GEOLOGICAL ASSOCIATION OF CANADA SPECIAL PAPER 29

THE CARSWELL STRUCTURE URANIUM DEPOSITS, SASKATCHEWAN

Edited by

**R. Lainé,
D. Alonso and
M. Svab**

R. Lainé, D. Alonso, and M. Svab
Amok Ltd.
817-825 45th Street
P.O. Box 9204
Saskatoon, Saskatchewan S7K 3X5

Canadian Cataloguing in Publication Data

Main entry under title:

The Carswell structure uranium deposits, Saskatchewan

(Geological Association of Canada special paper; 29)
Includes abstracts in French
ISBN 0-919216-27-7

1. Uranium ores — Athabaska, Lake, Region (Sask. and Alta.). 2. Geology — Athabaska, Lake, Region (Sask. and Alta.). I. Lainé, R. (Roger), 1946- . II. Alonso, D. (Daniel), 1944- . III. Svab, M. (Margaret), 1956- . IV. Geological Association of Canada. V. Series: The Geological Association of Canada special paper; 29.

QE390.2.U7C37 1985   553.4'932'0971241   C85-098697-4

All rights reserved. This book, or any part thereof, may not be reproduced in any form without the permission of the officers of the Geological Association of Canada.

Geological Association of Canada
Department of Earth Sciences
Memorial University of Newfoundland
St. John's, Newfoundland

Printed by: Johanns Graphics, Waterloo, Ontario

The GEOLOGICAL ASSOCIATION OF CANADA is Canada's national society for the geosciences. It was established in 1947 to advance geology and its understanding among both professionals and the general public. The GAC membership of 3,000 includes representatives of all geological disciplines from across Canada and many parts of the world employed in government, industry and academia. There are specialist divisions for environmental earth sciences, geophysics, mineral deposits, paleontology, Precambrian, sedimentology, tectonics, and volcanology. Regional sections of GAC have been set up in Victoria, Vancouver, Edmonton, Winnipeg, and St. John's, and there are affiliated groups in Toronto and the Maritimes.

GAC activities include the organization and sponsorship of conferences, seminars, short courses, field trips, lecture tours, and student and professional awards and grants. The Association publishes the quarterly journal *Geoscience Canada* and the quarterly newsletter *Geolog*, a *Special Paper* series, Short Course Notes, and several other continuing series. GAC also maintains liaison with other earth science societies and provides advice to government and the public on geological issues. The Association was incorporated under the Canada Corporations Act in January, 1984. For information contact: Geological Association of Canada, Department of Earth Sciences, Memorial University of Newfoundland, St. John's, Newfoundland, A1B 3X5, Canada.

CONTENTS

1. Geology and Mineralization in the Carswell Structure – A General Approach
   F. Tona, D. Alonso, M. Svab 1

2. Clay Mineral Stratigraphy of the Athabasca Group: Correlation Inside and Outside the Carswell Structure
   J. Hoeve, D. Quirt, D. Alonso 19

3. Geochronology of the Carswell Area, Northern Saskatchewan
   K. Bell 33

4. K-Ar Dating of Different Rock Types from the Cluff Lake Uranium Ore Deposits (Saskatchewan-Canada)
   N. Clauer, F. Ey, F. Gauthier-Lafaye 47

5. Petrographic and Geochemical Variations within the Carswell Structure Metamorphic Core and their Implications with Respect to Uranium Mineralization
   M. Pagel, M. Svab 55

6. Petrography and Geochemistry of the Earl River Complex, Carswell Structure, Saskatchewan – A Possible Proterozoic Komatiitic Succession
   K. Bell, A.D. Cacciotti, J.H. Schnessl 71

7. The Study of the Basal Athabasca Succession in the D, E, L, F, and S Areas of the Carswell Structure
   A. Pacquet, S. McNamara 81

8. The Carswell Formation, Northern Saskatchewan: Stratigraphy, Sedimentology, and Structure
   H.E. Hendry, K.L. Wheatley 87

9. Mineralogy and Metallogeny of Uraniferous Occurrences in the Carswell Structure
   F. Ruhlmann 105

10. A Uranium Unconformity Deposit: The Geological Setting of the D Orebody (Saskatchewan-Canada)
    F. Ey, F. Gauthier-Lafaye, F. Lillié, F. Weber 121

11. Mineralogical and Structural Aspects of the Dominique-Peter Uranium Deposit
    J.R. Blaise, E. Koning 139

12. Chemistry of Uranium Minerals in Deposits and Showings of the Carswell Structure (Saskatchewan-Canada)
    M. Pagel, F. Ruhlmann 153

13. A Chemical Study of the Carbonaceous Material from the Carswell Structure
    P. Landais, J.M. Dereppe 165

14. Geochemistry and Glacial Geology – Application to Exploration in the Carswell Structure
    J.S. Wilson 175

15. Case Histories of the Radon Tube Sampler in the Carswell Structure
    B. Powell 189

16. Geophysical Mapping of Gneiss Domes in the Carswell Structure and their Relationship to Uranium Mineralization
    B. Powell, E. Koning, R. Lainé 201

17. The Origin of the Carswell Circular Structure
    M. Pagel, K. Wheatley, F. Ey 213

18. Conclusion: The Carswell Uranium Deposits – An Example of Not So Unique Unconformity-Related Uranium Mineralization
    R. Lainé 225

FOREWORD

The Carswell structure is certainly one of the most conspicuous large diameter ring-type structures in Canada and its position within the flat-lying Athabasca Basin in northern Saskatchewan makes it even more obvious. This unique geological structure which could satisfy the dreams of any geologist seeking challenging research subjects, contains another characteristic besides its puzzling structure and history. It hosts some of the best uranium deposits in Canada, one of them having been for many years the world's richest deposit.

If we look back at the exploration, development, and mining adventures over the past 18 years, we may begin to understand the complexities which were encountered and which produced more questions than answers in the early years.

The story begins in the late 1960's when, following-up an airborne radiometric survey, the first uranium-bearing boulders were discovered at the north end of Cluff Lake. When geologists considered tracing these very high grade boulders to find the source area, they were confronted with a major problem: outcrop was very scarce, and they consequently had almost no geological background for the area, at least on a mining scale.

At that time, Amok had the choice between systematic, expensive and extensive drilling programs or more indirect approaches. The latter approach was selected, although quite unusual in the uranium industry at the time. Amok began using techniques such as radon surveys, airborne magnetic surveys, seismic refraction/reflection, geochemistry, overburden drilling, quaternary geology, geochronology, and magneto-telluric soundings. Innovation eventually proved to be the right choice for it is now known that the Athabasca high grade uranium deposits exhibit limited or poorly developed uranium haloes and that the classical method of radiometric ground surveys and drill hole gamma ray probing would not have so easily led to the present discoveries.

The majesty and regularity of the Carswell structure as a whole is as striking as its complexity in detail. Geological interpretation has progressed with uranium prospecting and drilling. The main uranium occurrences are clustered near the southern edge of the basement core where structural complexity is at its maximum, and it is understandable that the relationship between mineralization and various characteristics of the basement and the sedimentary cover could not be easily interpreted. The turning point in understanding the nature of the structure seems to have been reached in 1980 when the results of a high sensitivity airborne magnetic survey, flown on a very close spacing, were obtained and interpreted. The stratigraphic and structural relationships between the upper aluminous gneiss and the lower partially remobilized quartzo-feldspathic gneisses clearly appeared on the aeromagnetic map, leading to a fuller understanding of the ground resistivity results. With the assumption that the basement was not totally disrupted and retained most of its pre-Carswell event structure, the need for a synthesis of all exploration work done became apparent.

These interpretations were soon confirmed, and the discovery of the Dominique-Peter orebody at the end of the 1980 drilling campaign increased ore reserves significantly. The uranium mineralization was located precisely where the new concepts emerging among the geologists at Amok, Mokta, and Cogema (shareholders of Amok) indicated the uranium deposits should be, that is to say, at the stratigraphic contact between aluminous and quartzo-feldspathic gneisses along the flanks of magnetic domes of the latter. The major drilling program implemented to delineate and evaluate the Dominique-Peter area was a great source of information for the geologists. The role of mylonite zones and subsequent faulting as major structural controls for the mineralized veins became very evident.

This volume contains descriptions of the geology of the Carswell structure and the relationship between mineralization, geochemistry, geology, geochronology, geophysics, and tectonics. It includes valuable data regarding the Carswell event in which the structure was formed. It is left up to the reader to determine whether the origin of the structure is from a meteorite impact, or is the result of some other cryptoexplosion phenomenon.

It has taken over three years of study and reflection to achieve the synthesis provided in this volume. Our appreciation is extended to Cogema, Mokta, and Pechiney Cie., shareholders of Amok, for encouragement and criticisms of the manuscripts, and to SMDC (Saskatchewan Mining Development Corporation) for permission to publish the data. Much credit is due to the important contributions by earlier scientists. We would like to congratulate and thank, on behalf of Amok, all the present and past geologists who worked with us, and especially: Daniel Alonso, Bernard Blangy, Jacques Dardel, Jacques Dumas, Tony Durand, Roger Lainé, Andre Meunier, Kristo Tapaninen, Yves Thomas, and Frédéric Tona. The quality of their work is not clouded by the changes in interpretation of already well-documented facts. Now that this volume is complete, it is obvious that much still remains to be done, that a lot of problems remain unsolved. Fortunately, a lot of uranium will probably be discovered in the future around the existing deposits, providing a challenge for a future generation of geologists.

We wish to congratulate all of the authors in this special paper for their valuable contributions to this overview of the Carswell structure.

J.P. Slama
Executive Vice-President
AMOK LTD.

Ph. Artru
COGEMA
Responsable Régional
Amérique du Nord

PREFACE

Over the past several decades, information about the Carswell structure in the western part of the Athabasca Basin, northern Saskatchewan, has been accumulating in the form of unpublished reports, several graduate and undergraduate theses, and abundant uncollated facts and figures obtained in the course of uranium exploration. The decision to publish the data in a set of scientific papers and therefore to tie together all existing geological and related research on the 20 km wide circular basement inlier and the surrounding sedimentary rocks was made in early 1983, following a change in company policy which allowed the release of such information. The Geological Association of Canada offered the opportunity to publish the papers in a single volume.

This volume summarizes all aspects of the Carswell structure — geology, mineralogy, geochronology, geochemistry, stratigraphy, metallogeny, and geophysics. It also provides an avenue for some discussion regarding the controversial origin of such an unusual circular phenomenon. We, as editors of the volume, have seen many idaas evolve during the two year preparation of the individual manuscripts. Many of the conclusions arrived at would not be as advanced if not for the groundwork laid by earlier workers and more recently by the dialogue amongst the various authors in the course of writing and revisions.

The support and co-operation of John Kramers, Chairman of the Geological Association of Canada Publications Committee from the outset is most gratefully appreciated. The Geological Association of Canada acknowledges the generous support of the Natural Sciences and Engineering Research Council for providing a grant toward publication of this volume and to the Geological Survey of Canada and the Canadian Geological Foundation for their support of the Special Paper Series.

We also express our appreciation to Amok Ltd. and its shareholders for allowing us to act as editors and for the financing of the coloured map which accompanies the volume. Special thanks are due to Tom Sibbald, Sasktachewan Geolgoical Survey, and to Keith Bell, Carleton University, for their encouragement and guidance beyond that requested of a referee.

We also thank Daryl Geisbrecht, who drafted many of the figures in the volume and Lori-Lee McNally who typed most of the manuscripts. We are grateful to Maureen Dickson Czerneda, the Geological Association of Canada's Managing Editor for her advice and expertise concerning format and final printing.

The majority of the manuscripts contained in this Special Paper were written by scientists whose first language is not English. We appreciate the extra time and effort devoted by all referees who not only reviewed the papers for their scientific content but also edited the manuscripts, greatly improving their readability. The co-operation by the authors in making prompt revisions to their papers is also acknowledged.

The following referees provided reviews with a thoroughness and speed for which we are grateful: K. Bell, M.E. Bickford, F.R. Breaks, D.G. Brookins, M. Cameron-Schimann, M.E. Cherry, G.S. Clark, N. Clauer, G.L. Cumming, K.L. Currie, A.G. Darnley, J.A. Donaldson, W. Dyck, D. Francis, R.I. Grauch, Z. Hajnal, C.T. Harper, J. Hoeve, D.M. Kent, W.O. Kupsch, F.F. Langford, J.S. Leventhal, J.F. Lewry, G.S. Maciel, M. Mellinger, R.D. Morton, J.T. Nash, E. Nisbet, G. Parslow, P. Ramaekers, D.S. Robertson, D.H. Rousell, D. Roy, V. Ruzicka, J. Shaw, T.I.I. Sibbald, A.B. Siddans, M.R. Stauffer, H. Stolz, L.P. Tremblay, J. Wilson, D. York.

R. Lainé
D. Alonso
M. Svab
Editors

The Carswell Structure Uranium Deposits, Saskatchewan,
edited by R. Lainé, D. Alonso and M. Svab,
Geological Association of Canada Special Paper 29, 1985

GEOLOGY AND MINERALIZATION IN THE CARSWELL STRUCTURE — A GENERAL APPROACH

F. Tona
COGEMA BU/DRM, 2, Rue Paul Dautier, B.P. No. 4, 78141 Velizy Villacoublay Cedex, France

D. Alonso and M. Svab
Amok Ltd., 817-825 45th St. West, P.O. Box 9204, Saskatoon, Saskatchewan S7K 3X5

ABSTRACT

The Carswell circular structure, northern Saskatchewan, is located in the western part of the Athabasca Basin. It consists of a metamorphic core, similar in lithology to the Western Craton, and is encircled by Athabasca Group sedimentary rocks. The main uraniferous occurrences are concentrated along the S-SW perimeter of the metamorphic core and are hosted either by sandstones, as in the cases of the D orebody, or by basement gneisses, as in the case of the Dominique-Peter, Claude, N, and OP orebodies. Uranium deposition in the sandstones is controlled by the unconformity, a mylonite zone, and abundant faulting. Uranium deposition in the basement gneisses is controlled by a mylonite zone, extensive faulting, strong alteration, and an additional lithologic control from the interface between the Earl River feldspathic gneisses and the overlying Peter River aluminous gneisses. These gneissic assemblages form the two main units of the metamorphic core.

Four major paragenetic assemblages have been noted and dated in the Carswell structure: a chemical age from a uraninite-monazite association yields 1800 Ma; a polymetallic uraninite-Se-Te assemblage yields an age around 1100 Ma; a remobilization event around 900 Ma, and later around 380 Ma, produced a less complex assemblage of uraninite and simple sulphides. A fifth assemblage, dated around 1330 Ma, has been observed recently in some minor uraniferous occurrences which correlates well with ages derived elsewhere in the Athabasca Basin.

RÉSUMÉ

Situeé à l'ouest du Plateau Athabasca au Nord de la Province du Saskatchewan, la structure circulaire de Carswell se compose d'un noyau de roches métamorphiques assimilables à celles du Western Craton et d'une couverture annulaire de roches sédimentaires appartenant au groupe Athabasca. La structure actuelle dont l'origine, endogène (cryptoexplosion) ou exogène (impact météoritique), n'est pas éclaircie, s'est formée à l'Ordovicien.

Le noyau métamorphique est constitué par deux ensembles: 1) le complexe de l'Earl River qui pourrait représenter une ancienne série volcano-sédimentaire débutant par des komatiites et se terminant par des séquences arkosiques ou acides d'âge aphébien ou archéen, métamorphosée en gneiss quartzo-feldspathiques ou amphibolitiques; et 2) les gneiss de Peter River, ancienne série sédimentaire de type shales, d'âge aphébien, transformée en gneiss alumineux. Ces deux ensembles ont subi un métamorphisme subcatazonal (granulite faciès) à l'hudsonien ancien(?), avant d'être rétromorphosés au cours des phases hudsoniennes récentes (amphibolites faciès) conséquence d'une tectonique tangentielle NE-SW.

La couverture hélikienne est constituée d'une série basale volcano-sédimentaire acide, sédimentation résiduelle paléohélikienne, et de la série classique du groupe Athabasca complétée dans le secteur de Carswell par les formations de la Douglas et de Carswell.

L'orogenèse grenvillienne est marquée par l'intrusion de diabases au sud de la structure circulaire de Carswell. L'"événement" Carswell désorganise les structures et provoque une remontée du socle de plus de 1000m au travers des terrains Athabasca s'accompagnant par la mise en place de brêches à faciès "sub-volcaniques" (brêches de Cluff).

Les principaux gisements d'uranium se répartissent dans la partie sud du noyau métamorphique soit dans le socle, soit au contact grès-socle, mais toujours à proximité des grès Athabasca. Toutes les minéralisations se situent dans des zones tectonisées fortement altérées.

Plusieurs épisodes minéralisateurs ont été reconnus: un épisode hudsonien à uraninite-monazite (1800 Ma); un épisode majeur contemporain de la formation des autres gisements du Plateau Athabasca (1050 Ma): paragenèse à uraninite et sulfures polymétalliques (Se et Te); un épisode discret remobilisant la phase précédente (900 Ma): paragenèse à quartz, uraninite et sulfures; enfin un épisode caractéristique à hématite-pechblende (380 Ma) retrouvé surtout dans les grès Athabasca.

Le gisement D se situe à la discordance grès-socle, retournée lors de l'événement Carswell, sur la bordure sud du noyau métamorphique régolithisé. La majeure partie de la minéralisation très riche accompagne le ciment argileux d'une "zone à boules", zone de lenticulation tectonique développée dans les grès Athabasca à la suite du rejeu d'une zone mylonitique hudsonienne. Deux paragenèses à uraninite ont été identifiées: 1) uraninite-sulfures polymétalliques (Mo, Cu, Pb, Co, Ni), séléniures et tellurures, 2) uraninite-sulfures simples (Cu, Pb). La mise en place de l'uranium a été précédée d'une longue altération hydrothermale argileuse magnésienne.

Le gisement de Dominique-Peter situé à environ 1.5 km au NW de D, se développe entièrement dans le socle métamorphique. La minéralisation est contrôlée par une mylonite altérée, localisée préférentiellement à la limite des gneiss de la série de Peter River et du complexe de l'Earl River.

La minéralisation remplit deux systèmes de fractures à contre pendage de la mylonite, qui sont orientés N-S à pendage 50° W et N60° E à pendage NW. Des failles tardives NNE et N120° E recoupent les précédentes mais ne sont pas minéralisées. Un lambeau de grès et de socle métamorphique régolithisé a été découvert dans la zone de Dominique; il souligne la proximité de la discordance pré-Athabasca. Les paragenèses observées à D se retrouvent dans le gisement de Dominique-Peter.

Le gisement Claude, au NW de Dominique-Peter, est situé aussi dans les gneiss alumineux de Peter River. L'uranium est contrôlé par une zone mylonitique E-W où se développe une "zone à boules", qui sépare les gneiss altérés au nord des gneiss sains au sud. Des failles tardives N20° E donnent à l'enveloppe minéralisée la forme d'un "J".

Les gisements OP et N sont également dans le socle à proximité de la discordance. OP est constitué de trois structures N-S subverticales au sein des gneiss alumineux de Peter River immédiatement au nord de D. N se développe essentiellement dans les faciès quartzo-feldspathiques altérés du complexe de l'Earl River charriés sur les gneiss de Peter River. La minéralisation y est contrôlée soit par des "zones à boules" N-S soit par un filon vertical N40° E.

INTRODUCTION

The Carswell structure, northern Saskatchewan, is located approximately 60 km south of Lake Athabasca and 25 km east of the Alberta-Saskatchewan border (map in pocket). The roughly circular structure is lithostructurally located within the Western Craton in the Churchill Province, along the southwestern edge of the Canadian Shield. It is composed of a 20 km diameter metamorphosed basement core of mixed feldspathic and mafic gneisses of the Earl River complex overlain by aluminous Peter River gneiss, which is in turn unconformably overlain and surrounded by an 8 to 10 km wide ring of Helikian Athabasca Group sedimentary cover rocks. The inner ring of undifferentiated William River Subgroup Sediments is surrounded by pelitic sediments and carbonates of the Douglas and Carswell Formations. Altogether the structure has a diameter of approximately 40 km. A series of younger mafic dykes intrude the basement and the Athabasca cover to the south of the structure. The structure itself is intruded by dykes and veinlets of Ordovician Cluff breccia which resulted from the Carswell event when the structure was formed.

Uranium mineralization is found in both Athabasca Group sediments and the basement core. The controls for the mineralization in the sandstones include the presence of the unconformity, shear zones, and transverse faulting. The mined-out D orebody is the only example presently known which exhibits all of these features. Economic mineralization in the basement core is a function of lithologic variations, the proximity to the unconformity, the presence of mylonite zones, and transverse faulting. Examples of this type of occurrence include the Claude, Dominique-Peter, and N orebodies. Strong alteration in the vicinity of mineralization is pervasive in both geological settings.

Mineralization in the structure have given the following ages: an age of around 1800 Ma which appears to reflect the influence of Hudsonian deformation; an age between 1050 and 1150 Ma representing the most economic phase and probably corresponding to the main mineralization period; an age of around 900 Ma obtained from weaker mineralizations; and an age between 250 and 380 Ma. A U-Pb age of 1330 Ma has been recently obtained on a small amount of uranium and may reflect a separate phase of mineralization.

HISTORY OF EXPLORATION
Previous Work

The first known geological investigation in the Carswell Lake area was carried out in 1952 by D.A.W. Blake. Fahrig (1961) renamed Blake's "Trout Lake Limestone" the Carswell Formation. In 1958, an airborne magnetic and scintillometric survey was flown at a 1 mile spacing over the "Trout Lake Structure", outlining shallow magnetic anomalies and higher radiometric backgrounds at the centre of the dolomite ring (Ratcliffe, 1958). Field reconnaissance of the anomalies indicated the presence of basement gneisses in the area.

In 1964, M.J.S. Innes interpreted the Carswell structure as a meteorite impact crater after observing shock metamorphism features, striations resembling shatter cones, and unusual polymictic breccias in outcrops. Later, Currie (1969) examined the Carswell structure in detail and related the formation of the structure to crypto-explosion phenomena. A regional aeromagnetic map (Agarwal, 1965) also outlined the circular structure south of Carswell Lake. Herring (1976) and Harper (1982, 1983) mapped the metamorphic inlier in some detail.

In 1965, geologists from the French Atomic Energy Commission began working in northern Saskatchewan to evaluate the possibility of initiating a uranium exploration program. Mokta Canada Ltd. operated the project for the CEA (Commissariat à l'Energie Atomique), Mokta, and Pechiney Ltd. syndicate. After 1967, Amok Ltd./Ltée. replaced Mokta Canada Ltd., and the syndicate was joined in the mid and late seventies by Numac, Ontario Hydro, and SMDC (Saskatchewan Mining and Development Corporation). In 1976, COGEMA was created by the CEA and by 1979 assumed an active role in Amok.

Uranium Exploration

Two areas in northern Saskatchewan were investigated in the late 1960s: the metamorphic basement around Uranium City, and the unconformity at the base of the Athabasca Group in the Stony Rapids – Black Lake area where uranium occurrences had been reported. Airborne radiometric coverage of these chosen targets began in the 1966 field season. The airborne surveys over the Fond du Lac area were completed during the 1967 field season. Two other areas, the Carswell structure and the Cree Lake area were also covered by similar surveys during that season.

A SRAT scintillometer and recorder were mounted on a Cessna 180 and the recording was done without altimetric correction. The survey was of the total count type and lines were flown in an east-west pattern at a spacing of 1200 m. The Fond du Lac and Carswell Lake areas were the most anomalous. Due to the location of the field parties in that year, only the Fond du Lac anomalies were investigated on the ground, and uranium-bearing sandstone boulders were discovered in the area.

In the spring of 1968, a field party investigated an anomaly located at the northeast end of Cluff Lake. This led to the discovery of the "A" train boulders of uranium-bearing sandstone and massive pitchblende. A month and a half later, most of the metamorphic core of the Carswell structure and the surrounding sandstone had been staked (Fig. 1). A permit was obtained for the area at the end of the same year.

In September 1968, an airborne survey crew established camp at Cluff Lake and discovered the B, D, and E showings, as well as other smaller occurrences (Fig. 2). The B and E showings contained mineralization in sandstone, whereas the D showing was mainly yellow stains along fractures within highly altered hematized basement rocks.

In 1969 and 1970, airborne surveys were continued at closer spacings. Ground radiometry over the F, L, M, N, O, P, and Claude showings, geological mapping, as well as magnetic and radon surveys were also carried out. Although outcrops were scarce, geological mapping indicated that the area was highly tectonized and included overturned contacts and severely faulted basement and sandstone slabs. Hand-dug trenches revealed that the airborne radiometric anomaly over the N showing was due to a mineralized, flat-lying fault plane. In the showing, the radon anomaly corresponded to numerous pitchblende boulders in glacial till. The ground radiometric high at the SE edge of Claude Lake and at the NE end of an airborne anomaly consisted of radioactive reddish and greenish fault gouge and led to the discovery of the Claude orebody. A radon survey in the D area picked out a 10 times background anomaly over the D orebody and follow-up drilling with a Winkie diamond drill intersected mineralization. A trench opened on the showing revealed fractures that were filled by millimetre- to centimetre-wide veinlets of pitchblende and organic material.

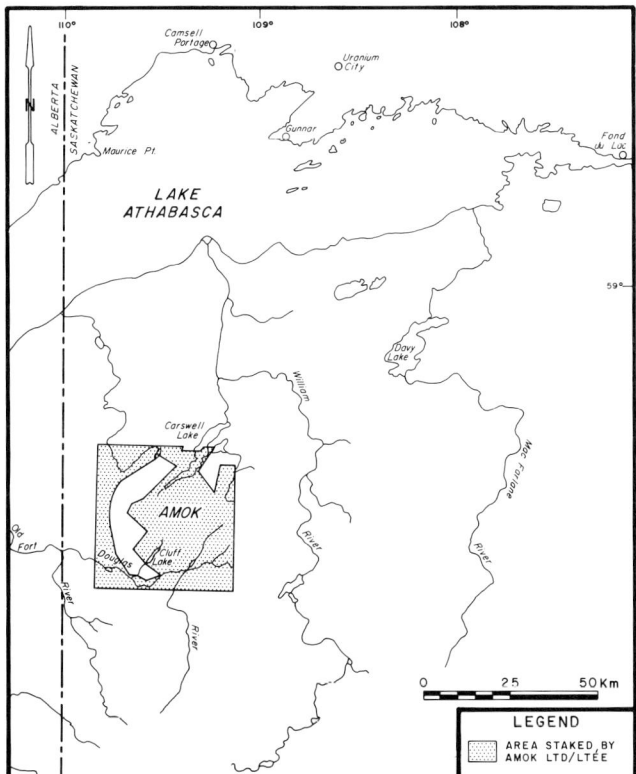

Figure 1. Location of areas staked in 1968 by Amok Ltd.

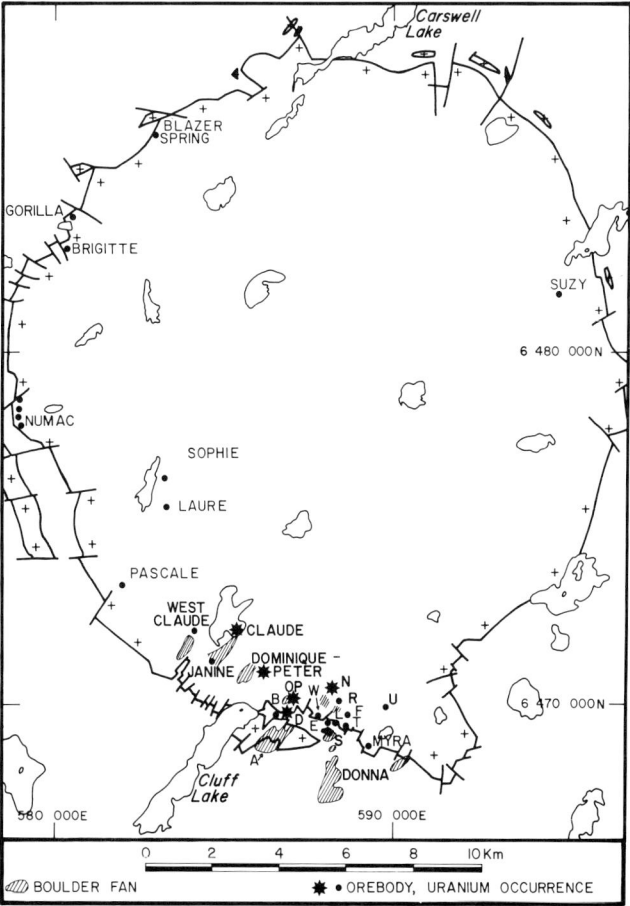

Figure 2. Location within the Carswell structure of all known uraniferous occurrences and orebodies.

By 1971, a follow-up detailed airborne radiometric survey was completed and detailed studies of the overburden in the Carswell structure were started to trace glacial events. A percussion drill program was carried out and probing techniques improved with the first prototype of the CEA Probing Unit using geiger, scintillometric, and neutron probes. During this more sophisticated drilling, three years after the first boulder discovery, it became clear that the D showing was a significant ore deposit.

Systematic diamond and percussion drilling around the D, Claude, Dominique, N, R, and OP mineralized areas from 1971 to 1974 outlined these orebodies. As new roads were created, radioactive boulders were also uncovered in what are now the R, S, T, U, W, Donna, and Janine areas. In 1974 percussion drilling was used for geochemical sampling in the overburden and bedrock. This technique helped to precisely delineate the basement-sandstone contacts and faulted slabs. As a result of this drilling, several thoriferous and uraniferous occurrences were found in the Suzy and West Claude areas.

From 1970 to 1982, magnetic, electromagnetic, high sensitivity magnetometer, and VLF/EM surveys were flown at various spacings over the Carswell structure. A tripole resistivity survey was also initiated to help geological mapping and was later replaced by EM-16 and EM-16R. The latter method was used to provide a blanket coverage along with geological mapping, radiometric prospecting, and radon surveys. In addition, various other geophysical methods were applied such as Turam EM, gravity, seismic refraction and reflection, magnetotellurics, Maxiprobe EMR-16, surface and borehole PEM, DEEPEM Crone, MaxMin, magnetometer, and IP.

In 1980, a fence of deep diamond drill holes was put down between the OP and Dominique mineralized zones to explore an area covered by thick overburden. This area is characterized by a magnetic low between two magnetic highs. Mineralization was intersected south of Peter River (SE of Dominique), and follow-up drilling outlined the Dominique-Peter orebody.

GEOLOGICAL SETTING OF THE CARSWELL STRUCTURE

Lithostructural Setting

The basement core of the Carswell structure lies within the Western Craton, the most westerly of the four broad crustal units (Fig. 3) which comprise the Canadian Shield in northern Saskatchewan (Lewry et al., 1978). Geophysical methods applied by Wallis (1970) delineated three domains in the area west of the Virgin River shear zone. Lewry et al. (1978) also describe three domains: from east to west they are the Western Granulite, Clearwater and Firebag Domains. The limits between the three domains are not well defined. Surface exposure of the Clearwater Domain is very poor, and its definition may only be a reflection of a Hudsonian mobile zone (Lewry et al., 1978). The Western Granulite Domain is composed of granulite facies layered gneisses, and its eastern boundary is marked by blastomylonite gneiss of the Virgin River shear zone. To the west, the Clearwater

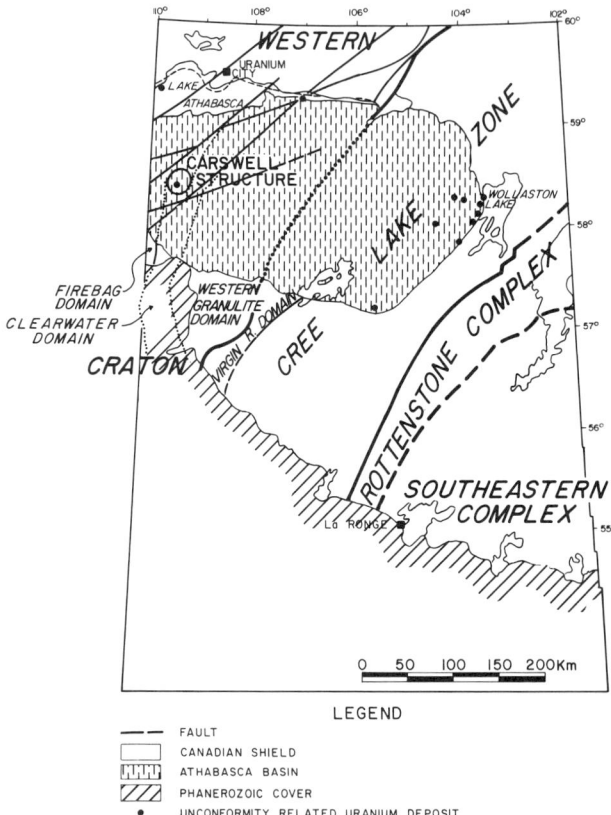

Figure 3. Location of the Carswell structure with respect to the lithostructural subdivisions of the western Canadian Shield (modified from Lewry et al., 1978).

Domain is composed mainly of granite and felsic gneisses, which are overlain by a younger metasedimentary unit called the Careen Lake Group (Scott, in press). The Firebag Domain, at the western edge of the Western Craton, appears to contain similar granulite gneisses as those in the Western Granulite Domain. In Alberta, Godfrey and Langenberg (1978) describe granodioritic rocks and mylonites as the major rock types in the Firebag Domain south of the Athabasca Basin.

The Firebag Domain exhibits magnetic lineations (Fig. 4) which trend between N140°E and N160°E and which are cut by N50°E right lateral faults such as the Black Bay and Deranger Creek Faults, by N-S faults, and by N70° to 80°E dextral faults. In the Western Granulite Domain, the magnetic lineations roughly parallel the trend in the Firebag Domain but swing sharply from N140°E to N60°E along the Virgin River shear zone. The magnetic lineations in the Clearwater Domain show a strong NE trend.

The Carswell structure lies just to the west of the Clearwater Domain, if the NE structural trend is extrapolated beneath the Athabasca Basin. Major faults within the structure itself are interpreted from diamond drilling and by geophysical methods and trend N-S, N60°E and N140°E, and to a lesser degree E-W (map in pocket). The N-S faults are best exposed along the eastern boundary of the structure, the

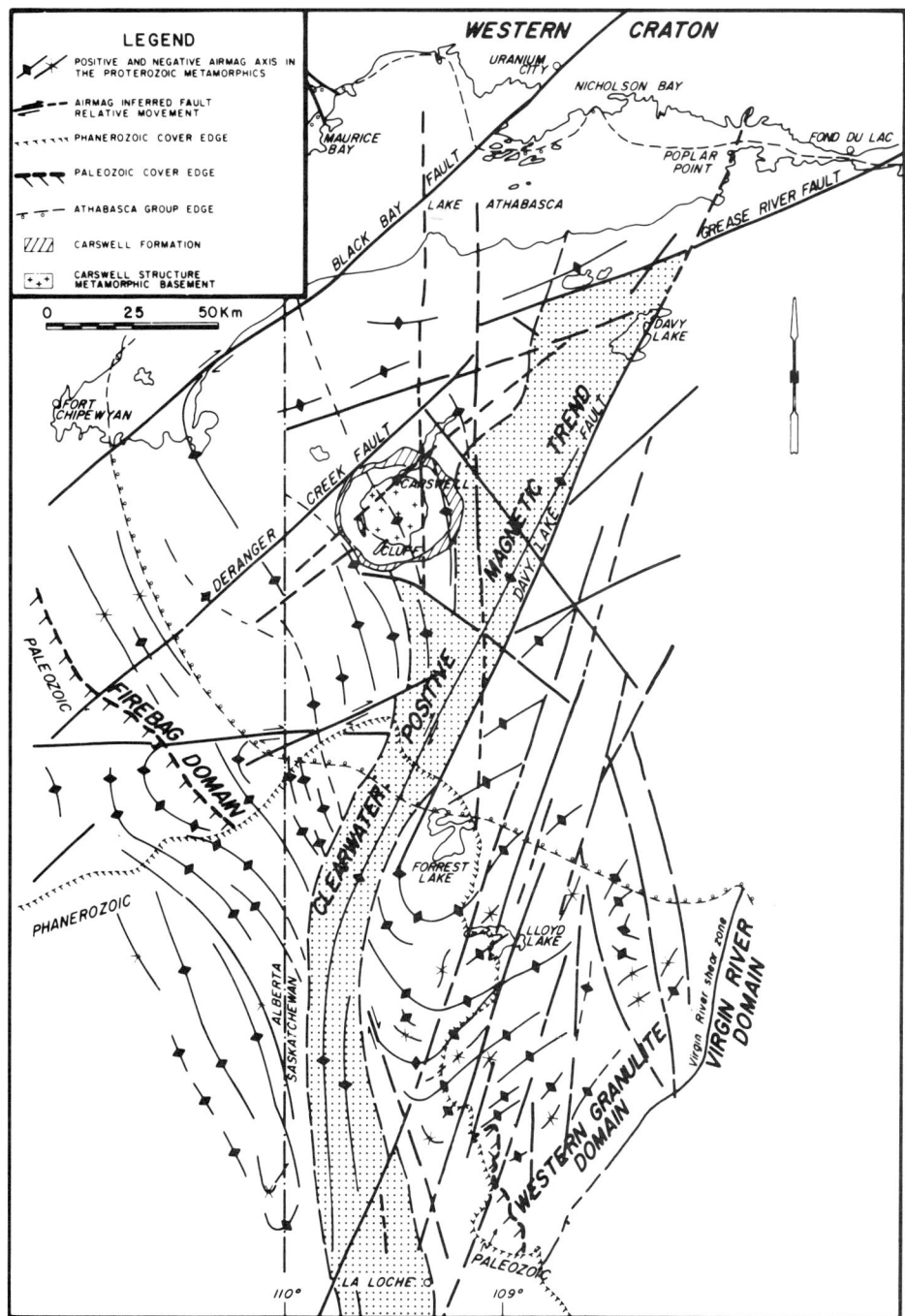

Figure 4. Structural setting of the Carswell structure.

N60°E faults are ubiquitous throughout and N140°E faults have been noted in the northern half of the structure as well as to the south where diabase dykes have intruded fractures along the same direction (Wanless *et al.*, 1979). The N60°E and N140°E fault directions are a result of late Hudsonian NE-SW compressive deformation.

Basement Geology

Geological and structural interpretation of the basement core of the Carswell structure has been based mainly on diamond drilling and geophysical methods. Bassaget and Camps (1973) distinguished four rock types: paragneisses, migmatites or granitic gneisses, granites and pegmatoids, and amphibolites. Herring (1976) subdivided the basement rocks into pelitic gneisses, quartzofeldspathic gneisses, mafic gneisses, pegmatoids, and cataclastic rocks. Harper (1982, 1983) subdivided the Carswell structure basement core into three main units: a granodiorite complex, quartzofeldspathic gneisses, and pelitic gneisses, all accompanied by metagabbros and granite pegmatites.

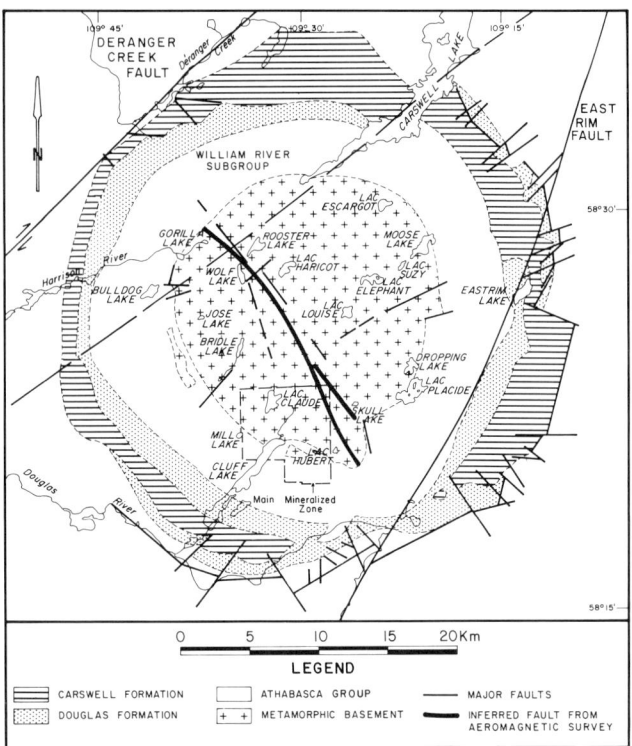

Figure 5. Bedrock geology of the Carswell structure.

Amok geologists have divided the basement core into two units: Earl River complex and Peter River gneiss.

Earl River Complex. Structurally underlying the Peter River gneiss, the Earl River complex is composed of feldspathic gneisses, mafic gneisses, amphibolitic rocks, granitoids, and pegmatoids. The feldspathic gneisses are medium-grained, well foliated, and composed mainly of potassium feldspar, plagioclase, quartz, and biotite. Frequently they contain aluminous bands of similar composition to the Peter River gneiss. Augen textures are locally well developed. A sedimentary origin for these gneisses has been suggested by Herring (1976), Harper (1983), and Pagel and Svab (1985). Whole-rock Rb-Sr ages for these gneisses reflect an event around 1800 Ma (Bell, 1985).

Granitoids are pink to grey, medium -to coarse-grained, massive to weakly foliated and are composed of potassium feldspar, plagioclase, and quartz. Garnet, minor cordierite, and sillimanite commonly occur as accessory minerals. The granitoids are thickest in the Earl River complex and are typically located near the contact with the Peter River gneiss. Pagel and Svab (1985) believe these granitoids represent meta-arkoses which have undergone anatexis. Bell *et al.* (1985) suggest an acid volcanic origin.

Mafic gneisses and amphibolites are generally intercalated with feldspathic gneisses and granitoids and are composed of variable amounts of hypersthene, clinopyroxene, biotite, and minor amphibole. They have been subjected to granulite facies metamorphism and later have been retrograded to upper amphibolite facies. These rocks appear to have a mafic volcanic origin, possibly komatiitic, and may form part of a larger volcanic succession (Bell *et al.*, 1985).

Peter River Gneiss. The Peter River gneiss is a distinct metasedimentary unit composed of aluminous gneisses. The most characteristic rock of this unit is a banded, well foliated, garnet-cordierite-sillimanite gneiss. Pyrite and graphite occur as disseminations and fracture fillings. Compositional banding may reflect the original sedimentary bedding or may be a result of remobilization during metamorphism. Small scale folding is fairly common. The percentage of garnets ranges up to 40%, however these minor bands of garnetite are usually less than 1 m thick and may have been originally iron-rich sedimentary layers or possibly represent a restite phase from partial anatexis. Granitic material is intimately interfoliated with these aluminous gneisses and varies in thickness from ten's of centimetres up to metres. According to Pagel and Svab (1985), the Peter River gneiss represents former shale deposits. Whole-rock Rb-Sr ages for the gneiss reflect an event around 1760 Ma (Bell, 1985).

The Peter River gneiss exhibits a well-defined geophysical response, particularly with the EM and gravity methods (Powell *et al.*, 1985), a property that has been of great benefit when interpreting the geology beneath the overburden.

Basement Rock Relationships. A series of negative magnetic anomalies define a broad NW-SE trending zone through the Carswell structure basement core (Fig. 5) and may represent a major shear zone. To the northeast of this

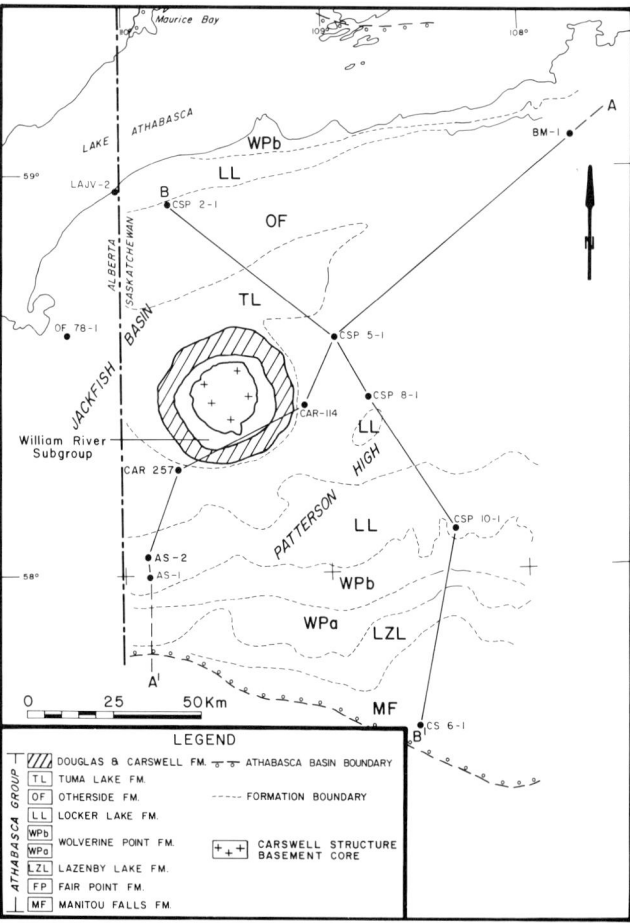

Figure 6. Athabasca Group in the western part of the Athabasca Basin (modified from Ramaekers, 1980) with cross section location.

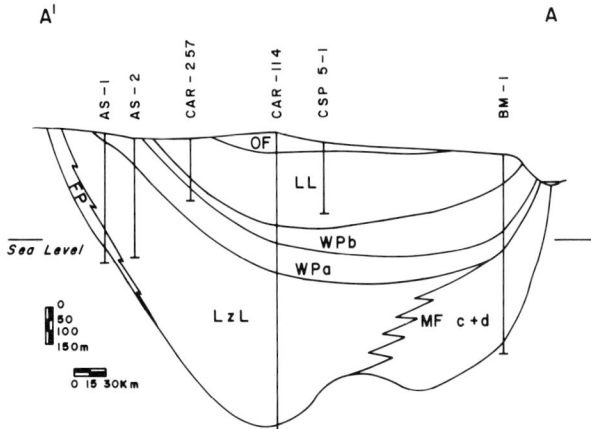

Figure 7. Clay mineral stratigraphy for the western part of the Athabasca Basin (modified from Hoeve et al., 1985). See Figure 6 for legend and cross section location.

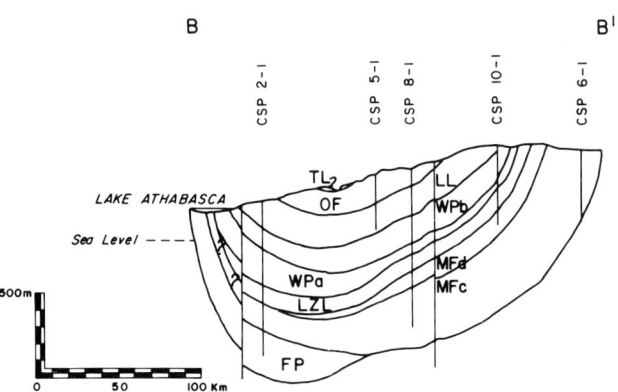

Figure 8. Cross section of the Athabasca Group in the western part of the Athabasca Basin (modified from Ramaekers, 1980). See Figure 6 for legend and cross section location.

zone, the predominant rock type is mafic-rich Earl River complex, while in the southwestern part of the core there are abundant Peter River gneisses which overlie the feldspathic gneiss of the Earl River complex. These observations suggest that the northeastern part of the core may be uplifted by faulting with respect to the southwestern part. A gravity survey has confirmed that the block of the Earl River complex in which the Gersendre and N areas are located is a faulted block which now overlies the Peter River gneiss (Map 1, in pocket). The mylonite zone developed at the contact between the Earl River complex and the Peter River gneiss is the typical expression of deformation at the interface. The gravity survey also outlined a type of mantle gneiss doming which is thought to have occurred during the late Hudsonian where the lighter Earl River complex ''intruded'' the denser Peter River gneiss and produced the type of dome structures presently observed (Powell et al., 1985).

Sedimentary Cover

Ramaekers (1979) developed a general stratigraphic section for the sedimentary rocks in the Athabasca Basin and defined six formations. From the base to the top they are: the Fair Point, Manitou Falls, Wolverine Point (a and b), Locker Lake, Otherside and Tuma Lake Formations. Later, the Lazenby Lake Formation was included which occurs stratigraphically below the Wolverine Point Formation (Ramaekers, 1980). All seven formations are present in the western half of the basin (Fig. 6).

The Douglas and Carswell Formations appear to conformably overlie the William River Subgroup and are the uppermost formations of the Athabasca Group. They are composed of pelitic sandstones and carbonates and are found only in the Carswell area where they form the outer ring of the structure (Hendry and Wheatley, 1985). Diamond drill hole CAR 114 is located several kilometres east of the Carswell structure, and provides a stratigraphic section for Athabasca Group sediments in the vicinity. Over 1200 m of sandstones were intersected prior to reaching basement rocks. Clay mineral interpretations (Hoeve et al., 1985) indicate that CAR 114 intersects the Otherside Formation at subsurface and that the base of the sandstone belongs to the Lazenby Lake Formation (Fig. 7). Figure 8 is a schematic cross section modified from Ramaekers (1981) to illustrate the relative formation thicknesses in the western half of the Athabasca Basin.

Early work within the Carswell structure (Currie, 1969) did not include any stratigraphic interpretation of the Athabasca Group which unconformably overlies and is often in faulted contact with the basement core. Ramaekers (1980) suggested the term William River Subgroup be applied to the undifferentiated sediments within the structure. Recent clay mineral studies for the general Athabasca Basin (Hoeve et al., 1981), illustrate that each formation typically has a distinctive clay mineral assemblage and that such information can be used to define formational boundaries (Table I). Studies by Tona and Alonso (1983) and Hoeve et al. (1983, 1985) on the William River Subgroup have resulted in a proposed stratigraphic section for all but the innermost 500 m of sediment adjacent to the basement core.

Hoeve et al. (1985) determined the clay mineralogy in sandstones that were located more than 500 m from the basement/sandstone contact. From the clay data, they identified the boundary between the Locker Lake Formation and the Otherside Formation. If the William River Subgroup contains the base of the Athabasca Group along the southern contact and also contains the top of the Athabasca Group less than 1 km from the basement core, the vertical difference is on the order of 1000 m.

Near the D orebody, Paquet and McNamara (1985) describe a 60 m thick sedimentary sequence which is preserved at the southern basement/sandstone contact over a distance of several kilometres. At the base of the sequence is a basal conglomerate, overlain by fine- to medium-grained red sandstones and pelites, with interbedded coarser grained and conglomeratic sandstones. The pelites contain evidence of paleovolcanism. A homogeneous variegated sandstone overlies the sequence. Each type of sandstone has a specific clay mineral assemblage. The fine- to medium-grained sandstones contain kaolinite and illite, the coarser grained sandstones contain chlorite and illite, while the variegated sandstones contain only illite. Ramaekers (1981) discussed the occurrence of late Aphebian unmetamorphosed redbed

TABLE I

SEDIMENTARY FORMATIONS AND CLAY ASSOCIATIONS WITHIN THE ATHABASCA GROUP
(AFTER HOEVE ET AL., (1981)

Formation	Clay Content
Otherside Fm.	Predominantly illite
Locker Lake Fm.	Predominantly illite
Upper Wolverine Point Fm.	Illite – chlorite association
Lower Wolverine Point Fm.	Illite – kaolinite association
Lazenby Lake Fm.	Predominantly illite
Manitou Falls Fm.	Chlorite (10%), illite (60%), kaolinite (30%)
Fair Point Fm.	Chlorite, illite, kaolinite (50 to 70%).

units which occur at the base of the Athabasca Group in small fault-bounded basins. The sedimentary sequence near the D orebody may be related to these small basins.

Recent age dating on the Douglas Formation (Clauer et al., 1985) yields dates of 1292 ± 27 Ma and 1220 ± 43 Ma (K-Ar on illites). Bell and Blenkinsop (1981) determined a whole-rock Rb-Sr age on the Wolverine Point (b) Formation of 1513 ± 24 Ma and suggest that it represents a probable maximum age for the deposition of the Athabasca Group sediments.

Diabase Dykes And Cluff Breccias

The Athabasca Basin is cut by diabase dykes. A WNW-ESE oriented dyke of similar composition which has K-Ar age of 949 ± 33 Ma (Wanless et al., 1979), is located south of the Carswell structure and is offset by NE-SW faults. There is no evidence of these younger mafic rocks in the basement core of the Carswell structure.

Cluff breccias occur ubiquitously in the area and are most frequently observed at the faulted sandstone/basement contacts and almost without exception along all major fault zones in the basement gneisses of the Carswell structure. They have been studied in detail by Currie (1969), Pagel (1975), and von Einseidel (1981) and are related to the Carswell event when the Carswell structure was formed. Currie (1969) postulated that the structure originated as a diapiric uplift, with probable concentration of volatiles at depth. Pagel (1975), however, used fluid inclusion studies to support a meteorite impact hypothesis. More detailed discussion on the origin of the structure and the Cluff breccias can be found elsewhere in this volume (see Pagel et al., 1985). Potassium-argon dates for the breccias (Currie, 1969; von Einseidel, 1981; Bell, 1985) range from 365 to 513 Ma.

Von Einseidel (1981) grouped the breccias into three varieties: volcanic-like, polymictic or classic variety, and tectonic-like. The most abundant variety of breccia that forms dykes and veinlets in the basement gneisses is the polymictic breccia. The volcanic-like breccia is most frequently observed along the edge of the basement core in faulted sandstone/basement contacts. Tectonic-like breccias are found near small fault zones within the basement gneisses.

MINERALIZATION

Uranium in the Carswell structure has been found in both basement gneisses and in the Athabasca Group cover rocks, concentrated along the perimeter of the basement core. All known economic deposits are found in the basement gneisses along the southern perimeter except for the D orebody which occurs in the Athabasca sandstones near the basement contact (Fig. 9). The controlling factors for uranium mineralization have previously been described by Tremblay (1982) in a synthesis of all uranium deposits related to the sub-Athabasca unconformity.

The existence of uranium in the sedimentary cover around the D orebody is mainly a function of three metallotects: the sandstone/basement unconformity, a mylonite zone, and transverse faults. All three features were important conduits for hydrothermal circulation (Ey et al., 1985). Although heavy mineral beds have been observed in the sandstones around the D orebody and are associated with uranium elsewhere in the Athabasca Basin (de Carle, 1980), their metallotectic role is not clearly understood (Ruhlmann, 1985).

Basement-hosted mineralization is commonly found near the contact between the Peter River gneiss and the Earl River complex. Mylonites and extensive fracturing are evidence of tectonic activity in this zone and provided channels for the circulation of uranium-rich fluids.

There are three main mineralizing episodes in the Carswell structure. The 1050 to 1150 Ma mineralization (U-Pb dated by Bellon et al., 1976; Gancarz, 1979; and Bell, 1985) is the most important economically (Ruhlmann, 1985). The Claude, N, and Dominique-Peter orebodies contain uranium of this age. The other two episodes observed in the orebodies (900 Ma and 250 to 380 Ma) are less significant economically. The 900 Ma and 1050 Ma episodes of uranium deposition have never been differentiated in the sandstones, however in the basement gneisses in the Dominique-Peter orebody, both episodes of mineralization have been observed. The 250 to 380 Ma mineralization episode has been observed in both sandstones and basement gneisses.

With the exception of the Hudsonian-age mineralization found in the Sophie area (Ruhlmann, 1985) which is presently only comparable with similar ages in the Uranium City area (Tremblay, 1972), the three main episodes of uranium mineralization are similar in age to deposits elsewhere in the Athabasca Basin which show approximate ages of 1200 Ma, 800 Ma, and 300 Ma. The 1050 Ma mineralization may be analogous to that at Rabbit Lake (Knipping, 1974; Cumming and Rimsaite, 1979), Midwest Lake (Worden et al., in press), Key Lake (de Carle, 1980; Wendt et al., 1978), and the Collins Bay B zone (Jones, 1980). Much of this uranium has been dated closer to 1200 Ma and occurs as uranium-arsenic-sulphide assemblages in the deposits in the eastern part of the basin. The deposits in the western part of the basin, the Carswell deposits and those in the Uranium City areas, contain selenides instead of arsenides. The difference may be explained by variations in the geology of the basement rocks, since the western deposits occur within the Western Craton, and the eastern deposits are generally located within the Wollaston Domain. Tremblay (1978) clas-

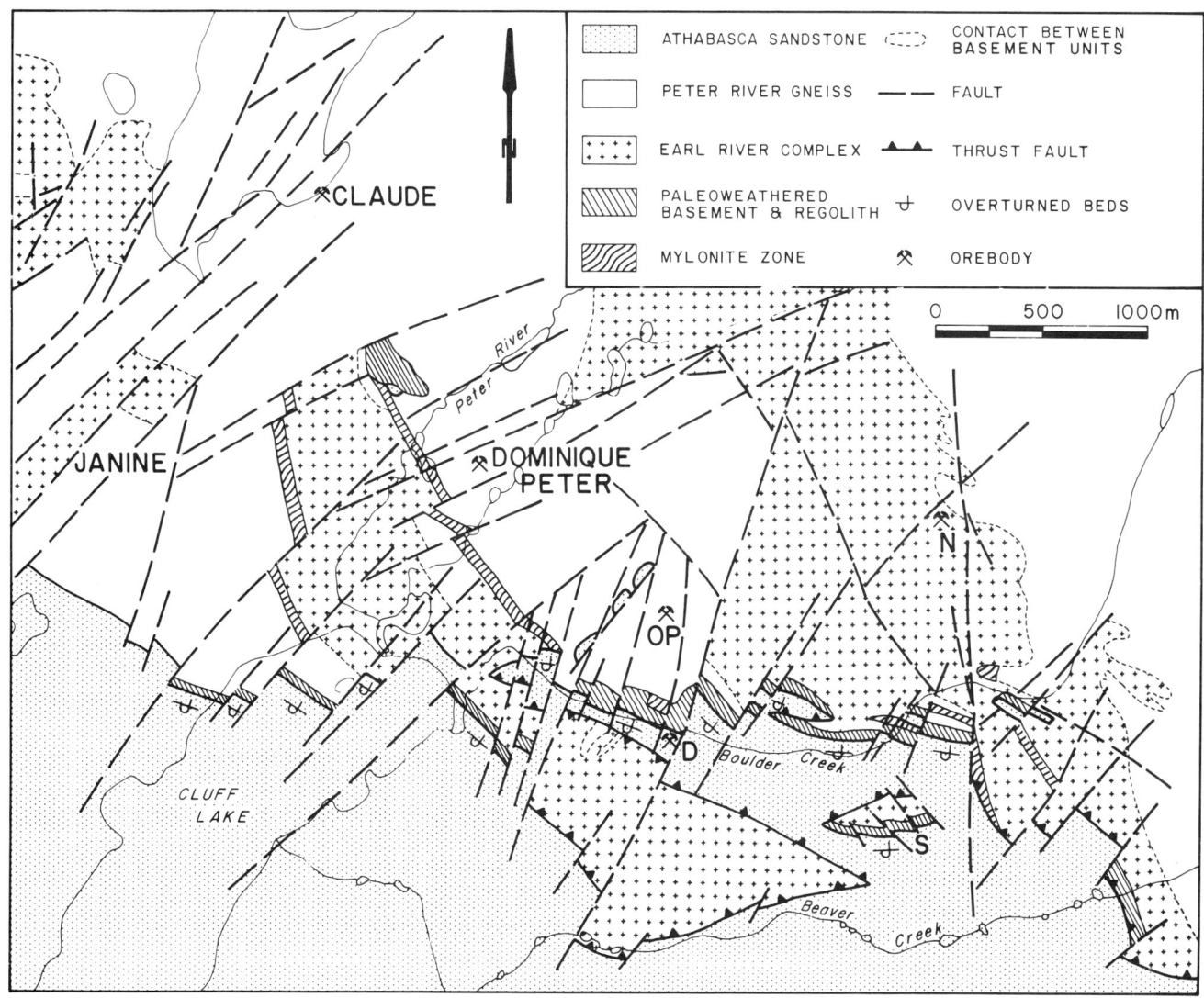

Figure 9. Subsurface geology and locations of ore deposits along the southern perimeter of the structure.

sified the basement-hosted deposits in the Carswell structure with the Beaverlodge District deposits.

The link between alteration and uranium mineralization has been widely studied in the Athabasca Basin deposits (for example: Macdonald, 1980; Sopuck et al., 1982; Hoeve, 1982; Tremblay, 1983). In the Carswell structure, Pagel and Svab (1985) have separated the alteration phases into retrograde (probably late Hudsonian in age), paleoweathering prior to (and diagenesis after) deposition of the Athabasca Group sediments, hydrothermal alteration during the main episode of uranium deposition, alteration linked with Cluff breccias formed during the Carswell event, and the youngest recorded alteration linked with the 380 Ma uranium mineralization in sandstones.

The following is a brief description of the individual orebodies (D, Dominique-Peter, Claude, N, and OP) and a few of the major showings in the Carswell structure.

D Orebody

The D orebody is located along the southern edge of the basement core at the unconformity with the overlying Athabasca Group sediments (Fig. 10). A highly altered mylonite zone cuts the Peter River gneiss and extends into the sediments giving rise to a "zone à boules" which is a lenticular shear zone of rotated blocks of sandstone enveloped in fault gouge (Ey et al., 1985). The Peter River gneiss, a regolith, and the Athabasca Group sediments were overturned and faulted up on bleached Earl River complex gneisses as a result of the Carswell event (Fig. 11). Ruzicka (1975) summarized the main features of the D deposit, and Ey et al. (1985) discuss the relationship between structure and uranium deposition in greater detail.

The orebody was mined by the open-pit method in 1979 and 1980. During that time, it yielded 110,000 tonnes at an average grade of 3.79% uranium. The mineralization formed discontinuous lenses over a 120 × 25 × 7 m area at the unconformity. Near the contact, uranium was also found in fractures within the basement gneiss.

Metallogenic studies on ore from the D orebody were conducted by Geffroy (1972) and Ruhlmann (1985). Massive

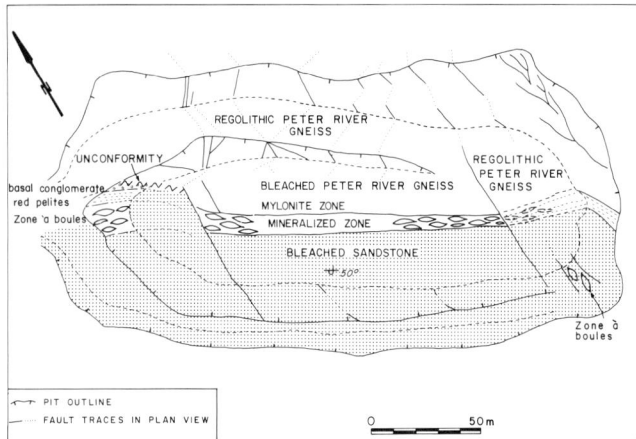

Figure 10. Plan view of the D open pit (modified from Ey et al., 1982).

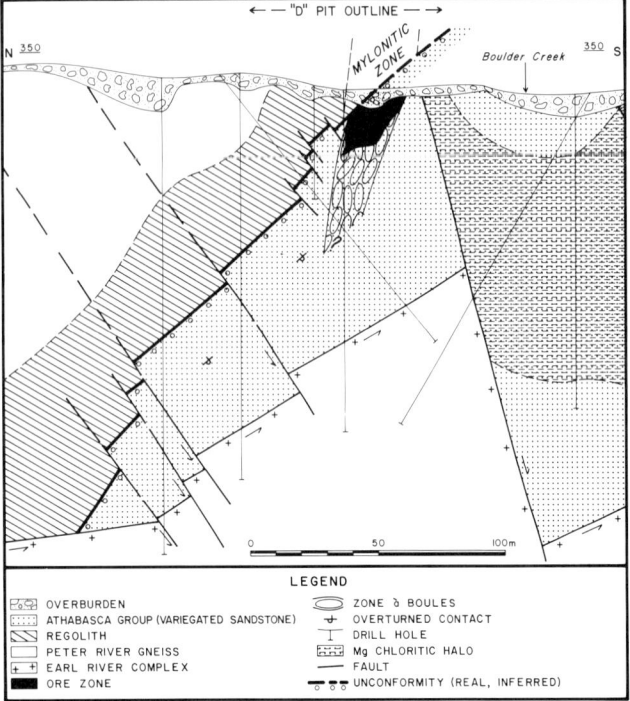

Figure 11. N-S cross section through the D open pit, showing alteration haloes and tectonic features (modified from Ey et al., 1982).

mineralization in the "zone à boules" contains uraninite and selenides (predominantly clausthalite), gold, tellurides, sulphur-nickel arsenides, jordisite, and minor pyrite and chalcopyrite. Organic material deposited later was studied by Disnar and Rouzaud (1980), and Landais and Dereppe (1985) and their role in uranium deposition was found to be a minor one. Coffinite is a secondary depositional phase.

Uraninite has also been observed in the hanging wall of the D orebody. In the sandstones, there is a narrow lens of abundant detrital titanium oxides and zircons. Uraninite occurs in this environment along the cleavages in the altered titanium minerals.

In the vicinity of the D orebody, but not in the open-pit itself, sporadic mineralization of a hematite-pitchblende association has also been noted. The type of occurrence and association is similar to that of the Donna area sandstone boulders and is part of the same depositional episode (Ruhlmann, 1982).

The emplacement of the main mineralization was accompanied by pervasive hydrothermal alteration that affects the basement as well as the sedimentary cover. Similar alteration episodes are noted in the east part of the Athabasca Basin (Hoeve et al., 1981; Hoeve, 1982; Sopuck et al., 1982) and the Northern Territory, Australia (Gustafson and Curtis, 1983).

In the D orebody, the deformation due to the Carswell event is very complex, however, a principal alteration halo characterized by magnesium-rich chlorite and dravite is still evident (Fig. 11). Near and inside the orebody, the clays are composed of both magnesium and aluminous chlorites. Secondary dravite is often observed as overgrowths on detrital tourmalines. Illite is the primary alteration clay outside of the deposit.

Alteration phenomena were observed in 70 diamond drill holes from the D orebody. The results, summarized on Figure 12, are projections of these observations to the bedrock surface. The dip of the overturned unconformity is 50° to 60° to the north, and therefore, the real extension of the haloes is somewhat exaggerated. There are three main alteration haloes in the orebody:

1) A mineralized halo (Fig. 12a): All significant radioactive occurrences are outlined by a halo. One may observe that faults F3 and F4 seem to deflect the contour lines. Below the lowest contour, the background uranium in the sandstone is less than 5 ppm. At the level of the orebody, anomalous radioactivities are apparent over a maximum area 40 m wide by 153 m long.

2) A reduction halo (Fig. 12b): To the north, the regolith is composed of a red and a red-green zone. All facies which are grey-green to olive green, or green to pale green are considered here to be related to a reducing phenomenon irregardless of the clay mineralogy. To the south, in the sediments, the same criteria was used. The haloes appear to be less extensive in the basement (25 m by 148 m maximum) than in the sediments (40 m by 185 m). The K-Ar age of illites from this alteration halo is 1293 ± 36 Ma (Clauer et al., 1985), and therefore the emplacement of the mineralization (1050 to 1150 Ma) post-dates the alteration phase.

3) A hematite halo (Fig. 12c): An overprinting hematite halo exists close to the edge of the reduced zone in the sandstones. Located one to several metres above the zone, this halo may enclose some mineralization. It varies in colour from earthy hematitic to brown to buff to reddish or greenish brown. It is sometimes silicified and is located over a maximum area of 138 m by 25 m in plan view. This hematite may be the same phase linked with the 380 Ma pitchblende-hematite episode (Ruhlmann, 1985).

Beyond the reduced and hematized zones, the sandstones are bleached and are usually grey, white, or pale pink. They are commonly vuggy, fractured, and occasionally show quartz dissolution over an area extending 50 to 60 m to the south of the orebody. Due to the marked contrast in the sandstones away from the ore zone, these haloes, although disturbed by the Carswell event, could by used as a guide for other D-type orebodies. This method is presently being tested along the southern rim of the basement core between the D and Myra areas (Fig. 2).

Devillers and Nordmann (1974) determined ages of 1150 ± 25 Ma ($^{206}Pb/^{238}U$ and $^{207}Pb/^{235}U$) on a selenide-bearing, massive pitchblende and uraninite sample. Gancarz (1979) derived an age of 1050 ± 30 Ma ($^{206}Pb/^{207}Pb$) on massive mineralization in the "zone à boules".

Dominique-Peter Orebody

The Dominique-Peter orebody is located northeast of Cluff Lake and approximately 1 km north of the southern boundary of the basement/sandstone contact. Uranium mineralization is contained entirely within the basement gneisses. The controlling factors for the deposition are a combination of structure, lithology, degree of alteration, and possibly the proximity to the unconformity.

The Peter River gneisses and the Earl River complex in the area are commonly separated by a mylonite (Fig. 13), probably Hudsonian in age. The northwest-trending mylonite zone at the interface between these two series is usually highly altered to a mixture of sericite and chlorite. This zone is the same mylonite zone described in the basement gneisses of the D orebody.

Pagel and Svab (1985) observed a variation in the chlorites towards the mineralized veins, becoming Mg-rich close to mineralization. These chlorites become more Fe-rich outside of the mineralization and are replaced by illite far away from the mineralization.

A wedge of overturned sandstone and regolithic basement occurs northwest of Dominique-Peter (Fig.9). Although true regolithic alteration is absent in the Dominique-Peter ore zone itself, the proximity to the unconformity may be another controlling factor for the mineralization, as is suggested by the presence of the sedimentary wedge.

The area is tectonically affected by three sets of faults trending N60°E, N-S and N120°E, all of which displace the main mylonite zone. Mineralization is found mainly in the N60°E and the N-S faults and occurs as fracture fillings, veinlets, and disseminations. This structurally controlled mineralization in the Peter River gneiss is usually located less than ten's of metres from the mylonite zone (Blaise and Koning 1985). In the Earl River complex, mineralization is not as rich and is more fracture controlled. Mechanical reworking of mineralization by Cluff breccia has been observed on a local scale. The average tonnage calculated for the Dominique-Peter orebody (both indicated and inferred) is 1,761,000 tonnes at an average grade of 0.66% uranium.

Two main paragenetic assemblages have been observed in the Dominique-Peter orebody: a uraninite-polymetallic sulphide assemblage and a uraninite-dravite-simple sulphide assemblage. Bell (1985) has determined U-Pb ages ranging

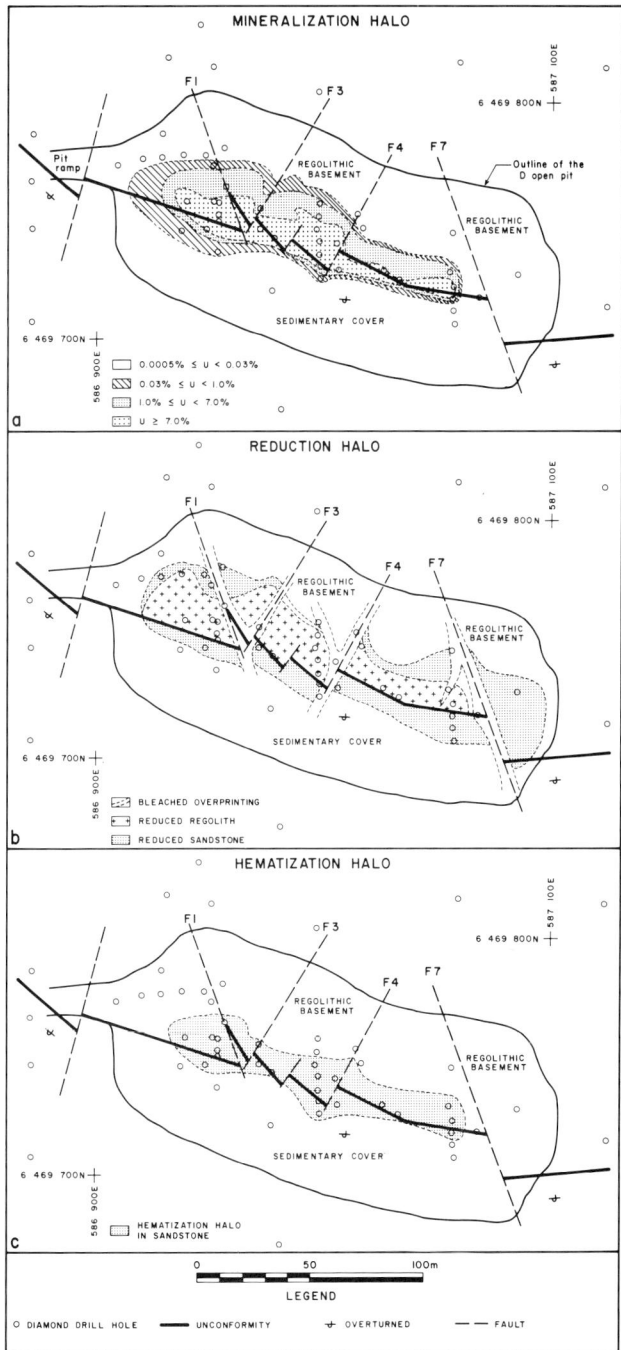

Figure 12. a) Mineralization, b) reduction and c) hematization haloes around the D orebody.

between 950 and 1050 Ma for uranium mineralization of the polymetallic association from the deposit. By analogy with the OP deposit, where Bell (1985) dated a uraninite-dravite assemblage ranging between 820 and 890 Ma, it can be inferred that the similar assemblage at Dominique-Peter is also the younger mineralization.

Tectonic breccia zones are also strongly altered and often contain a secondary mineralizing episode where coffinite is the principal uranium mineral.

Claude Orebody

The Claude orebody is located north of Dominique and is also completely hosted by basement gneisses. Contrary to the N and Dominique-Peter mineralization, the principal control is a tectonic zone with rotated fault blocks which are analogous to the "zone à boules" described in the D orebody. In Claude, the main mineralization is contained in the E-W vertical tectonic zone, but some younger NE-SW faults are also mineralized. This mineralization was traced in a decline in 1979, and then by an experimental pit in 1982 (Fig. 14). The average tonnage calculated for the Claude orebody is 640,000 tonnes at an average grade of .35% uranium.

The host rocks are composed of Peter River gneiss, and several interfoliated granitoids outcrop at the northern end of the pit. The gneiss is strongly altered around the mineralization, and thin section observations indicate abundant argillization of the silicates. Chloritization is ubiquitous and overprints a variable sericitization. Garnets are generally chloritized but still exhibit fresh fragments. The sericitization and chloritization of the biotite is accompanied by the formation of aggregates of anatase (Reyx, 1980). Tourmaline may be present locally.

Lillié (1982) observed a N-S trending S_1 foliation attributed to regional metamorphism, overprinted by an E-W trending S_2 schistosity which reflects a later ductile deformational event. This second schistosity may be associated with the development of the mylonite zones in the Dominique-Peter and N areas. Two northeasterly-trending shear zones intersect the E-W zone and may be due to an even later tectonic event.

A series of late N20°E faults have displaced the E-W mineralization to the east. Despite the extreme complexity on a small scale, the envelope containing the mineralization is uniform. The southern part of the E-W pod associated with the "zone à boules" dips subvertically from the experimental pit to the 50 m level in the decline. The N20°E shear zone plunges gently to the west and can be clearly traced in both the pit and the decline.

Mineralization is located in a variety of different structures, but is principally contained in the "zone à boules." It was deposited in fault gouge between the "boules" and has a banded appearance. Other mineralization was found in NE-SW clay fillings, in a few fractures in fresh rock, and in some Cluff breccias (Fig. 15).

Mineralization was studied by Geffroy (1972) who observed uraninite, uranium-titanium minerals, clausthalite, galena, pyrite, chalcopyrite, sphalerite, and jordisite. Uraninite-polymetallic sulphide mineralization has been dated using $^{206}Pb/^{238}U$ and $^{207}Pb/^{235}U$ at 1050 ± 65 Ma (Devillers, 1974), and occurs as impregnations in the fault gouge. The crystal form of uranium depends on host minerals:

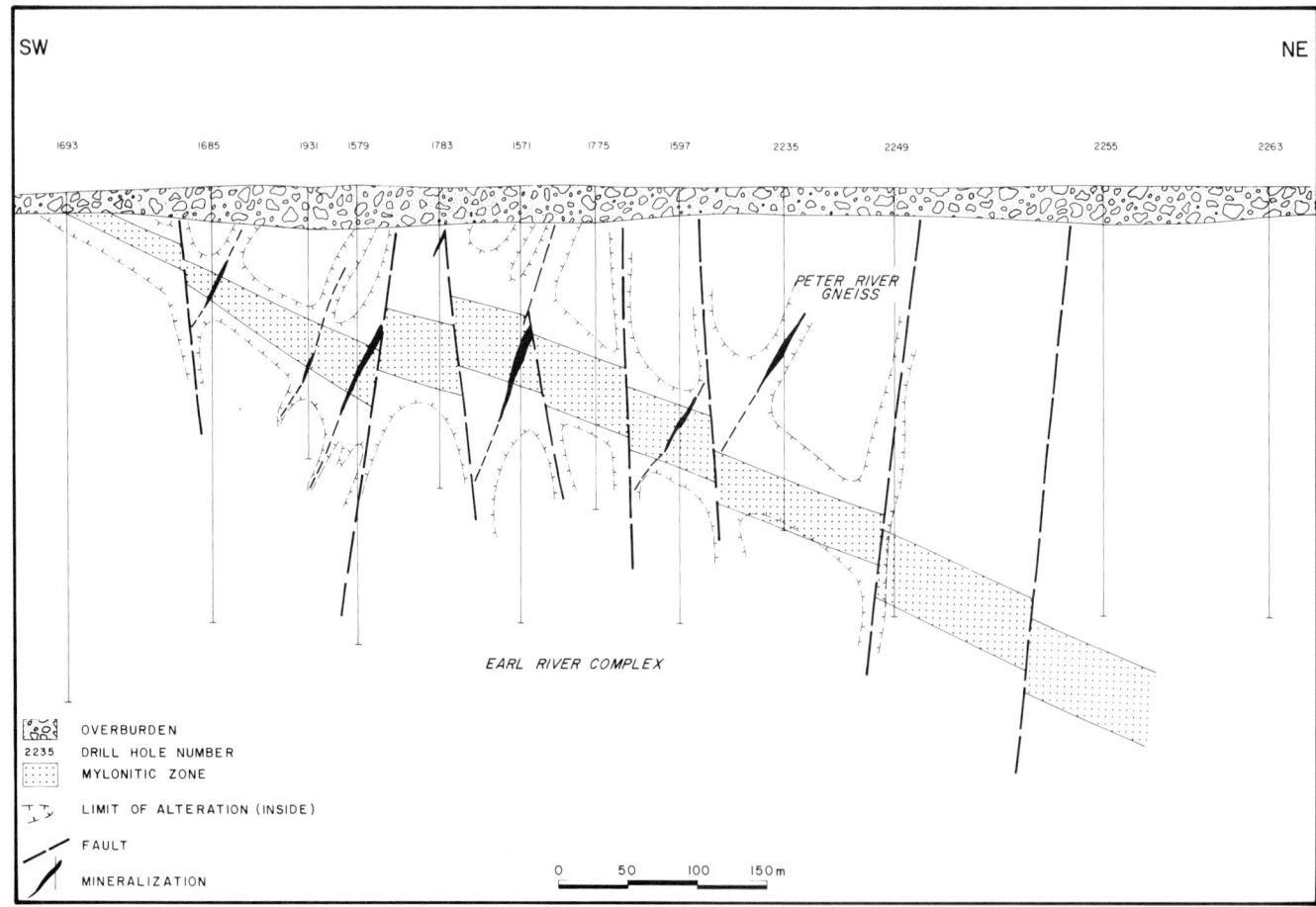

Figure 13. NE-SW cross section through the Dominique-Peter deposit (Blaise and Koning, 1985).

pitchblende occurs in the clays parallel to cleavage; and euhedral uraninite occurs in aureoles around titanium oxide minerals. Coffinite, covellite, digenite, hematite, and limonite have also been observed.

Organic matter associated with the mineralization is younger than the main mineralization. Uraniferous inclusions in the organic material are leached phases from pre-existing mineralization (Landais and Dereppe, 1985).

N Orebody

The N orebody is located within strongly altered mixed gneisses and granitoids of the Earl River complex which overlie Peter River gneiss (Leguere and Chauvet, 1982). It is composed of a series of discontinuous mineralized fractures trending N-S and dipping 45° to the west (Fig. 16.). These fractures are cut in the north by a N40° fault which contains a mineralized vein-type breccia. The steeply dipping N40° fault shows dextral movement. A third set of faults, N135° to N150°E are conformable with the foliation.

Geffroy (1972), and more recently, Harper (1983) distinguished an old illite-kaolinite alteration associated with regolith, and a later hydrothermal chlorite alteration. Regolithic alteration has been observed in the N area and is again an indication of proximity to the Athabasca sediments as in the Dominique-Peter deposit.

The N40° vein-like structures are filled with bleached fault gouge. Cluff breccias were later injected into the structure, however, the walls remained unaltered. The mineralization

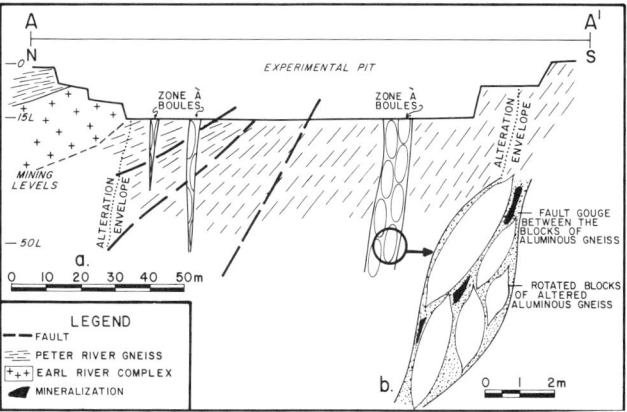

Figure 15. a) N-S cross section (A-A') through the Claude orebody. b) Detailed view of mineralization in the "zone à boules".

is observed as impregnations, as veinlets, or in the matrix of the mineralized breccia (Fig. 17a). It is mainly a mechanical remobilization of a pre-existing mineralization.

The N-S mineralization (Fig. 17b) is contained in an altered zone that is controlled by fractures dipping 45°W. This envelope is a mixture of altered tectonic breccias and fresh rocks of the Earl River complex. The configuration of these blocks of fresh rock resembles that of the "zone à boules" in the D and Claude orebodies. Mineralization is found in the altered zones between the blocks. Cluff breccias contain fragments of displaced mineralization. The average tonnage in the N orebody from both major structures has been calculated to be 505,000 tonnes at an average grade of 0.34% uranium.

The gangue is essentially chlorite with some illite. Uraninite is the principal uranium mineral and occurs along the phyllite cleavages or in veinlets. Galena, clausthalite, chalcopyrite, covellite, pyrite, and jordisite are accessory to the uranium in the 1050 Ma mineralization (Geffroy, 1972). Coffinite haloes and veinlets of quartz-pitchblende could be evidence of 900 Ma mineralization. Late organic material (thucholite and bituminous material) is also present. There is a late transformation of uraninite to uranophane near surface.

OP Orebody

The OP orebody is located to the east and northeast of the O and P showings, respectively. These two showings were highly mineralized boulders discovered in areas to the north and northwest of D. Drilling through resistivity contacts and conductors intersected discontinuous mineralization, and an exploratory decline was opened during 1980 and 1981.

At the beginning of the decline, a slab of fresh Peter River gneiss striking E-W and dipping 30°N is thrust over a slab of sandstone, conglomerate, and pelite (Fig. 18). Further east, an overturned slab of regolith and sediments are thrust along a N65°E fault over other basement gneisses. The O mineralization occurs at the junction of these two thrust faults. The Peter River gneiss comprises the main rock type for the remainder of the decline. Mineralization strikes N150° to 170°E, dipping 45° to 60°E, with minor fractures oriented

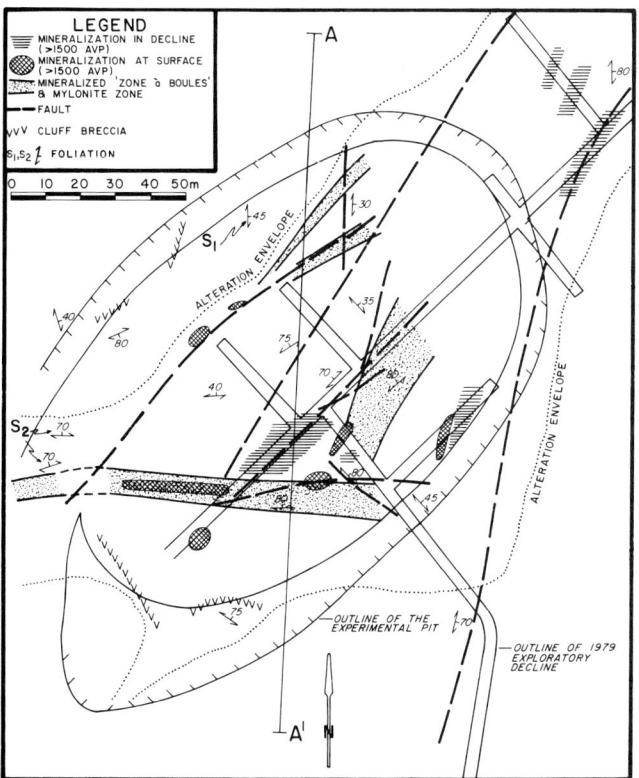

Figure 14. Plan view of the Claude orebody with cross section location. Structural interpretation and mineralized areas at surface and in the decline (modified from Lillié, 1982).

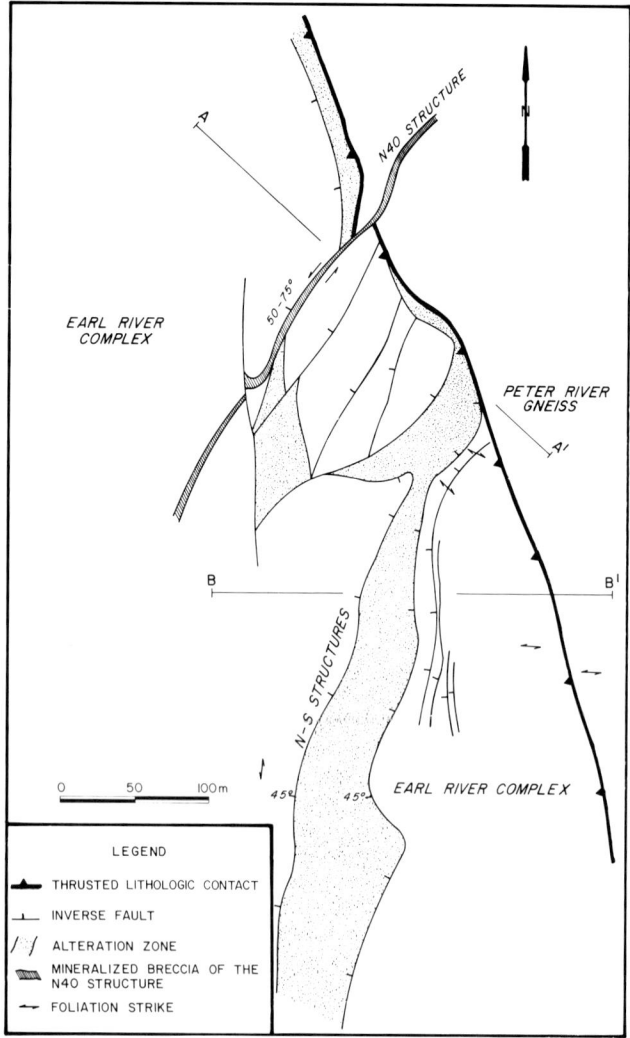

Figure 16. Plan view and lithostructural interpretation of the N orebody with cross section location (modified from Leguéré, 1982).

N30° to 40/45°W. The relatively fresh rocks are crosscut by narrow, discontinuous NE-trending, mineralized quartz veins. Garnetite, aluminous gneiss, granitoid, and minor mafic gneisses are the major rock types in this area. Cluff breccias are ubiquitous and have often crosscut the mineralization causing both mechanical and chemical remobilization.

Three main subvertical mineralized N-S structures have been observed in the decline from east to west. Discontinuous, thin mineralization occurs with vein quartz in these fractures. Uraninite is deposited on quartz crystals, along with chalcopyrite, galena, clausthalite, and pyrite. Magnesium-rich tourmaline with a second quartz phase is commonly crystallized after the uraninite. Coffinite and magnesium chlorite have also been observed in contact with the uraninite mineralization (Ruhlmann, 1985) and alteration zones around the veins are very narrow. Uranium in quartz veins has also been observed in Dominique-Peter and is related to similar hydrothermal processes.

Gancarz (1979) dated some of the mineralization from OP at 800 Ma ($^{207}Pb/^{206}Pb$). He considered it to be totally independent of the 1050 Ma mineralizing episode. Similar ages have been described at Key Lake (918 Ma, U-Pb, Wendt et al., 1978). In the Carswell structure, this episode affects all known mineralized areas and is particularly well developed in the OP orebody, where Bell (1985) obtained U-Pb ages from a uraninite-dravite assemblage between 820 and 890 Ma.

The average calculated tonnage for the OP orebody is 55,000 tonnes with an average grade of 0.28% uranium.

Other Uranium Occurrences

In the Carswell structure, hematite-pitchblende and carbonate-pitchblende assemblages have been classified as the youngest known uranium mineralization. These assemblages occur elsewhere in the Athabasca Basin and appear to originate from remobilization of the 1050 Ma mineralization event (Hoeve et al., 1981). This type of mineralization is described at Key Lake (270 to 250 Ma, Wendt et al., 1978), Midwest Lake (334 ± 16 Ma, Worden et al., 1981), and in the Beaverlodge District (300 Ma, Tremblay, 1972). In the Carswell area, it has been identified in samples from the Donna boulder train.

The Donna boulder train lies 1.5 km SE of the D orebody in a north-south orientation (Fig. 2). The sandstone boulders are generally fine-to medium-grained and moderately to strongly hematized. Hematite commonly forms the cement of these sandstones and may also be present along microfractures. Epigenetic uranium mineralization is found along fractures or in brecciated areas of the boulders. Vein-type textures have been observed in which the center is filled with hematite and pitchblende. In the more highly radioactive boulders, uranium minerals replace the matrix, quartz grains are corroded and sometimes dissolved completely. Gold and galena are found along with limonite and goethite, all in association with hematite and pitchblende (Ruhlmann, 1985). U-Pb ages on a hematite-pitchblende sandstone boulder yield 380 ± 5 Ma (Bell, 1985).

Basement boulders were also discovered in the Donna boulder train. They are highly radioactive and stained with yellow products. The host for the mineralization may be either pegmatoid or Peter River gneiss. The pegmatoid boulders are altered and contain minor biotite and garnets. Accessory minerals include allanite, zircon, and titanium minerals. The biotite-rich Peter River gneiss boulders are highly chloritized and sericitized.

The R occurrence was discovered by diamond drilling and is located in Peter River gneiss to the SE of the N orebody. A set of flat-lying NE fractures plunging NW control a large box-shaped alteration zone. Uranium mineralization is discontinuous and consists of uraninite (sometimes altered to coffinite), pyrite, neodigenite, and jordisite (Geffroy, 1972). Organic material is also present. Gangue is composed of sericite and chlorite, and in a recent study, Reyx (pers. commun., 1983) found relics of titanium minerals associated with uraninite and molybdenum sulphides. Minor chalcopyrite, galena, and pyrite have been observed. This mineralization could be compared to the mineralization in the N orebody.

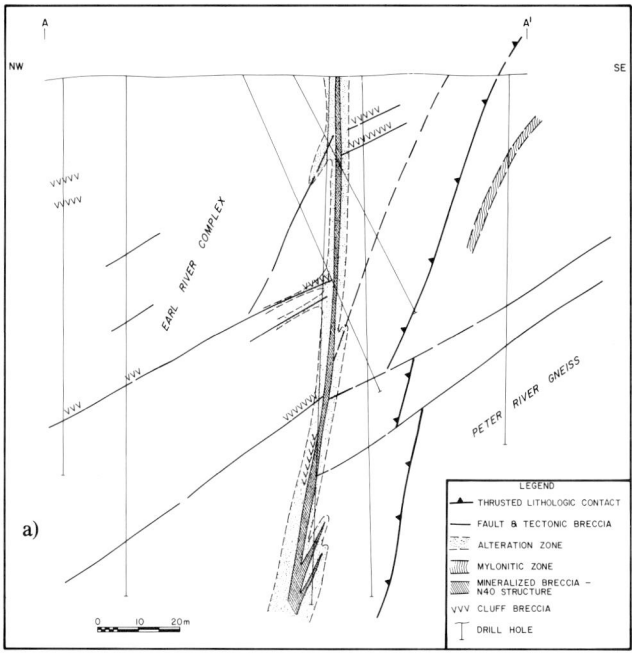

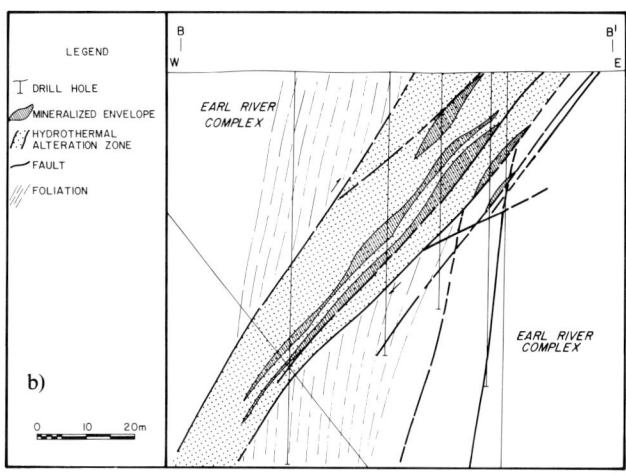

Figure 17. a) NW-SE cross section (A-A') through the N orebody, showing the N40° mineralized structures. b) E-W cross section (B-B') showing N-S mineralized structure.

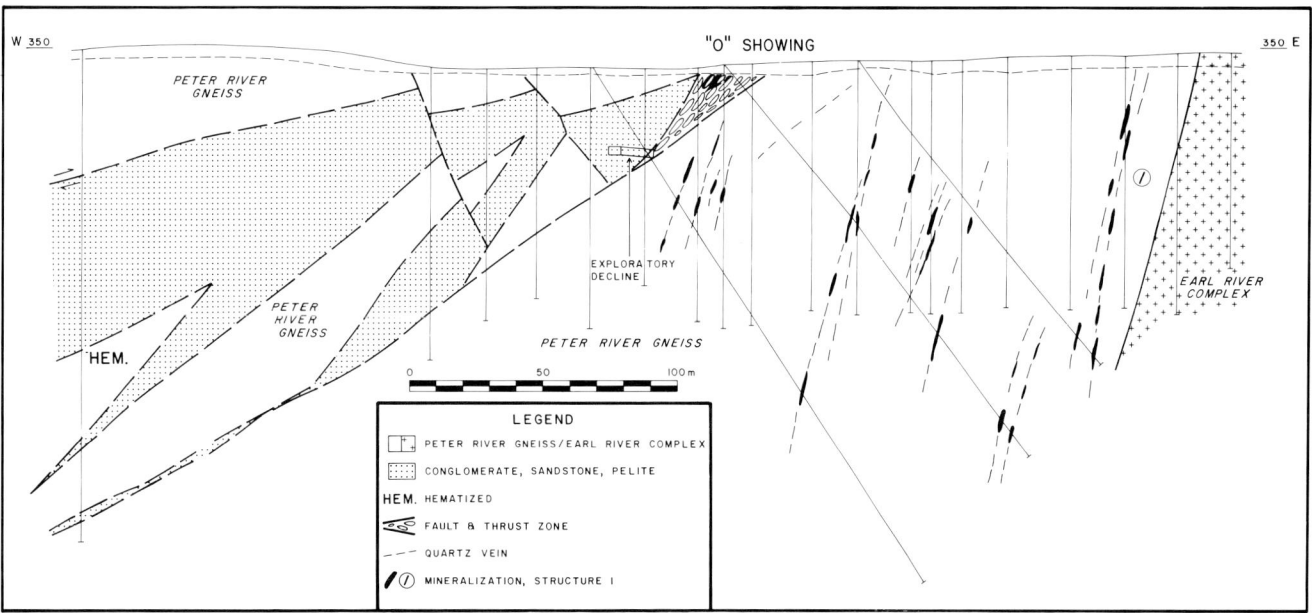

Figure 18. E-W cross section through the OP orebody.

CONCLUSION

The Carswell circular structure is a complex assemblage of metamorphic rocks which form the basement, and Helikian sedimentary rocks which form the cover. Two major units can be observed in the basement core: the Earl River complex and the Peter River gneiss, which underwent granulite facies metamorphism with retrograde metamorphism during late Hudsonian. The retrograde metamorphism had a primary role in the liberation of uranium from phyllitic minerals. All age dating performed on basement core samples from the Carswell area yield Aphebian dates.

The basement rocks were affected by paleoweathering prior to deposition of the Helikian sediments of the Athabasca Basin. A pocket of unusual red sedimentary rocks directly overlie the basement rocks in the Cluff Lake area and could be paleo-Helikian in age, since it is very difficult to correlate these sediments with any found elsewhere in the inner sedimentary ring or even outside of the structure.

The Ordovician Carswell event created a new tectonic pattern in the area which overprinted all earlier linear patterns and also produced the vein and dyke-like intrusives of Cluff breccia.

All presently known economic uranium is located along the southern perimeter of the basement core, however, minor occurrences have been observed throughout the structure.

The deposits can be separated into two types: mineralization not apparently associated with the unconformity, and mineralization directly associated with the unconformity. Strict identification of Hudsonian-age uranium is difficult. A chemical age of 1800 Ma is calculated on uranium at a contact between mafic gneiss and pegmatoid in the Laure area. The Numac, Sophie, Brigitte, and Pascale areas could also contain relict mineralization associated with Hudsonian retrograde metamorphism since these occurrences are very similar to the Laure area. In the Sophie area a uraninite-monazite assemblage hosted by feldspathic gneisses and pegmatoids can be compared with assemblages in the Beaverlodge district belonging to a 2000 Ma mineralizing episode (Tremblay, 1982). These minor occurrences are considered to be pre-concentrations of uranium.

The mineralization directly associated with the unconformity has three different ranges of dates: the major 1050 Ma episode, a lesser 900 Ma episode, and the latest 380 Ma episode.

The 1050 Ma episode produced a uraninite-polymetallic sulphide assemblage. Variations in the mineralogy are related to the type of host rock. Peter River gneiss contains uraninite, brannerite, Mo-sulphides, selenides, tellurides, gold, silver, and bismuth. The mixed gneisses of the Earl River complex typically contain uraninite, Pb and Cu sulphides, and selenides. The Claude, Dominique-Peter, N, Claude (West Claude), and D orebodies contain this uranium assemblage.

The controls for this major mineralization have been divided into four metallotects. The first is the interface between the Peter River gneiss and the Earl River complex. The interface is not only a former stratigraphic contact but also marks the limit of anatexis which has taken place in the Earl River complex (mantle gneiss doming?). The pyrite and graphite within the Peter River gneiss form a good reducing environment. The second metallotect is a mylonite zone at the same contact which can be traced in the Carswell structure from the Claude orebody in the NW to the Myra area in the SE and passes through or near to the Dominique-Peter, OP, D, and N orebodies and the R occurrence. The D orebody occurs at the intersection of the mylonite and the Athabasca unconformity. The third metallotect is the transverse faulting through the mylonite which opened fractures (N-S and N60°E) as a result of strong NE-SW compression. All of the above structural metallotects created significant channelways for multiple phases of hydrothermal circulation. The fourth metallotect is essential for uranium deposition and is definitely related to the pre-Athabasca unconformity. The overlying sediments, besides acting as a protective cover, provide an ideal fluid source and may also be a minor source for uranium.

Mineralization dated at around 900 Ma contains the same metal assemblage as the 1050 Ma mineralization but without significant amounts of selenides and tellurides. This vein-type mineralization is very well displayed in the OP orebody and is hosted by Peter River gneiss. It is probably a remobilization of the 1050 Ma episode.

The 380 Ma mineralization observed in the boulders from the Donna boulder train, S area, and the southern wall of the D orebody is a hematite-pitchblende assemblage which is probably also a remobilization of the major 1050 Ma episode.

Recognition of the various similarities and metallotects which characterize the Carswell deposits and their correlation with deposits elsewhere in the Athabasca Basin should be considered as valuable guides for further exploration in northern Saskatchewan.

ACKNOWLEDGEMENTS

The authors are grateful to Amok Ltd. for support, encouragement, and unrestricted access to data. This paper was read, and valuable criticisms were made, by three anonymous reviewers. The english text was greatly improved by K. Wheatley and D.M. Burton. Drafting was done by D. Geisbrecht, and the manuscript was typed by L.L. McNally. The authors wish to recognize that without the diligent and often unacknowledged work by numerous past exploration scientists in the Carswell area, this paper could not have been written.

REFERENCES

Agarwal, R.G., 1965, Regional Correlation of Geological and Geophysical Data in the Wollaston Lake Area, Northern Saskatchewan: Saskatchewan Department of Mineral Resources Report no. 82.

Bassaget, J.P., and Camps, P., 1973, Quadrant Nord-Ouest du Noyau de Socle, Géologie et Prospection (Claimblocks 2325, 2326, 2327): Amok Internal Report.

Bell, K., 1985, Geochronology of the Carswell Area, Northern Saskatchewan: in Lainé, R., Alonso, D., and Svab, M., eds., The Carswell Structure Uranium Deposits, Saskatchewan: Geological Association of Canada Special Paper 29.

Bell, K., and Blenkinsop, J., 1981, Saskatchewan Shield Geochronology Project: in Summary of Investigations 1980, Saskatchewan Geological Survey Miscellaneous Report 80-4, p. 18.

Bell, K., Cacciotti, A.D., and Schnessl, J.H., 1985, Petrography and Geochemistry of the Earl River Complex, Carswell Structure, Saskatchewan − A Possible Komatiitic Succession: in Lainé, R., Alonso, D., and Svab, M., eds., The Carswell Structure Uranium Deposits, Saskatchewan: Geological Association of Canada Special Paper 29.

Bellon, H., Devillers, C., Hagemann, R., and Touray, J.C., 1976, Dater les Minéralisations: Mémoire Hors Série de la Société Géologique de France, No. 7, p. 265-268.

Blaise, J.R., and Koning, E., 1985, Mineralogy and Structural Aspects of the Dominique-Peter Uranium Deposit: in Lainé, R., Alonso, D., and Svab, M., eds., The Carswell Structure Uranium Deposits, Saskatchewan: Geological Association of Canada Special Paper 29.

Blake, D.A.W., 1952, Geological Notes on the Region South of Lake Athabasca and Black Lake, Saskatchewan and Alberta: Geological Survey of Canada Paper 55-53, 12 p.

de Carle, A.L., 1980, Geology of the Key Lake Deposits, Saskatchewan: Canadian Institute of Mining, District Annual Meeting, September, 1980, Flin Flon, Manitoba.

Clauer, N., Ey, F., and Gauthier-Lafaye, F., 1985, K-Ar Dating of Different Rock Types from the Cluff Lake Uranium Ore Deposits (Saskatchewan-Canada): in Lainé, R., Alonso, D., and Svab, M., eds., The Carswell Structure Uranium Deposits, Saskatchewan: Geological Association of Canada Special Paper 29.

Cumming, G.L., and Rimsaite, J., 1979, Isotopic Studies of Lead-Depleted Pitchblende, Secondary Radioactive Minerals and Sulphides from the Rabbit Lake Uranium Deposit, Saskatchewan: Canadian Journal of Earth Sciences, v. 16, p. 1702-1715.

Currie, K.L., 1969, Geological Notes on the Carswell Circular Structure, Saskatchewan (74K): Geological Survey of Canada Paper 67-32, 60 p.

Devillers, C., 1974, Datation de l'Echantillon "Claude" par la Méthode U/Pb. Commissariat à l'Energie Atomique, Internal Report.

Devillers, C., and Nordmann, F., 1974, Datation du Minerai Uranifère de Cluff – Premiers Résultats. Commissariat à l'Energie Atomique, Internal Report.

Disnar, J.R., and Rouzaud, J.N., 1980, Etude du Kérogène Provenant du District Uranifère de Cluff (Saskatchewan), Protocole Experimental et Premiers Résultats: Amok Internal Report.

von Einsiedel, C.A., 1981, Petrography and Geochemistry of the Cluff Lake Breccias, Carswell Structure, Northern Saskatchewan: Unpublished B.Sc. Thesis, Carleton University, Ottawa, 44 p.

Ey, F., Lillié, F., Gauthier-Lafaye, F., and Weber, F., 1985, A Uranium Unconformity Deposit: The Geological Setting of the D Orebody (Saskatchewan-Canada): in Lainé, R., Alonso, D., and Svab, M., eds., the Carswell Structure Uranium Deposits, Saskatchewan: Geological Association of Canada Special Paper 29.

Fahrig, W.F., 1961, The Geology of the Athabasca Formation: Geological Survey of Canada Bulletin 68, 41 p.

Gancarz, A.J., 1979, Chronology of the Cluff Lake Area Uranium Deposit, Saskatchewan, Canada (Abstract): International Uranium Symposium on the Pine Creek Geosyncline, N.T., Australia, Extended Abstracts, p. 91-94.

Geffroy, J., 1972, Données Préliminaires sur le District Uranifère du Lac Cluff (Saskatchewan): Cadre Géologique, Pétrographique et Paragénèses, Amok Internal Report.

Godfrey, J.D., and Langenberg, C.W., 1978, Metamorphism in the Canadian Shield of Northeastern Alberta: in Metamorphism in the Canadian Shield: Geological Survey of Canada Paper 78-10, p. 129-138.

Gustafson, L.B., and Curtis, L.B., 1983, Post-Kombolgie Metasomatism at Jabiluka, Northern Territory, Australia and its Significance in the Formation of High Grade Uranium Mineralization in Lower Proterozoic Rocks: Economic Geology, v. 78, p. 26-56.

Harper, C.T., 1982, Geology of the Carswell Structure, Central Part (Parts of NTS Areas 74K-5, -6, -11, -12): Saskatchewan Mineral Resources Report 214, 6 p.

———, 1983, The Geology and Uranium Deposits of the Central Part of the Carswell Structure, Northern Saskatchewan, Canada: Unpublished Ph.D. Thesis, Colorado School of Mines, Golden, Colorado, 337 p.

Hendry, H.E., and Wheatley, K.L., 1985, The Carswell Formation, Northern Saskatchewan: Stratigraphy, Sedimentology, and Structure: in Lainé, R., Alonso, D., and Svab, M.; eds., The Carswell Structure Uranium Deposits, Saskatchewan: Geological Association of Canada Special Paper 29.

Herring, B.G., 1976, The Metamorphism and Alteration of the Basement Rocks in the Carswell Circular Structure, Saskatchewan: Unpublished M.Sc. Thesis, University of British Columbia, Vancouver, 134 p.

Hoeve, J., 1982, Host Rock Alteration and its Application as an Ore Guide at the Midwest Uranium Deposit, North Saskatchewan: Saskatchewan Research Council Publication, p. 1-44.

Hoeve, J., Rawsthorn, K., and Quirt, D., 1981, Uranium Metallogenic Studies: Clay Mineral Stratigraphy and Diagenesis in the Athabasca Group: in Summary of Investigations 1981: Saskatchewan Geological Survey Miscellaneous Report 81-4, p. 76-89.

Hoeve, J., Quirt, D., and Alonso, D., 1983, Uranium Metallogenic Studies: Clay Mineral Distribution in the Athabasca Group, Southwest Part of the Basin: in Summary of Investigations 1983: Saskatchewan Geological Survey Miscellaneous Report 83-4, p. 89-95.

———, 1985, Clay Mineral Stratigraphy of the Athabasca Group: Correlation Inside and Outside the Carswell Structure: in Lainé, R., Alonso, D., and Svab, M., eds., The Carswell Structure Uranium Deposits, Saskatchewan: Geological Association of Canada Special Paper 29.

Innes, M.J.S., 1964, Recent Advances in Meteorite Crater Research of the Dominion Observatory: Meteoritics, v. 2, no. 2, p. 230-234.

Jones, B.E., 1980, The Geology of the Collins Bay Uranium Deposit, Saskatchewan: Canadian Institute of Mining Bulletin, v. 73, no. 818, p. 84-90.

Knipping, H.D., 1974, The Concepts of Supergene versus Hypogene Emplacement of Uranium at Rabbit Lake, Saskatchewan, Canada: in Formation of Uranium Ore Deposits: International Atomic Energy Agency, Vienna, p. 531-548.

Landais, P., and Dereppe, J.M., 1985, A Chemical Study of the Carbonaceous Material from the Carswell Structure: in Lainé, R., Alonso, D., and Svab, M., eds., The Carswell Structure Uranium Deposits, Saskatchewan: Geological Association of Canada Special Paper 29.

Leguere, J., and Chauvet, J.F., 1982, Rapport de Synthèse du Gisement N: Amok Internal Report.

Lewry, J.F., Sibbald, T.I.I., and Rees, C.J., 1978, Metamorphic Patterns and their Relation to Tectonism and Plutonism in the Churchill Province in Northern Saskatchewan: in Metamorphism in the Canadian Shield: Geological Survey of Canada Paper 78-10, p. 139-154.

Lillié, F., 1982, Analyse Tectonique du Gisement Claude (Cluff Lake, Saskatchewan): Amok Internal Report.

Macdonald, C.C., 1980, Mineralogy and Geochemistry of a Precambrian Regolith in the Athabasca Basin: Unpublished M.Sc. Thesis, University of Saskatchewan, Saskatoon, 151 p.

Pacquet, A., and McNamara, S., 1985, A Study of the Basal Athabasca Succession in the D, E, L, F, and S Areas of the Carswell Strucutre: in Lainé, R., Alonso, D., and Svab, M., eds., The Carswell Structure Uranium Deposits, Saskatchewan: Geological Association of Canada Special Paper 29.

Pagel, M., 1975, Cadre Géologique des Gisements d'Uranium dans la Structure Carswell (Saskatchewan, Canada): Etude des Phases Fluides: 3è Cycle Docteur de Spécialité, Université de Nancy, 157 p.

Pagel, M., and Ruhlmann, F., 1985, Chemistry of Uranium Minerals in Deposits and Showings of the Carswell Structure (Saskatchewan-Canada): in Lainé, R., Alonso D., and Svab, M., eds., The Carswell Structure Uranium Deposits, Saskatchewan: Geological Association of Canada Special Paper 29.

Pagel, M., and Svab, M., 1985, Petrographic and Geochemical Variations within the Carswell Structure Metamorphic Core and their Implications with Respect to Uranium Mineralization: *in* Lainé, R., Alonso, D., and Svab, M., eds., The Carswell Structure Uranium Deposits, Saskatchewan: Geological Association of Canada Special Paper 29.

Powell, B., Koning, E., and Lainé, R., 1985, Geophysical Mapping of Gneiss Domes in the Carswell Structure and their Relationship to Uranium Mineralization: *in* Lainé, R., Alonso, D., and Svab, M., eds., The Carswell Structure Uranium Deposits, Saskatchewan: Geological Association of Canada Special Paper 29.

Ramaekers, P., 1979, Stratigraphy of the Athabasca Basin: *in* Summary of Investigations 1979, Saskatchewan Geological Survey Miscellaneous Report 79-10, p. 154-160.

———, 1980, Stratigraphy and Tectonic History of the Athabasca Group (Helikian) of Northern Saskatchewan: *in* Summary of Investigations 1980: Saskatchewan Geological Survey Miscellaneous Report 80-4, p. 99-106.

———, 1981, Hudsonian and Helikian Basins of the Athabasca Region, Northern Saskatchewan: *in* Campbell, F.H.A., ed., Proterozoic Basins of Canada: Geological Survey of Canada Paper 81-10, p. 219-233.

Ratcliffe, J.H., 1958, Preliminary Report on an Aeromagnetic Survey of the Trout Lake Area, Saskatchewan, for W.S. Kennedy (1958) Grubstake: Unpublished Report from Lundsberg Explorations Limited.

Reyx, J., 1980, Travaux Miniers de Reconnaissance de la Minéralisation Claude: Amok Internal Report.

Ruhlmann, F., 1982, Etudes Minéralogiques et Métallogèniques de Quelques Occurrences Uranifères de l'Anneau de Carswell: Amok Internal Report.

———, 1985, Mineralogy and Metallogeny of Uraniferous Occurrences in the Carswell Structure: *in* Lainé, R., Alonso, D., and Svab, M., eds., The Carswell Structure Uranium Deposits, Saskatchewan: Geological Association of Canada Special Paper 29.

Ruzicka, V., 1975, Some Metallogenic Features of the "D" Uranium Deposit at Cluff Lake, Saskatchewan: *in* Report of Activities, Part C: Geological Survey of Canada Paper 75-1C, p. 279-282.

Scott, B.P., in press, Geology of the Upper Clearwater River Area (Southeast Part of 74-F): Saskatchewan Geological Survey Report 202.

Sopuck, V.J., de Carle, A., Wray, E.M., and Cooper, B.R., 1982, Application of Lithogeochemistry in Locating Unconformity-Type Uranium Deposits: *in* 9th International Geochemical Exploration Symposium, Program With Abstracts, p. 6-7.

Tona, F., and Alonso, D., 1983, The Athabasca Group in the Cluff Lake Area: Study of Lithogeochemistry and Clay Minerals: Amok Internal Report.

Tremblay, L.P., 1972, Geology of the Beaverlodge Mining Area, Saskatchewan: Geological Survey of Canada Memoir 367, 265 p.

———, 1978, Uranium Sub-Provinces and Types of Uranium Deposits in the Precambrian Rocks of Saskatchewan: *in* Current Research, Part A: Geological Survey of Canada Paper 78-1A, p. 427-435.

———, 1982, Geology of the Uranium Deposits Related to the Sub-Athabasca Unconformity, Saskatchewan: Geological Survey of Canada Paper 81-20, 56 p.

———, 1983, Some Chemical Aspects of the Regolithic and Hydrothermal Alterations Associated with the Uranium Mineralization in the Athabasca Basin, Saskatchewan: *in* Current Research, Part A: Geological Survey of Canada Paper 83-1A, p. 1-14.

Wallis, R.H., 1970, A Geological Interpretation of Gravity and Magnetic Data, Northwest Saskatchewan: Canadian Journal of Earth Sciences, v. 7, p. 858-868.

Wanless, R.K., Stevens, R.D., Lachance, G.R., and Delabio, R.N., 1979, Age Determinations and Geological Studies, K-Ar Isotopic Ages: Geological Survey of Canada Paper 79-2, 67 p.

Wendt, I., Hohndorf, A., Lenz, H., and Voultsidis, V., 1978, Radiometric Age Determinations on Samples of the Key Lake Uranium Deposits: *in* Short Papers of the 4th International Conference, Geochronology, Cosmochonology, and Isotope Geology: United States Geological Survey Open File Report 71-701, p. 448-449.

Worden, J.M., Cumming, G.L., and Baadsgaard, H., 1981, Geochronological Setting and Mineralization Ages of the Midwest Uranium Deposit, Northern Saskatchewan: Canadian Institute of Mining Uranium Symposium, September 8-13, 1981, Saskatoon, Technical Program, p. 10.

The Carswell Structure Uranium Deposits, Saskatchewan,
edited by R. Lainé, D. Alonso and M. Svab,
Geological Association of Canada Special Paper 29, 1985

CLAY MINERAL STRATIGRAPHY OF THE ATHABASCA GROUP: CORRELATION INSIDE AND OUTSIDE THE CARSWELL STRUCTURE

J. Hoeve and D. Quirt
Saskatchewan Research Council, 30 Campus Drive, Saskatoon, Saskatchewan S7N 0K1

D. Alonso
Amok Exploration Ltd., P.O. Box 9204, Saskatoon, Saskatchewan S7K 3X5

ABSTRACT

Using previously established criteria, the clay mineral stratigraphy of the Athabasca Group in the area around the Carswell structure proves to be similar to that elsewhere in the Athabasca Basin. The Lazenby Lake Formation, however, is much thicker than realized before and is recognized to form a third clastic wedge, stratigraphically equivalent to the Fair Point and Manitou Falls Formations. The eastern limit of the Fair Point Formation is now recognized to pass through the Carswell structure, trending north-northeast. In this location the Lazenby Lake Formation underlies and interfingers with the Fair Point Formation. The presence of diaspore within siltstone layers intercalated with the sandstones of the Lazenby Lake and Locker Lake Formations indicates a terrestrial depositional environment for these lithostratigraphic units.

Correlation of sequences inside and outside the Carswell structure are based upon identification of the boundary between the Locker Lake and Otherside Formations in both areas. Detailed correlations within the structure are, however, hampered by a shortage of drill holes intersecting stratigraphic boundaries.

RÉSUMÉ

La composition des argiles des roches du Groupe Athabasca a été étudiée aux rayons X, à l'intérieur de la structure de Carswell, pour caractériser stratigraphiquement les différentes formations. Les résultats obtenus sont les suivants:

1) La Formation Fair Point est composée presque exclusivement de kaolinite, l'illite n'apparaissant qu'en faibles proportions. 2) La Formation Manitou Falls renferme kaolinite et illite dans les mêmes proportions; toutefois, la kaolinite est légèrement dominante dans le membre inférieur de la série et inversement l'illite dans le membre supérieur. 3) Les argiles de la Formation Lazenby Lake sont essentiellement illitiques; la kaolinite n'y est observable qu'à l'état de traces. 4) Dans la partie inférieure de la Wolverine Point on observe surtout de l'illite associée à la kaolinite et un peu de chlorite; dans sa partie supérieure, les argiles incluent une forte proportion de chlorite et d'illite, la kaolinite y est absente. 5) La Formation Locker Lake est principalement illitique et se caractérise par un niveau kaolineux à la base. 6) La Formation Otherside renferme surtout de l'illite et un peu de chlorite; elle est également soulignée par de nombreux niveaux kaolineux. 7) Les pélites et les silts de la Formation Douglas présentent un assemblage d'illite et chlorite semblable à celui observé dans la partie supérieure de la Formation Wolverine. 8) Les dolomies de la Formation de Carswell, pauvres en argiles ne renferment qu'un peu de chlorite, d'illite et de kaolinite.

Cette stratigraphie des minéraux argileux du groupe Athabasca aux environs de la structure de Carswell est semblable à celle observée par ailleurs dans le Bassin Athabasca. Néanmoins, la Formation Lazenby Lake, beaucoup plus épaisse que ne pouvaient le laisser prévoir les études antérieures, semble former une langue de dépôts stratigraphiquement équivalente aux formations Fair Point et Manitou Falls qui se déposent en provenance du Sud-Sud Ouest dans le sous-bassin Mirror.

La présence de diaspore dans les silts intercalaires des formations Lazenby Lake et Locker Lake indique un environnement continental au moment du dépôt de ces unités. La corrélation des séquences à l'intérieur et à l'extérieur de la structure Carswell est basée sur l'identification de limites entre les Formations Locker Lake et Otherside. Toutefois, une étude plus détaillée de l'intérieur de la structure est limitée par un manque de sondages profonds.

INTRODUCTION

The Athabasca Basin, underlying an area of approximately 100,000 km² in northern Saskatchewan, is filled with red bed sandstone, conglomerates, and minor shales and dolomites of the middle Proterozoic Athabasca Group (nomenclature after Ramaekers, 1979, 1980, 1981). The area was mapped by Fahrig (1961) who considered the red beds to consist of ultramature, polycyclic orthoquartzites. Since then the discovery of unconformity-type uranium deposits in the basin has sparked renewed interest in the stratigraphy and sedimentology of the sedimentary rocks of the basin. Ramaekers (1979, 1980, 1981) distinguished several lithostratigraphic formations and interpreted the clastics to represent first-cycle sediments which owe their apparent maturity to peculiarities of the vegetationless Proterozoic sedimentary environment, and to extensive diagenetic alteration.

On the basis of sedimentological and stratigraphic analysis, Ramaekers concluded that the Athabasca Group was deposited in a tectonically active sedimentary basin, whose development was initiated by the formation of three northeast-southwest trending sub-basins (Fig. 1), controlled by wrench movements along major Hudsonian faults in the metamorphic basement. These were quickly filled in with detritus and coalesced to form the larger Athabasca Basin. Nonetheless, the faults remained active during sedimentation and controlled, to some extent, later basin development and stratigraphic differences between the sub-basins.

The basal part of the lithostratigraphic column, as defined by Ramaekers (1979, 1980, 1981) comprises two clastic wedges, the Manitou Falls Formation (MF) in the Cree and Mirror Basins, and the Fair Point Formation (FP) in the Jackfish Basin. The former, subdivided into lower and upper members (MFa + b, MFc + d respectively), consists of fluviatile sandstones and conglomerates which were transported from the east. The Fair Point Formation (FP), stratigraphically equivalent to the lower Manitou Falls Formation (MFa + b), is interpreted as a marine sandstone deposited in a high-energy, near-shore environment. These two formations are overlain by a series of nearshore to shallow-shelf, marine sandstones which comprise several transgressive-regressive cycles related to repeated uplift in the source

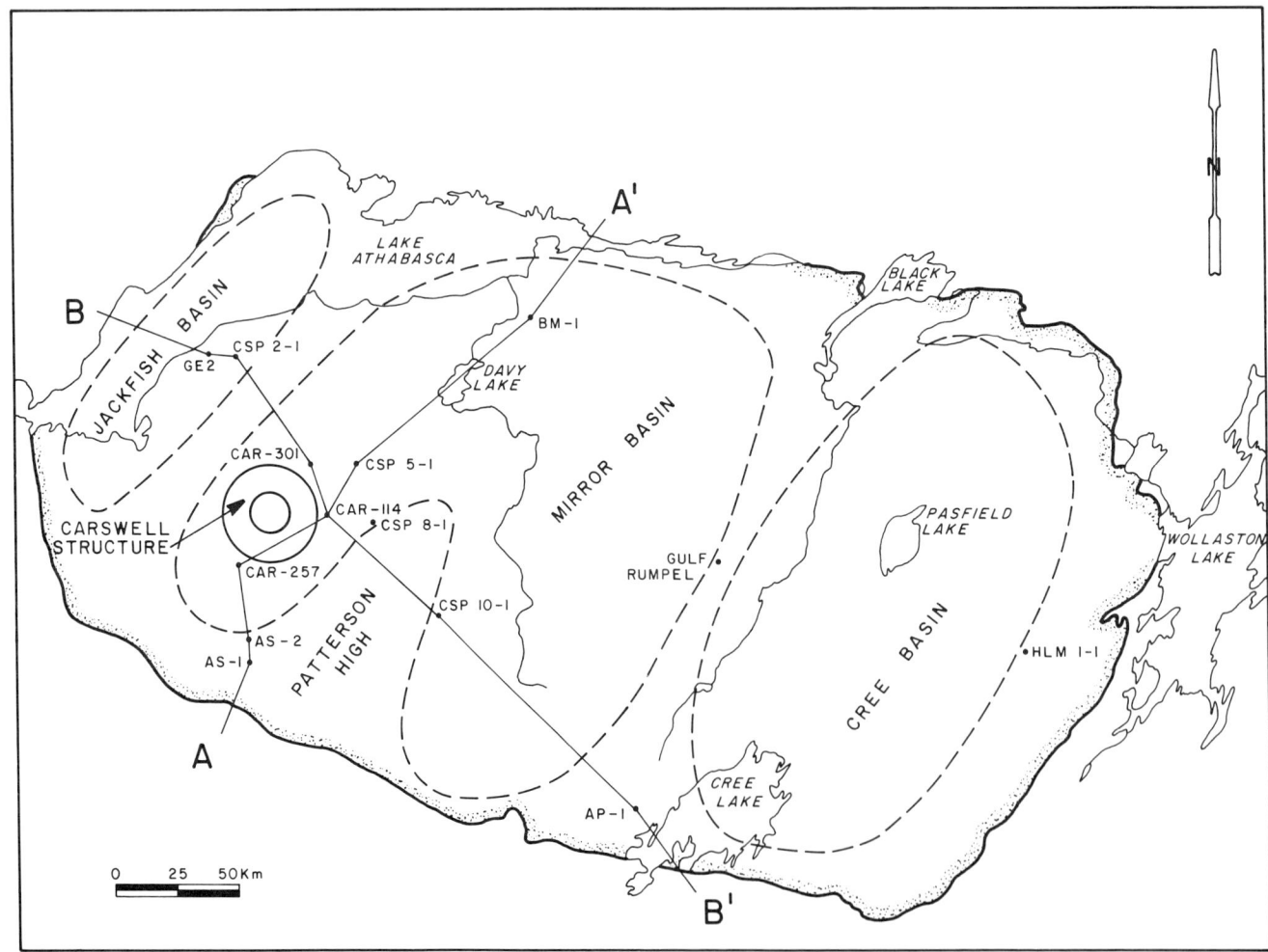

Figure 1. Location of drill holes and cross-sections discussed in text. Configuration of sub-basins modified after Hobson and MacAulay (1969) and Ramaekers (1981).

areas: Lazenby Lake (LzL, restricted to the western and central sub-basins), lower and upper Wolverine Point (WPa and WPb, respectively), Locker Lake (LL), Otherside (OF), and Tuma Lake (TL) Formations. The uppermost part of the sedimentary pile consists of shales and stromatolitic dolomites of the Douglas (DG) and Carswell (CW) Formations, which are only preserved within the Carswell structure (Figs. 1 and 2). Contacts of the Douglas with the underlying Tuma Lake and overlying Carswell Formations are gradational. The overall stratigraphic succession, of fluviatile, to near-shore marine sandstones, to marine stromatolitic dolomites, is similar to that of the Thelon Formation in the Northwest Territories (Cecile, 1973) with which the Athabasca Group is often correlated.

The preserved part of the stratigraphic column is at least 1500 m thick (Rumple Lake stratigraphic drill hole, Fig. 1 and Hoeve et al., 1981), however, fluid inclusion evidence (Pagel, 1975a, 1975b) suggests an original thickness of the sediment cover in the order of 5 km, which is consistent with clay mineral studies indicating deep burial and high grade diagenesis (Hoeve et al., 1981).

The authors have previously studied the clay mineral distribution within the Athabasca Group by means of quantitative XRD-techniques and demonstrated the use of clay mineral assemblages in stratigraphic interpretation (Hoeve et al., 1981). The present contribution will discuss the clay mineral stratigraphy in the little known Southwestern portion of the basin and attempt to correlate tectonically disturbed sequences inside the Carswell Structure with undisturbed sequences of the Athabasca basin.

Approximately 600 core samples were collected at 5 to 10 m intervals from drill holes located outside (AS-1, AS-2,

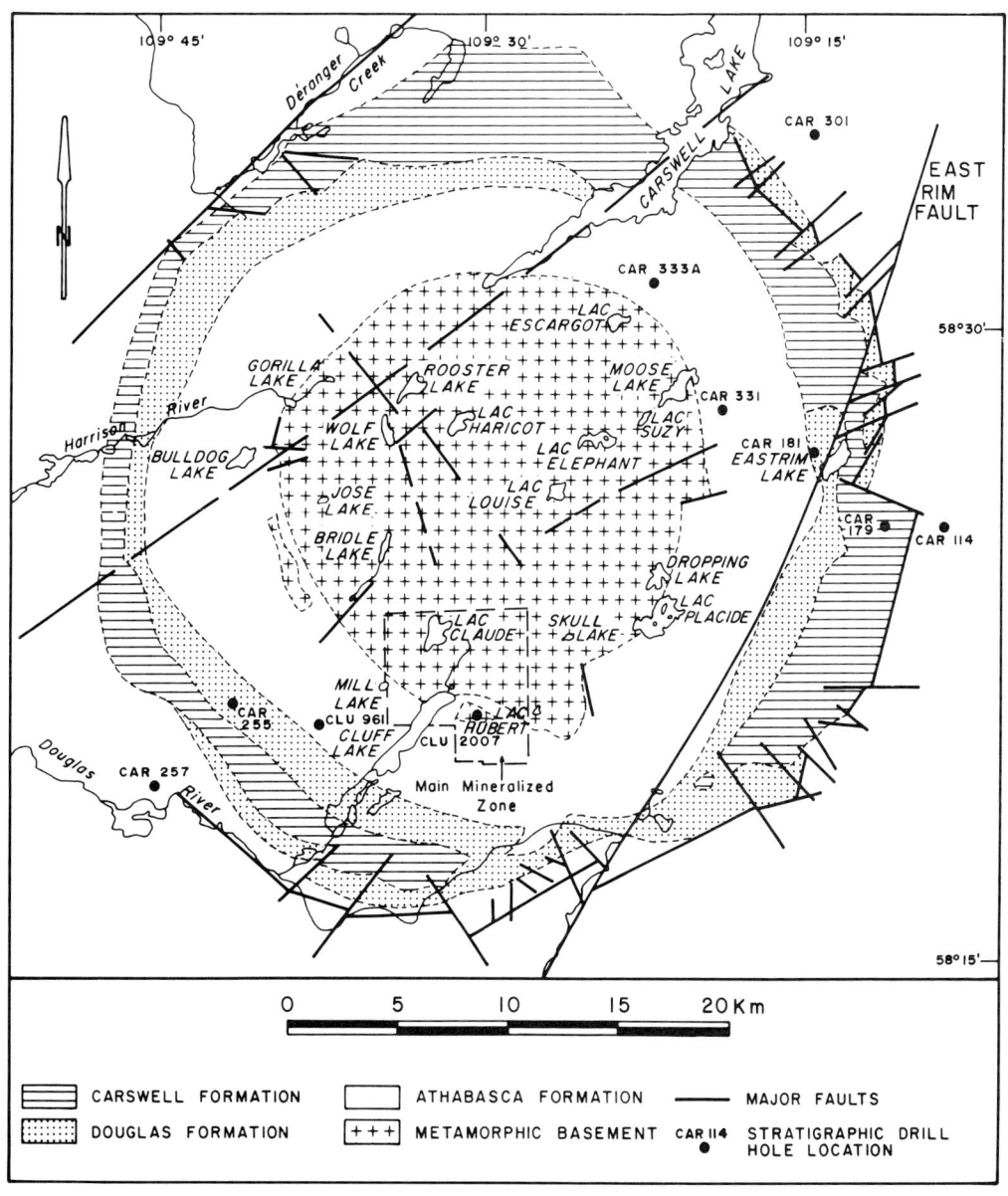

Figure 2. Geology of the Carswell structure and location of stratigraphic drill holes (after Tona et al., 1985).

CAR-257, CAR-114, CAR-301, and BM-1) and inside (CAR-333A, CAR-331, CLU-2007, CLU-961, CAR-181, CAR-255, and CAR-179) the structure, and analyzed for clay minerals by means of automated, quantitative XRD techniques. Sample preparation methods and analytical procedures are described elsewhere (Hoeve et al., 1981). Clay mineral compositions in the following tables and diagrams are expressed as relative weight percentages of clay mineral species present. The sampling was started by Tona and Alonso in 1983 in an attempt to clarify stratigraphic relationships in and around the Carswell Structure using Hoeve et al. (1981)'s procedures. The present study is a follow-up and expansion of their initial work. A brief field study by the authors, P. Ramaekers, and Amok geologists in the summer of 1984, resulted in corroborating the identification of Athabasca Group Formations within the Carswell structure and helped with the final sample selection.

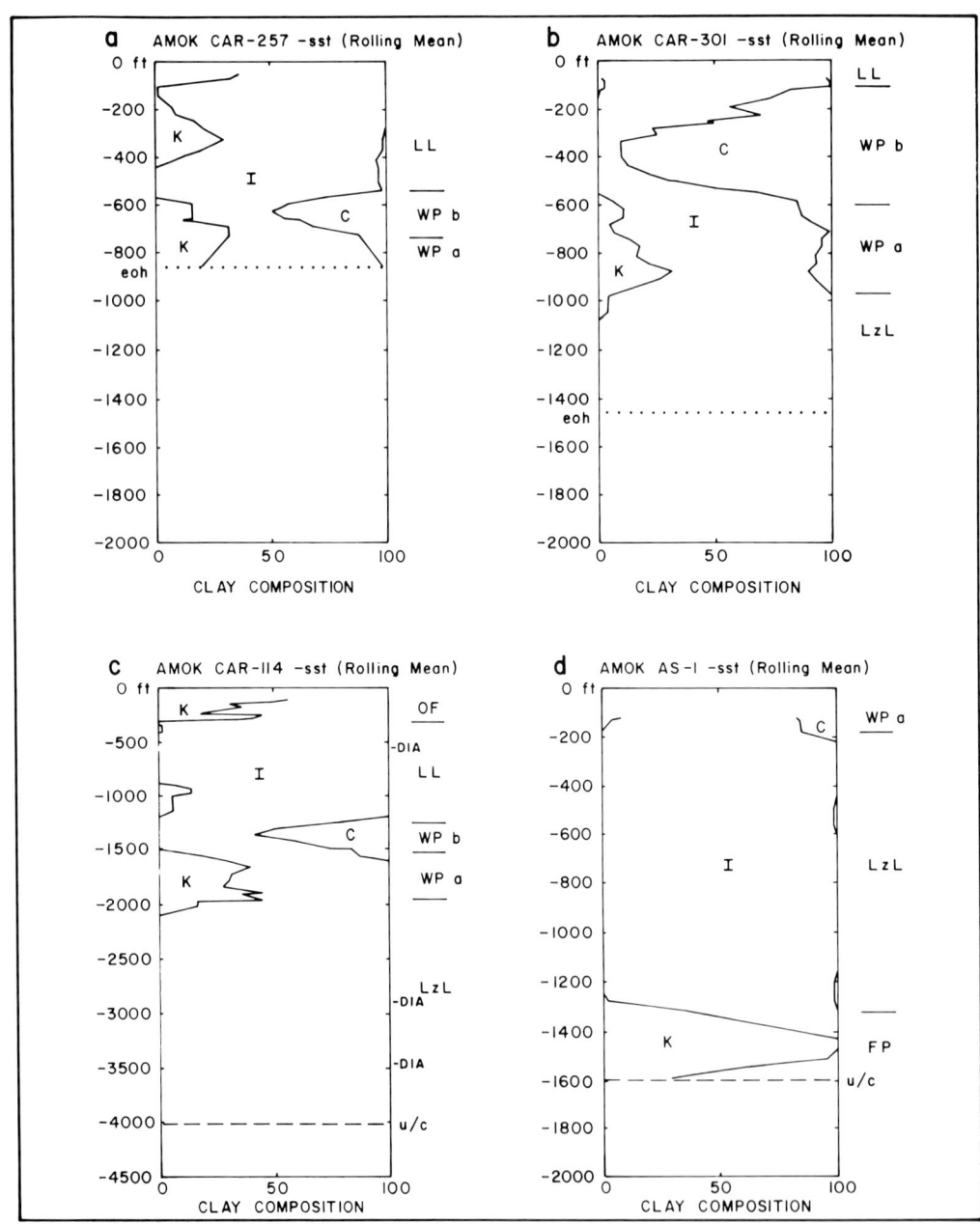

Figure 3. Clay mineral logs of drill holes CAR 257, CAR 301, CAR 114, and AS-1. Clay composition = K, kaolinite; I, illite; C, chlorite; DIA, diaspore; DOL, dolomite. Athabasca Group stratigraphy: FPa, b, Fair Point Formation lower and upper members; MFa + b, c + d, Manitou Falls Formation lower and upper members; LzL, Lazenby Lake Formation; WPa, b, Wolverine Point Formation lower and upper members; LL, Locker Lake Formation; OF, Otherside Formation; DG, Douglas Formation; CW, Carswell Formation.

CLAY MINERAL STRATIGRAPHY OUTSIDE THE STRUCTURE

The samples consist of the same clay mineral assemblages as those reported (Hoeve et al., 1981) for other parts of the basin, i.e., varying proportions of kaolinite, illite, and chlorite. No smectites or mixed-layer clay minerals were noted. Illite has a high crystallinity (Kubler Index (Ki) - 4-5) and is of the 2M polymorph. Chlorite is of the Mg-Al-rich ditrioctahedral variety sudoite (nomenclature according to Bailey, 1980). In addition, diaspore was identified in several thin intercalated silt layers at various depth levels of drill hole CAR-114 (Fig. 3c). The total clay content (weight per cent of the $\leq 2\ \mu$ fraction) ranges between 3 to 36%.

Using previously established criteria (Hoeve et al., 1981), the clay mineral logs for each of the drill holes (Figs. 3, 4, 5, and 6) can be subdivided into depth intervals that correlate with distinct lithostratigraphic units. The Fair Point Formation in the western portion of the basin is characterized by a clay mineral assemblage of almost exclusively kaolinite (± illite). The Manitou Falls Formation in the eastern and central part of the basin contains kaolinite and illite in approximately equal proportions, the lower member (MFa + b) being slightly kaolinite-dominated and the upper member

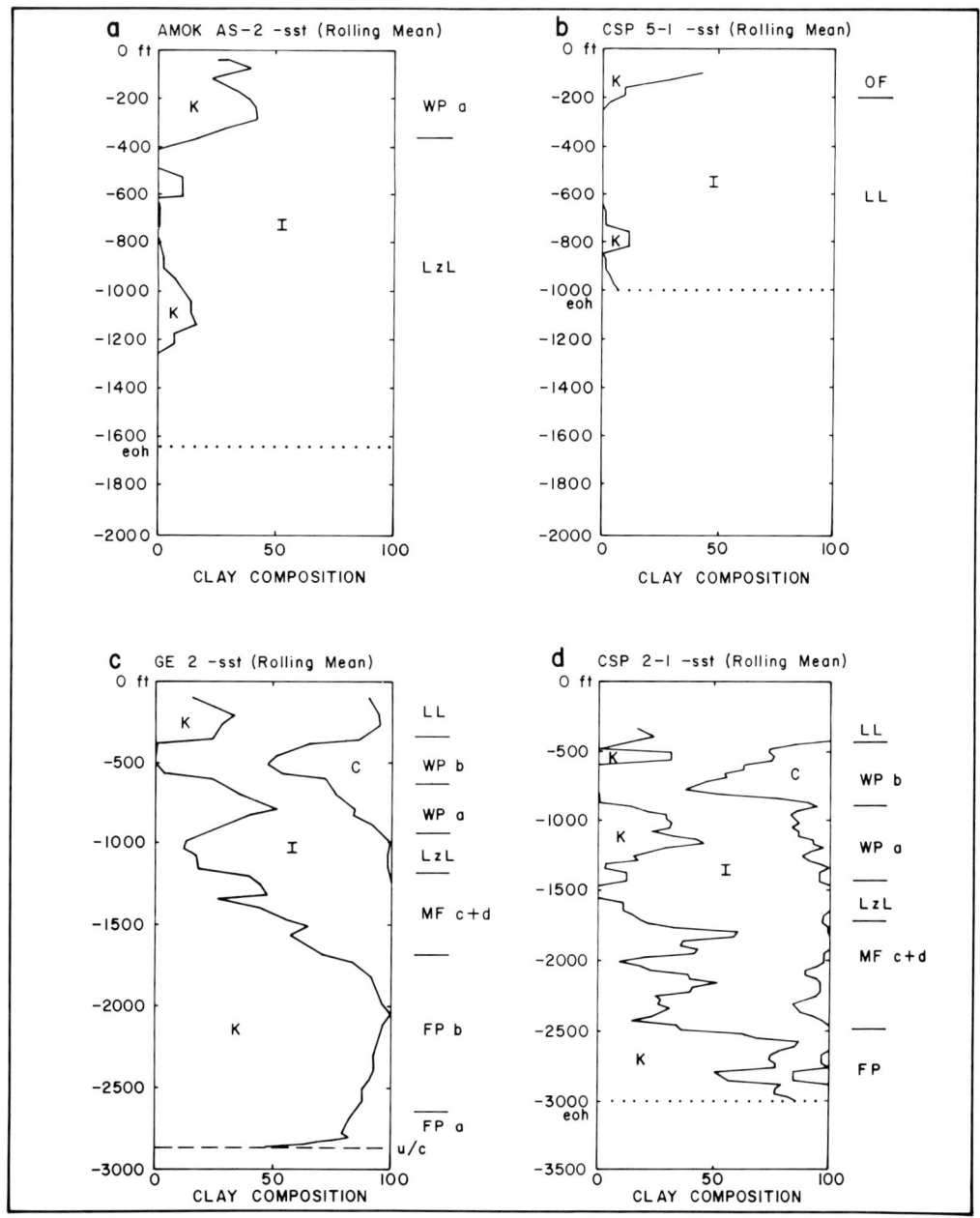

Figure 4. Clay mineral logs of drill holes AS-2, CSP 5-1, GE 2, and CSP 2-1. See Figure 3 for explanation of symbols.

(MFc + d) being slightly illite-dominated. The Lazenby Lake Formation (LzL) is marked by a clay mineral assemblage consisting of illite and traces only of kaolinite and by a low total clay content, generally less than 8%. The overlying lower Wolverine Point Formation (WPa) shows predominance of illite with intermediate kaolinite and accessory chlorite, whereas the upper Wolverine Point Formation (WPb) is distinguished by high proportions of both chlorite and illite, combined with a virtual absence of kaolinite. The Locker Lake Formation is characterized by extreme illite predominance with the boundary against the upper Wolverine Point Formation being identified by a characteristic "kaolinite hump" on the clay mineral logs (e.g., Figs. 3c and 5d). The clay mineral signature of the Otherside Formation (OF) is incompletely known, owing to insufficient drill hole coverage. The base of the formation is, however, marked by the appearance of another "kaolinite hump" on the logs (e.g., Figs 3c, 4b, and 6a). The Tuma Lake Formation (TL) has not been identified in any of the analyzed drill cores.

Statistical data on mean composition and compositional range of the various formations are presented in Table I. Correlation of clay mineral intervals from one drill hole to another along section A-A' (Fig. 1) renders a coherent clay mineral stratigraphy (Fig. 7a), which bears no surprises as

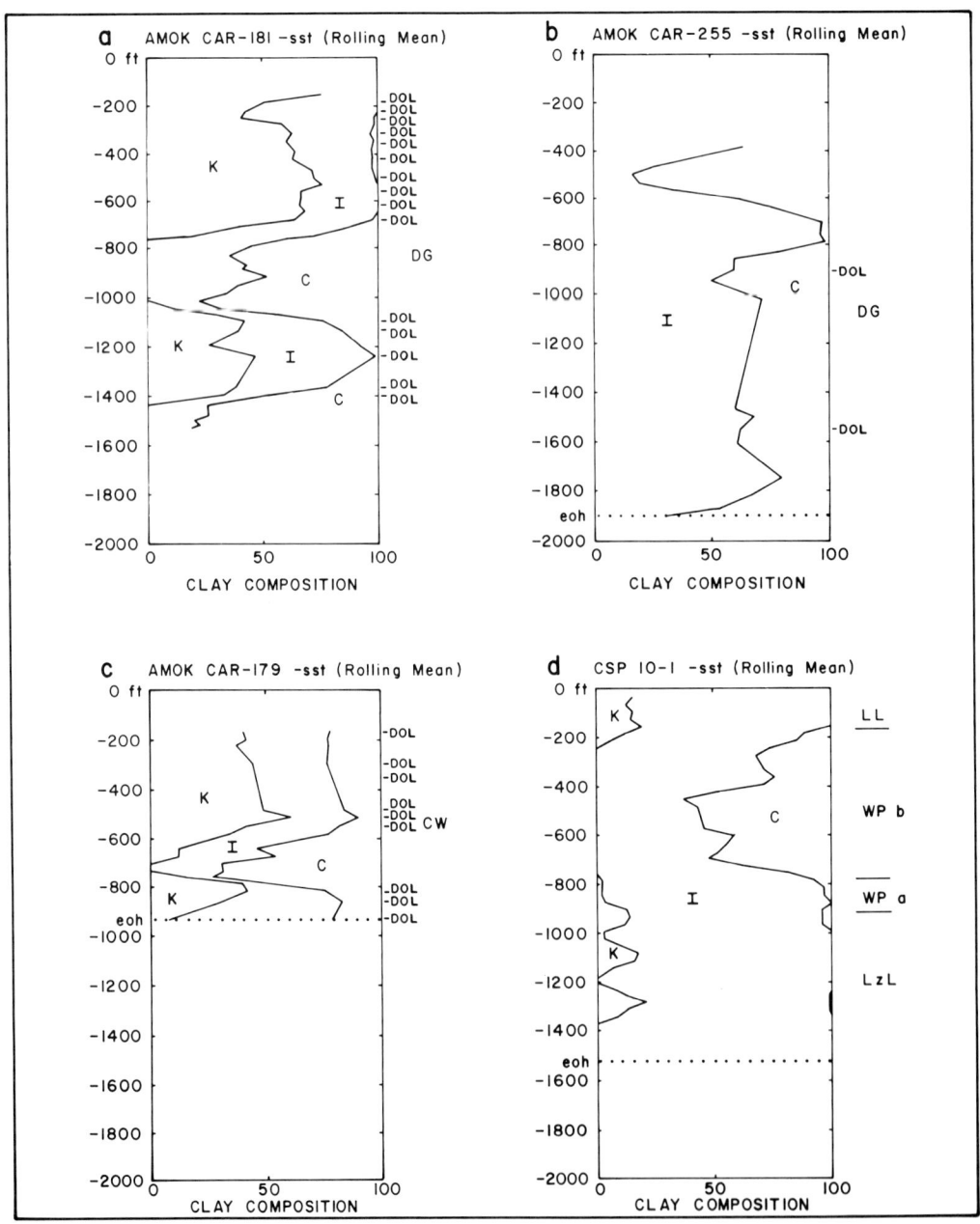

Figure 5. Clay mineral logs of drill holes CAR 181, CAR 255, CAR 179, and CSP 10-1. See Figure 3 for explanation of symbols.

compared to the results of previous studies elsewhere in the basin, except for the unexpected development of the Lazenby Lake Formation. According to Ramaekers (1979, 1980, 1981), this formation represents a thin (150 m) marine sandstone, which in the western and central sub-basins overlies the Fair Point and Manitou Falls Formations. Consistent with this interpretation, the Lazenby Lake Formation is identified in drill holes (GE-2 and CSP 2-1, Figs. 4c and 4d) as a 100 m interval between the lower Wolverine Point (WPa) and upper Manitou Falls (MFc+d) Formations. Yet, in drill hole CAR-114 (Fig. 3c) located immediately east of the Carswell structure, the Manitou Falls and Fair Point Formations both appear to be absent, and 600 m of Lazenby Lake Formation is found instead between the basement and the lower Wolverine Point Formation. In drill hole AS-1, to the south of the structure, 350 m of Lazenby Lake Formation is sandwiched between the Wolverine Point Formation and a thin (100 m) intersection of Fair Point Formation, whereas AS-2, though not reaching basement, intersects 400 m of Lazenby Lake Formation before terminating in this unit. In drill hole BM-1, located 100 km to the northeast of the Carswell structure, the Lazenby Lake Formation is absent and the lower Wolverine Point Formation directly overlies the upper Manitou Falls Formation.

Section A-A' is recast in Figure 7b which shows the drill holes according to their stratigraphic intersection, taking the base of the lower Wolverine Point Formation (WPa) as the level of reference.

Similar relationships are noted in section B-B' (Figs. 1 and 8), which shows the Lazenby Lake Formation to be located between the Fair Point and Manitou Falls Formations, against which units it onlaps to the west and east respec-

TABLE I

CLAY MINERAL DATA FOR ATHABASCA GROUP STRATIGRAPHIC SUBDIVISIONS
SANDSTONE SAMPLES

	LL + OF	WPb	WPa	LzL	MF c+d	MF a+b	FPb	FPa
Kaolinite								
n	180	153	83	201	207	70	34	38
X (%)	9.49	3.67	35.09	2.49	34.88	55.62	89.96	72.00
95% conf. int. (%)	±2.623	±2.634	±4.671	±1.047	±2.404	±4.239	±3.885	±5.947
σ_{n-1} (%)	17.78	16.44	21.26	7.51	17.50	17.65	10.97	17.85
RSD (%)	187.47	447.88	60.60	301.14	50.17	31.74	12.19	24.80
Range (%)	0.0 - 98.0	0.0 - 93.4	0.0 - 94.8	0.0 - 50.0	0.0 - 90.7	0.0 - 91.7	32.0 - 100.0	0.0 - 100.0
Chlorite								
n	180	153	83	201	207	70	34	38
X (%)	1.26	46.04	4.62	0.23	1.52	1.56	0.14	1.53
95% conf. int. (%)	±0.747	±4.501	1.645	±0.204	±0.860	±1.728	±0.206	±2.564
σ_{n-1} (%)	5.06	28.09	7.49	1.46	6.26	7.20	0.58	7.70
RSD (%)	403.25	61.01	161.99	624.68	411.14	460.35	411.96	503.40
Range (%)	0.0 - 51.2	0.0 - 100.0	0.0 - 25.8	0.0 - 14.6	0.0 - 75.0	0.0 - 50.0	0.0 - 2.8	0.0 - 46.8
Illite								
n	180	153	83	201	207	70	34	38
X (%)	89.26	50.41	60.29	97.27	63.60	42.68	9.90	26.47
95% conf. int. (%)	±2.745	±4.457	±4.876	±1.146	±2.465	±3.558	±3.891	±4.642
σ_{n-1} (%)	18.61	27.81	22.20	8.22	17.95	14.81	10.99	13.94
RSD (%)	20.85	55.17	36.82	8.45	28.22	34.71	111.00	52.65
Range (%)	2.0 - 100.0	0.0 - 100.0	5.2 - 100.0	48.5 - 100.0	8.1 - 100.0	8.3 - 93.5	0.0 - 68.0	0.0 - 74.1
I_{002}/I_{001}								
n	165	90	75	192	197	63	5	29
X (%)	0.40	0.43	0.39	0.39	0.36	0.34	0.33	0.38
95% conf. int. (%)	±0.018	±0.031	±0.023	±0.011	±0.014	±0.028	±0.149	±0.028
σ_{n-1} (%)	0.12	0.15	0.10	0.08	0.10	0.11	0.11	0.07
RSD (%)	29.11	34.11	25.40	20.08	28.16	32.14	32.57	18.99
Range (%)	0.23 - 0.88	0.13 - 0.85	0.18 - 0.70	0.17 - 0.75	0.15 - 0.77	0.16 - 0.67	0.19 - 0.45	0.24 - 0.57
Kübler Index								
n	174	144	80	199	200	70	25	37
X (%)	5.03	8.78	5.02	5.26	5.02	5.45	5.59	4.45
95% conf. int. (%)	±0.160	±0.496	±0.245	±0.139	±0.151	±0.363	±0.730	±0.346
σ_{n-1} (%)	1.07	3.00	1.09	0.99	1.08	1.51	1.73	1.03
RSD (%)	21.20	34.19	21.80	18.80	21.53	27.75	30.99	23.05
Range (%)	2.6 - 9.0	3.2 - 15.8	3.2 - 8.2	1.8 - 8.2	2.1 - 7.6	2.5 - 10.7	3.0 - 9.9	3.1 - 7.6

where: n = number of samples σ_{n-1} = standard deviation
        X = mean value   RSD = relative standard deviation

tively. The kaolinitic intervals within the Lazenby Lake Formation, intersected in drill hole CSP 10-1 (Fig. 5d), may represent interfingering with the Manitou Falls Formation towards the east. Illitic intervals within the upper Manitou Falls Formation, intersected in drill holes CSP 2-1 and GE-2, may be similarly interpreted.

CLAY MINERAL STRATIGRAPHY WITHIN THE STRUCTURE

The clay mineral logs for drill holes located within the Carswell structure are presented in Figures 5 and 6. Correlations between these drill holes are less reliable than for those outside the structure on account of a shortage of drill holes intersecting formational boundaries and the tectonic complexity of the structure.

The dolomites of the Carswell Formation, intersected in drill hole CAR 179 which did not reach the underlying Douglas Formation, are marked by a clay mineral assemblage of kaolinite, illite, and chlorite, whereas intercalated shaley intervals show illite and chlorite only (Fig. 5c). The shales of the Douglas Formation, in drill holes CAR 181 and CAR 255, consist of an assemblage of illite and chlorite; dolomite bearing intervals (Fig. 5a, CAR 181) also contain kaolinite, i.e., these have a clay mineral signature similar to that of the Carswell Formation. Correlations between

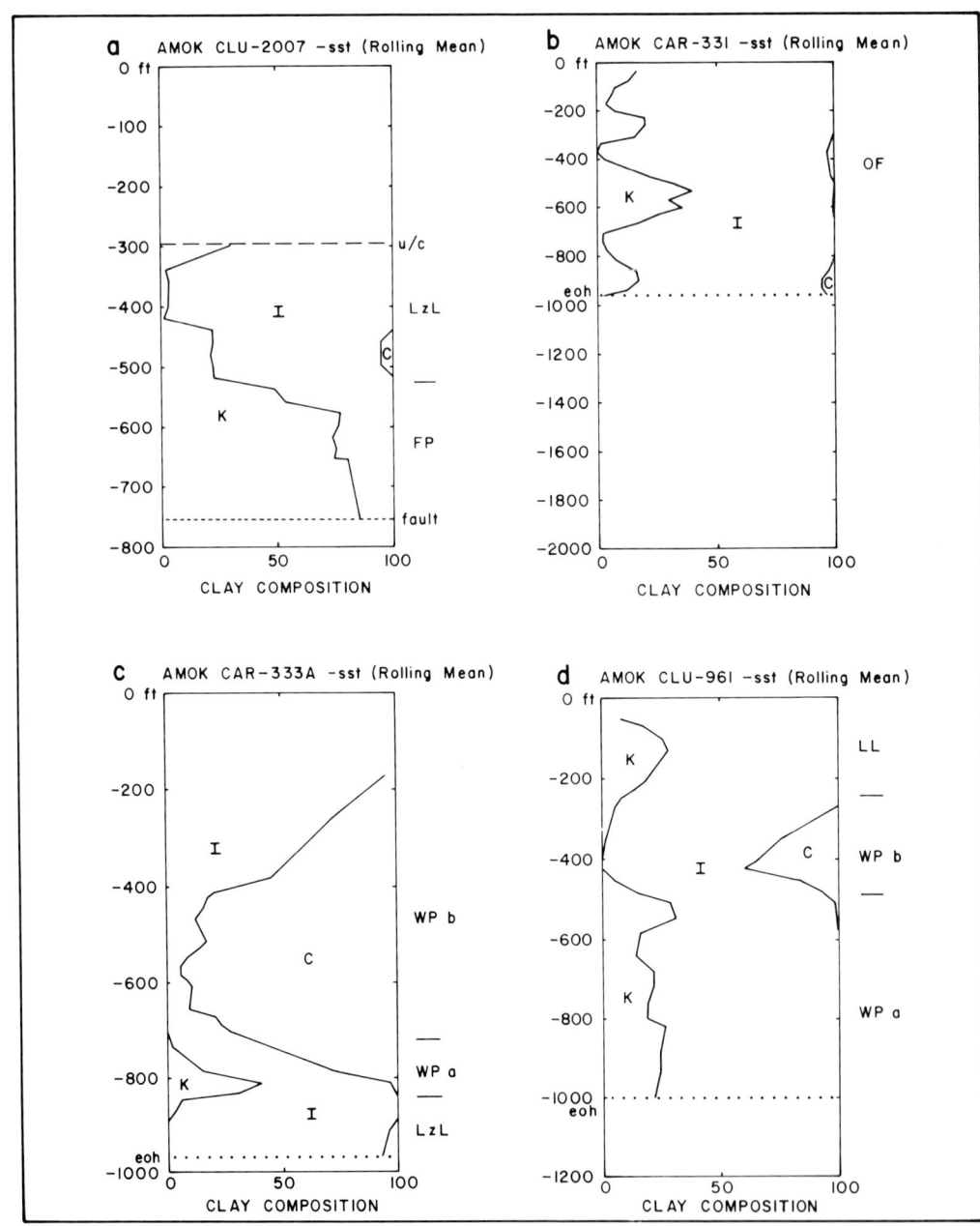

Figure 6. Clay mineral logs of drill holes CAR 2007, CAR 331, CAR 333A, and CLU 961. See Figure 3 for explanation of symbols.

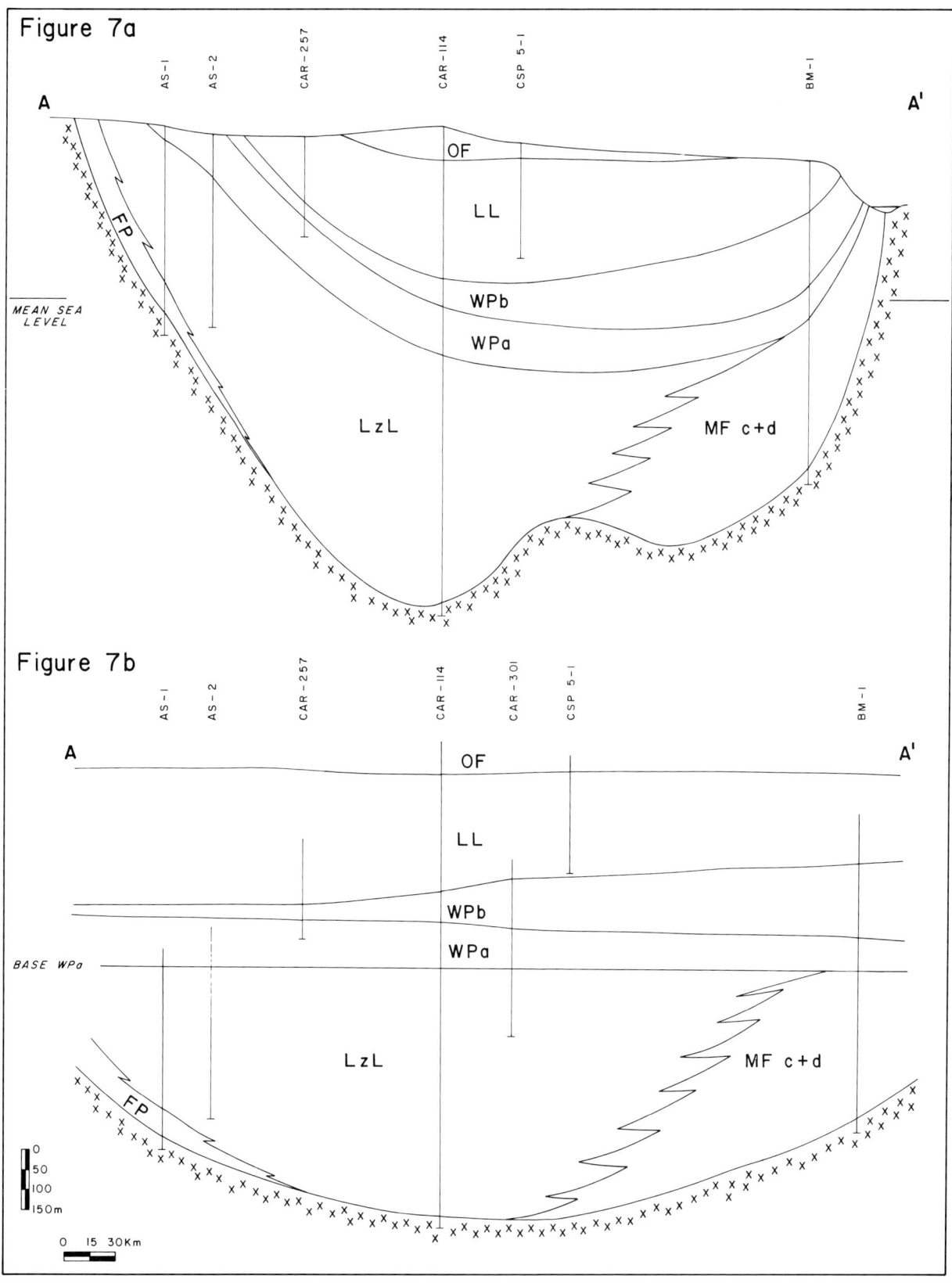

Figure 7. Interpreted clay mineral stratigraphy, section A-A'; (a) reference horizon = mean sea level, (b) reference horizon = base WPa.

drill holes CAR 179, CAR 181, and CAR 255 are in part based upon lithological descriptions indicating gradational contacts between the Carswell and Douglas Formations via a downward decrease of dolomitic intervals and a concomitant increase in shaley layers.

The Tuma Lake Formation, described by Ramaekers (1980, 1981) as a pebbly sandstone of only 80m thickness which underlies the Douglas Formation and has gradational contacts with this unit was not identified in any of the drill holes. It is included here with the Otherside Formation.

The Otherside Formation, as intersected in drill hole CAR 331 (Fig. 6b), is characterized by predominant illite, minor chlorite, and multiple kaolinitic intervals. Correlations within the Otherside Formation between drill holes are uncertain due to lack of distinct marker horizons. This could mean that the kaolinitic intervals are of limited lateral extent.

Within the structure, the contact between the Otherside Formation and the underlying Locker Lake Formation is not intersected in the available suite of drill holes. The stratigraphic thickness within the structure of the Locker Lake and Lazenby Lake Formations (Fig. 9) are taken from the log of drill hole CAR-114. The maximum stratigraphic thickness of the Otherside Formation outside of the structure (350 m) was obtained from Ramaekers (1979) and is used here as a minimum within the structure. The lower portion of the Locker Lake Formation appears in drill hole CLU-961 (Fig. 6d) as do both the upper and lower Wolverine Point Formations. Drill hole CAR-333A (Fig. 6c) also intersects the Wolverine Point Formations as well as the top of the Lazenby Lake Formation. The clay mineral signatures of these formations are identical to the equivalent units outside of the structure.

The Lazenby Lake Formation, as the basal unit, is intersected by a number of drill holes collared along the eastern flank of the basement plug (Fig. 10) while the Fair Point Formation is similarly intersected along the western flank. Both formations appear along the southern margin of the plug, as illustrated by drill hole CLU-2007 (Fig. 6a) which is located in the overturned "S" basement slab. At this location the Lazenby Lake Formation underlies and interfingers with a thin unit of Fair Point Formation. This relationship has been noted over an E-W distance of approximately 3 km.

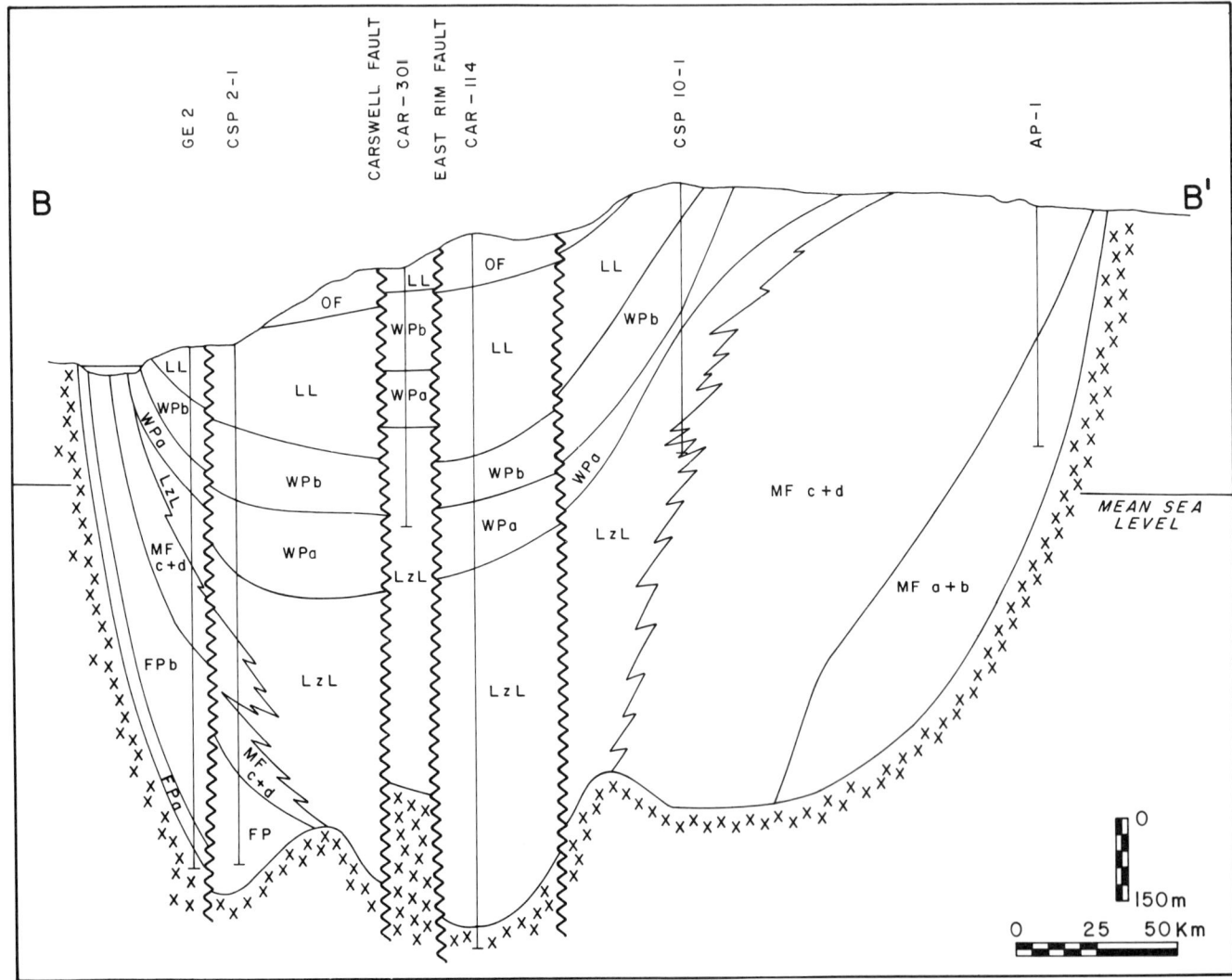

Figure 8. Interpreted clay mineral stratigraphy, section B-B'; reference horizon = mean sea level.

A composite cross-section from the centre to the rim of the Carswell structure is presented in Figure 9, which shows the position of the drill holes according to their radial distance from the central basement plug. Note that drill hole CLU-961 appears to be in an upthrown block. Drill hole CAR-114, located outside the structure, is added for reference. The drill holes are plotted according to their interpreted stratigraphic intersection using the base of the lower Wolverine Point Formation (WPa) as the level of reference.

Correlations between drill holes inside and outside the Carswell structure are based upon drill holes CLU-961 and CAR-114, which both intersect the boundary between the upper Wolverine Point and Locker Lake Formations and drill holes CLU-2007 and AS-1, which both intersect the boundary between the Lazenby Lake and Fair Point Formations (Figs. 6d 3c, 6a and 3d). It appears that within the structure a stratigraphically higher portion of the Otherside Formation is preserved than anywhere else in the basin; however, correlations within this formation are tenuous. Although the upper and lower boundaries of the Douglas Formation are not intersected in any of the drill holes, on the basis of lithological descriptions we believe that gaps in coverage between CAR-179, CAR-181, and CAR-255 as well as between CAR-255, CAR-331 and CLU-961 are small.

INTERPRETATION

Clay mineral evidence indicates that the Lazenby Lake Formation forms a third clastic wedge, stratigraphically equivalent to the Manitou Falls and Fair Point Formations, which enters the basin from the southwest and distally shows interfingering relationships with these two units. Such an interpretation is consistent with the observation of increasing grain size to the southwest, pointing to sediment influx from that direction (Ramaekers, 1979, 1980, 1981).

Based upon the geometric configuration of the Lazenby Lake Formation and in accordance with the results of a seismic study by Hobson and McAulay (1969), we have modified, from Ramaekers (1980, 1981), the configuration of the Mirror Basin (Fig.1). The Patterson High, instead of forming a broad high separating the Jackfish and Mirror Ba-

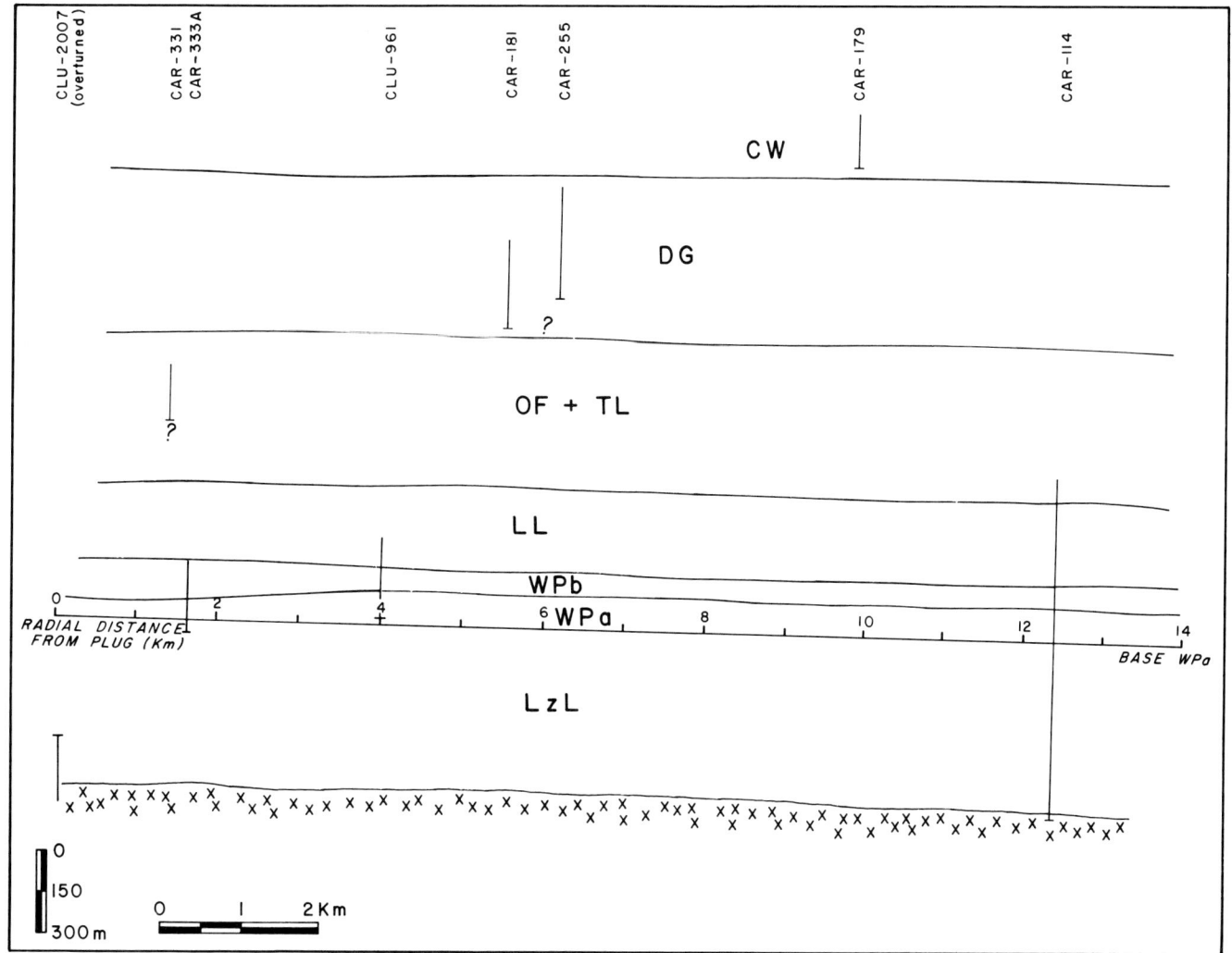

Figure 9. Interpreted clay mineral stratigraphy, composite section within Carswell structure; reference horizon = base WPa.

sins, is reduced to a ridge which divides the latter into a western and an eastern arm. The Lazenby Lake Formation was initially restricted to the western arm, and it was only after the Patterson High had been overcome that the formation began to spread out eastwards and to interact with the approaching clastics of the Manitou Falls Formation.

From the drill hole coverage shown in Figure 10, it is apparent that the eastern limit of the Fair Point Formation is located in the vicinity of the Carswell basement plug. This boundary appears to trend approximately NNE through the plug, probably running from around Thompson Bay on Lake Athabasca toward Bourassa Creek along the southern edge of the Basin. This is illustrated by the thicknesses of Fair Point material found in drill holes CSP 2-1 and AS-1, its absence in drill holes BM-1 and CAR 114 and the presence of both Fair Point and Lazenby Lake Formations in drill hole CLU 2007.

The identification of diaspore at several levels within drill hole CAR 114 is of significance to the interpretation of the depositional environment. The mineral was found in the paleoweathering profile at the top of the metamorphic basement and in three silt layers intercalated within the overlying sandstones. Two of these are within the Lazenby Lake Formation, one within the Locker Lake Formation. Its presence within the sub-Athabasca paleoweathering profile adds support to earlier interpretations of the lateritic nature of this

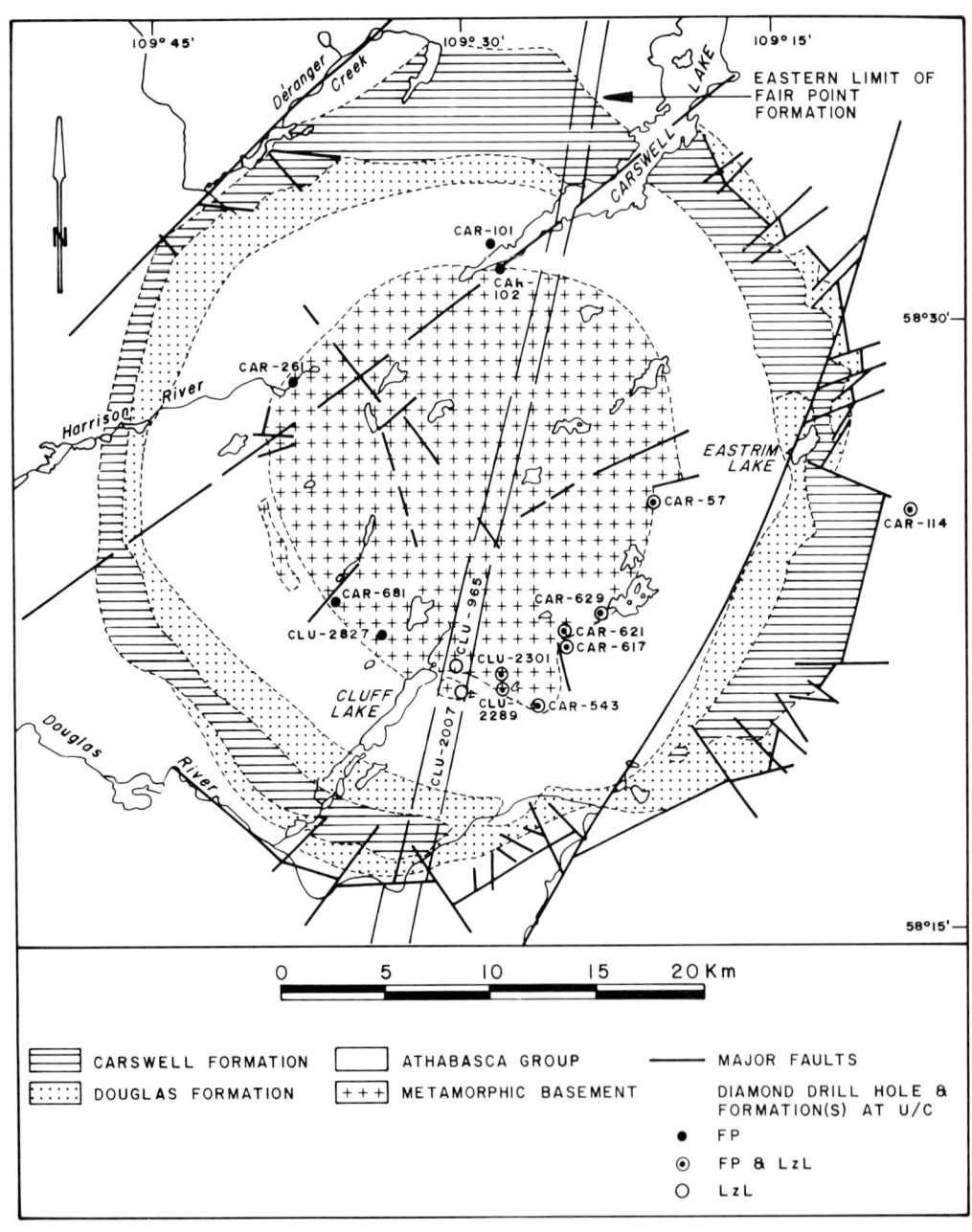

Figure 10. Location of the eastern limit of the Fair Point Formation and illustrative drill holes.

profile (Hoeve and Sibbald, 1978; MacDonald, 1980). The occurrence of diaspore at several stratigraphic levels within the sandstones may suggest intervals of non-deposition accompanied by intensive weathering, or alternatively, may represent detrital material washed in from a nearby weathering profile. In either case, the mineral is indicative of a strong terrestrial influence which may cast doubt on previous interpretations of the Lazenby Lake and Locker Lake Formations as having a marine origin (Ramaekers, 1980, 1981).

CONCLUSIONS

The clay mineral stratigraphy of the Athabasca Group in the southwestern part of the Athabasca Basin proves to be similar to that studied elsewhere in the basin. The Lazenby Lake Formation is, however, much thicker than realized before and is now recognized to form a third clastic wedge, which is stratigraphically equivalent to the Manitou Falls and Fair Point Formations. The presence of diaspore within the Lazenby Lake and Locker Lake Formations indicates a terrestrial rather than a marine depositional environment for these formations.

The presence of Fair Point Formation along the western margin of the Carswell structure basement plug, the presence of Lazenby Lake Formation along the eastern margin, and the presence of both units along the southern edge indicates that the eastern limit of the Fair Point Formation passes through the structure. The presence of both formations adjacent to the plug also confirms that a nearly complete Athabasca Group stratigraphic sequence is preserved within the structure.

Within the Carswell structure, clay minerals offer great promise in stratigraphic and structural interpretations, although their potential is currently hampered by insufficient coverage of critical intersections. Nevertheless, sequences inside and outside the structure can be correlated because the boundaries between the Locker Lake and upper Wolverine Point Formations and the Fair Point and Lazenby Lake Formations are identified in both areas.

ACKNOWLEDGEMENTS

This study was made possible by Mr. Frederic Tona, Exploration Manager for Amok Ltd. at the time (now with COGEMA), who initiated the sampling program. Discussions with Dr. Paul Ramaekers of the Saskatchewan Mining and Development Corporation and with the Amok geologists M. Svab, P. Honarvar, and C. McAleenan during a field trip to Carswell proved most helpful in illucidating uncertainties in details of stratigraphic correlation. Bonn Energy Corporation gracefully gave permission to sample drill core from its properties in the area south of the Carswell Structure and allowed the results to be incorporated in this paper (drill hole AS1 and AS2). We gratefully acknowledge Amok Ltd. and the Saskatchewan Mining Development Corporation for permission to publish the results of this study.

REFERENCES

Bailey, S.W., 1980, Summary of Recommendations of AIPEA Nomenclature Committee on Clay Minerals: American Mineralogist, v. 65, p. 1-7.

Cecile, M.P., 1973, Lithofacies Analysis of the Proterozoic Thelon Formation, N.W.T.: Unpublished M.Sc. Thesis, Carleton University, Ottawa.

Fahrig, W.F., 1961, The Geology of the Athabasca Formation: Geological Survey of Canada Bulletin 68, 41 p.

Hobson, G.D., and MacAulay, H.A., 1969, A Seismic Reconnaissance Survey of the Athabasca Formation, Alberta and Saskatchewan (Part of 74): Geological Survey of Canada Paper 69-18, 23 p.

Hoeve, J., and Sibbald, T.I.I., 1978, On the Genesis of Rabbit Lake and Other Unconformity-Type Uranium Deposits in Northern Saskatchewan, Canada: Economic Geology, v. 73, p. 1450-1473.

Hoeve, J., Rawsthorne, K., and Quirt, D., 1981, Uranium Metallogenic Studies: Clay Mineral Stratigraphy and Diagenesis in the Athabasca Group: Summary of Investigations 1981, Saskatchewan Geological Survey Miscellaneous Report 81-4, p. 76-89.

Hoeve, J., Quirt, D., and Alonso, D., 1983, Uranium Metallogenic Studies: Clay Mineral Distributions in the Athabasca Group, Southwest Part of the Basin: Summary of Investigations 1983, Saskatchewan Geological Survey, Miscellaneous Report 83-4, p. 89-95.

Macdonald, C.C., 1980, Mineralogy and Geochemistry of a Precambrian Regolith in the Athabasca Basin: Unpublished M.Sc. Thesis, University of Saskatchewan, Saskatoon, 151 p.

Pagel, M., 1975a, Cadre Géologique de Gisements d'Uranium dans la Structure Carswell (Saskatchewan, Canada): Etudes des Phases Fluides; Thèse d'Etat, Université de Nancy, France, 157 p.

———, 1975b, Détermination des Conditions Physicochimiques de la Silicifications Diagénétique des Grès Athabasca (Canada) au Moyen des Inclusions Fluides: Comptes Rendus de l'Académie des Sciences Paris, t. 280, Série D., p. 2301-2304.

Ramaekers, P., 1979, Stratigraphy of the Athabasca Basin: Summary of Investigations 1979, Saskatchewan Geological Survey, Miscellaneous Report 79-10, p. 154-160.

———, 1980, Stratigraphy and Tectonic History of the Athabasca Group (Helikian) of Northern Saskatchewan: Summary of Investigations 1980, Saskatchewan Geological Survey Miscellaneous Report 80-4, p. 99-106.

———, 1981, Hudsonian and Helikian Basins of the Athabasca Region, Northern Saskatchewan: in Campbell, F.H.A., ed., Proterozoic Basins of Canada, Geological Survey of Canada Paper 81-10, p. 219-233.

Tona, F., Alonso, D., and Svab, M., 1985, Geology and Mineralization in the Carswell Structure – A General Approach: in Lainé, R., Alonso, D., and Svab, M., eds., The Carswell Structure Uranium Deposits, Saskatchewan: Geological Association of Canada Special Paper 29.

The Carswell Structure Uranium Deposits, Saskatchewan,
edited by R. Lainé, D. Alonso and M. Svab,
Geological Association of Canada Special Paper 29, 1985

GEOCHRONOLOGY OF THE CARSWELL AREA, NORTHERN SASKATCHEWAN

Keith Bell
Ottawa-Carleton Centre for Geoscience Studies, Department of Geology, Carleton University, Ottawa, Ontario K1S 5B6

ABSTRACT

New isotopic data from the basement complex of the Carswell structure, northern Saskatchewan, outline a complex history for this part of the Hudsonian orogen. Rb-Sr whole-rock, and U-Pb zircon dates from quartzofeldspathic units indicate events at 2320 Ma, 2130 Ma, 2000 Ma and 1880 Ma ago. A much younger event at about 1750 Ma, indicated by whole-rock Rb-Sr data from some metapelites, is similar to the age normally associated with the culmination of the Hudsonian orogeny. Archean events have yet to be documented from the Carswell basement.

New U-Pb data from the unconformity-type uranium deposits indicate that mineralization and/or reworking of the deposits took place over a considerable period of time. The U-Pb data are discordant and most suggest recent Pb loss. Uranium mineralization associated with the Dominique-Peter ore zone probably occurred between 950 and 1050 Ma, at about the same time as or slightly later than the formation of the D deposit. Uraninite data from the OP deposit indicate ages of between 820 and 890 Ma, while uraninite data from the Donna boulder train yield a significantly younger date of 380 Ma. Events at perhaps 1330 Ma, 1060 Ma and 820 Ma are suggested by new U-Pb data from the Numac showing.

K-Ar laser-fusion dates from highly-altered samples associated with the Dominique-Peter zone range from 1650 Ma to 500 Ma, an observation consistent with long-lived, fluid activity. Laser-fusion dates also indicate that one of the most recent events within the Carswell structure was the formation of the Cluff breccias sometime during the Palaeozoic.

RÉSUMÉ

De nouvelles datations Rb-Sr sur roche totale et U/Pb sur zircons du socle de la structure de Carswell donnent aux événements des âges allant de 2320 à 1750 Ma.

Les événements les plus anciens, datés à 2320 Ma, 2130 Ma, 2000 Ma, et 1880 Ma sont mesurés dans les gneiss du complexe de Earl River. Les métapélites de Peter River, stratigraphiquement plus jeunes, ont donné des âges Rb-Sr sur roche totale de 1750 Ma, indiquant que soit les gneiss de Peter River font partie d'une séquence sédimentaire beaucoup plus récente reposant sur un socle pré-1880 Ma, soit les métapélites ont été rajeunies pendant une phase tardive de l'orogénie Hudsonienne. Tous les événements pré-Athabasca observés à l'intérieur de la structure de Carswell ont été observés dans les roches du socle à l'extérieur de la structure. Néanmoins, aucun âge Archéen n'a été observé.

Presque toute la minéralisation uranifère est tardive par rapport au dépôt de sédiments Athabasca. La minéralisation uranifère s'est probablement mise en place à (1) 1100 ± 50 Ma (D, Dominique-Peter), (2) 820 - 890 (OP), (3) autour de 380 Ma (train de galets de Donna). Des épisodes possibles à 1330 Ma, 1060 Ma, et 820 Ma sont suggérés par les résultats isotopiques sur des minéralisations en provenance de Numac. Le nombre d'épisodes distincts, enregistrés par chaque gisement, indique que les conduits utilisés par les fluides minéralisateurs sont restés ouverts pendant des périodes considérables. Bien qu'il soit difficile de corréler les minéralisations uranifères avec des événements rencontrés ailleurs dans la Saskatchewan, il est intéressant de noter que les âges obtenus sur les gisements sont semblables à ceux de l'activité orogénique dans les provinces de Grenville et des Appalaches plus à l'Est. Les datations au Ar^{40}/Ar^{39} obtenues par la technique de fusion au laser sur les minéraux d'altération sont difficiles à interpréter. Aucune n'est plus vieille que 1650 Ma mais aucune n'est plus jeune que 500 Ma. Les datations par fusion au laser, des brèches de Cluff, vont de 515 Ma à 365 Ma, âges pas incompatibles à ceux obtenus à partir de certaines minéralisations uranifères. Si les brèches sont le résultat de l'impact d'une météorite, alors cet épisode doit être post-365 Ma et l'écart dans les dates doit être attribué à des degrés différents de rétention de l'argon. Si les brèches ne sont pas le résultat d'un impact alors l'écart dans les dates pourrait indiquer une activité continue (fluide?) durant plus de 150 Ma.

Les faits principaux de cette étude sont : (1) le socle sub-Athabasca a été soumis à des épisodes magmatiques et/ou métamorphiques à 2320 Ma, 2130 Ma, 2000 Ma, 1880 Ma, et 1750 Ma; 2) la minéralisation uranifère et/ou la remobilisation se sont probablement déclenchées dès 1330 Ma et terminées il y a quelques 380 Ma. De nombreux épisodes à 1330 Ma, 1100 Ma, 850 Ma, et 380 Ma ont été observés; et 3) une activité post-600 Ma est soulignée par la formation des brèches de Cluff et quelques minéralisations uranifères. La minéralisation est en tout cas plus jeune que les estimations connues de dépôt des sédiments Athabasca.

INTRODUCTION

Although a great deal is known about the geochronology of the Churchill Province in Saskatchewan, little published information is available concerning the nature and age of basement rocks in the Carswell structure. With less than 1% surface exposure, almost all of the known geological relationships are based on drill core data.

The circular Carswell structure, 39 km in diameter, sits in the western part of the Athabasca Basin and consists of outer and inner rings of deformed sedimentary rocks, and a central core of high-grade metamorphic rocks (see map in pocket, and Fig. 1). An outer rim of deformed sedimentary rocks includes dolomites of the Carswell Formation, underlain by siltstones, mudstones and fine-grained sandstones of the Douglas Formation. Between the outer ring and the basement is an annular zone of deformed Athabasca conglomerates and sandstones.

Comparison with areas outside the structure suggests that the basement gneisses of the Carswell structure are either Archean and /or Aphebian in age. Although heterogeneous, the basement rocks can be broadly divided into: (i) granitoids and granitoid gneiss (mainly quartzofeldspathic), (ii) mafic gneisses (amphibolite and pyroxene gneisses), and (iii) pelitic metasedimentary rocks (sillimanite-cordierite-garnet schists and gneisses). Evolution of the Carswell basement involved extensive polyphase deformation, granulite facies metamorphism, retrogression to amphibolite facies, and subsequent alteration by low-temperature hydration (Herring, 1976). Chloritization (mainly associated with uranium mineralization), sericitization, and silicification are common.

Sometime after deposition of the Athabasca sediments, polyphase hydrothermal uranium mineralization produced major unconformity-type deposits at the sandstone-basement contact. Deposits are hosted within either the crystalline basement close to the contact of the Peter River gneiss and Earl River complex, e.g., the Dominique-Peter orebody, or the overlying sandstone, e.g., the D orebody (Ruhlmann, 1985). Later episodes superimposed minor mineralization on some of the earlier deposits. Argillized shear zones and mylonites in both basement and cover are associated with one of the hydrothermal events (Ey et al., 1985).

The formation of the Carswell structure resulted in a series of overturned, isoclinal anticlines and synclines, restricted mainly to the outer parts of the circular feature. Because of tectonic slicing and thrusting and the complex ordering of the slices, there is difficulty in correlating individual units, even from one drill hole to another.

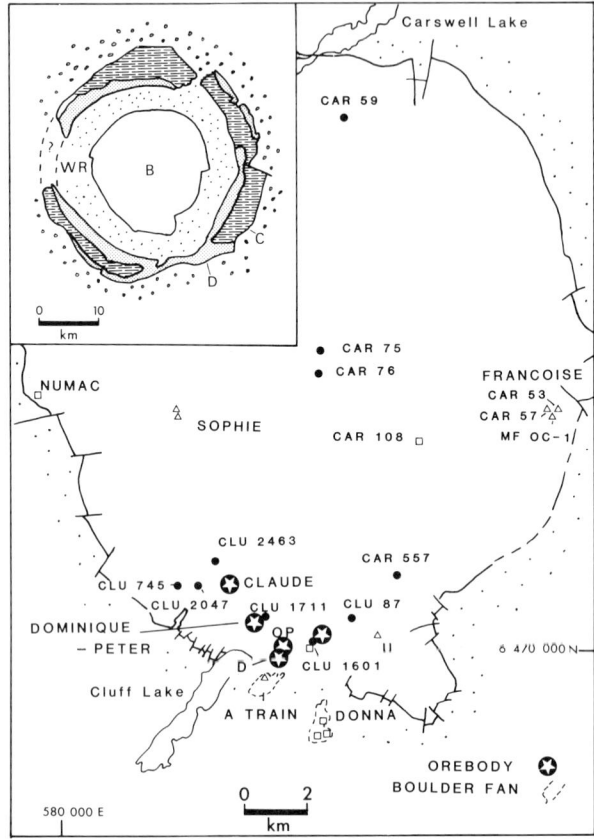

Figure 1. Sample localities in the Carswell basement. Inset shows the regional geology of the Carswell structure. B = basement, WR = William River Subgroup, D = Douglas Formation, C = Carswell Formation (after Ramaekers, 1983). Suites collected for Rb-Sr analysis marked by dots, those for U-Pb work by open squares. Cluff breccia localities denoted by open triangles.

Both basement and Athabasca rocks are cut pervasively by Cluff breccias — a series of post-metamorphic polymict breccias thought to be related to the same event that produced the Carswell dome structure. These randomly distributed breccias are thought, by some, to be the result of meteorite impact (Innes, 1964; Harper, 1978). Other features considered as evidence of shock metamorphism include shatter cones, planar deformation lamellae in quartz, and split and re-cemented conglomerate pebbles (see Tapaninen, 1976).

A simplified sequence of events within the structure involves: (i) formation of granitoid and pelitic basement rocks, (ii) metamorphic activity, (iii) deposition and diagenesis of the Athabasca Group sediments, (iv) uranium mineralization and associated alteration, and (v) formation of the Carswell structure and the Cluff breccias.

Geochronological studies undertaken to date in Saskatchewan mainly concern areas well outside the Carswell structure. Within the last decade U-Pb zircon and whole-rock Rb-Sr dates have supplemented existing K-Ar mineral data leading to a framework that can be used for further geochronological work (Bell, 1981; Bell and Macdonald, 1982). The new geochronological studies on the Carswell area summarized here shed some light on: (i) the age of the basement rocks, (ii) the time of regional metamorphisms, (iii) the age of the uranium mineralization, (iv) the timing of alteration events, and (v) the age of the Cluff breccias. Sample localities are shown in Figure 1.

PREVIOUS WORK

The handful of previously published dates from the Carswell structure suggests a complex history for the area, indicating events from about 2300 Ma through to 234 Ma ago. Although several dates are quoted in the literature, in many cases the analytical data are not given so that it is impossible to re-calculate some of the dates using the most recent decay constants. Whole-rock Rb-Sr data from granitoid rocks collected from surface exposures by members of the Saskatchewan Geological Survey suggest a minimum age of at least 2300 Ma for some of the basement rocks, although "scatterchrons" rather than true isochrons are indicated (Blenkinsop and Bell, 1981); an event at about 1745 Ma is also suggested (Bell and Blenkinsop, 1982). A K-Ar biotite date of 1973 Ma (Tapaninen, 1976) from a garnet-cordierite gneiss suggests that parts of the basement have remained relatively undisturbed since the early Proterozoic while a K/Ar date from sericite of 988 Ma showed that hydrothermal alteration took place after the deposition of the Athabasca sediments (Tapaninen,1976). A gabbro dyke that cut the Athabasca succession, a short distance to the south of the external rim of the Carswell structure, yields a K-Ar whole-rock date of 950 ± 35 Ma (Wanless et al., 1979), and places a younger limit on the Athabasca sediments. This is consistent with Rb-Sr data from Athabasca rocks (Ramaekers and Dunn, 1976; Blenkinsop and Bell, 1981). U-Pb dates from both the D and Claude orebodies (Devillers and Nordmann, 1974; Gancarz, 1979) indicate a series of events that extended from about 1150 to 800 Ma ago for the mineralization and alteration, with subsequent disturbance of these systems at or later than 234 Ma ago. Pb isotopic ratios from rocks outside of the D ore body, yet within the Carswell structure, yield a model age of 1330 ± 30 Ma (Gancarz, 1979). Although these rocks are radioactive, they contain high concentrations of thorium, with little or no uranium (Lainé, pers. commun., 1984). Bellon et al. (1976) mention intersection U-Pb dates of 1150 ± 25 Ma and 250 ± 20 Ma from Cluff Lake. Although no details are given of either sample localities or analytical data, the analysed material is from the D orebody. A K-Ar date of about 1130 Ma, recorded by Wanless et al. (1979) from a chloritic schist that hosts the D orebody probably marks the time of alteration, but the uncertainty of 450 Ma is so large that the data are of limited value. The lower Palaeozoic date for the Cluff breccia of about 480 Ma, based on two K-Ar determinations by the Geological Survey of Canada, was considered by Wanless et al. (1968) to be the age of ultra-mylonitization produced by faulting during the late stages of the development of the Carswell structure.

In spite of the fact that most of the data consist of single dates from isolated outcrops, suggested are: (i) a complex history for the pre-Athabasca basement that probably extended back to at least 2300 Ma, (ii) major uranium mineralization between 1150 Ma and 800 Ma, and (iii) formation of the Cluff breccias about 480 Ma ago. Also indicated are later disturbances of the U-Pb systems.

ANALYTICAL DETAILS

All core samples for Rb-Sr work were first cleared of their outer skin with a tungsten carbide drill and then crushed to less than 100 mesh in a Bleuler Mill. Rb and Sr abundances were determined by X-ray fluorescence using the method of Norrish and Chappell (1967) in which the mass-absorption coefficients are measured directly. The analytical uncertainties for Rb and Sr abundances are $\pm 2\%$ at the 2 sigma level. Some of the Sr isotopic ratios were measured as the nitrate using a double filament technique on a multi-collector MAT-261 solid-source mass spectrometer. The multi-collection technique measures the ^{88}Sr, ^{87}Sr and ^{86}Sr ion beams simultaneously. The reproducibility of the $^{87}Sr/^{86}Sr$ ratios measured by the multi-collector method is considered good to 0.00003 at the 2 sigma level. All $^{87}Sr/^{86}Sr$ ratios are adjusted to be compatible with a value for the Eimer and Amend Sr standard of 0.70800. Isochrons were calculated using the York 2 method (York, 1969).

Samples for argon work were analyzed using the $^{40}Ar/^{39}Ar$ method, at the Department of Physics, University of Toronto. Crumb-sized pieces were fused, after irradiation, using a continuous, variable-power (15 watt maximum) argon ion laser (Spectra-Physics Model 171-08) in broadband mode. A laser probe diameter of about 200-300 μm was used. ^{39}Ar and ^{40}Ar were measured, after sample fusion, by conventional methods.

U-Pb analyses were obtained from Geospec Consultants Ltd. U/Pb ratios are known to better than 1% while the Pb isotopic ratio measurements are considered precise to less than 0.1% per mass unit. All analytical data have been corrected for mass discrimination by comparison to NBS standard reference material.

All new dates have been calculated using the decay constants recommended by Steiger and Jäger (1977) and all errors are quoted at the 2 sigma level.

WHOLE-ROCK Rb-Sr RESULTS

Ten suites of core material were collected from the Carswell basement for whole-rock Rb-Sr work. The basement rocks are divided into a younger group of pelitic metasedimentary rocks, the Peter River gneiss, and an older heterogeneous group of pelites, quartzofeldspathic rocks and mafic gneisses and amphibolites known collectively as the Earl River complex (see Pagel and Svab, 1985). The rocks analysed in the present work include two suites of pelitic metasediments from the Peter River gneiss (suites CLU 87, and CLU 1711; the numbers refer to individual holes) and seven suites of quartzofeldspathic rocks (suites CAR 59, CAR 75, CAR 76, CAR 557, CLU 745, CLU 1601 and CLU 2047). An additional suite (CLU 2463) contains both pelitic and quartzofeldspathic material. The data are given in Table I and summarized in Table II. The data for all samples, other than suites CAR 557 and CAR 59, are shown in Figures 2-9. Thin section examination showed that the samples available for analysis were not the most suitable for isotopic study.

TABLE I
Rb-Sr ANALYTICAL DATA

No.	Sample No.	Rb (ppm)	Sr (ppm)	^{87}Sr/^{86}Sr (atomic)	^{87}Rb/^{86}Sr (atomic)
	CAR 59				
1	238	267	105	1.01849	7.57
2	334	200	136	0.85104	4.31
3	448	173	260	0.77465	1.94
4	451	243	175	0.84248	4.07
5	475	197	180	0.80918	3.18
	CAR 75				
6	52	228	189	0.8309	3.53
7	77	173	268	0.7823	1.88
8	88	262	175	0.8387	4.39
9	195	244	125	0.8975	5.75
10	198	223	104	0.9087	6.32
	CAR 76				
11	31	132	310	0.76038	1.24
12	73	91.5	340	0.74360	0.781
13	112	105	379	0.74228	0.804
14	140	93.6	361	0.74298	0.752
15	208	200	293	0.78049	1.99
16	239	264	224	0.82007	3.45
	CAR 557				
17	25	179	243	0.77854	2.15
18	109	186	58	1.05921	9.57
19	231	177	158	0.83154	3.28
20	234	209	176	0.82793	3.48
	CLU 87				
21	75.6	189	69.7	0.9206	8.01
22	81.75	240	75.4	0.9587	9.43
23	87.5	152	20.0	1.2554	23.2
24	104	147	76.4	0.8627	5.64
25	112	220	41.3	1.1307	16.1
	CLU 745				
26	78	119	55.5	0.8860	6.30
27	105	155	55.8	0.9380	8.22
28	265	138	70.0	0.8646	5.79
29	280	132	70.5	0.8603	5.49
30	300	120	71.3	0.8461	4.93
	CLU 1601				
31	133	205	218	0.7884	2.74
32	168	157	312	0.7532	1.46
33	197	206	288	0.7695	2.08
34	278	224	272	0.7781	2.40
35	339	165	194	0.7770	2.48
	CLU 1711				
36	142	205	67.5	0.9475	9.00
37	311	267	102	0.9160	7.70
38	338	156	81.5	0.8654	5.61
39	383	188	108	0.8501	5.10
40	448	193	94.5	0.8755	6.00
	CLU 2047				
41	78	113	86	0.81921	3.83
42	81.3	131	73	0.85669	5.28
43	86.5	160	130	0.81676	3.60
44	88	106	136	0.77839	2.27
	CLU 2463				
45	72	272	73	1.01433	11.1
46	87.5	248	76	0.97310	9.64
47	88.5	267	85	0.96699	9.25
48	89	226	110	0.87681	6.09
49	98	299	73	1.03721	12.2
50	100.5	285	108	0.91358	7.82
51	294	155	103	0.83467	4.45
52	301.5	140	97	0.83208	4.24

^{87}Sr/^{86}Sr ratios adjusted to be compatible with a value of 0.70800 for the Eimer and Amend Standard.
All Sr values normalised to a ^{86}Sr/^{88}Sr ratio of 0.1194.

All samples were affected by one or more of the following features: chloritization, sericitization, hematization, silicification, fracturing, and comminution of quartz and feldspar.

Certain data points had to be arbitrarily excluded from the calculation of the isochrons in order to obtain an acceptable MSWD (mean square of weighted deviates), a statistical parameter indicating that the data lie on an acceptable straight line (Brooks et al., 1972). Any MSWD less than 2 in the present work suggests that the data form an acceptable straight line, i.e., an isochron. Table II shows that most of the MSWD's for the calculated dates are about 3 or less, other than those associated with suites CLU 2047 and CLU 2463 (all samples). There was no apparent correlation between the extent of alteration seen in thin section and the degree of scatter of the data. The data for suites CAR 557 and CAR 59 are not plotted because of pronounced scatter. Both sets of data have MSWD's greater than 26.

The two pelitic suites (CLU 87 and CLU 1711), both quartz-cordierite-garnet-sillimanite schists, give dates of 1775±55 Ma and 1730±80 Ma (see Figs. 2 and 3), figures in reasonable agreement with those generally quoted for the culmination of the Hudsonian metamorphism (e.g., Stockwell, 1982). The average of these two dates, of about 1750 Ma, is slightly less than the date of 1760 Ma quoted elsewhere in this volume; the latter was based on earlier data. Initial ^{87}Sr/^{86}Sr ratios of 0.718±0.006 and 0.725±0.007 are consistent with the involvement of older material.

The data from the quartzofeldspathic suites, in general, indicate dates significantly greater than those obtained from the metasedimentary rocks. The dates shown in Table II can be divided into two groups, one corresponding to an event ca. 2000 Ma and the other to events at ca. 1870 Ma (Figs. 4-8). Most of the initial ^{87}Sr/^{86}Sr ratios are high enough to suggest involvement with older material.

The dates of 1730 Ma and 1775 Ma from the pelitic material can be interpreted in two ways. They can reflect either a much younger sedimentary sequence overlying an older quartzofeldspathic basement, or differences in the ways that these two different rock types have adjusted during high-grade regional metamorphism. In an attempt to choose between these alternatives, a bore-hole was chosen (CLU

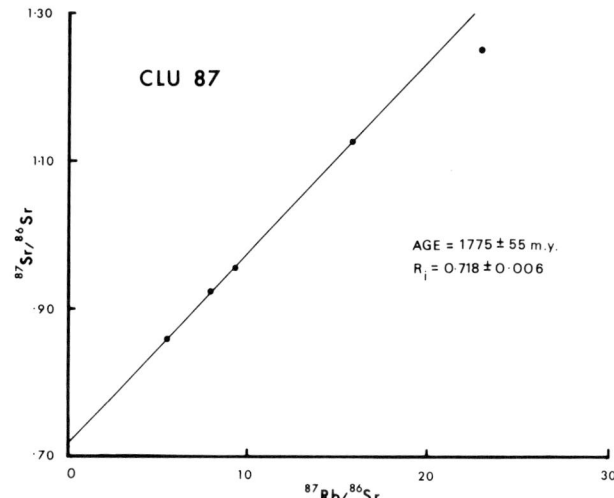

Figure 2. Rb-Sr isochron plot - suite CLU 87.

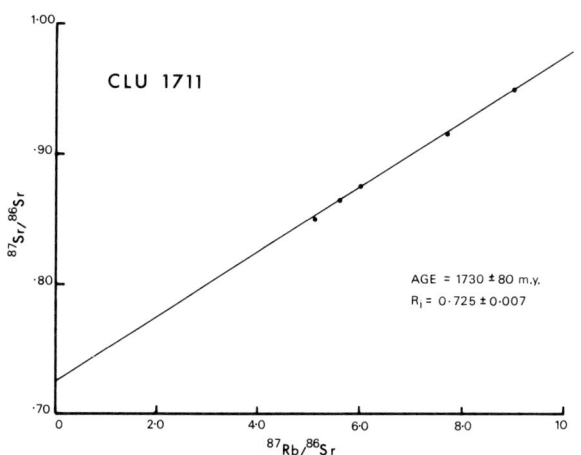

Figure 3. Rb-Sr isochron plot - suite CLU 1711.

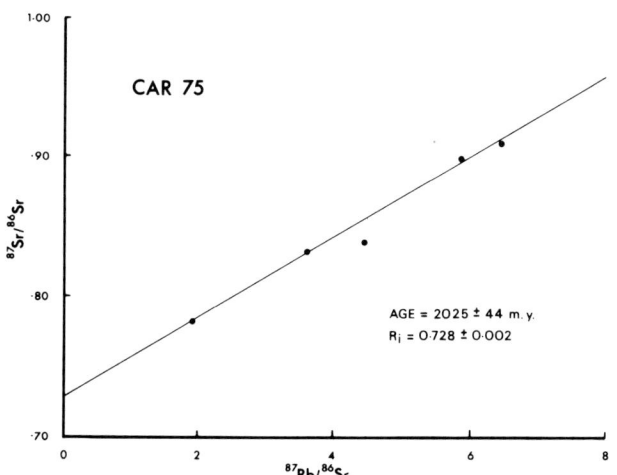

Figure 4. Rb-Sr isochron plot - suite CAR 75.

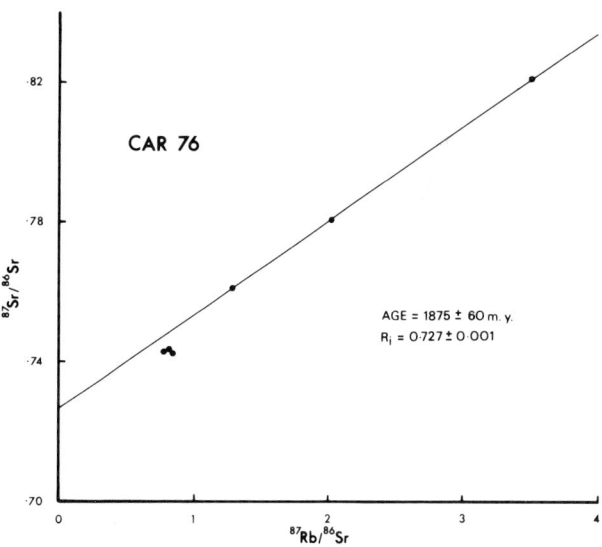

Figure 5. Rb-Sr isochron plot - suite CAR 76.

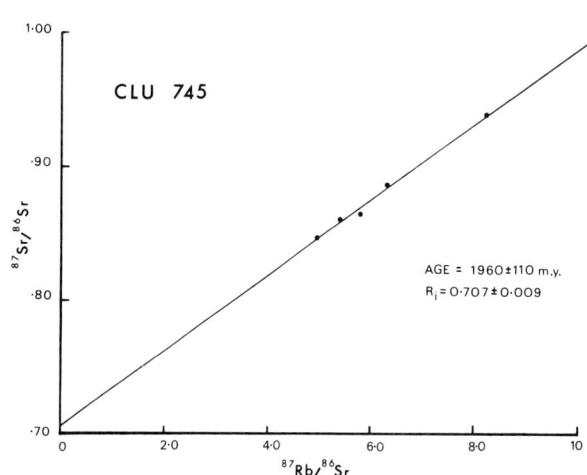

Figure 6. Rb-Sr isochron plot - suite CLU 745.

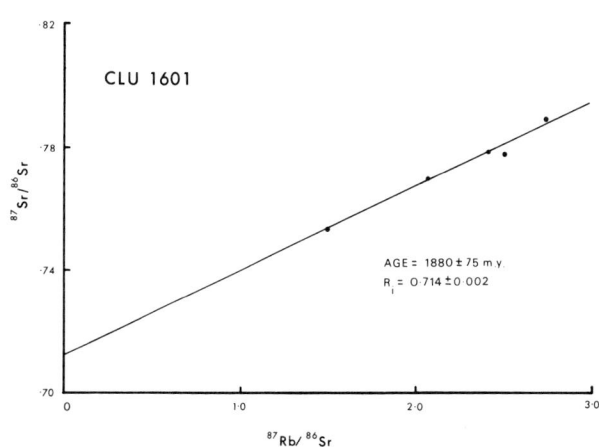

Figure 7. Rb-Sr isochron plot - suite CLU 1601.

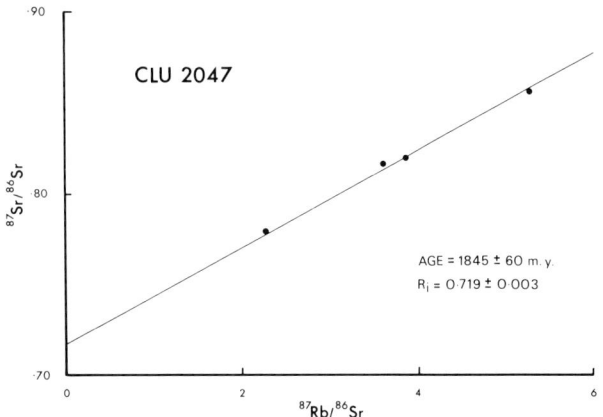

Figure 8. Rb-Sr isochron plot - suite CLU 2047.

2463) that cuts through both the quartzofeldspathic and pelitic sequence. The data from five quartzofeldspathic gneisses and three pelitic rocks are included in Table I and the data are plotted on an isochron diagram in Figure 9. Data from all eight samples (pelites + granitoids) yield a scatterchron that corresponds to a "date" of 1825 ±35 Ma and an initial ^{87}Sr/^{86}Sr of 0.718±0.003 (see Table II). If only the data from the quartzofeldspathic gneisses (samples 72, 87.5, 88.5, 294 and 301.5) are used in the calculation then the date corresponds to a value of 1860±35 Ma and the MSWD drops from 6.4 to 1.6. Using only the analytical data from the metapelitic rocks a figure of 1820±80 Ma (MSWD 26) is obtained, but this is much too imprecise to draw any firm conclusions about isotopic rehomogenization during regional metamorphic activity.

U-Pb DATING

Two quartzofeldspathic suites were collected for zircon dating. The analytical data are shown in Table III. Sample CAR 108-56, from the Earl River complex south of Dominique-Peter, contains honey-coloured, rounded zircons with a length to width ratio of up to 3. The data from both magnetic and non-magnetic fractions are discordant and yield upper and lower intercept dates of 2320±20 Ma and 168±17 Ma, respectively, with about 20% discordance (see Fig. 10). The event at 2320 Ma is one of the oldest so far detected in this part of the Churchill Province. Sample CLU

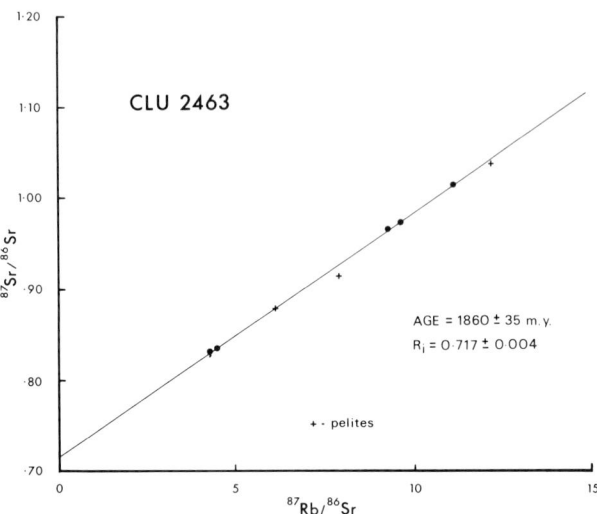

Figure 9. Rb-Sr isochron plot - suite CLU 2463. Data from pelites marked by crosses, dots represent quartzofeldspathic material. Isochron shown based on data from quartzofeldspathic material only.

1601-337.5 yielded sufficient material for three different magnetic fractions. The zircons are lighter in colour than those separated from CAR 108-56 and not as cloudy. The three data points yield a discordia (Fig. 10) with intercepts of 2130±22 Ma and 628±66 Ma. This date is significantly younger than the one obtained from CAR 108-56. The data show about 30% discordance.

Four mineralized areas were sampled for U-Pb dating. These were (i) the Donna boulder train, that lies 1.5 km southeast of the D orebody, (ii) the Dominique-Peter zone, northeast of Cluff Lake and approximately 1 km north of the southern boundary of the basement-sandstone contact, (iii) the OP decline, southeast of Dominique-Peter, and (iv) the Numac showings at the central, western margin of the basement. The analytical data are given in Table IV. Attempts were made to obtain isotopic data from samples that contained only one generation of U-bearing minerals, but this proved to be extremely difficult.

Samples from the Donna boulder train consist of hematite-uraninite associated with gold, galena, limonite and goethite. Data from three of the four Donna samples (sam-

TABLE II

SUMMARY OF ISOCHRON DATA

No.	Suite No.	No. of points	Age (Ma)	^{87}Sr/^{86}Sr (initial)	MSWD
1	CAR 75	4	2025 ± 44	0.728 ± 0.002	2.5
2	CAR 76	3	1875 ± 60	0.727 ± 0.001	0.1
3	CLU 87	4	1775 ± 55	0.718 ± 0.006	0.6
4	CLU 745	4	1960 ± 110	0.707 ± 0.009	0.8
5	CLU 1601	4	1880 ± 75	0.714 ± 0.002	0.9
6	CLU 1711	5	1730 ± 80	0.725 ± 0.007	1.1
7	CLU 2047	4	1845 ± 60	0.719 ± 0.003	5.5
8	CLU 2463	8	1825 ± 35	0.718 ± 0.003	6.4
9	CLU 2463 (granitoids only)	5	1860 ± 35	0.717 ± 0.004	1.6

ples 2, 3, and 4, Table IV) are co-linear within experimental error (see line B, Fig 11) and the intersection with the concordia gives an upper intercept value of 380±5 Ma. The discordia passes through zero and suggests recent Pb loss.

A single uraninite sample from Dominique-Peter (CLU 1571-5) yielded a $^{207}Pb/^{206}Pb$ date of 960 Ma, a figure that represents a minimum age for this deposit assuming recent Pb loss. In an attempt to obtain data sufficient to define a discordia, sample CLU 1571-5 was divided into three pieces. The four data points obtained, including the initial results,

TABLE III
U-Pb ZIRCON DATA

No.	Sample No.	$^{207}Pb/^{206}Pb$ (atomic)	$^{208}Pb/^{206}Pb$ (atomic)	$^{204}Pb/^{206}Pb$ (atomic)	U(%) (weight)	Pb(%) (weight)	Common Pb(%) (atomic)	$^{207}Pb/^{235}U$ (atomic)	$^{206}Pb/^{238}U$ (atomic)
1	CAR 108-56 NM	0.17830	0.16608	0.00246	0.06	0.03	12.07	7.0991	0.3520
2	CAR 108-56 M	0.15760	0.11189	0.00094	0.05	0.02	4.60	6.2817	0.3135
3	CLU 1601-337.5 1M	0.13236	0.12529	0.00047	0.10	0.04	2.32	5.4312	0.3125
4	CLU 1601-337.5 3M	0.12620	0.11113	0.00023	0.12	0.04	1.11	4.8896	0.2879
5	CLU 1601-337.5 5M	0.12344	0.10614	0.00023	0.20	0.06	1.11	4.3933	0.2648

M = magnetic fraction
NM = non-magnetic fraction

TABLE IV
U-Pb DATA

No.	Sample No.	$^{207}Pb/^{206}Pb$ (atomic)	$^{208}Pb/^{206}Pb$ (atomic)	$^{204}Pb/^{206}Pb$ (atomic)	$^{208}Pb/^{204}Pb$ (atomic)	U(%) (weight)	Pb(%) (weight)	Common Pb(%) (weight)	$^{207}Pb/^{235}U$ (atomic)	$^{206}Pb/^{238}U$ (atomic)
1	DON 49	0.05565	0.00494	0.00013	39.2	7.11	0.32	0.86	0.3642	0.0491
2	DON 37	0.05550	0.00372	0.00008	43.8	22.89	1.28	0.58	0.4605	0.0616
3	DON 16	0.05570	0.00379	0.00009	41.0	67.56	2.19	0.63	0.2664	0.0356
4	DON 10	0.05448	0.00128	0.00002	52.9	43.78	5.55	0.17	1.0433	0.1398
5	OP F-1-1	0.08343	0.04008	0.00102	39.2	55.05	5.54	6.46	0.9693	0.1023
6	OP F-1-2	0.08428	0.04073	0.00107	38.2	40.27	3.32	6.72	0.7940	0.0835
7	OP F-1-3	0.07469	0.02026	0.00052	39.3	18.20	1.41	3.34	0.7589	0.0818
8	OP F-1-4	0.09956	0.08404	0.00214	39.2	3.21	2.93	12.85	8.1667	0.8619
9	CLU 1571-5	0.07178	0.00186	0.00004	42.8	56.59	5.30	0.29	0.9955	0.1015
10	CLU 1571-5(A)	0.07092	0.00094	0.00002	41.9	44.94	5.25	0.15	1.2330	0.1267
11	CLU 1571-5(B)	0.07177	0.00208	0.00005	40.6	40.85	4.78	0.34	1.2419	0.1268
12	CLU 1571-5(C)	0.07311	0.00218	0.00005	41.8	78.83	8.99	0.35	1.2298	0.1233
13	3B-OP (A)	0.06841	0.00176	0.00004	43.1	28.23	2.19	0.27	0.7867	0.0841
14	3B-OP (B)	0.06877	0.00528	0.00013	39.7	113.8	10.07	0.88	0.8812	0.0956
15	3B-OP (C)	0.06942	0.00771	0.00020	38.8	43.82	3.45	1.31	0.7773	0.0847
16	TR-16 (A)	0.07026	0.01580	0.00043	37.0	0.34	1.92	2.20	53.890	6.0990
17	TR-16 (B)	0.10593	0.11904	0.00296	40.2	0.08	0.15	0.32	17.030	1.9586
18	TR-16 (C)	0.06883	0.00680	0.00017	39.7	51.79	6.71	1.06	1.2790	0.1398
19	TR-16 (D)	0.06702	0.00478	0.00012	40.1	58.22	8.88	0.78	1.4870	0.1652
20	TR-16 (F)	0.06404	0.00938	0.00023	40.5	3.55	13.78	1.36	35.080	4.1925
21	TR-16 (G)	0.06444	0.00956	0.00023	41.8	4.91	14.96	1.43	27.720	3.2886
22	TR-16 (H)	0.06679	0.00451	0.00011	40.4	52.57	8.26	0.72	1.5298	0.1702
23	TR-18-1(A)	0.27231	0.58551	0.01381	42.4	44.94	3.96	51.96	0.4341	0.0418
24	TR-18-1(B)	0.26922	0.57738	0.01367	42.2	0.56	0.42	51.47	3.6852	0.3612
25	TR-18-1(C)	0.26696	0.55555	0.01331	41.7	0.12	0.23	49.89	9.9157	0.9263
26	TR-18 (A)	0.23186	0.46164	0.01099	42.0	0.98	0.47	45.04	2.7612	0.2667
27	TR-18 (B)	0.27799	0.58006	0.01364	42.5	0.63	0.52	49.91	4.7157	0.3962
28	TR-18 (C)	0.26814	0.57759	0.01354	42.6	4.87	4.25	51.28	4.3253	0.4186
29	TR-18 (D)	0.28327	0.59540	0.01408	42.3	1.83	1.46	51.20	4.4233	0.3778
30	TR-18 (E)	0.30380	0.66045	0.01553	42.5	7.05	2.28	53.92	1.6688	0.1418
31	TR-18 (F)	0.28168	0.61962	0.01451	42.7	2.00	0.97	53.06	2.2867	0.2222

Assumed common Pb isotopic composition for DON samples: $^{206}Pb/^{204}Pb$ = 18.056, $^{207}Pb/^{204}Pb$ = 15.628, $^{208}Pb/^{204}Pb$ = 37.972; for OP and CLU samples $^{206}Pb/^{204}Pb$ = 17.342, $^{207}Pb/^{204}Pb$ = 15.571, $^{208}Pb/^{204}Pb$ = 37.150; for Numac: TR-16, TR-18 (A,C,F), and TR-18-1, $^{206}Pb/^{204}Pb$ = 17.067, $^{207}Pb/^{204}Pb$ = 15.544, $^{208}Pb/^{204}Pb$ = 36.840; TR-18, (B,D,E), $^{206}Pb/^{204}Pb$ = 16.503, $^{207}Pb/^{204}Pb$ = 15.476, $^{208}Pb/^{204}Pb$ = 36.220.
Error on some *absolute concentrations* can be extremely high because of low sample weights.

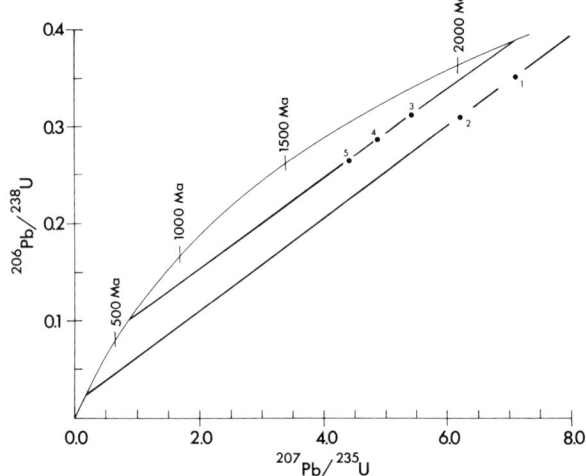

Figure 10. Concordia diagram. Lower discordia defined by data from CAR 108-56, upper by data from CLU 1601-337.5. Sample numbers refer to analyses in Table III.

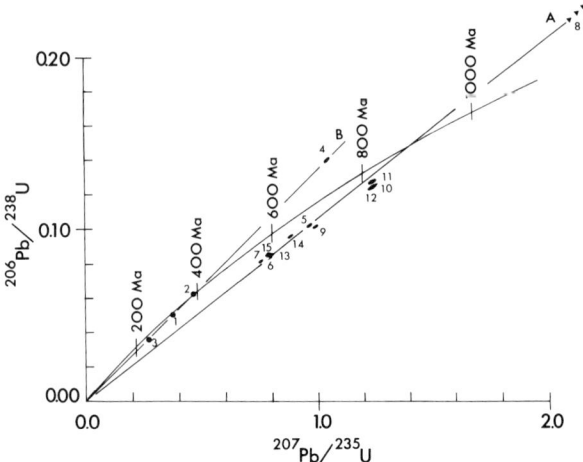

Figure 11. Concordia diagram showing data from the Dominique-Peter, Donna and OP deposits. Discordia A based on data from OP samples, discordia B from Donna samples. Sample numbers refer to analyses in Table IV.

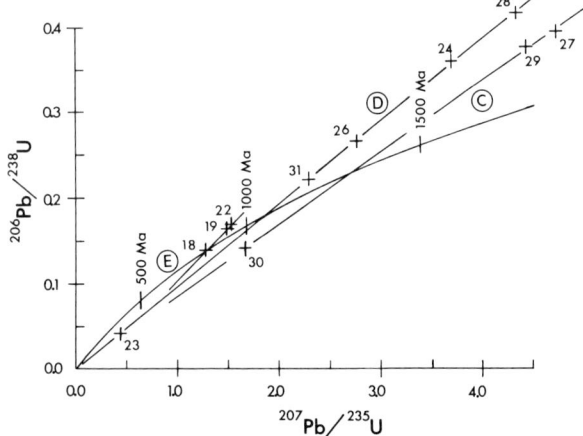

Figure 12. Concordia diagram showing 'near concordant' data from Numac samples. Discordia C calculated using three points from TR-18, discordia D using five points from TR-18 and TR-18-1, and discordia E using the three least discordant samples from TR-16. Sample numbers refer to analyses in Table IV.

are discordant (points 9, 10, 11, 12, Fig. 11) and do not define a straight line. The oldest date was obtained by combining the data from two of the points (analyses 9 and 10, Table IV) to yield an upper intersection age of about 1050 Ma. The $^{207}Pb/^{206}Pb$ dates for all four samples fall between 995 Ma and 945 Ma.

A single specimen (OP-F) of material from the OP deposit, containing coffinite, quartz, and tourmaline, was divided into four pieces and each piece was considered as an individual sample. Data from all four samples are discordant on the concordia diagram but data from three are co-linear (see line A, Fig. 11) within experimental uncertainty. The discordia intersects the concordia at 892±5 Ma and at zero. An additional sample from the OP body (3B-OP), mainly uraninite, was divided into three pieces. Data from these, together with the data from the original sample form a poorly-defined discordia with a range of intercepts that vary from about 820 Ma to 890 Ma. One sample from the OP body and one from Donna lie above the concordia (reverse discordance) which suggests either Pb gain or U loss from the system.

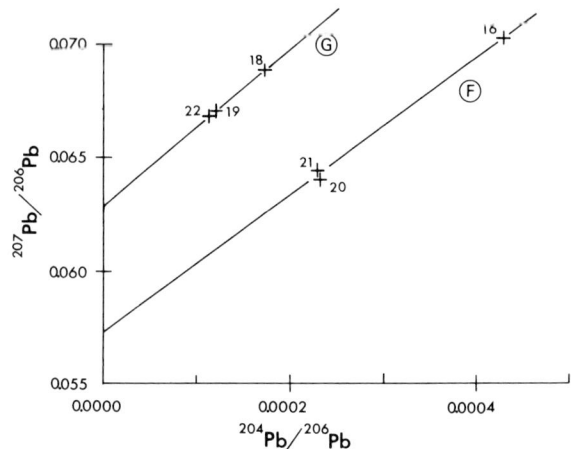

Figure 13. Plot of $^{207}Pb/^{206}Pb$ vs $^{204}Pb/^{206}Pb$ for the data from the TR-16 samples. The two groupings yield intercepts at 0.057 (500 ±50 Ma) and 0.063 (705±2 Ma).

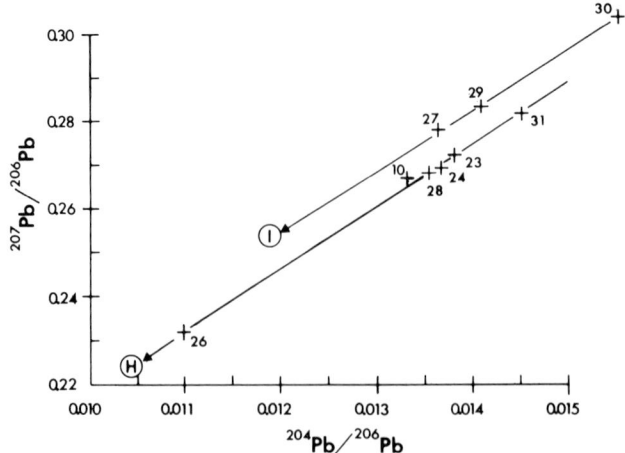

Figure 14. Plot of $^{207}Pb/^{206}Pb$ vs $^{204}Pb/^{206}Pb$ for the data from the TR-18 and TR-18-1 samples. The two groupings yield intercepts of 0.076 (1095±95 Ma) and 0.090 (1425±115 Ma).

Three samples (TR-16, TR-18 and TR-18-1) from the Numac area were analysed for U and Pb. Seven TR-16, six TR-18 and three TR-18-1 separates were analyzed. Because of extensive alteration (kasolite?, uranophane?) it was difficult to separate pitchblende from any of the samples. All separates of TR-16 and TR-18-1 were treated as pitchblende and were dissolved overnight in warm, dilute HNO_3. The remaining separates were dissolved in concentrated HNO_3 containing a few drops of HF. Only in a few cases were the samples completely dissolved.

The U-Pb data for the Numac samples are given in Table IV. Features worthy of note are: (i) only those samples that dissolved completely show U and Pb concentrations characteristic of pitchblende, (ii) all separates from TR-18 and TR-18-1 contain about 50% common Pb, even though the total Pb contents vary between 0.2% to 4.0%, and (iii) the $^{208}Pb/^{204}Pb$ ratio is fairly consistent for the high common Pb samples and lies close to a value of 42.

Almost all of the Numac data show reverse discordance on a concordia diagram. Plotted on Figure 12 are the data that lie close to the concordia. All seven TR-16 samples show a crude linearity and a large negative intersection with the concordia. If the three least concordant samples are used (line E, Fig. 12), the intersections correspond to dates of 823 ± 5 Ma and -324 ± 61 Ma. Because the intercept at 823 Ma is well-defined by points near the concordia, the date is considered to be somewhat reliable in spite of the "negative" lower intercept. The reasons for the latter are presently unknown.

Samples with high common Pb contents can sometimes yield useful information (see Gancarz, 1979) when the $^{207}Pb/^{206}Pb$ ratio is plotted against the $^{204}Pb/^{206}Pb$ ratio. A linear regression line through the data gives an intersection that approximates the $^{207}Pb/^{206}Pb$ ratio that existed when the common Pb content (non-radiogenic component) was zero. This ratio can be used to calculate a model $^{207}Pb/^{206}Pb$ age. Implied in this approach is that the common Pb added to the system had a uniform Pb isotopic composition. The data for the TR-16 samples, shown in Figure 13, can be used to generate two lines with intercepts at 0.057 (500 Ma) and 0.063 (705 Ma). In general, the more discordant samples on the concordia diagram tend to have the lower $^{207}Pb/^{206}Pb$ intercepts.

Five points from separates of TR-18 and TR-18-1 are nearly co-linear (line D, Fig. 12). The line intersects the concordia at 1062 ± 10 Ma and at -8 ± 11 Ma. A regression line through six of the data points on the $^{207}Pb/^{206}Pb$ vs $^{204}Pb/^{206}Pb$ plot (see line H, Fig. 14) yields an intercept age of 1095 ± 95 Ma which is in good agreement with the concordia intersection. This date probably corresponds to a real event.

Three of the data points from TR-18 are roughly co-linear. The discordia (line C, Fig. 12) intersects the concordia at 1327 ± 26 Ma and at essentially 0 Ma. These three points on a $^{207}Pb/^{206}Pb$ vs $^{204}Pb/^{206}Pb$ diagram (see line I, Fig. 14) yield a date of 1425 ± 115 Ma, in general agreement with the concordia interpretation.

K-Ar DATES

Chloritization and sericitization, fairly widespread throughout the structure, are associated with either diagenesis or hydrothermal activity. Nine altered basement samples were provided by M. Pagel (CREGU) from four holes drilled into the Dominique-Peter zone. Some samples contain chlorite (MPCL 81-22, -26, -46 and -60) and some contain potassic micas (MPCL 81-13, -22, -23, -26, -42 and -84). One sample (MPCL 81-24) was obtained from an interlayered argillaceous horizon developed in a zone of uranium mineralization.

Because the alteration products are intimately intermixed and hence fairly difficult to separate, small whole-rock aliquots were analyzed using the laser-fusion technique in an attempt to evaluate the degree of argon loss. This approach enables clusters of small grains or even individual grains to be partly or completely fused under vacuum. In most cases the alteration products are of the order of a micron or tens of microns in size (Clauer, pers. commun., 1984) so that several grains were probably analysed using the laser fusion technique.

Such data may place constraints on (i) the age of the last thermal event to affect the mineralized zones, and (ii) the scale of isotopic re-setting. The wide range of dates from 1615 ± 5 Ma through to 500 ± 50 Ma, shown in Table V, show a crude correlation with the geology observed in each hole although further work on separated mineral phases is needed before such relationships can be accepted with any degree of

TABLE V

$^{40}Ar/^{39}Ar$ DATA FROM ALTERATION PRODUCTS

No.	Sample No.	K/Ca	$^{40}Ar^*/^{39}Ar$	% atmos.	Age[+] (Ma)
1	MPCL 81-60 (1549)	12.6	56.74	8.6	1175 ± 15
2	MPCL 81-13 (1571)	44.3	89.58	3.7	1615 ± 5
3	MPCL 81-84 (1589)	4.9	67.29	15.0	1325 ± 10
4	MPCL 81-22 (1607-260)	26.7	83.58	16.9	1540 ± 25
5	MPCL 81-23 (1607-276)	7.4	72.61	24.3	1400 ± 50
6	MPCL 81-24 (1607-280)	6.4	19.89	80.6	500 ± 50
7	MPCL 81-26 (1607-294)	34.5	74.22	2.9	1420 ± 5
8	MPCL 81-42 (1607-495)	24.2	45.89	13.1	1000 ± 10
9	MPCL 81-46 (1607-548)	16.8	83.71	62.9	1540 ± 30

numbers in parentheses refer to hole numbers.
+ rounded off to nearest 5 Ma.

confidence. Sample 81-13, from hole CLU 1571, lies within the main mineralized zone and close to a regolith zone in the metamorphic succession (see Blaise and Koning, 1985). This latter feature, along with the similarity of the 1615±5 Ma date to a whole-rock Rb-Sr isochron date of 1630±30 Ma (Fahrig and Loveridge, 1981) from deeply-weathered basement rocks, may simply reflect processes related to formation of the regolith. Samples MPCL 81-22, -23, -24, -26, -42 and -46 were collected from a single hole (CLU 1607) that cuts through a variety of rocks. Between depths of 276′ and 294′ aluminous gneisses are intercalated with arenaceous material, logged as Cluff breccia. The date of 500±50 Ma obtained from sample 81-24 from a depth of 280′ is consistent with previous figures of 475 Ma and 495 Ma obtained from the Cluff breccia (Wanless et al., 1968). Pagel (1982) has noted that between depths of 276′ and 307′ in hole CLU 1607 Mg-tourmalines are associated with mineralization. The 1400±50 Ma and 1420±5 Ma dates are restricted to this section of core, are not found elsewhere in the structure, and could conceivably be associated with either mineralization and/or the formation of the Mg-tourmalines. Sample 81-42, collected from a highly-mineralized section of quartzofeldspathic gneisses at a depth of 495′, yields a date of 1000±10 Ma similar to many of the U-Pb dates obtained from ore material elsewhere in the structure. Sample 81 84, from hole CLU 1589, has had significant additions of boron and fluorine during alteration (Pagel, pers. commun., 1983). Open-system behaviour such as this may be reflected in the date of 1325±10 Ma. The two dates of 1540 Ma are from two highly-altered aluminous gneisses from hole CLU 1607. The samples represent the deepest and shallowest collected for isotopic work from hole CLU 1607. These dates are similar to the estimated depositional age for the Athabasca sediments. Although the correlations between the laser fusion dates and the geological features observed at different levels within the hole are made on a somewhat *ad hoc* basis, the ^{40}Ar/^{39}Ar dates do, nevertheless, suggest that disturbances are fairly localized, with several events documented over a relatively short distance within this particular hole.

Eight samples of the fine-grained, volcanic-like Cluff breccia (see von Einsiedel, 1982), selected on the basis of thin section examination, were crushed to a 1 to 2 mm size fraction. Aphanitic matrix material was hand-picked from the crushed samples and care was taken to avoid any visible fragmental material. Breccia localities are shown in Figure 1. Only two of the samples, CAR 53-187 and CAR 57-87, were from core material; the remainder were from surface outcrops. The ^{40}Ar/^{39}Ar laser fusion analyses are summarized in Table VI.

Earlier K-Ar work by Wanless et al. (1968) on the Cluff breccias suggested some degree of concordancy of the dates at about 485 Ma. However, the dates shown in Table VI, between 515 Ma and 365 Ma, are somewhat discordant, although most fall between 500 Ma and 440 Ma.

Additional information from the two core samples yielded some interesting results. Both samples were analysed twice using a laser spot of about 150 μm. Whereas the two runs on sample CAR 53-187 were from separate spots, sample CAR 57-87 did not yield enough radiogenic ^{40}Ar for single spot analysis. This problem was overcome by fusing a tight cluster of four spots for each of the two runs. Repeat runs were made on the same chunks that were used for the first spots. The additional spots were within 2 mm of the original fusion sites. The results from sample CAR 53-187 agree fairly closely within the 2 sigma error limits, while the two dates from sample CAR 57-87 are discordant by 75 Ma. The date of 365 Ma from CAR 57-87 was obtained from the results of the first of the two fusion experiments, so that it seems unlikely that the differences are due to argon loss during the heating procedures. Such discordances could have been produced by continuous argon leakage over a considerable period of time, variable degrees of argon loss during a single event, or repeated argon loss at discrete intervals during the period 515 Ma to 365 Ma.

INTERPRETATION OF DATA

The U-Pb zircon dates of 2320 Ma and 2130 Ma clearly document the oldest events yet detected in the Carswell structure. The basement chronology based on whole-rock, Rb-Sr data involved events at ca. 2000 Ma, ca. 1880 Ma and another at about 1750 Ma. Although some of these events are indicated by only "scatterchrons" they do agree fairly well with events found in basement rocks outside of the Carswell structure. For example, granitoid activity ca. 2000 Ma has been found in the Beaverlodge area (Bell and Blenkinsop, 1982). The 1880 Ma figure, a weighted-mean based on some

TABLE VI

^{40}Ar-^{39}Ar DATA FROM CLUFF BRECCIAS

No.	Sample No.	K/Ca	^{40}Ar-^{39}Ar	% atmos.	Age+ (Ma)
1	Sophie 1-1	36.8	14.52	24.2	415 ± 10
2	Sophie OC-1	34.7	18.44	51.0	515 ± 10
3	'A' Train OC-2B	14.5	17.23	45.7	485 ± 10
4	'A' Train OC-1R	20.6	17.92	36.1	500 ± 10
5	UAR OC-1 ('U'-area)	17.1	16.26	48.3	460 ± 10
6	MF. OC.	11.9	15.80	51.5	450 ± 20
7	53-187 (1)	18.8	13.43	41.2	430 ± 5
8	53-187 (2)	19.3	14.90	8.1	470 ± 40
9	57-87 (1)	22.7	11.23	32.7	365 ± 5
10	57-87 (2)	25.5	13.80	6.9	440 ± 5

+ rounded off to nearest 5 Ma.

of the data shown in Table II, is also similar to whole-rock Rb-Sr (Blenkinsop et al., 1983) and U-Pb zircon dates (Ray and Wanless, 1980) from the Watharnan batholith in the Rottenstone Domain, data that almost certainly indicate a magmatic event.

The 1750 Ma event recorded by the metapelites corresponds to the culmination of Hudsonian diastrophism but it is not clear whether this reflects an upgrading of an older succession or sedimentary deposition at that time. If the latter interpretation is accepted then an unconformity must exist somewhere within the Cluff succession. It is interesting to note that the 1750 Ma event is not recorded in any of the data from the granitoid units in spite of the fact that some re-setting of the Rb-Sr systems has taken place. The data from hole CLU 1601 highlight this problem. The U-Pb zircon date of 2130±20 Ma differs markedly from the whole-rock Rb-Sr isochron date of 1880±75 Ma, a feature that can be attributed to disturbances brought about by the 1880 Ma event. It is still, however, puzzling that this and some of the other quartzofeldspathic suites have not been reset by the later 1750 Ma event.

In spite of the new U-Pb data from the Carswell uranium deposits there are still no signs of major mineralization that pre-dates the deposition of the Athabasca sediments. Precambrian uranium mineralization in Saskatchewan was broadly divided into three age groups (Bell, 1981): those ca. 1880 Ma (syngenetic uraninite in pegmatites), those ca. 1740 Ma (the vein-type deposits of the Beaverlodge area) and those between 1300 Ma and 800 Ma. The last-mentioned group includes most of the unconformity-type deposits and almost all of the Precambrian mineralization within the Carswell structure.

Attempts to correlate the isotopic data from the uranium deposits with the mineralogical and geochemical classification of Pagel and Ruhlmann (1985) are made difficult because of open-system behaviour for almost all of the U-bearing minerals. Although uranium and/or lead loss or gain make interpretation of the isotopic data far from straightforward, three, perhaps as many as four, events are indicated at ca. 1300 Ma, 1100 ± 50 Ma, ca. 850 Ma, and 380 ± 5 Ma. The oldest event is hinted at by the data from the Numac showing, while the 1100 Ma event includes formation of both the D and Claude deposits (Devillers and Nordmann, 1974; Bellon et al., 1976; Tapaninen, 1976; Gancarz, 1979) and possibly Dominique-Peter and parts of the Numac showing. The ca. 850 Ma group includes the OP and Numac bodies, while the 380 Ma event includes the mineralization associated with the Donna boulder train. Tona et al. (1985) consider the 1100

TABLE VII

SUMMARY OF NEW GEOCHRONOLOGICAL DATA FROM THE CARSWELL STRUCTURE

Isotopic Event	Geochronometric Details	Geologic Event	Proterozoic (Pre-Athabasca) events elsewhere in Saskatchewan (after Bell and Macdonald, 1982)
ca. 380 Ma	U-Pb, discordant uranium ores	uranium mineralization	
ca. 515-365 Ma	^{40}Ar-^{39}Ar, laser-fusion, whole-rock	formation of Cluff breccias	
ca. 1330-850 Ma	U-Pb, discordant uranium ores	uranium mineralization and associated alteration	
	K-Ar, alteration minerals		ca. 1450 Ma deposition of Athabasca sediments
			ca. 1580 Ma uplift and localized remobilization
			ca. 1670 Ma metamorphism (Glennie Lake Domain)
ca. 1750 Ma	Rb-Sr, whole-rock (metapelites)	"Main" Hudsonian metamorphism (uplift?)	ca. 1740 Ma widespread "main" Hudsonian metamorphism plus magmatism. Epigenetic uranium mineralization.
ca. 1880 Ma	Rb-Sr, whole-rock (quartzofeldspathic gneisses)	granitoid magmatism and/or metamorphism	ca. 1880 Ma granitoid magmatism (Watharnan batholith) plus metamorphism. Syngenetic uranium mineralization.
ca. 2000 Ma	Rb-Sr, whole-rock (quartzofeldspathic gneisses)	granitoid magmatism and/or metamorphism	
ca. 2130 Ma	U-Pb, discordant zircons (quartzofeldspathic gneiss)	granitoid magmatism and/or metamorphism	ca. 2180 Ma granitoid magmatism (Donaldson Lake event)
ca. 2320 Ma	U-Pb, discordant zircons (quartzofeldspathic gneiss)	granitoid magmatism and/or metamorphism	

Ma event to be the most important within the Carswell structure.

The range of K-Ar dates shown in Table V from the altered samples is difficult to interpret, but may reflect either different degrees of argon loss brought about by a single unique event, or alteration, or re-setting of older material by diagenesis or episodic mineralization. Possible disturbances could involve the migration of fluids during weathering of the basement, deposition of the Athabasca sediments, hydrothermal activity, and formation of the Cluff breccias. A fairly widespread event ca. 1200-1300 Ma has now been documented by Clauer et al. (1985) that is correlated with intensive hydrothermal activity that affected both basement and sedimentary cover within the Carswell structure.

There is now little doubt that the Cluff breccias formed fairly late in the evolution of the Carswell structure. If the Cluff breccias are the result of meteorite impact, then the dates listed in Table VI imply a post-365 Ma impact event, in which case the variations would have to be attributed to different degrees of argon retention during impact. On the other hand, the spread in dates, rather than reflecting different argon retentivities, could simply be the result of repeated or continuous argon loss during the period 515 to 365 Ma ago. In this context it is interesting to note that this spread is similar to dates obtained from the post-800 Ma uranium deposits.

CONCLUSIONS

The most significant findings of this paper are: (i) The sub-Athabasca basement was affected by magmatic and/or metamorphic events during the Proterozoic at 2320 Ma, 2130 Ma, 2000 Ma, 1880 Ma and 1750 Ma ago. Archean ages have yet to be detected. (ii) Uranium mineralization started perhaps as early as 1300 Ma ago and ended some 380 Ma ago. Possible events at ca. 1330 Ma (Numac?), 1100 Ma (D, Numac), ca. 850 Ma (OP, Numac?) and 380 Ma (Donna) are indicated. (iii) Post-600 Ma activity involved formation and subsequent evolution of the Cluff breccias, as well as uranium mineralization and/or reworking.

The new data from the Carswell structure can be accommodated into the geochronometric framework established by Bell and Macdonald (1982) for the Hudsonian orogen in Saskatchewan. Comparison is made in Table VII between the chronology established, so far, within the Carswell structure and a slightly revised version of the calibration proposed by Bell and Macdonald (1982). Almost all of the Proterozoic events documented in other parts of the province are recorded by the Carswell basement rocks.

On a regional scale, the dates of uranium mineralization from the Carswell structure are similar to those found elsewhere in the Athabasca Basin. Dates similar to those from Carswell (in the range 1350 Ma to 380 Ma) have been documented from other deposits elsewhere in Saskatchewan (Fahrig, 1961; Knipping, 1974; Little, 1974; Wendt et al., 1978; Cumming and Rimsaite, 1979; Gatzweiler et al. 1979; Jones, 1980), but the isotopic data suggest that events ca. 1350 Ma were more important outside of the structure (see Tremblay, 1982) than the main 1100 ±50 Ma mineralization event associated with the D and Claude deposits. An interesting feature to emerge from this present study, particularly from the D orebody and Numac showing, is that the isotopic data suggest that the channelways for mineralizing solutions seem to have remained open during considerable periods of time.

To correlate the mineralizing events within the Carswell structure with any of the major thermotectonic events known elsewhere in Saskatchewan is difficult. The similarity in age between the oldest grouping of mineralization and the Cree Lake dyke swarm has already been pointed out (Bell, 1981) but although mineralization and emplacement of diabase dykes appear to have been active at roughly the same time, this probably reflects different responses to a much more widespread phenomenon. It is tempting to relate the ages at 1300 to 800 Ma and 600 to 380 Ma to large-scale events outside the province, the former to activity further to the east in the Grenville Province, and the latter to orogenic activity in the Appalachians. Although such correlations are complicated by the enormous distances involved, there nevertheless appears to be a crude agreement between the ages of uranium mineralization at Carswell and some of the major orogenies in Canada.

ACKNOWLEDGEMENTS

Thanks are extended to Amok Ltd. for financial support of this study. Discussions of the geology of the area with D. Alonso, R.T. Lainé, M. Svab, K. Tapaninen and F. Tona added greatly to the overall interpretation of the isotopic data. D. York and C.M. Hall of the University of Toronto (Geophysics) kindly carried out the laser-fusion K-Ar analyses. J. Blenkinsop, J.W. Card, N. Clauer, G.L. Cumming, and R. Macdonald kindly reviewed the manuscript and offered numerous helpful suggestions. J. Blenkinsop is also thanked for help with some of the analytical work.

REFERENCES

Bell, K., 1981, A Review of the Geochronology of the Precambrian of Saskatchewan — Some Clues to Uranium Mineralization: Mineralogical Magazine, v. 44, p. 371-378.

Bell, K., and Blenkinsop, J., 1982, Saskatchewan Shield Geochronology Project: in Summary of Investigations 1982: Saskatchewan Geological Survey Miscellaneous Report 82-4, p. 16.

Bell, K., and Macdonald, R., 1982, Geochronological Calibration of the Precambrian Shield in Saskatchewan: in Summary of Investigations 1982: Saskatchewan Geological Survey Miscellaneous Report 82-4, p. 17-22.

Bellon, H., Devillers, C., Hagemann, R., and Touray, J.-C., 1976, Dater les Minéralisations: Mémoire Hors Série de la Société Géologique de France, no. 7, p. 265-268.

Blaise, J.R., and Koning, E., 1985, Mineralogical and Structural Aspects of the Dominique-Peter Uranium Deposit: in Lainé, R., Alonso, D., and Svab, M., eds., The Carswell Structure Uranium Deposits, Saskatchewan: Geological Association of Canada Special Paper 29.

Blenkinsop, J., and Bell, K., 1981, Saskatchewan Shield Geochronology Project: in Summary of Investigations 1981: Saskatchewan Geological Survey Miscellaneous Report 81-4, p. 25.

Blenkinsop, J., Bell, K., Cole, T.J.S., Menagh, D.P., and Ray, G.E., 1983, The Wathaman Batholith – a Comparison Between U-Pb and Rb-Sr Dating: Geological Association of Canada/Mineralogical Association of Canada Annual Meeting, Program with Abstracts, v. 8, p. A7.

Brooks, C., Hart, S.R., and Wendt, I., 1972, Realistic Use of Two-Error Regression Treatments as Applied to Rubidium-Strontium Data: Reviews of Geophysics and Space Physics, v. 10, p. 551-577.

Clauer, N., Ey, F., and Gauthier-Lafaye, F., 1985, K-Ar Dating of Different Rock Types from the Cluff Lake Uranium Ore Deposits (Saskatchewan-Canada): in Lainé, R., Alonso, D., and Svab, M., eds., The Carswell Structure Uranium Deposits, Saskatchewan: Geological Association of Canada Special Paper 29.

Cumming, G.L., and Rimsaite, J., 1979, Isotopic Studies of Lead-Depleted Pitchblende, Secondary Radioactive Minerals, and Sulphides from the Rabbit Lake Uranium Deposit, Saskatchewan: Canadian Journal of Earth Sciences, v. 16, p. 1702-1715.

Devillers, C., and Nordmann, F., 1974, Datation du Minerais Uranifère de Cluff - Premiers Résultats: Commissariat à l'Energie Atomique, Internal Report.

von Einsiedel, C.A., 1982, Petrography and Geochemistry of the Cluff Lake Breccias, Carswell Structure, Northern Saskatchewan: unpublished B.Sc. Thesis, Carleton University, Ottawa, 44 p.

Ey, F., Lillié, F., Gauthier-Lafaye, F., and Weber, F., 1985, A Uranium Unconformity Deposit: The Geological Setting of the D Orebody (Saskatchewan-Canada): in Lainé, R., Alonso, D., and Svab, M., eds., The Carswell Structure Uranium Deposits, Saskatchewan: Geological Association of Canada Special Paper 29.

Fahrig, W.F., 1961, The Geology of the Athabasca Formation: Geological Survey of Canada Bulletin 68, 41 p.

Fahrig, W.F., and Loveridge, W.D., 1981, Rb-Sr Isochron Age of Weathered Pre-Athabasca Formation Basement Gneiss, Northern Saskatchewan: Rb-Sr and U-Pb Isotopic Age Studies, Report 4: in Current Research, Part C: Geological Survey of Canada Paper 81-1C, p. 127-129.

Gancarz, A.J., 1979, Chronology of the Cluff Lake Uranium Deposit Canada (Abstract): International Uranium Symposium on the Pine Creek Geosyncline, N.T., Australia, Extended Abstracts, p. 91-94.

Gatzweiler, R., Lehnert-Thiel, K., Clasen, D., Tan, B., Voultsidis, V., Strnad, J.G., and Rich, J., 1979, The Key Lake Uranium-Nickel Deposits: Canadian Mining and Metallurgical Bulletin, v. 72, no. 807, p. 73-79.

Harper, C.T., 1978, The Geology of the Cluff Lake Uranium Deposits: Canadian Mining and Metallurgical Bulletin, v. 71, no. 800, p. 68-78.

Herring, B.G., 1976, The Metamorphism and Alteration of the Basement Rocks in the Carswell Circular Structure, Saskatchewan: Unpublished M.Sc. Thesis, University of British Columbia, Vancouver, 136 p.

Innes, M.J.S., 1964, Recent Advances in Meteorite Crater Research at the Dominion Observatory, Ottawa: Meteoritics, v. 2, p. 219-241.

Jones, B.E., 1980, The Geology of the Collins Bay Uranium Deposit, Saskatchewan: Canadian Mining and Metallurgical Bulletin, v. 73, no. 818, p. 84-90.

Knipping, H.D., 1974, The Concepts of Supergene Versus Hypogene Emplacement of Uranium at Rabbit Lake, Saskatchewan, Canada: in Formation of Uranium Ore Deposits: International Atomic Energy Agency 374, p. 531-549.

Little, H.W., 1974, Uranium in Canada: in Report of Activities, Part A: Geological Survey of Canada Paper 74-1A, p. 137-139.

Norrish, K., and Chappell, B.W., 1967, X-ray Fluorescence Spectrography: in Zussman, J., ed., Physical Methods in Determinative Mineralogy: London and New York, Academic Press, p. 161-214.

Pagel, M., 1982, Les Altérations Liées Aux Zones Minéralisées de la Structure Carswell, la Métasomatose Magnésienne: Internal Report, Centre de Recherches sur la Géologie de l'Uranium, 27 p.

Pagel, M., and Svab, M., 1985, Petrographic and Geochemical Variations within the Carswell Structure Metamorphic Core and their Implications with Respect to Uranium Deposition: in Lainé, R., Alonso, D., and Svab, M., eds., The Carswell Structure Uranium Deposits, Saskatchewan: Geological Association of Canada Special Paper 29.

Pagel, M., and Ruhlmann, F., 1985, Chemistry of Uranium Minerals in Deposits and Showings of the Carswell Structure (Saskatchewan-Canada): in Lainé, R., Alonso, D., and Svab, M., eds., The Carswell Structure Uranium Deposits, Saskatchewan: Geological Association of Canada Special Paper 29.

Ramaekers, P., 1983, Geology of the Athabasca Group, NEA/IAEA Athabasca Test Area: in Cameron, E.M., ed., Uranium Exploration in Athabasca Basin, Saskatchewan, Canada: Geological Survey of Canada Paper 82-11, p. 15-25.

Ramaekers, P., and Dunn, C.E., 1979, Geology and Geochemistry of the Eastern Margin of the Athabasca Basin: in Dunn, C.E., ed., Uranium in Saskatchewan: Saskatchewan Geological Society Special Publication ng. 3, p. 297-322.

Ray, G.E., and Wanless, R.K., 1980, The Age and Geological History of the Wollaston, Peter Lake, and Rottenstone Domains in Northern Saskatchewan: Canadian Journal of Earth Sciences, v. 17, p. 333-347.

Ruhlmann, F., 1985, Mineralogy and Metallogeny of Uraniferous Occurrences in the Carswell Structure: in Lainé, R., Alonso, D., and Svab, M., eds., The Carswell Structure Uranium Deposits, Saskatchewan: Geological Association of Canada Special Paper 29.

Steiger, R.H., and Jäger, E., 1977, Subcommission on Geochronology: Convention on the Use of Decay Constants in Geo- and Cosmochronology: Earth and Planetary Science Letters, v. 36, p. 359-362.

Stockwell, C.H., 1982, Proposals for Time Classification and Correlation of Precambrian Rocks and Events in Canada and Adjacent Areas of the Canadian Shield. Part 1. A Time Classification of Precambrian Rocks and Events: Geological Survey of Canada Paper 80-19, 135 p.

Tapaninen, K., 1976, Cluff Lake Area: in Trigg, C.M., and Woolett, G.N., eds., Uranium Deposits of Northern Saskatchewan: Geological Association of Canada/Mineralogical Association of Canada Guidebook, Edmonton '76, p. 50-71.

Tona, F., Alonso, D., and Svab, M., 1985, Geology and Mineralization in the Carswell Structure – A General Approach: in Lainé, R., Alonso, D., Svab, M. eds., The Carswell Structure Uranium Deposits, Saskatchewan: Geological Association of Canada Special Paper 29.

Tremblay, L.P., 1982, Geology of the Uranium Deposits Related to the Sub-Athabasca Unconformity, Saskatchewan: Geological Survey of Canada Paper 81-20, 56 p.

Wanless, R.K., Stevens, R.D., Lachance, G.R., and Edmonds, C.M., 1968, Age Determinations and Geological Studies, K-Ar Isotopic Ages, Report 8: Geological Survey of Canada Paper 67-2, Part A, 141 p.

Wanless, R.K., Stevens, R.D., Lachance, G.R., and Delabio, R.N., 1979, Age Determinations and Geological Studies, K-Ar Isotopic Ages, Report 14: Geological Survey of Canada Paper 79-2, 67 p.

Wendt, I., Höhndorf, A., Lenz, H., and Voultsidis, V., 1978, Radiometric Age Determination on Samples of the Key Lake Uranium Deposit: *in* Zartman, R.E., ed., Short Papers of the Fourth International Conference, Geochronology, Cosmochronology and Isotope Geology: United States Geological Survey Open File Report 78-701, p. 448-449.

York, D., 1969, Least-Squares Fitting of a Straight Line with Correlated Errors: Earth and Planetary Science Letters, v. 5, p. 320-324.

K-Ar DATING OF DIFFERENT ROCK TYPES FROM THE CLUFF LAKE URANIUM ORE DEPOSITS (SASKATCHEWAN-CANADA)

N. Clauer, F. Ey, and F. Gauthier-Lafaye
Centre de Sédimentologie et Géochimie de la Surface, (CNRS), 1 rue Blessig, 67084 Strasbourg-Cedex (France)

ABSTRACT

K-Ar data were obtained on clay fractions (<2 μm) of different rock types, originating mainly from the D ore deposit near Cluff Lake.

A tectonic event is recognized at about 1293 ± 36 Ma in the basement and the sedimentary cover. It is clearly related to intensive hydrothermal activity which is responsible, to some degree, for the mineralization near the shear zone, and whose effects are observed throughout the entire sequence of the Athabasca Group. The clay minerals of these untectonized units also record an isotopic homogenization at 1222 ± 52 Ma, which can be considered to be contemporaneous with the tectonic events inasmuch as the isotopic ages agree within the limits of experimental error.

The results of this study also confirm that the Carswell event, at around 480 Ma, had no influence on the uranium concentration or on the K-Ar isotopic system of the clay minerals.

RÉSUMÉ

Les fractions argileuses <2 μm des différents types de roches rencontrées au gisement D près du Lac Cluff, grès stérile, socle hématitique, socle blanchi et "zone à boules" ont été soumises à la datation isotopique K-Ar. Quelques échantillons de grès prélevés en dehors de la structure de Carswell ont également été analysés pour comparaison.

L'évènement tectonique, qui s'est traduit dans le socle et la couverture sédimentaire par respectivement une zone mylonitique décolorée (socle blanchi) et un faciès "à boules", se place vers 1293 ± 36 Ma. Il est associé à une intense activité hydrothermale partiellement responsable des minéralisations d'uranium et qui s'est propagée dans l'ensemble de la séquence gréseuse de l'Athabasca et de la formation de Douglas placée au-dessus. Les minéraux argileux de ces formations ont, en effet, enregistré une remise à zéro de leur système K-Ar à la même époque, à l'erreur de mesure près, vers 1222 ± 52 Ma. Cet âge n'est donc pas à relier à l'époque de la sédimentation; il n'a pas de valeur stratigraphique et ne constitue qu'un repère minimal. Le socle hématitique n'a pu être daté; ses fractions argileuses semblent constituées de composants hétérogènes de différentes générations.

Les datations obtenues ici permettent également de dissocier clairement les effets de l'impact de Carswell des minéralisations d'uranium. Aucune remobilisation tardive n'a pu être décelée sur le système K-Ar des minéraux argileux; les échantillons extérieurs à la structure présentent rigoureusement le même comportement que ceux pris à l'intérieur.

INTRODUCTION

Structural analysis of the uranium deposits at Cluff Lake (northwestern Saskatchewan, Canada; Fig. 1) and petrographic observations of the D orebody (Ey et al., 1985), have clearly shown that the uranium mineralization is closely related to major tectonic structures of a mylonitic type and to intense hydrothermal activity.

In order to further delineate the chronology of the diagenetic, tectonic, and hydrothermal events, K-Ar isotopic measurements were made on clay fractions from 32 samples of the major rock types associated with the uranium deposits at Cluff Lake.

SAMPLE DESCRIPTION

Most of the samples were collected from the D orebody (Fig. 2) where the overall structural, petrographic and metallogenic picture has been examined in detail (Ey et al., 1983; Ruhlmann, 1983). Four units have been distinguished on the basis of petrographic and structural criteria.

1) The untectonized, barren sandstones of the Athabasca Group contain an illitic matrix with occasional traces of chlorite (Table I). Among the eight samples of this facies, dated by the K-Ar method, five are from the D orebody. Two other samples (682 and 684) belong to drill hole CAR 114, located outside of the Carswell structure and were taken at depths of 1633' and 2801', respectively. The eighth sample (691) is from drill hole CAR 283, taken at a depth of 455' in the Douglas Formation. These three last samples were added to those from the D orebody in order to provide contrast with material that was not touched by the event responsible for the circular structure, as well as to obtain an evaluation of the relationship between the Athabasca sandstones and the sedimentary rocks of the overlying Douglas Formation.

2) The bleached basement is characteristic of alteration in the immediate vicinity of the D orebody. It represents a highly sheared zone that was entirely modified by an intense hydrothermal event. This event pervasively altered the parent rocks which now are composed exclusively of clay minerals: illite and hyperaluminous chlorite, the latter becoming more abundant towards the mineralized zone. The samples contain at least as much illite as chlorite (Table I); they were specifically chosen to avoid the potential problems related to the dating of pure chlorite.

3) The "zone à boules" represents the mylonitization of the sedimentary cover which was affected by the same tectonic and hydrothermal events as the bleached basement. As

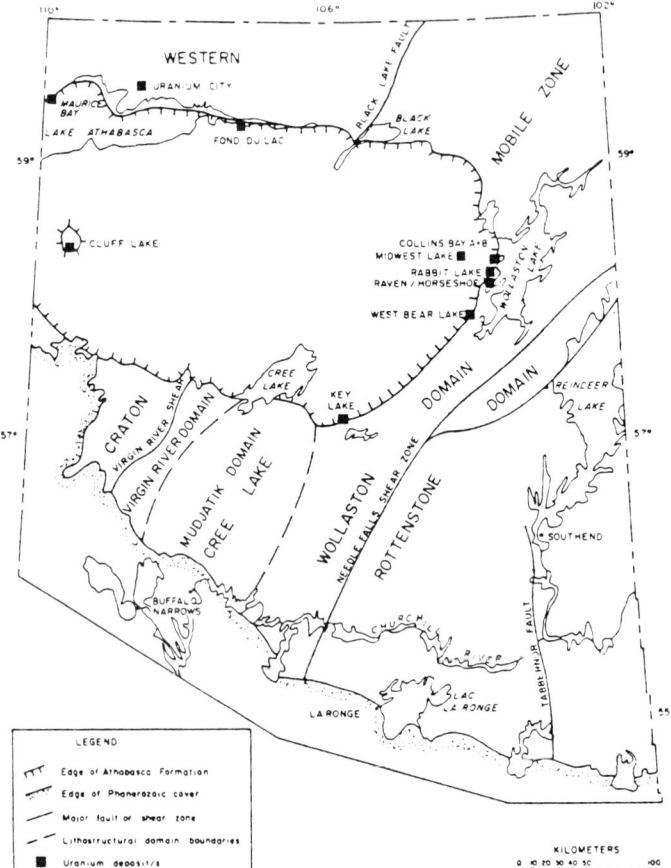

Figure 1. Location of the Cluff Lake uranium deposit in the Athabasca Basin (modified from Sibbald *et al.*, 1977).

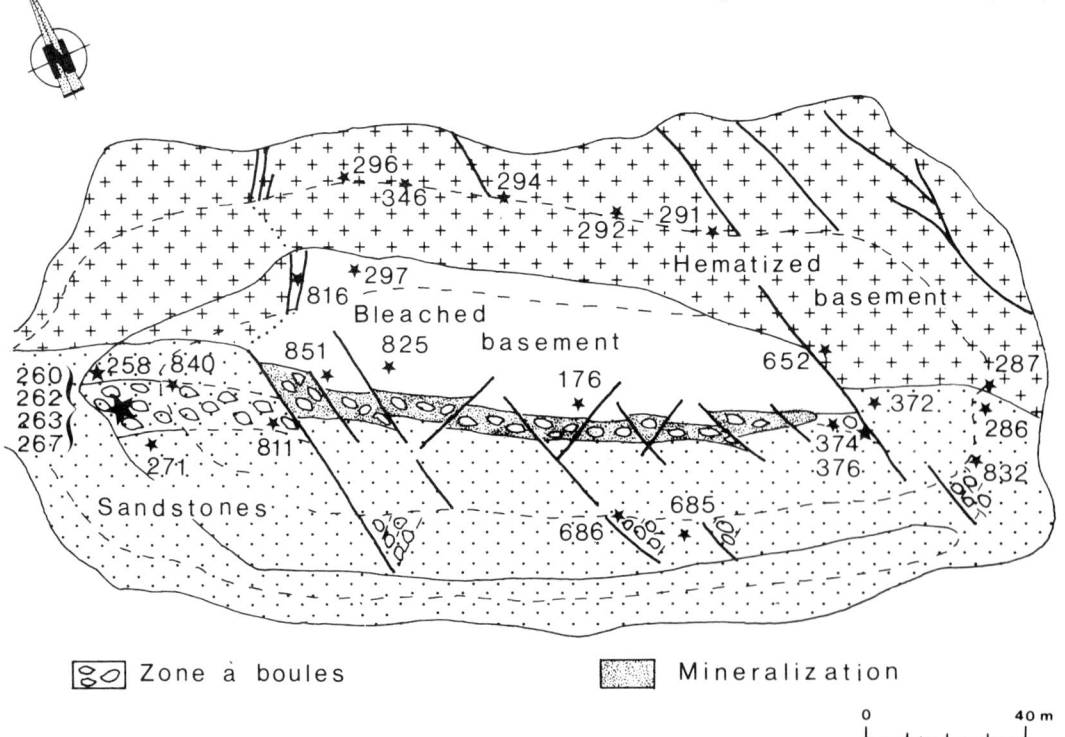

Figure 2. Location of the samples dated by the K-Ar method within the D ore deposit.

TABLE I
ANALYTICAL DATA ON THE CLAY FRACTIONS <2 μm.

SAMPLE	LOCALITY	MINERALOGY			ISOTOPIC DATA				AGE Ma
		NATURE & SPECIES PROPORTION < 2 μm	ILLITE POLYTYPES	CRYSTALLINITY INDEX	K_2O	%Ar*	$^{40}Ar^* \cdot 10^{-6}$ cc	$^{40}Ar/^{36}Ar \times 10^3$	$^{40}K/^{36}Ar \times 10^{-3}$	
ATHABASCA SEDIMENTS		NOT MINERALIZED								
258	D	ILLITE 9, Al CHLORITE 1	IM	4.0	8.31	99.69	447.24	95.322	0.965	1181 ± 25
271	D	ILLITE 8, Al CHLORITE 2	IM	4.0	6.46	99.38	364.03	47.661	0.463	1221 ± 26
286	D	ILLITE 7, Al CHLORITE 3		4.0	5.20	98.75	306.20	23.640	0.219	1261 ± 28
372	D	ILLITE 6, CHLORITE 4		3.5	5.44	97.68	313.75	12.737	0.120	1242 ± 27
682	CAR 114	KAOLINITE 10			1.87	94.64	116.93	5.513	0.046	1316 ± 33
684	CAR 114	ILLITE 10		3.0	4.08	95.78	222.71	7.002	0.068	1194 ± 28
832	D	ILLITE 10	IM – 2M	4.5	7.58	99.79	448.03	140.714	1.359	1265 ± 27
DOUGLAS FORMATION										
691	CAR 283	ILLITE 10		6.5	7.00	99.07	425.59	31.774	0.287	1292 ± 27
"ZONE À BOULES"										
260	D	ILLITE 8, Al CHLORITE 2		4.0	6.74	99.69	352.17	95.323	0.996	1155 ± 25
262	D	ILLITE 7, Al CHLORITE 3		4.0	6.95	99.65	370.25	84.429	0.884	1172 ± 25
263	D	ILLITE 8, Al CHLORITE 2		4.0	7.96	99.41	449.64	50.085	0.490	1223 ± 26
267	D	ILLITE 5, Al CHLORITE 5	IM – 2M	3.5	4.24	99.06	252.92	31.436	0.289	1272 ± 28
310	OP	ILLITE 10	3T	4.5	9.16	99.62	530.29	77.763	0.751	1245 ± 26
315	OP	ILLITE 10	IM	4.0	9.09	99.33	516.19	44.104	0.425	1228 ± 26
374	D	ILLITE 9, Al CHLORITE 1	IM – 2M	4.0	8.44	99.91	537.99	328.33	2.748	1335 ± 28
376	D	ILLITE 10	IM – 2M	4.0	8.82	99.91	549.32	328.33	2.822	1314 ± 28
685	D	ILLITE 10	IM – 2M	4.5	8.36	99.80	525.08	147.750	1.313	1322 ± 28
686	D	ILLITE 9, Al CHLORITE 1	IM – 2M	4.5	8.79	99.84	522.63	184.688	1.711	1270 ± 26
811	D	ILLITE 10	IM – 2M – 3T	5.0	4.05	99.91	246.40	328.333	3.000	1292 ± 29
840	D	ILLITE 8, Al CHLORITE 2	IM – 2M – 3T	5.0	8.05	99.72	483.29	105.536	0.959	1280 ± 27
HEMATIZED BASEMENT										
287	D	ILLITE 8, Al CHLORITE 2	IM	3.5	6.18	98.45	315.00	19.065	0.205	1134 ± 25
291	D	ILLITE 9, Al CHLORITE 1	IM – 2M	4.0	8.01	99.54	423.67	64.239	0.676	1167 ± 25
292	D	ILLITE 7, Al CHLORITE 3	IM	4.0	5.53	98.87	325.04	26.150	0.244	1259 ± 28
294	D	ILLITE 9, Al CHLORITE 1		3.5	4.72	99.43	513.84	51.842	0.264	1902 ± 42
296	D	ILLITE 10	2M	3.5	7.90	99.34	523.45	44.773	0.373	1372 ± 89
346	D	ILLITE 9, Al CHLORITE 1	IM	3.5	7.61	99.40	476.28	49.250	0.434	1317 ± 28
652	D	ILLITE 9, Al CHLORITE 1	IM – 2M	4.0	8.18	99.80	570.09	147.750	1.202	1421 ± 30
BLEACHED BASEMENT										
176	D	ILLITE 5, Al CHLORITE 5		4.0	7.40	99.39	298.27	48.443	0.662	949 ± 20
297	D	ILLITE 6, Al CHLORITE 4	IM	3.5	5.05	99.27	307.46	40.479	0.368	1291 ± 28
825	D	ILLITE 5, CHLORITE 5		3.5	7.67	99.16	490.06	35.179	0.302	1337 ± 29
851	D	ILLITE 8, CHLORITE 2	IM – 2M	4.5	7.56	99.46	507.00	54.722	0.450	1383 ± 29
F_1 FAULT CLAY										
816	D	ILLITE 9, Al CHLORITE 1	IM – 2M	4.0	7.93	99.85	468.90	109.444	1.014	1266 ± 27

a result, the sandstones are extensively argillized to illite and minor hyperaluminous chlorite. Illite is characterized by its 3T polymorphic form which is typical of high pressure environments (Frey et al., 1983) and which was never observed in the equivalent untectonized horizons. When uranium mineralization occurs in this "zone à boules", the clay fraction becomes a pure Mg-chlorite. Among the twelve samples dated, five were collected from barren facies on the western wall of the D pit, relatively far from the uranium mineralization. Three other samples (374, 376, and 811) were taken from the footwall of the ore deposit (Fig. 2). Two additional samples (685 and 686) were also collected from the footwall of the mineralization but further away. The two last samples (310 and 315) were taken from a barren "zone à boules" in the OP exploration decline which is located about 500 m north of D.

4) The hematized basement also underwent intense hydrothermal activity, consequently only a few residual components such as quartz and hematized garnets could be identified in the clay matrix. This matrix is mainly illitic with subordinate chlorite and was probably generated during the hydrothermal event. The entire set of samples belonging to this facies was collected in the D open pit.

ANALYTICAL PROCEDURE AND RESULTS

The K-Ar isotopic dating method has been applied to sediments for about 25 years (Wasserburg et al., 1956). Since this pioneer attempt, studies have concentrated on systematic dating of glauconitic minerals (Odin, 1982). Different aspects and problems of its application to other sediments have also been discussed recently (Clauer, 1981; Bonhomme, 1982a), and encouraging results were obtained on clay minerals directly related to ore deposits (Bonhomme, 1982b). The general use of the latter application seems premature, however, since the results are still relatively scattered, for reasons still unknown. This may be due to a complex behaviour of the K-Ar system of clay minerals in hydrothermal environments.

The classical use of the K-Ar dating method requires the assumption that the $^{40}Ar/^{36}Ar$ initial ratio was sealed in the minerals during their crystallization. The individual apparent ages of the samples are calculated using, initially, the value of the present-day atmospheric $^{40}Ar/^{36}Ar$ ratio which is 295.5 (Nier, 1950). This assumption is commonly accepted for all geological material. The amounts of "initial" Ar are very low in the clay minerals dated here (Table I); some of this Ar could be due to contamination by adsorption of present-day atmospheric Ar during preparation and analysis. The amounts of low "initial" plus "contamination" Ar are, on the other hand, consistent with the low periodically controlled Ar residuals of the extraction line and the mass-spectrometer, and with the fact that present-day Ar contamination does not increase when the size of the analyzed material decreases (Liewig et al., 1981). These technical controls allow: 1) the assumption that the total non-radiogenic Ar measured in the series of samples is "initial" in origin, and 2) the use of $^{40}Ar/^{36}Ar$ vs $^{40}K/^{36}Ar$ diagrams for elucidation of the Ar isotopic data (McDougall et al., 1969). This graphic treatment is worthy because the calculation of the age and initial $^{40}Ar/^{36}Ar$ ratios of the minerals is made without any assumption. The fact that almost no present-day Ar contamination could be detected offered an opportunity to directly obtain a value for the $^{40}Ar/^{36}Ar$ ratio of the tectonic and hydrothermal environments responsible for the clay recrystallization. In the case of older material such as the Cluff Lake samples, this treatment has a drawback: the analytical error on the initial $^{40}Ar/^{36}Ar$ ratios may be relatively important and consequently obscure their meaning. For that reason, we have compared the average of the individual K-Ar apparent ages assuming an initial ratio of 295.5 to the graphic interpretation. The regression calculations used do not take into account the uncertainties of the individual isotopic ratios. The errors on the slope and intercept of the lines depend, therefore, only on the fitting of the data, and another least square calculation by plotting for instance ($^{36}Ar/^{40}Ar$ vs $^{40}K/^{40}Ar$) is useless. This calculation adapted from a proposal by Roddick et al. (1980), would virtually erase the correlation between the errors of the abcissas and the ordinates in a traditional plot $^{40}Ar/^{36}Ar$ vs $^{40}K/^{36}Ar$, especially when the errors are dominated by a large uncertainty in the determination of very small amounts of ^{36}Ar.

The analyses were made on clay fractions smaller than 2 μm that were separated from the crushed rocks by a classical sedimentation technique in distilled water. The Ar determinations were made by an extraction method described by Bonhomme et al. (1975). Prior to Ar extraction, the samples were heated under vacuum at 120°C for at least 4 hours. Ar isotopic abundances were calculated using an Ar tracer, and K concentrations were obtained with a precision of ± 2% by flame spectrometry. The ages were calculated using the constants proposed by Steiger and Jäger (1977). Errors of the

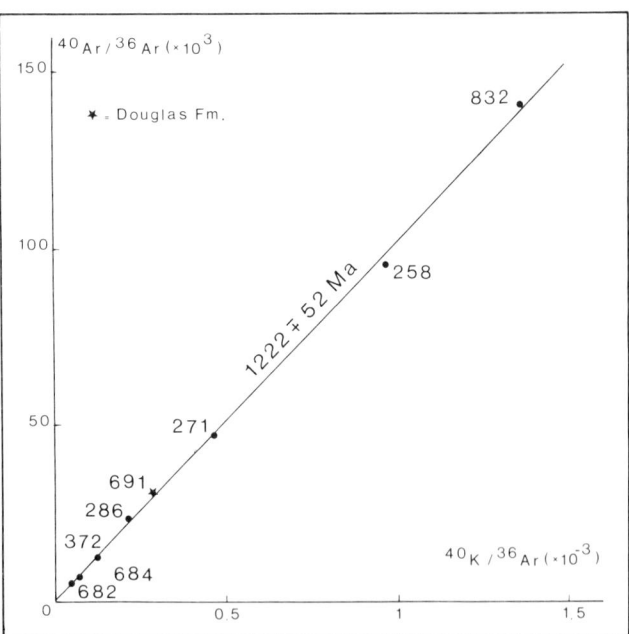

Figure 3. ($^{40}Ar/^{36}Ar$; $^{40}K/^{36}Ar$) isochron diagram with the samples of the barren sandstones from the Athabasca Group and Douglas Formation. The intercept of the isochron is about 792; it has to be considered after restoring to the original scale of data.

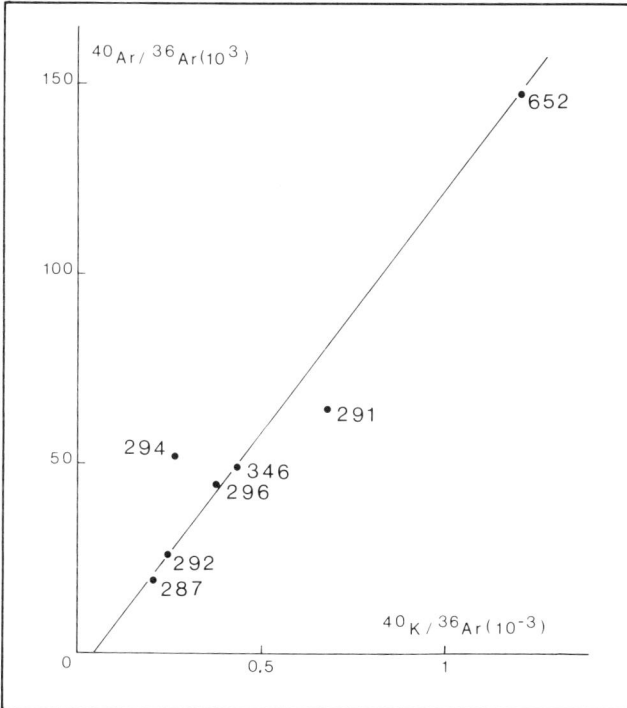

Figure 4. (^{40}Ar/^{36}Ar; ^{40}K/^{36}Ar) isochron diagram with the samples of the hematized basement. No values for the age and initial ^{40}Ar/^{36}Ar are intentionally given; see text for explanation.

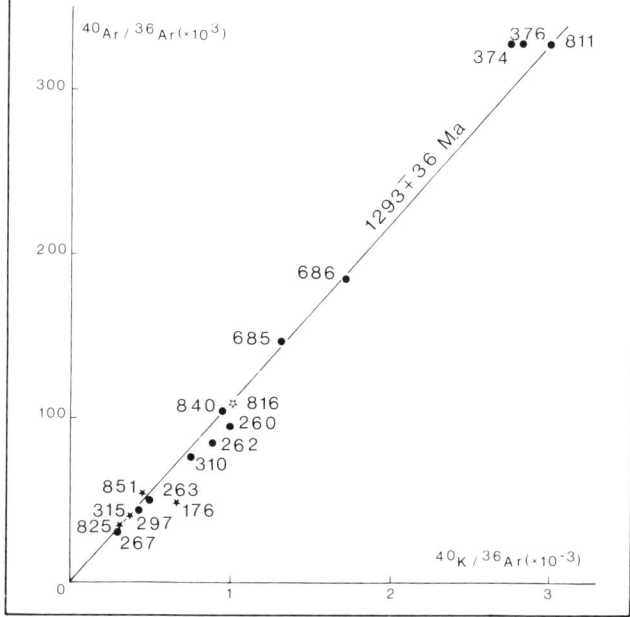

Figure 5. (^{40}Ar/^{36}Ar; ^{40}K/^{36}Ar) isochron diagram with the samples of the bleached basement (full stars), the "zone à boules" (full points) and the F1 fault (open star). The intercept of the isochron is about 1027; it has to be considered after restoring to the original scale of data.

ages are estimated at ± 3% based on replicate measurements of the standard mineral GLO (8 measurements during the present study gave an average value of 24.97 ± 0.36 (2σ) × 10^{-6} cm^3/g STP of radiogenic ^{40}Ar).

The eight fractions smaller than 2 μm extracted from barren sandstones define a unique line on an isochron diagram (Fig. 3) with a slope of 1222 ± 52 Ma for an intercept approximately at 792. The mean K-Ar apparent age of these samples is 1247 ± 47 Ma, which is comparable, within analytical uncertainty, to the one obtained by linear regression. The error of the intercept of the isochron is so large that it includes the ^{40}Ar/^{36}Ar ratio of the present-day atmospheric Ar. This explains why the mean value is similar to that obtained by the graphic method and prevents, as expected, discussion on the meaning of the slightly high initial ^{40}Ar/^{36}Ar of the clay minerals.

The K-Ar apparent ages of the seven fractions representative of the hematized basement range widely between 1134 ± 25 Ma and 1902 ± 42 Ma. Plotted on a similar isochron diagram (Fig. 4), the fractions remain scattered; five of them fit on a line whose intercept is clearly negative.

The four clay fractions representative of the bleached basement and the twelve fractions from the "zone à boules" have been plotted on the same type of diagram (Fig. 5) as they were taken from rocks having undergone the same tectonic event (Ey et al., 1985). Eleven of the sixteen samples define a line with a slope of 1293 ± 36 Ma for an intercept approximately at 1027. Two illites (samples 374 and 376) and two mixtures of illite and Al chlorite (samples 260 and 262) from the "zone à boules", as well as one mixture of illite and chlorite from bleached basement in the D orebody (sample 176) lie off the line. Analysis of a clay fraction from the F1 fault in the D orebody fits the same line. The overall mean K-Ar age of the eleven samples is again similar to the former value: 1284 ± 46 Ma. The analytical uncertainty of the intercept of the isochron again includes the present-day atmospheric ^{40}Ar/^{36}Ar ratio, and a useful discussion is not possible on a hypothetical initial excess of ^{40}Ar in the clay fractions.

DISCUSSION

The data for most of the samples from the pervasively altered shear zone, including those of the clay fraction sampled on the F1 fault of the D orebody, fall on a unique line. This certainly argues in favour of an isochron whose age of 1293 ± 36 Ma can be assumed to be characteristic of this combined event. This interpretation is further supported by the crystallinity indices and the polymorphic forms of the illites which are representative of such an event. Although they were chosen on the basis of the same criteria as the other samples, six of the samples from the shear zone do not fit the isochron, four belonging to the sedimentary cover and two to the bleached basement. This behaviour is not well understood; it certainly is related to the hydrothermal activity but needs to be examined in more detail.

Data for the other samples of the Athabasca sandstones and Douglas Formation which were collected away from the tectonic environment and its associated hydrothermal activity define another line with a parallel slope. The crystallinity indices of the illites and the occurrence of 2M polymorphic forms together with 1M forms may be indicative of detrital clay fractions. However in such a case, a scatter of the isotopic data is expected as discussed by Bonhomme (1982a). The unique isotopic value obtained here is more probably representative of a diagenetic-to-metamorphic event which reset the "K-Ar clock" of the clay minerals with a trend towards the time of the event. The diagenetic event recorded in the sediments at 1222 ± 52 Ma must be considered, within the analytical uncertainty, to be contemporaneous with the tectonic activity at 1293 ± 36 Ma.

The intense hydrothermal activity related to the shearing at the unconformity extended to the neighbouring facies, away from the tectonic environment, so much so that it could be detected inside and outside the Carswell structure. This interpretation not only explains the crystallographic parameters of the illites, but also the high temperatures determined by Pagel (1975) in the sandstones of the eastern part of the Athabasca Group without having to assume significant burial of the sediments or an exceptionally high geothermal gradient. It also suggests that the Douglas Formation underwent the same but weaker alteration as the lower Athabasca sandstones, even if it was never deeply buried.

The K-Ar data of the samples from hematized basement are scattered. The individual apparent ages range widely, and the line joining four of the six analysed samples cannot be considered an isochron because the intercept is negative. This intercept usually represents the initial $^{40}Ar/^{36}Ar$ sealed in the minerals during crystallization and for obvious chemical reasons it cannot have a negative value. Control of the analytical procedure excludes any technical uncertainties. The graphic method is useful, therefore, in identifying K-Ar data that have no geological relevance. Langley (1978) once obtained a line with a negative intercept while analysing sediments, but did not interpret it. Neither gain nor loss of one of the isotopes can explain a negative initial $^{40}Ar/^{36}Ar$ as it is difficult to consider that these alterations can modify the $^{40}Ar/^{36}Ar$ ratio sealed in the minerals during crystallization. The only reasonable explanation is to suggest that the line is characteristic of a mechanical mixture in the clay fractions smaller than 2 μm of at least two inhomogeneous components that have different origins. One of these components could belong to the basement and the second could have formed during the hydrothermal activity.

In summary, a tectonic event and a concommitant intense and extensive hydrothermal activity, to which the uranium mineralization is clearly related, have been dated at about 1260 Ma. Among the geochronological data available in the literature on the Athabasca Basin and its uranium ore deposits, only a few were obtained on rocks from Cluff Lake. According to Bellon et al. (1976) and Gancarz (1979), uranium mineralization of the D ore deposit ranges between 1150 and 1050 Ma; the petrographic observations show, however, several generations of pitchblende (Ruhlmann, 1983). Wanless et al. (1968), Currie (1969), and Bell (1981) have dated breccias related to the Carswell structure and found ages around 480 Ma. This value clearly separates the effects of the formation of the circular structure from hydrothermal events dated in this study at about 1260 Ma. In Collins Bay and Key Lake, where ore deposits similar to that at Cluff Lake were described, Wendt et al. (1978) dated pitchblende at 1270 Ma, Gatzweiler et al. (1979) published a date of 1228 Ma and Jones (1980) one of 1238 Ma, all on pitchblende. Hohndorf et al. (1981) reported on three generations of uranium mineralization, one of which probably formed around 1200 Ma. Furthermore, Stevens et al. (1982) determined K-Ar ages of about 1280 Ma on micaceous minerals from weathered basement. Other data available on these regions either are relative to the older basement or appear unreliable because they are obtained on minerals not clearly characterized or that are difficult to date, such as chlorites. Two observations arise from this compilation and comparison: 1) the hydrothermal event of circa 1260 Ma, which induced a mineralizing episode and has been described in the eastern part of the Athabasca Basin, also exists within the Carswell structure, and 2) the actual dating of the uranium minerals gives systematically lower ages between 1150 and 1050 Ma in the Carswell structure.

CONCLUSIONS

The K-Ar dating of clay fractions from different rock types of the D ore deposit in the Carswell structure, as well as from sandstones outside of this structure, yields the following conclusions:

1) The tectonic event responsible for the "zone à boules" in the sedimentary cover and for bleaching in the basement associated with a mylonitic zone occurred at 1293 ± 36 Ma.

2) The tectonic event is correlated with an intense hydrothermal activity responsible, to some degree, for the uranium mineralization; this hydrothermal activity extended widely over the entire sequence of the Athabasca Group, including the Douglas Formation.

3) The hydrothermal overprinting was recognized on a large scale because the clay fractions of the untectonized sandstones from inside and outside of the Carswell structure give concordant K-Ar isotopic ages of 1222 ± 52 Ma.

4) The widespread hydrothermal activity is one of the oldest events at about 1260 Ma; it is directly associated, in the shear zone of the Carswell structure, with the uranium mineralization.

5) The formation of the Carswell structure is clearly a late event which had no influence on the clay minerals; the samples taken outside of it recorded the same history as those collected within it.

ACKNOWLEDGEMENTS

The authors would like to thank Amok Ltd. for help, interest, and financial support during the study. Special thanks are given to Dr. R.T. Lainé (Amok Ltd.), Dr. K. Bell (Carleton University), Dr. J.R. O'Neil (U.S. Geological Survey, Menlo Park), and Dr. D. York (University of Toronto) for important improvements to an earlier draft of the manuscript, as well as to R. Winkler and R.A. Wendling (Centre de Sédimentologie et Géochimie de la Surface, Strasbourg) for technical assistance.

REFERENCES

Bell, K., 1981, Cluff Lake Geochronological Program 1981: Amok Internal Report.

Bellon, H., Devillers, C., Hagemann, R., and Touray, J.C., 1976, Dater les Minéralisations: Mémoire Hors Série de la Société Géologique de France, no. 7, p. 265-268.

Bonhomme, M., 1982a, The Use of Rb-Sr and K-Ar Dating Methods as a Stratigraphic Tool Applied to Sedimentary Rocks and Minerals: Precambrian Research, v. 18, p. 5-26.

_____, 1982b, Age Triasique et Jurassique des Argiles Associées aux Minéralisations Filoniennes et aux Phénomènes Diagénétiques Tardifs en Europe de l'Ouest. Contexte Géodynamique et Implications Génétiques: Comptes Rendus Académie des Sciences, Paris, t. 294, Série D, p. 521-524.

Bonhomme, M., Thuizat, R., Pinault, Y., Clauer, N., Wendling, A., and Winkler, R., 1975, Méthode de Datation Potassium-Argon. Appareillage et Technique: Notes Techniques de l'Institut de Géologie, Université Louis Pasteur, Strasbourg, v. 3, 53 p.

Clauer, N., 1981, Rb-Sr and K-Ar Dating of Precambrian Clays and Glauconites: Precambrian Research, no. 3-4, v. 15, p. 331-352.

Currie, K.L., 1969, Geological Notes on the Carswell Circular Structure, Saskatchewan (74K): Geological Survey of Canada Paper 67-32, 69 p.

Ey, F., Gauthier-Lafaye, F., Lillié, F., Schumacher, F., and Weber, F., 1983, Le Gisement d'Uranium de Cluff D (Saskatchewan-Canada), Etude Structurale et Pétrographique: in Pagel, M., éd., Les Gisements d'Uranium Lié Spatialement aux Discordances: Géologie et Géochimie de l'Uranium, Mémoire 1, Nancy, p. 97-113.

Ey, F., Gauthier-Lafaye, F., Lillié, F., and Weber, F., 1985a, A Uranium Unconformity Deposit: The Geological Setting of the D Orebody (Saskatchewan-Canada): in Lainé R., Alonso, D., and Svab, M., eds., The Carswell Structure Uranium Deposits, Saskatchewan: Geological Association of Canada Special Paper 29,

Frey, M., Hunziker, J.C., Jäger, E., and Stern, W.B., 1983, Regional Distribution of White K-Mica Polymorphs and their Phengite Content in the Central Alps: Contributions to Mineralogy and Petrology, No. 83, p. 185-197.

Gancarz, A.J., 1979, Chronology of the Cluff Lake Uranium Deposit, Canada (Abstract): in International Uranium Symposium on the Pine Creek Geosyncline, N.T., Australia, Extended Abstracts, p. 91-94.

Gatzweiler, R., Lehnert-Thiel, K., Clasen, D., Voultsidis, V., Strnad, J.G., and Rich, J., 1979, The Key Lake Uranium Nickel Deposits: Bulletin of the Canadian Institute of Mining and Metallurgy, v. 70, no. 807, p. 73-97.

Hohndorf, A., Voultsidis, V., and von Pechmann, E., 1981, U/Pb Isotopic Investigations of the Maurice Bay Uranium Deposit, Lake Athabasca (Preliminary Results): Canadian Institute of Mining Uranium Symposium, September 8-13, 1981, Saskatoon, Technical Program, p. 23-24.

Jones, B.E., 1980, The Geology of the Collins Bay Uranium Deposit, Saskatchewan: Canadian Institute of Mining Bulletin, v. 73, no. 818, p. 84-90.

Langley, K.M., 1978, Dating Sediments by a K-Ar Method: Nature, v. 276, p. 56-57.

Liewig, N., Caron, J.M., and Clauer, N., 1981, Geochemical and K-Ar Isotopic Behaviour of Alpine Sheet Silicates during Polyphase Deformation: Tectonophysics, v. 78, p. 273-290.

McDougall, I., Polach, H.J.A., and Stipp, J.J., 1969, Excess Radiogenic Argon in Young Subaerial Basalts from the Aukland Volcanic Field, New Zealand: Geochimica et Cosmochimica Acta, v. 33, p. 1485-1520.

Nier, A.O., 1950, A Redetermination of the Relative Abundance of the Isotopes of Carbon, Nitrogen, Oxygen, Argon, and Potassium: Physics Review, v. 77, p. 789-793.

Odin, G.S., 1982, How to Measure Glaucony Ages: in Odin, G.S., ed., Numerical Dating in Stratigraphy: New York, Wiley and Sons, p. 387-403.

Pagel, M., 1975, Détermination des Conditions Physicochimiques de la Silicification Diagénétique des Grès Athabasca (Canada) au Moyen des Inclusions Fluides: Comptes Rendus Académie des Sciences, Paris, tome 280, Série D, p. 2301-2304.

Roddick, J.C., Cliff, R.A., and Rex, D.C., 1980, The Evolution of Excess Argon in Alpine Biotites — $^{40}Ar/^{39}Ar$ Analysis: Earth and Planetary Science Letters, v. 48, p. 185-208.

Ruhlmann, F., 1983, Caractéristiques Minéralogiques des Principales Occurrences Uranifères de l'Anneau de Carswell (Cluff Lake, Athabasca): in Pagel, M., éd., Les Gisements d'Uranium Liés Spatialement aux Discordances: Géologie et Géochimie de l'Uranium, Mémoire 1, Nancy, p. 117-122.

Sibbald, T.I.I., Munday, R.J.C., and Lewry, J.F., 1977, The Geological Setting of Uranium Mineralizations in Northern Saskatchewan: in Dunn, ed., Saskatchewan Geological Survey Special Publication No. 3, p. 51-92.

Steiger, R.H., and Jäger, E., 1977, Subcommission on Geochronology: Convention on the Use of Decay Constants in Geo- and Cosmochronology: Earth and Planetary Science Letters, v. 36, p. 359-362.

Stevens, R.D., Delabio, R.N., and Lachance, G.R., 1982, Age Determinations and Geological Studies, K-Ar Isotopic Ages, Report 15: Geological Survey of Canada Paper 81-2, p. 34-35.

Wanless, R.K., Stevens, R.D., Lachance, G.R., and Rimsaite, J.Y.H., 1968, Age Determinations and Geological Studies, K-Ar Isotope Ages, Report 6: Geological Survey of Canada Paper 65-17, p. 65-77.

Wasserburg, G.I., Hayden, R.I., and Iensen, K.J., 1956, $^{40}Ar/^{40}K$ Dating of Igneous Rocks and Sediments: Geochimica et Cosmochimica Acta, v. 10, p. 153-165.

Wendt, I., Hohndorf, A., Lenz, H., and Voultsidis, V., 1978, Radiometric Age Determinations on Samples of the Key Lake Uranium Deposits: in Short Papers of the 4th International Conference, Geochronology, Cosmochronology, and Isotope Geology: United States Geological Survey Open File Report 71-701, p. 448-449.

The Carswell Structure Uranium Deposits, Saskatchewan,
edited by R. Lainé, D. Alonso and M. Svab,
Geological Association of Canada Special Paper 29, 1985

PETROGRAPHIC AND GEOCHEMICAL VARIATIONS WITHIN THE CARSWELL STRUCTURE METAMORPHIC CORE AND THEIR IMPLICATIONS WITH RESPECT TO URANIUM MINERALIZATION

M. Pagel
Centre de Recherches sur la Géologie de l'Uranium, B.P. 23, 54501 - Vandoeuvre-Lès-Nancy Cédex, France

and

Centre de Recherches Pétrographiques et Géochimiques, B.P. 20, 54501 - Vandoeuvre-Lès-Nancy Cédex, France

M. Svab
Amok Ltd., 817 - 825 45th St. W., P.O. Box 9204, Saskatoon, Saskatchewan S7K 3X5

ABSTRACT

Two major lithologic units have been recognized in the Carswell structure metamorphic core. The Earl River complex, composed of mixed feldspathic gneisses and mafic gneisses, is overlain by aluminous Peter River gneisses. Granitoids and pegmatoids are ubiqitous in both rock units and are, to a great extent, the result of in-situ anatexis of the metasedimentary gneisses. These gneisses are interpreted to be a normal detrital succession of arkoses and greywackes at the base, forming the Earl River complex, and shales at the top which make up the Peter River gneisses. Metamorphism reached granulite facies in the presence of fluids rich in cardon dioxide, methane, and nitrogen. Compared with other granulite facies terrains, the basement is relatively rich in in uranium and thorium. The metamorphic core has undergone at least six alteration phases: retrograde metamorphism, surface weathering, alteration related to diagenesis in the overlying Athabasca Group, alteration related to the episode of the main uranium deposition, alteration associated with the formation of the Carswell structure, and finally alteration associated with a later episode of uranium deposition and remobilization. The main mineralizing phase is associated with the circulation of brines and is accompanied by the formation of aluminous and magnesian chlorite, magnesian chlorite, illite, and alkali poor Mg-tourmalines.

RÉSUMÉ

Deux unités lithologiques principales ont été reconnues dans la partie méridionale du socle de la structure Carswell. Le complexe d'Earl River est surmonté par les gneiss de Peter River. Les gneiss de Peter River sont caractérisés par l'assemblage grenat-cordiérite-sillimanite-biotite tandis que les gneiss quartzofeldspathiques de l'Earl River, fréquemment riches en biotite, sont souvent lités. Des gneiss basiques à hypersthène-clinopyroxène-biotite ou à hornblende-biotite sont intercalés dans cette dernière formation. Des granitoïdes et des pegmatoïdes sont fréquents dans les deux formations, résultant au moins en grande partie de l'anatexie in situ des métasédiments. Ces formations métasédimentaires sont interpretées comme résultant du métamorphisme de sédiments détritiques avec à la base essentiellement des arkoses et des grès (complexe Earl River) passant à des shales (gneiss de Peter River). Si les shales présentent des traces de confinement, il est toutefois à noter qu'aucune trace de sédimentation évaporitique n'a été reconnue. Ces sédiments ont atteint les conditions du faciès granulite, de haute temperature et moyenne pression, en présence d'un fluide riche en gaz carbonique, méthane, et azote. Par comparaison avec de nombreuses zones ayant été soumises au faciès granulite, le socle est relativement riche en uranium et thorium. Il a été affecté par de nombreuses phases d'altération qui sont par ordre de succession temporelle: l'altération rétrograde contemporaine des dernières phases de l'orogénèse hudsonienne, l'altération de surface entre 1700 et 1500 millions d'années, en grande partie reprise par une altération liée à la diagénèse de la formation Athabasca, l'altération liée aux principales phases de dépôt des minéralisations, l'altération liée à la formation de la structure Carswell, et enfin les épisodes qui depuis 400 millions d'années environ ont abouti à la remobilisation des minéralisations uranifères. Les phases principales de minéralisation, provoquées par la circulation de saumures, sont accompagnées par la formation de chlorite magnésienne, de chlorite alumineuse, d'illite, et de tourmaline magnésienne déficitaire en alcalins. Dans la zone Dominique-Peter, à côté des éléments minéralisateurs, les zones d'altération ont des teneurs anomaliques en F et B tandis que l'anomalie en Li observée dans le gisement D est absente.

INTRODUCTION

The Carswell structure, in northern Saskatchewan, is a roughly circular core of Precambrian basement rocks, about 20 km in diameter, surrounded by a ring of faulted Helikian Athabasca Group sediments (Fig. 1). The youngest formations of the Athabasca Group, the Carswell and the Douglas Formations, outcrop in a fairly continuous ring around the outside of the structure. The total structure is approximately 39 km in diameter.

Surface exposure of the basement rocks is very poor in the area, and lithologic and stratigraphic interpretations have been largely produced from diamond drill core and geophysical information. As a result, there is abundant data from diamond drilling in the main mineralized zone along the southern margin of the core, however, information is rather sparse from the less extensively drilled northern edge.

The basement core of the Carswell circular structure was briefly noted by Fahrig (1961), and Currie (1969) later mapped the core in some detail. Herring (1976) and Harper (1982, 1983) mapped the core in greater detail as more diamond drill core became available for interpretation.

The basement rocks which form the Carswell structure are composed of aluminous, feldspathic, and mafic gneisses, all of which are intercalated with granitoids and pegmatoids. Rb-Sr isochrons (Bell, 1985) for the aluminous and feldspathic gneisses reflect an event around 1800 Ma, an age which is usually associated with Hudsonian orogenic activity. No events of Archean age have yet been documented in the Carswell structure. The metamorphic grade reached granulite facies with later retrograde metamorphism to amphibolite facies (Herring, 1976; Harper, 1983). Major and trace element geochemistry was used in determining the original rock types and indicated a metasedimentary origin for the feldspathic and aluminous gneisses and a volcanic origin for the mafic gneisses. Bell et al. (1985) interpret these mafics to be a high MgO (komatiitic?) suite.

Deformation in the basement is very complex and includes minor folding, abundant faulting, and widespread cataclasis (Tona et al., 1985). The alteration that is associated with the deformation, and also related to the Athabasca Group unconformity, can be divided into six phases: 1) retrograde metamorphism, 2) paleosurface-related alteration, 3) alteration related to diagenesis of the overlying Athabasca Group sediments, 4) alteration related to the major episode of uranium deposition, 5) alteration related to the formation of the Carswell structure, and 6) alteration related to episodes of later uranium deposition and remobilization around 400 Ma and later.

Fluid inclusion studies provide compositions of solutions present during some of the above-mentioned alteration phases. Results indicate that the solutions were primarily brines during major episodes of uranium deposition.

Chemical analyses of various alteration minerals within and near the zones of uranium deposition reveal a variety of both aluminous and magnesian chlorite, magnesium-rich tourmaline, and fine-grained potassium mica. These minerals form very localized haloes around the uranium deposits.

BASEMENT LITHOLOGY

The Carswell structure basement rocks have been divided into two major units (see map in pocket): metapelitic Peter River gneisses which overlie interlayered feldspathic and mafic gneisses of the Earl River complex. Table I contains representative geochemical analyses of the major rock types in the basement core.

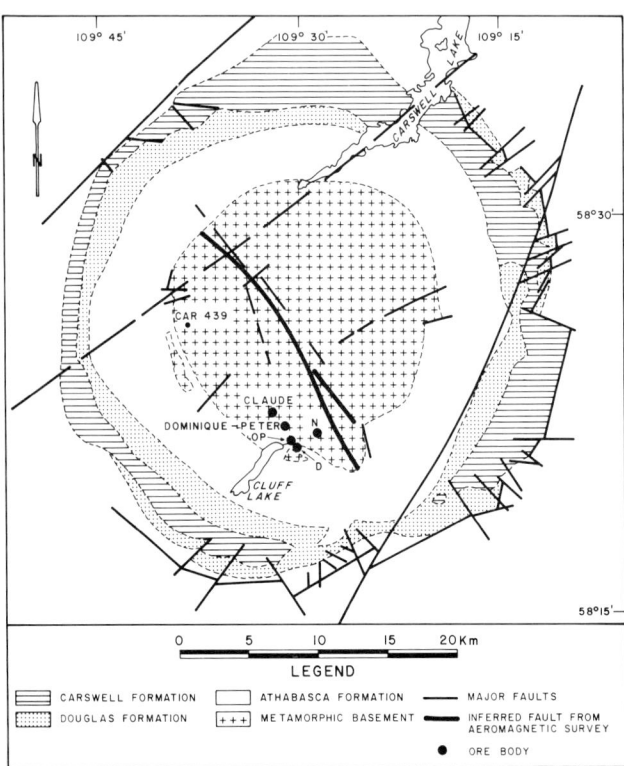

Figure 1. Geology of the Carswell structure and location of areas cited in text (modified from Tona et al., 1985).

Figure 2. Two-pyroxene granulite (Earl River complex). Hypersthene-augite-plagioclase-biotite gneiss. Note the irregular plagioclase twinning (crossed nicols, bar scale equals 1 mm).

TABLE I

REPRESENTATIVE MAJOR ELEMENT ANALYSES FOR THE FOUR MAJOR ROCK TYPES IN THE CARSWELL STRUCTURE BASEMENT CORE

	Peter River Gneiss	Feldspathic Gneiss	Granitoid/ Pegmatoid	Mafic[1] Gneiss
SiO_2	62.57	68.13	73.75	47.40
Al_2O_3	16.77	14.54	14.82	12.27
CaO	0.31	1.27	0.77	5.83
Fe_2O_3 (tot)	8.26	4.14	1.12	10.40
K_2O	3.96	5.35	4.45	2.70
MgO	2.49	2.04	0.38	15.20
MnO	0.09	0.04	0.03	0.15
P_2O_5	0.12	0.26	0.18	0.24
Na_2O	0.94	1.99	3.28	0.79
S	0.57	N.D.	0.11	N/A
TiO_2	0.47	0.33	0.03	1.14
L.O.I.	3.24	1.43	0.97	3.02
Total	99.79	99.52	99.89	99.14
	n=8	n=8	n=6	n=14

N.D. = Non-Detectable
N/A = Not Analysed

1 = Arithmetic means of 14 analyses from the Dominique-Peter area (Bell et al., 1985).

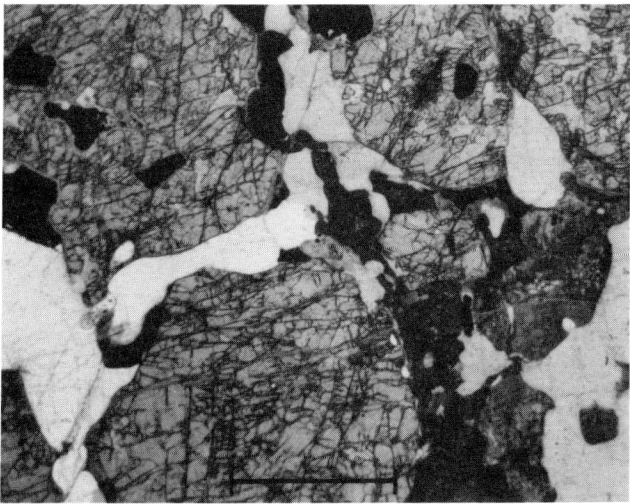

Figure 3. Garnetite. The dark cloudy minerals are stained K-feldspars. Garnet-biotite-quartz-K-feldspar gneiss (plane light, bar scale equals 1 mm).

Earl River Complex

The mafic gneisses are a relatively minor rock type in the southern part of the Carswell structure, however, they are the predominant rock type in much of the northern half of the core. Herring (1976), Bruneton (1981), and Bell et al. (1985) subdivided the mafic gneisses into two suites: a banded, two-pyroxene granulite, and a more massive amphibole-pyroxene gneiss.

The two-pyroxene granulite is a hypersthene-clinopyroxene-biotite-amphibole-plagioclase (An $_{50-60}$)-K-feldspar gneiss. Sulphides and carbonates are usually present in minor amounts. Hypersthene occurs as optically oriented lobate grains (Fig. 2) and almost always shows good exsolution textures. Clinopyroxene has been identified as belonging to the diopside-hedenbergite series (Herring, 1976; Bruneton, 1981), however augite has been noted in some samples (Bell et al., 1985; Amok Ltd., unpublished data). Biotite is found either interstitial to, or included in, the pyroxenes and feldspars, and defines the foliation in the granulites. Amphibole (green hornblende) is a minor mineral in the granulites and occurs as small interstitial grains or as rounded inclusions within pyroxenes.

Plagioclase (An $_{50-60}$) is the most abundant feldspar in the mafic gneisses. It is sometimes twinned, although twinning planes are always bent. It is more commonly poorly twinned and clouded, and in the northern part of the Carswell structure in particular, this clouding forms highly irregular zoning in the grains. Minor K-feldspar and quartz can occur as small intergranular grains. In some zones within the Earl River complex, K-feldspar and quartz may form centimetre-thick leucocratic bands.

The massive amphibole-pyroxene gneiss contains a similar assemblage as the two-pyroxene granulite except that there is abundant hornblende and an overall coarser grain size. There are only very minor bands of true amphibolite in the basement core. The massive mafic gneisses that have been called "amphibolites" may represent a gradational change in mineralogy and texture from the granulites (Bell et al., 1985).

The mixed feldspathic gneisses of the Earl River complex occur structurally between the Peter River gneisses and the mafic gneisses. Classification of these banded gneisses is dependent upon the mineralogy of the melanocratic component and the lithologic setting. Feldspathic gneisses near Peter River gneisses typically contain garnet, cordierite, and biotite in amounts up to 20% by volume, whereas feldspathic gneisses located near mafic gneisses typically contain pyroxene, amphibole, and biotite. These feldspathic gneisses are interpreted to be a result of migmatization and anatexis of original feldspar-rich sediments.

Peter River Gneiss

The metapelitic rocks of the Peter River gneiss are garnet-cordierite-sillimanite-biotite-K-feldspar-plagioclase-quartz gneisses. The garnet is almandine and is typically porphyroblastic, highly fractured, and anhedral. In the well-foliated gneisses, garnet is usually elongate parallel to foliation and is often poikiloblastic, or may exhibit "fish-net" and snowball textures (Spry, 1969). In the more massive gneisses, the garnet is more equant, generally smaller and less poikiloblastic. High concentrations of equant garnets form metre-thick zones of garnetite (Fig. 3), typically at a contact with granitoids. Concentrations have also been observed in magnetite-hematite-rich horizons and may reflect primary iron-rich sedimentary bedding.

Cordierite is also porphyroblastic and may be poikiloblastic, and is usually flattened parallel to foliation. Fresh cordierite exhibits irregular cyclic twinning, but more frequently the mineral is strongly pinitized. Cordierite often encloses fragments of garnet (Fig. 4) and is usually associated with garnet and sillimanite.

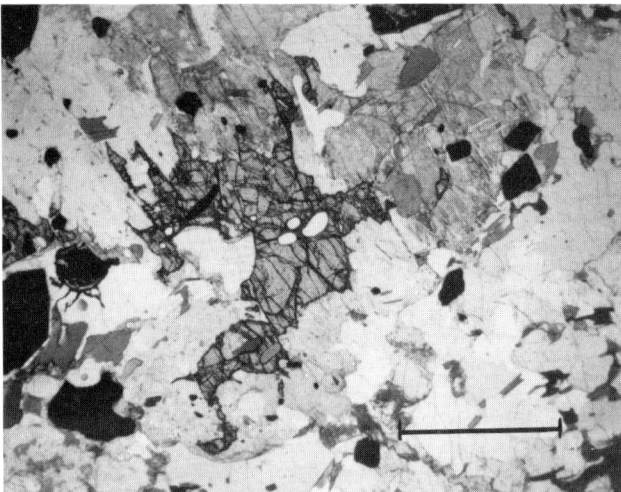

Figure 4. Peter River gneiss. Garnet within a cordierite porphyroblast. Both grains overgrow biotite. The opaque minerals are predominantly sulphides (plane light, bar scale equals 1 mm).

Figure 5. Peter River gneiss. Typical garnet-cordierite-sillimanite gneiss. Note the abundance of K-feldspar (plane light, bar scale equals 1 mm).

Prismatic sillimanite co-exists with both cordierite and garnet in the Peter River gneisses, and aggregates of sillimanite often form bands parallel to foliation (Fig.5). It occasionally occurs as inclusions within garnet and cordierite but more commonly envelopes the larger minerals.

Biotite is always present in minor amounts. It forms small, ragged grains parallel to foliation and usually contains exsolved titanium oxides along cleavage planes. Biotite is often overgrown by garnet and cordierite but has also been observed crystallized in pressure shadows near the porphyroblasts, including two generations of crystallization.

Felsic minerals of the Peter River gneisses constitute from 30% to 80% of the rock volume. Plagioclase (An_{30}), microcline, and quartz may be homogeneously distributed or may form distant bands in the gneiss. Microcline often occurs with cordierite and forms small anhedral aggregates. Plagioclase and quartz are more uniformly distributed. Herring (1976) observed albite in some of these metapelitic gneisses, however he attributed this association to the presence of nearby granitic veins.

Zircon, monazite, sphene, apatite, graphite, and sulphides are common accessory minerals, although in some zones, graphite and sulphides can become major constituents of the Peter River gneiss.

Granitoids and Pegmatoids

Granitoids and pegmatoids are ubiquitous in the basement, however they often appear to increase in volume with depth, and particularly near the contact between the Peter River gneiss and the Earl River complex. They are most abundant within the Earl River complex. Quartz and feldspar often define a lineation, usually in proximity to shear zones. This texture led some earlier geologists to use the term "leptynite" in describing these rocks. The authors in this study apply the terms granitoids and pegmatoids to all rocks of granitic composition in the Carswell structure.

Microcline is usually microperthitic (Fig. 6) and twinning is often bent. Twinned plagioclase (An_{30}) is commonly bent

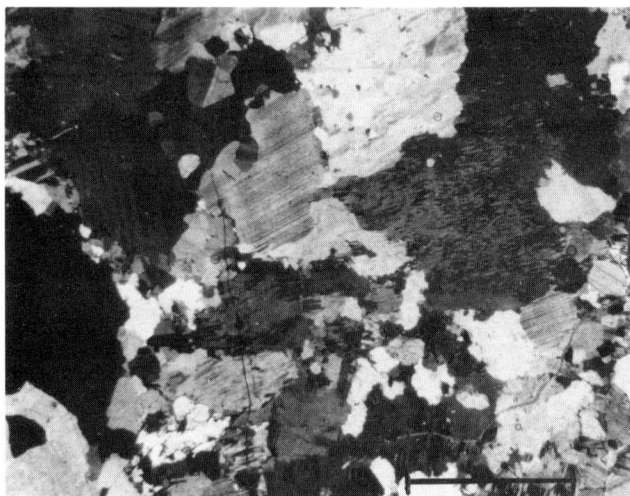

Figure 6. Pegmatoid. Fresh pegmatoid showing strong perthitic texture in the K-feldspars (plane light, bar scale equals 1 mm).

and recrystallized. Antiperthitic grains have frequently been noted in pegmatoids in the Earl River complex.

Accessory minerals in these rocks vary according to mineralogy of the country rock, but they often contain minor amounts of zircon, monazite, biotite, apatite, and sulphides. It is more usual to observe garnet, cordierite, and sillimanite in minor amounts in those granitoids and pegmatoids which occur within the Peter River gneiss, and to note an increase in biotite content and the occurrence of pyroxenes in those same rocks within the Earl River complex.

GEOCHEMISTRY OF THE METASEDIMENTARY SEQUENCE

Numerous diagrams have been developed for the determination of the origin of metamorphic rocks, particularly for metasediments, using major element geochemistry (e.g., de La Roche, 1968; Moine, 1974; Jarousse et al., 1978). The

TABLE II

RANGE AND MEAN VALUES OF SELECTED TRACE ELEMENTS FOR ALUMINOUS GNEISSES, FELDATHIC GENEISSES, AND BIOTITE GNEISSES

	ALUMINOUS GNEISS		FELDSPATHIC GNEISS	BIOTITE GNEISS
	CLU 87[1] 8 Samples	CLU 1711[1] 8 Samples	CLU 1601[1] 8 Samples	CAR 439[2] 2 Samples
C org %	0.02-0.70 (0.20)	0.02-0.20 (0.07)	N.D.-0.12 (0.01)	–
C tot %	0.16-1.00 (0.37)	0.10-0.50 (0.22)	0.02-0.30 (0.14)	–
F ppm	600-1100 (830)	290-910 (560)	470-910 (650)	–
Cl ppm	100-800 (260)	N.D.-100	N.D.-150	–
B ppm	30-100 (60)	15-70 (40)	10-30 (20)	8-10
S ppm	5800-18000 (10600)	2800-13100 (5500)	N.D.-600	–
Cu ppm	37-280 (91)	8-90 (42)	3-22 (13)	24-77
Pb ppm	5-15 (7.5)	4-9 (6)	3-4 (3.5)	20-60
Zn ppm	19-86 (39)	12-27 (20)	23-32 (29)	46-51
V ppm	112-208 (153)	79-158 (111)	51-57 (55)	40-50
Ni ppm	27-74 (53)	25-53 (39)	9-12 (10)	5-15
Co ppm	12-40 (23)	7-19 (14)	4-8 (6)	–
Mo ppm	3-6 (4)	2-6 (4)	N.D.-2	–
Th ppm	14-49 (33)	16-35 (25)	42-57 (51)	7-8
*U ppm	N.D.-1 (.70)	0.2-1 (0.6)	1-10 (4.0)	–
As ppm	N.D.-22 (3.25)	N.D.-4 (1.4)	N.D.	–

[1] CLU 87, CLU 1711, and CLU 1601 have been analyzed by Bondar-Clegg, Canada.
[2] CAR 439 has been analyzed by COGEMA-Fontenay aux Roses, France.
N.D. = non detectable, – = not analyzed
*U – analysed fluorometrically by partial extraction (this is not total uranium).

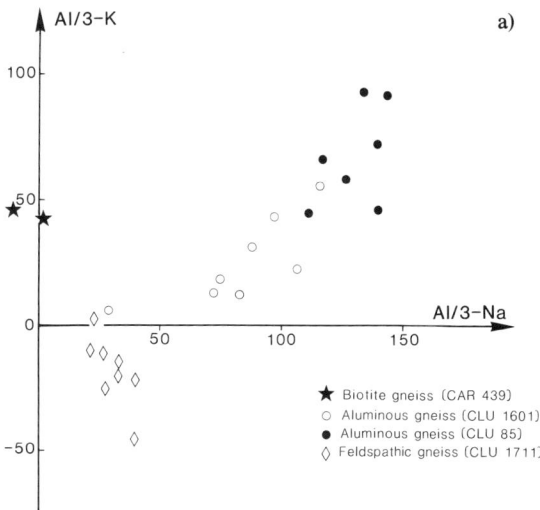

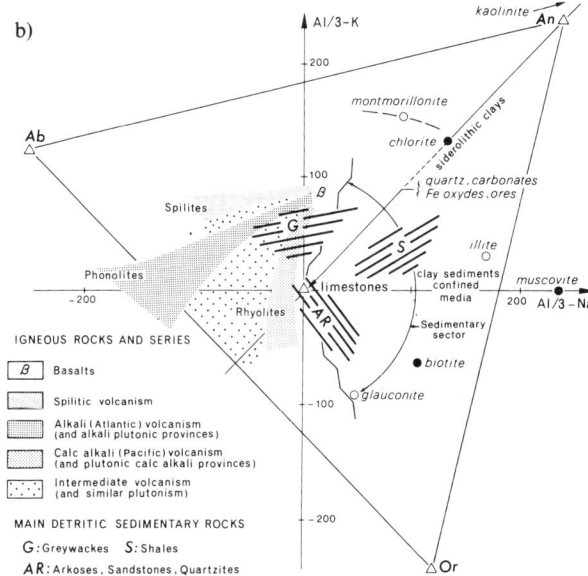

Figure 7. a) Distributions of the metasedimentary rocks of the Carswell structure in an Al/3-Na versus Al/3-K diagram. b) General distribution of metasedimentary rocks and minerals in an Al/3-Na versus Al/3-K diagram.

Peter River aluminous gneisses and the feldspathic gneisses of the Earl River complex are the two main types of metasedimentary rocks in the basement core. Figure 7a (de La Roche, 1968) is a graphical representation of the geochemical diversity of igneous and sedimentary rocks. Geochemical data from four diamond drill holes were plotted on the diagram. Diamond drill holes CLU 1711, located near the Dominique-Peter orebody, and CLU 87, located near the N orebody both plot in the "shale" field. Diamond drill hole CLU 1601, in the same vicinity as CLU 87, plots in the "arkose" field, and diamond drill hole CAR 439, from the western edge of the core (Fig. 1), plots in the "greywacke" field.

Table II contains selected trace elements from all four of the diamond drill holes. The aluminous gneisses are richer in Cu, Pb, Mo, Ni, Co, As, V, and organic carbon than the feldspathic gneisses; these metashales show a geochemical trend (Fig. 7a) which is also characterized by an enrichment in metals (Cu, Mo, Co, and V) coupled with an enrichment in B, S, and organic carbon. The trace element contents are higher in CLU 87 than in CLU 1711. The organic carbon enrichment is related to an increase in clay minerals, specifically to chlorite or chlorite and kaolinite. There is also an increase in the Mg content of the metashales coupled with the increase in trace elements. It is apparent from Figure 7b that the enrichment is not due to illite.

Table III shows the average abundances of selected minor and trace elements in the Peter River metashales and shales from outside of the Carswell structure. Compared with other Aphebian shales from the Canadian Shield (Cameron and Garrels, 1980), Peter River gneisses show slightly lower average metal values although the average shield values fall well within the range of Peter River gneiss element concentrations. With respect to shales of a different age, the average for all types of shales (Turekian and Wedepohl, 1961) are closer to Peter River gneiss compositions than true black shales (Vine and Tourtelot, 1970).

The two samples from the biotite gneiss in diamond drill hole CAR 439 plot in the "greywacke" field in Figure 7a. Their major element compositions are given in Table IV and compare well with an average greywacke of Pettijohn (1963). The Na_2O/K_2O ratios in each of the samples are notably similar.

Reconstruction of the original sedimentary sequence from major and trace element data shows a normal detrital sequence with arkosic sediments at the base and shales at the top. Locally, some greywackes are present. There is no indication of carbonate sedimentation and specifically no evidence of evaporites. In the metashales, there is evidence of a transition to a more restricted sedimentation, reflected in the higher organic carbon, sulphur, and trace element values.

GEOCHEMISTRY OF PEGMATOIDS AND GRANITOIDS

Mineralogical compositions of pegmatoids and granitoids are highly dependent on their wall-rock lithology, except in the case of large masses. Major element geochemistry of fresh pegmatoid samples (Table V) show that two major groups can be distinguished using de La Roche (1980) diagrams (Fig. 8a). Only general observations can be made using these diagrams; more detailed geochemistry is necessary to identify reasons for the wide variations.

The pegmatoids of Group I are potassic and suggest an enrichment in K-feldspar, whereas the plagioclase composition is more albitic than the plagioclases of Group II (Fig. 8b). In the pegmatoids of Group II, plagioclase has a generally higher abundance and more calcic composition. The higher Fe and Mg contents are correlated with the increase in the anorthite component, and Group II pegmatoids are also richer in U, Th, Zn, V, and Pb.

Granitoids plot, in part, in the same field as the pegmatoids (Figs. 9a and b), although occasionally they are slightly richer in quartz and orthoclase. Major element analyses are comparable with Group I pegmatoids (Table V).

The origin of these granitoids seems to be associated with in situ mobilization of predominantly feldspathic but also some aluminous gneisses. Different stages of mobilization can be observed in small quartzo-feldspathic pods in the

TABLE III

AVERAGE ABUNDANCES OF SELECTED MINOR AND TRACE ELEMENTS IN THE PETER RIVER GNEISSES AS COMPARED TO SHALES FROM THE LITERATURE

ELEMENT	PETER RIVER GNEISS (Arithmetic mean)	CONCENTRATION RANGE	No. of SAMPLES	CANADIAN SHIELD[1] (Weighted Average)	AVERAGE BLACK SHALE[2]	AVERAGE SHALE[3]
Ag (ppm)	.20	.2 – 1.90	197	.41	1	0.07
As (ppm)	4	0 – 85	125	23	N/A	N/A
B (ppm)	40	10 – 150	122	25	50	100
Ba (ppm)	875	0 – 1000	8	492	300	580
Co (ppm)	10	0 – 54	221	22	10	19
Cu (ppm)	38	0 – 490	227	75	70	45
Mo (ppm)	4	0 – 63	225	4.5	10	3
Ni (ppm)	39	3 – 330	226	57	50	68
Pb (ppm)	14	2 – 266	219	22	20	20
Sr (ppm)	80	0 – 150	8	69	200	300
V (ppm)	127	5 – 780	110	188	150	130
Y (ppm)	9	1 – 20	8	25	30	41
Zn (ppm)	38	2 – 512	220	114	300	95
U (ppm)	3.25	.2 – 76	126	5.9	N/A	N/A
Th (ppm)	28	1 – 250	86	N/A	N/A	N/A
C (organic) %	0.25	0 – .70	174	1.6*	3.2	0.65
C (mineral) %	0.21	0 – 4.70	122		0.33	1.62

N/A = Not Analysed

* = Total Carbon

[1] Weighted average of 326 samples of Aphebian shales from the Canadian Shield (Cameron and Garrels, 1980, p. 188-189).

[2] Median of the median of 20 sets of black shale from the United States. Ages of the shales range from Cambrian to Eocene (Vine and Tourtelot, 1970, p. 261).

[3] Average shales (arithmetic means) from numerous sources. Ages and number of samples are not given (Turekian and Wedepohl, 1961, Table 2).

TABLE IV

CHEMICAL COMPOSITIONS (%) OF TWO BIOTITE GNEISSES (1 AND 2) FROM CAR 439 AND COMPARISON WITH AN AVERAGE GREYWACKE (3) (PETTIJOHN 1963)

	1	2	3
SiO_2	68.87	70.23	66.7
Al_2O_3	15.94	13.79	13.5
Fe_2O_3 (tot)	3.72	4.63	5.5
MnO	0.05	0.08	0.1
MgO	0.84	1.78	2.1
CaO	2.26	2.02	2.5
Na_2O	3.59	2.76	2.9
K_2O	2.74	2.31	2.0
TiO_2	0.34	0.38	0.6
Total	98.35	97.98	95.9

TABLE V

GEOCHEMICAL VARIATIONS AMONG THE GRANITOIDS AND PEGMATOIDS IN THE CARSWELL STRUCTURE BASEMENT CORE (SEE FIGS. 8 AND 9)

	PEGMATOID (Group I)	PEGMATOID (Group II)	GRANITOIDS	GRANITOIDS (CLU 745)
	n = 8	n = 4	n = 19	n = 6
SiO_2	71.87	69.75	72.84	73.75
Al_2O_3	15.06	15.12	14.18	14.82
Fe_2O_3 (total Fe)	1.10	2.50	1.03	1.12
MnO	0.03	0.03	0.05	0.03
MgO	0.47	1.34	0.46	0.38
CaO	0.54	1.06	0.48	0.77
Na_2O	2.95	5.50	2.18	3.28
K_2O	6.12	2.55	6.78	4.45
TiO_2	0.11	0.16	0.07	0.03
TOTAL	98.25	98.01	98.07	98.63

pressure shadows of garnets, in the development of quartzo-feldspathic augens, in felsic bands along foliation, and also by invasion of quartzo-feldspathic material disrupting foliation and forming veins which may or may not be conformable with the foliation. There is some problem with the strontium data obtained on granitoids from diamond drill hole CLU 745 from the Claude area (Fig.1) which shows a low initial strontium ratio of 0.707 ± 0.009 (Bell, 1985). This value is not compatible with the direct formation of granitoids from aluminous or feldspathic gneisses. However, the possible existence of different generations of granitoids should not be disregarded. Samples from CLU 745 do not plot exactly in the field of other basement core granitoids in an AA-CC-MM diagram (Fig. 9b).

URANIUM AND THORIUM GEOCHEMISTRY

Detailed and accurate knowledge of the distribution of uranium and thorium in fresh basement rocks of the Carswell structure is very important for the understanding of petrogenetic processes and also for the evaluation of the metallogenic potential of uranium deposits which are spatially related to the area. At present only partial data is available. Fresh samples are extremely rare, since the rocks have been affected by one or more alteration processes. Many samples have been collected very near to the uranium deposits, and therefore some contamination is possible. Analyses over the years have been performed by a variety of laboratories using partial extraction methods, with very few analyses of total uranium. A detailed geochemical and mineralogical study is in progress to resolve this problem (Halter and Pagel, in progress).

Despite these uncertainties, several general observations can be made:

1) The average uranium content of the basement rocks is significantly high. From the histograms in Figure 10, it appears that the basement rocks from the Carswell structure have average uranium contents which are slightly higher than the averages in the upper crust (U = 2.7 ppm, Taylor, 1964). Weighted uranium averages calculated from radiometric analyses for the Canadian Shield (Fahrig and Eade, 1968) and Canadian Shield averages (Shaw *et al*., 1967) are 1.5 and 3.45 ppm respectively. Cuney (1981), and Barbey and Cuney (1982) have shown that uranium is not systematically de-

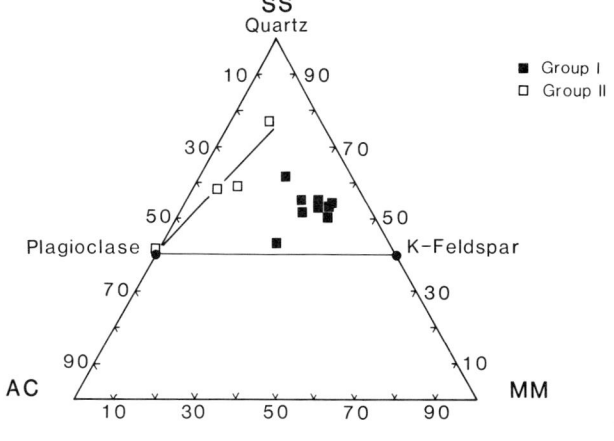

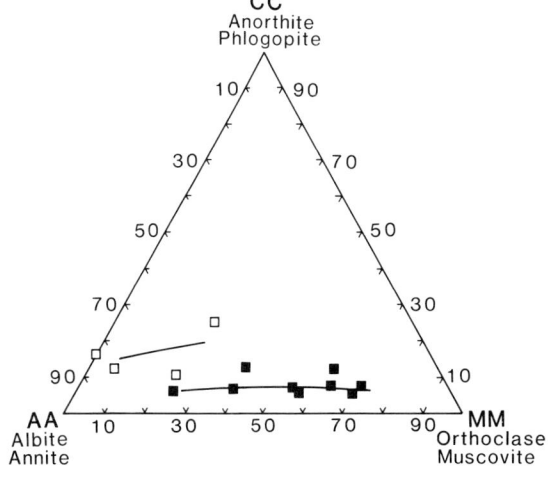

Figure 8. a) Distributions of pegmatoids in the Carswell structure in an SS-AC-MM ternary diagram (de La Roche, 1980). b) Distributions of pegmatoids in the Carswell structure in a CC-AA-MM ternary diagram (de La Roche, 1980).

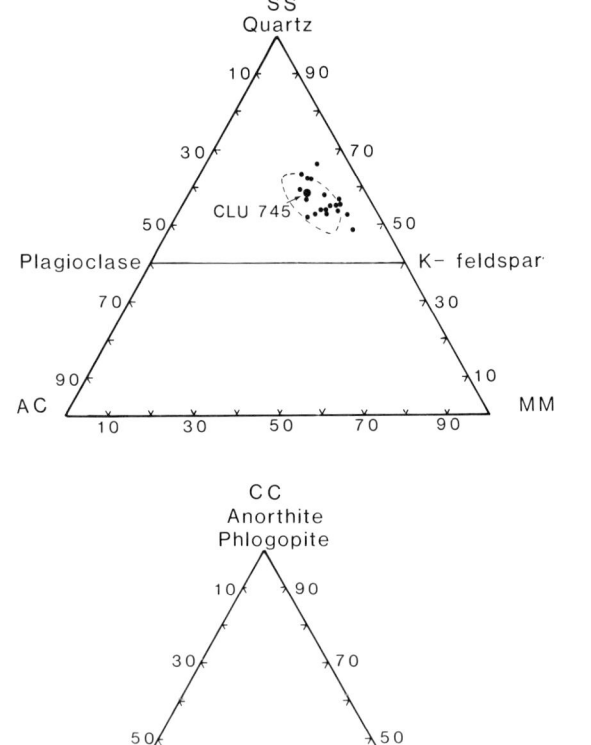

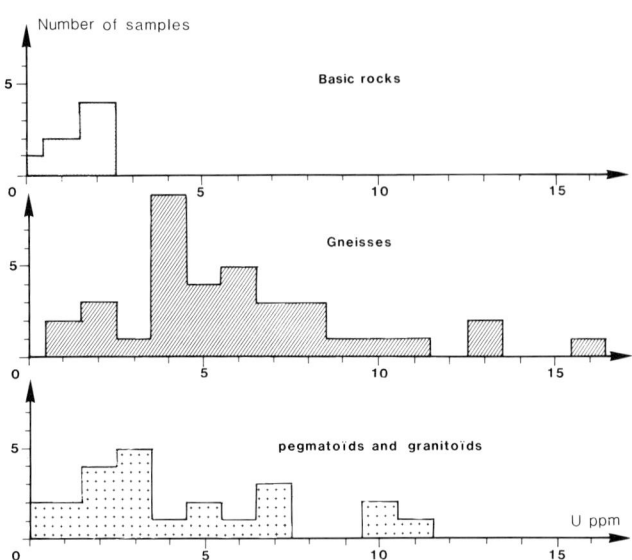

Figure 9. a) Distribution of granitoids in the Carswell structure in an SS-AC-MM ternary diagram (de La Roche, 1980). The dashed line indicates the field of Group 1 pegmatoids. b) Distributions of granitoids in the Carswell structure in a CC-AA-MM ternary diagram (de La Roche, 1980). The dashed line indicates the field of Group 1 pegmatoids.

Figure 10. Histograms of uranium contents (ppm) of the main lithologies in the Carswell structure. Data was compiled from Bruneton (1981 unpublished), Schnessl (1983), Amok Ltd. (unpublished), and this work.

pleted in granulitic terrains, as first proposed by Lambert and Heier (1968). Widespread occurrences of graphite in gneisses, which represent low fO_2 during metamorphism, could be responsible for the low solubility of uranium in metamorphic fluids (Nguyen and Poty, 1977; Cuney, 1981). Further evidence is found in the fluid inclusions which contain methane and have been related to prograde metamorphism.

2) The thorium content of the gneisses is also quite high with most values between 10 and 50 ppm. East of Cluff Lake, there are small areas showing anatectic mobilization which are characterized by high thorium contents up to 3500 ppm. Similar contents are known in migmatized areas related to pegmatoids (Cuney, 1981; Barbey and Cuney, 1982). Breaks (1982) noted high Th/U ratios in S-type granites from metasedimentary migmatites in Ontario relative to uranium-enriched I-type potassic granitoids.

3) Although there is considerable overlap, each lithologic unit in the basement gneisses displays distinctive U-Th characteristics (Fig. 11). Mafic gneisses from the Dominique-Peter zone and other areas are poor in uranium and thorium. Cacciotti (1983) has obtained, using neutron activation analysis, values below 0.5 ppm for uranium and below 8 ppm for thorium. Other mafic gneisses have uranium contents up to 2.5 ppm, and up to 11 ppm for thorium (Bruneton, 1981; this work) indicating that the Th/U ratios for mafic gneisses are highly variable. Obviously, there are several types of basic rocks in the Carswell structure.

Feldspathic gneisses have the highest thorium contents of the typical rocks except for pegmatoids, with thorium values up to 60 ppm and variable uranium values up to 13 ppm. These variations are not totally understood since there are obvious traces of mobilization in these rocks which may be attributed either to fractionation effects in anatectic melts or to similar erratic variations in the protolith.

Aluminous gneisses have lower thorium contents than feldspathic gneisses, but show the same variations in uranium content. Since these aluminous gneisses are interpreted to be metamorphosed shales, this interpretation is emphasized by their enrichment in uranium compared to the average shale of Rogers and Adams (1969), where U = 3.7 ppm and a similar thorium enrichment compared with average shales of Adams and Weaver (1958) where thorium averages 12 ppm. Garnetite has a significantly lower thorium content than the aluminous gneisses, while the uranium content (2.4 to 10.8 ppm) is similar to that of uranium in the aluminous gneisses.

Granitoids and pegmatoids show extremely variable uranium and thorium values (Fig. 11). There are obviously different generations of granitoids and pegmatoids, however further work is needed with respect to age determinations,

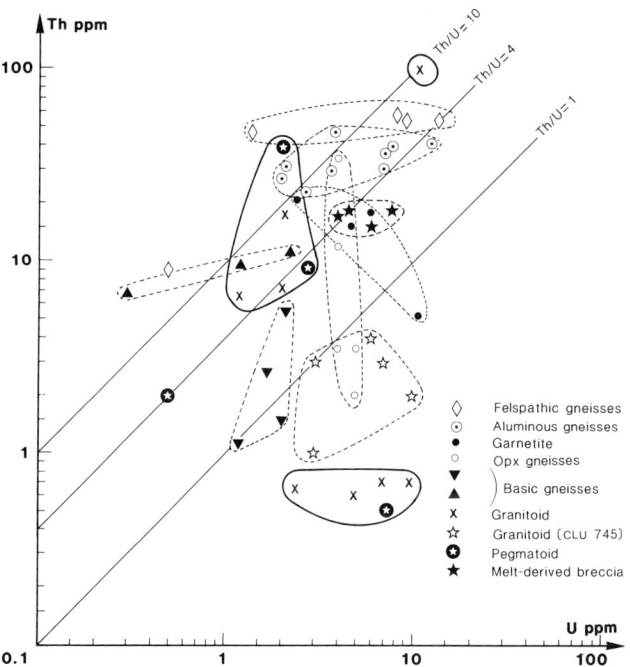

Figure 11. Selected U versus Th plot for the Carswell structure basement rocks.

defining wall rock lithology and determining physico-chemical conditions during crystallization. Granitoids from CLU 745, which give a Rb-Sr age of around 1960 ± 110 Ma (Bell, 1985), show a distinct U-Th field from other granitoids which are intrusive into the aluminous or feldspathic gneisses. Pegmatoids, very rich in U-Th, or both, have been observed in the Carswell structure.

BASEMENT ALTERATION

Six main types of alteration have been recognized in the Carswell structure basement based on their spatial distribution, petrographic characteristics, paragenetic association, and mutual relationships: 1) retrograde alteration during the later stage of the Hudsonian orogeny; 2) alteration due to exposure at surface around 1700 to 1500 Ma; 3) diagenetic alteration just below the Athabasca sedimentary cover; 4) hydrothermal alteration related to the main uranium mineralization process; 5) alteration related to the formation of the Carswell structure; and 6) late alteration due to mineralization or remobilization processes around 400 Ma and later. Detailed chemical analyses have been carried out to characterize the hydrothermal alteration related to the main uranium mineralization processes.

Retrograde Metamorphism

Retrograde metamorphism in the basement is pervasive, and fresh rocks are rare. At the present time, no dates have been calculated for the retrograde assemblage. There is, however, visible overprinting by pre-Athabasca surface alteration, which would place the retrograde event between 1800 Ma, the age of Hudsonian metamorphism and 1500 to 1550 Ma, the approximate age of the deposition of the Athabasca Group sediments. The development of secondary amphibole, K-micas, Fe-rich chlorite, carbonate, epidote, titanium oxides, pyrite, and hematite has been observed. Veins containing an epidote-chlorite-quartz-carbonate-sulphide assemblage are also found. Locally, there are processes in the basement in which quartz is dissolved, with the accompanying development of an albite-chlorite-K-mica-carbonate assemblage. This is comparable to the episyenitization processes described in leucogranites from France (Leroy, 1978), and also similar to features described in the Beaverlodge district, northern Saskatchewan (Tremblay, 1972). These alteration processes correspond to important stages of redistribution of uranium in the basement and further quantification of the process is in progress.

Surface Alteration and Diagenesis

Surface alteration and diagenesis are included under one heading because the regolith and paleoweathering in the Athabasca Basin has been overprinted by a recrystallization phenomenon after the deposition of the Athabasca Group.

Beneath the Athabasca sediments, there is a pervasive alteration which forms distinct red and green zones. The actual extent of the paleoweathering profile is difficult to reconstruct since the formation of the Carswell structure has modified the configuration and the relationship between the basement and the sedimentary cover. The red zone is present directly beneath the basement-sandstone contact, whereas the green zone is most abundant in the deeper part of the profile.

The red zone in the area of the D orebody (along the southern edge of the basement core) contains a combination of illite and chlorite, quartz and hematite. In the zone near the deposit, Ey *et al.* (1985) have observed 1M illite with a crystallinity index (Dunoyer de Segonzac, 1969) between 3.5 and 4.0. Macdonald (1980) studied cases in the Athabasca Basin in which the actual crystallinity index of illite and the clay paragenesis indicate that paleoweathering has been overprinted by a late, higher temperature phenomenon which produced a new paragenesis. It should be noted that elsewhere along the basement-sandstone contact, kaolinite is often present in the red zone. In the green zone, hematite is absent, and the predominant clay mineral is chlorite.

Locally, just beneath the sedimentary cover, there are irregularly distributed bleached zones where hematite is scarce and occasionally absent. In these bleached zones, quartz corrosion is developed along the borders and also along narrow bands within the zone.

Hydrothermal Alteration Related to the Main Uranium Mineralization Processes

There is intense hydrothermal alteration around uranium veins in the basement, where sometimes quartz is the only mineral preserved. Plagioclase and ilmenite are the first minerals to be intensely altered. Further away from the mineralization, illite, and occasionally carbonate are formed. Ilmenite destabilizes to form iron and titanium oxides. Magnetite remains quite stable; its breakdown is synchronous with biotite alteration.

Near the mineralized veins, all host rock minerals are altered. Illite and chlorite develop, and quartz is sometimes corroded. Acicular dravite is common and is often localized in open fractures in association with quartz. The dravite appears late in paragenesis. Fluorite has been observed only in one sample, in the Dominique-Peter deposit. Sulphides, essentially pyrite and chalcopyrite, are common in the altered zones near mineralization. Carbonate is absent in the intensely altered zones but may be present within the vein. Close to pitchblende, there is a development of a fine-grained, accordion-pleated mineral which may possibly be a mixed-layer clay. Veinlets of apparently later kaolinite have been observed within the alteration halo but their relationship to uranium deposition is not known.

Alteration Related to the Formation of the Carswell Structure

Little alteration appears to be directly related to the formation of the Carswell structure. However, chlorite, illite, and hematite have formed in the breccias which are a result of tectonic events. Vesicles or dolomite cement within the Cluff breccias have also been observed.

Late Alteration Due to Mineralization or Remobilization Processes Around 400 Ma and Later

Alteration producing chlorite, hematite, carbonate, kaolinite, and other clay minerals is associated with the later remobilization processes. No detailed analyses are presently available to separate the different events, particularly that which is related to the main uranium deposition (Pagel and Ruhlmann, 1985).

CHEMICAL VARIATIONS AROUND MINERALIZED ZONES

Major Element Distribution

Due to widespread anatexis, special attention has been focused on the granitoids and pegmatoids in the Carswell structure; these rocks are less complex and can be used easily in discriminant diagrams. A detailed study has been conducted in the Dominique-Peter area. Table VI contains the chemical analyses of granitoids. Chemical variations between the fresh and altered granitoids are quite consistent. There is an important increase in water due to the development of clay minerals and extensive loss of Na_2O and lesser depletion of CaO which corresponds to the alteration of plagioclase. After the initial increase in K_2O and MgO, K_2O decreases in the presence of high MgO near uranium veins. There is a slight increase in iron during potassium metasomatism and a corresponding decrease during intense magnesium metasomatism. Although it is not certain, there is an apparent small decrease in silica for an apparent increase in alumina. Even if aluminum is assumed to be constant, the above characteristics remain consistent. Tremblay (1983) comments that similar trends are present in almost all unconformity-related deposits in Saskatchewan.

For rocks of primary granitic compositions, de La Roche (1980) diagrams discriminate between altered and unaltered samples, and between illite-bearing and chlorite-bearing rocks. Because of the mineralogy of the granitoids, the AA-CC-MM diagram is most suitable, since it permits a good assessment of the feldspar development (Fig. 12). Three

TABLE VI

CHEMICAL COMPARISON BETWEEN FRESH AND ALTERED GRANITOIDS FROM DRILL HOLES 1589 AND 1609 (DOMINIQUE-PETER AREA)

	FRESH GRANITOIDS					ALTERED GRANITOIDS			
DRILL HOLE	(1589)			(1609)		(1589)			
DEPTH	731'	999'	1038.7'	696.6'	876.3'	281'	341'	519'	631.6'
SiO_2	73.5	74.5	74.5	73.0	72.5	72.5	71.0	71.5	72.0
Al_2O_3	14.0	13.8	14.7	15.0	14.4	13.6	15.3	15.6	14.5
Fe_2O_3 (total)	1.0	1.2	0.7	1.2	1.4	1.7	1.0	0.7	1.5
CaO	0.45	0.95	0.65	0.40	1.10	0.20	0.30	0.20	0.15
MgO	0.15	0.35	0.25	0.90	1.15	2.40	3.50	3.85	2.20
Na_2O	2.30	2.7	2.5	2.1	2.4	0.3	0.3	0.2	0.4
K_2O	8.10	5.1	5.7	6.3	5.4	5.8	3.0	2.9	6.6
MnO	0.01	0.01	0.01	0.01	0.03	0.01	0.01	0.01	0.01
TiO	0.10	0.15	0.10	0.10	0.05	0.10	0.10	0.10	0.10
P_2O_5	0.21	0.19	0.21	0.17	0.20	0.22	0.22	0.17	0.16
S	0.02	0.02	0.02	0.02	0.02	0.02	0.02	0.02	0.02
L.O.I.	0.64	0.91	1.12	1.02	1.27	2.60	4.40	4.60	2.35
TOTAL	100.46	99.86	100.44	100.20	99.90	99.43	99.13	99.82	99.97
Th	10	9	14	8	7	N.D.	2	N.D.	20
U	1	2	2	1	2	13	3	0.8	3
Mo	4	6	6	2	2	9	5	1	3
B	N.D.	N.D.	10	N.D.	N.D.	30	100	50	20
F	30	110	100	40	55	210	310	330	280
As	N.D.	N.D.	N.D.	N.D.	N.D.	2	7	N.D.	N.D.

N.D. = Non-Detectable
L.O.I. = Loss On Ignition

characteristics emerge: 1) a primary fractionalization in granitoids which involves albitic plagioclase and potassium feldspar; 2) a trend due to the development of illite, and 3) a trend due to the development of Mg-chlorite. These features are in agreement with petrographic observations. Such diagrams could be useful during prospecting to identify the most favourably altered zones.

Trace Element Distribution

Trace element contents of the host rocks within the main alteration zone can be directly compared to those associated with the mineralization itself. In the Dominique-Peter area, there is an increase in boron and fluorine during metasomatism. At high concentrations, these elements may be expressed mineralogically as tourmaline and fluorite. Although in the vicinity of the D orebody, there are high lithium values up to 550 ppm, the lithium content in the Dominique-Peter area is always below 200 ppm, and there are no clear indications of any enrichment in altered zones.

Near the mineralized veins, there may be anomalies in Cu, Pb, Zn, As, Mo, Co, Au, Ni, V, and Y, however, not all the elements are anomalous in each sample, as is illustrated in Figure 13. The following trace element associations are recognized: As-Ni, As-Ni-V-(Co), As-Cu-(Mo)-(Ni)-(Co)-(Zn)-(V), Pb-Cu-Mo-(V), and Cu-Mo-V. The Pb-Cu-Mo-(V) association is the most important for uranium mineralization and forms a uraninite-sulphide assemblage (Ruhlmann, 1985).

FLUID INCLUSIONS RELATED TO THE HYDROTHERMAL ALTERATION IN THE BASEMENT DURING MINERALIZATION

During hydrothermal alteration in the mineralized zones, there is development of chlorite, illite, sericite, mixed-layer clays, and occasionally dissolution of quartz. These are not favourable environments for the study of fluid inclusions, however, some information has been obtained from quartz-uranium oxide-dravite-quartz veins from the OP area, quartz-uranium oxide veinlets from the Dominique-Peter deposit, and an altered aluminous gneiss from the D deposit.

Quartz-Uranium Oxide-Dravite-Quartz Veins from OP

In the euhedral quartz grains that have been deposited prior to uranium oxides, there are extremely rare fluid inclusions. Three phase inclusions have been observed: liquid, gas bubble, and cubic crystal. The cubic mineral dissolves during heating, inferring that it is probably halite. In the later quartz phase which is associated with dravite, there are either two-phase (liquid + gas) or three-phase (liquid + gas + halite cube) inclusions. They yield homogenization temperatures, considering individual inclusion planes, of up to 300°C for inclusions of variable salinity, from dilute solutions

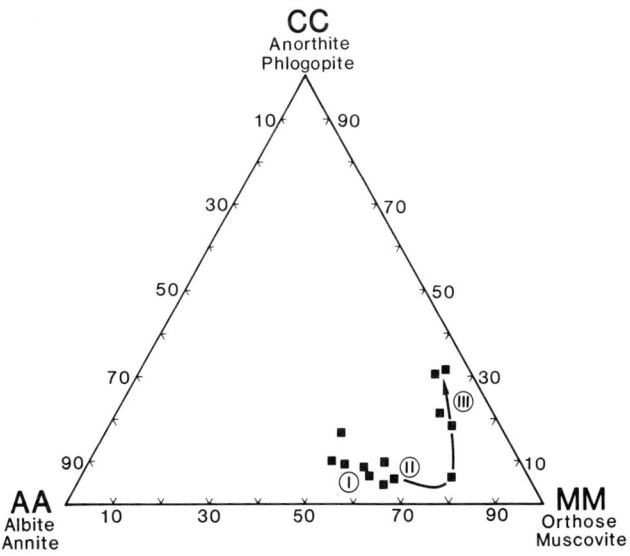

Figure 12. Characteristics of hydrothermal alteration around uraniferous zones in the Dominique-Peter area, as plotted on a CC-AA-MM diagram (de La Roche, 1980). I are fresh granitoids, II is the path indicating the development of illite, and III is the path indicating the development of chlorite.

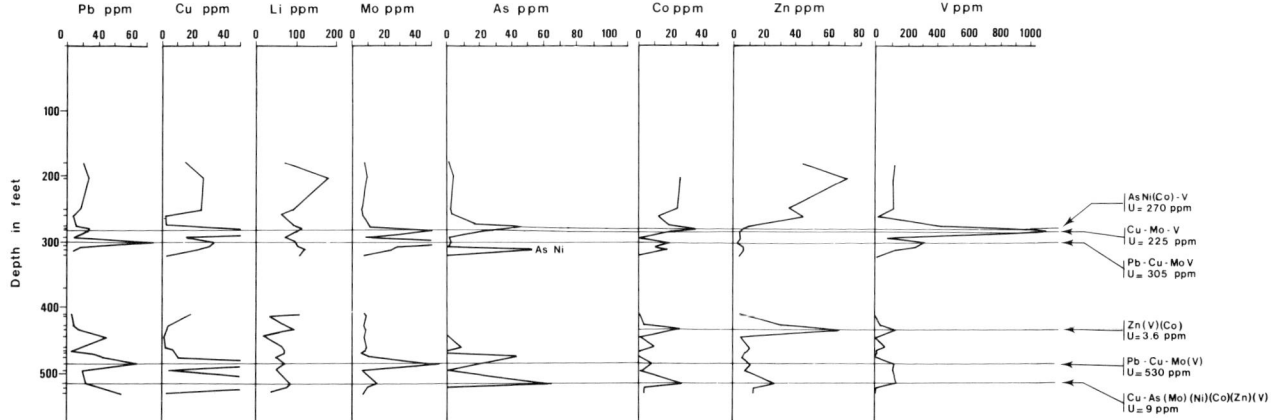

Figure 13. Variations of selected elements (Pb, Cu, Li, Mo, As, Co, Zn, and V) with depth in diamond drill hole CLU 1607, Dominique-Peter area. Twenty-seven samples were analysed by atomic absorption, CREGU, France.

to supersaturated brines. Leaching after crushing of the automorphic quartz has proven that NaCl is the principal component; the K/Na ratio is very low (approximately 0.0085).

Quartz-Uranium Oxide Veinlets from Dominique-Peter

In one sample from Dominique-Peter, fluid inclusions within the quartz that were deposited just before uranium oxide are very similar to those which have been observed at Rabbit Lake, Saskatchewan (Pagel, 1977; Pagel and Jaffrezic, 1977). These inclusions have a liquid phase, a gas phase in which the bubble is occasionally flattened, a cube of halite, and a few needles of an unknown birefringent compound.

Altered Aluminous Gneiss from D

In the D deposit, there is a widespread greenish alteration which affects the gneissic basement. Metamorphic quartz shows strong dissolution features with abundant two-phase fluid inclusions. These inclusions may contain a hexagonal reddish material, possibly hematite. They have melting temperatures from -11.7°C to -47.4°C, and a maximum homogenization temperature of 172°C.

The above data illustrates a number of common factors: 1) Brines are present in all Precambrian deposits where fluid inclusions have been observed in the Carswell structure, with NaCl dominant. The implication of their presence in the genesis of unconformity-related depostis has also been demonstrated at Rabbit Lake (Pagel, 1977; Pagel and Jaffrezic, 1977), and Nabarlek and Jabiluka, Australia (Ypma and Fujikawa, 1980). 2) Minimal formation temperatures are variable, based on deductions from homogenization temperatures. 3) Post-mineralization events in the OP deposit give temperatures as high as 300°C. Geochronological studies (Bell, 1985) indicate that this event does not correspond to the main mineralization process, but is later, at about 820-890 Ma. The presence, in the same quartz crystal, of fluid inclusions containing dilute and supersaturated brine solutions with inclusions of intermediate compostion is interpreted to be a mixing of two aqueous solutions rather than boiling because in the latter case, dilute inclusions would become homogeneized in a gas phase at the same temperature as brine inclusions in a liquid phase.

A CHEMICAL STUDY OF ALTERATION MINERALS IN MINERALIZED ZONES

It has already been stated that chlorite, illite, sericite, and other clay minerals are widespread in the altered wall rocks of mineralized veins from the Dominique-Peter deposit. A detailed electron microprobe study has been performed on the different types of clay minerals in an attempt to relate chemical variations to the proximity of the veins. The presence of Mg-tourmaline is also significant, and its chemical composition has been determined.

Chlorite

Several types of chlorite are microscopically discernible in altered zones from the Dominique-Peter deposit. Structural

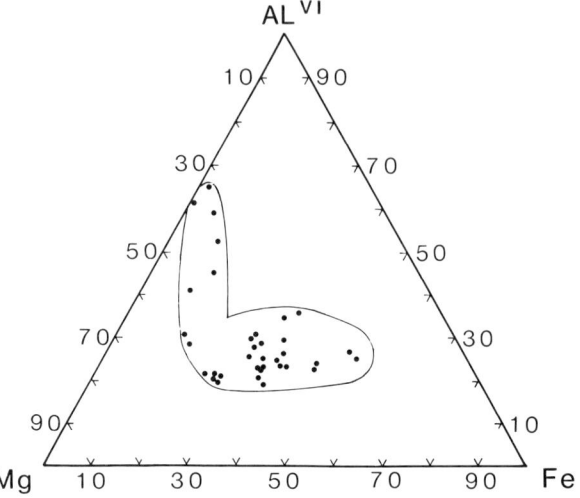

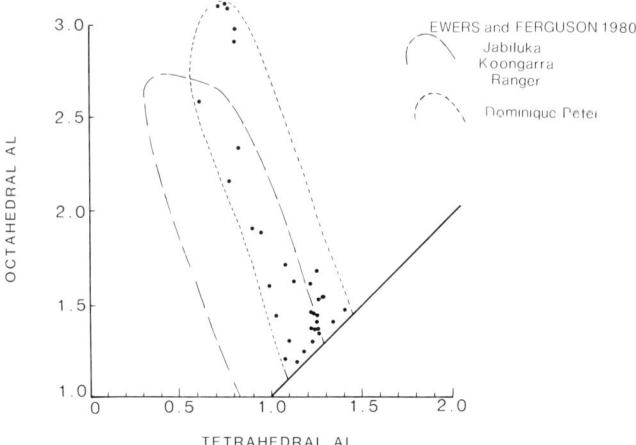

Figure 14. Graphic representation of chlorites from the Dominique-Peter area: a) Tetrahedral Al versus octahedral Al. The dashed line is the contour for chlorite from the Jabiluka, Koongarra, and Ranger deposits (Ewers and Ferguson, 1980). b) Al(VI)-Mg-Fe ternary plot.

formulae calculated from electron microprobe analyses and Figure 14 yield the following observations: 1) With proximity to mineralization, there is an increase of octahedral Al, whereas there is a small decrease in tetrahedral Al. The increase in octhedral Al is followed by a decrease of octahedral site occupancy (Table VII). Similar data have been found for the Jabiluka, Koongarra, and Ranger deposits, Australia (Ewers and Ferguson, 1980).

The relationship between total Al, Fe, and Mg indicates a continuous evolution (Fig. 14b). Representative analyses are given in Table VII. There is aluminous chlorite (sudoite) in samples where uranium oxides have been observed. Aluminous and magnesian chlorites are most abundant in altered zones around uranium veins. More rarely, there are iron-rich chlorites in garnetiferous basement rocks.

From these observations, indications of the proximity of uranium concentrations appear to be characterized by aluminous and magnesium chlorite, whereas mineralized samples are characterized by hyperaluminous chlorite. This

TABLE VII

REPRESENTATIVE ELECTRON MICROPROBE ANALYSES (WEIGHT %) OF TYPICAL CHLORITE FROM THE ALTERED ZONES IN THE DOMINIQUE-PETER DEPOSIT

	1	2	3
SiO_2	35.15	27.52	23.37
Al_2O_3	35.58	20.14	22.16
FeO*	1.72	18.52	33.78
MnO	–	–	–
MgO	11.57	20.51	8.23
CaO	0.08	0.07	0.02
Na_2O	0.04	–	–
K_2O	0.49	0.07	–
TiO_2	–	–	0.06
TOTAL	84.63	86.83	87.62
Si (4+)	3.230	2.816	2.585
Al (IV)	0.760	1.184	1.415
Al (VI)	3.085	1.245	1.475
Ti (4+)	–	–	0.005
Fe (2+)	0.132	1.585	3.125
Mg (2+)	1.585	3.128	1.357
Mn (2+)	–	–	–
Ca	0.008	0.008	0.002
Na	0.007	–	–
K	0.057	0.009	–

* = Total FeO

1. Sample MP-CL-81-41, DDH 1607, 485' – Uranium oxide is present.
2. Sample MP-CL-81-46, DDH 1607, 548' – Neoformation of chlorite at the expense of potassium feldspar.
3. Sample MP-CL-81-53, DDH 1433, 617' – Narrow zone of chlorite in garnet.

trend appears to be very characteristic of Proterozoic unconformity-related uranium deposits. It has not been observed in other geological environments where brines are also involved; for example, the uraninite-albite veins from Mistamisk, Canada (Kish and Cuney, 1982) or uraninite concentrations from Shaba, Africa (Audeoud, 1983).

K-Micas

The development of K-micas is widespread in altered zones around mineralized veins. Four groups have been microscopically differentiated in an attempt to correlate the chemical composition of K-mica with the proximity to mineralization.

In non-mineralized samples, a coarse K-mica, a fine K-mica which coexists with the coarse mica, and a ubiquitous fine K-mica have been differentiated. In mineralized samples, only a fine K-mica has been observed.

Large variations in Fe/Fe+Mg (X_{Fe}) and total aluminum occur (Fig 15). With a strengthening of the magnesian character of the K-mica, there is a corresponding increase in aluminum. If a paragonitic substitution is considered and comparison made with data of Leroy and Cathelineau (1982) for hydrothermally developed K-micas, the low paragonite content (1% to 3%) indicates a temperature of formation below 250°C, which is in agreement with fluid inclusion data and geological context.

Other Clays

Other clays have been observed near uranium mineralization. Microprobe analyses show variable chemical compositions. Five analyses on two samples from the Dominique-Peter zone show that they are rich in magnesium and aluminum. They contain high MgO contents (6.1% to 10.2%) for a low FeO content (0.4% to 2.8%) and K_2O contents between 3.8 and 5.2%. They are also aluminum-rich (Al_2O_3 = 28.2 to 32.8%) with SiO_2 from 41.3% to 45.6%.

Tourmaline

Fine, colourless needles of tourmaline are locally abundant in the altered zones from the D, OP, Claude, and Dominique-Peter deposits. Due to the grain size and intimate mixture of secondary quartz, it has been difficult to obtain good microprobe analyses. Table VIII contains representative analyses, showing that the tourmalines are magnesium-rich. Iron is virtually absent in all analyses. The tourmalines are further characterized by alkali deficiencies.

A typical approximate structural formula for a sample from the Dominique-Peter zone is:

$(Na_{0.21} K_{0.04} Ca_{0.01}) (Mn_{0.05} Mg_{2.09} Al_{0.85}) (Al_6) (Si_{5.93} Al_{0.07}) (BO_3)_3 (OH_4)$

This formula is close to alkali-poor dravite (Rosenberg and Foit, 1975):

$(Mg_2Al) Al_6 (BO_3)_3 Si_6O_{18} (OH)_4$

There is a lack of experimental and mineralogical data to explain this particular tourmaline composition. Similar compositions for tourmalines have been determined in the Rabbit Lake deposit, Canada (Rimsaite, 1979) and the Jabiluka deposit, Australia (Durak, in prep).

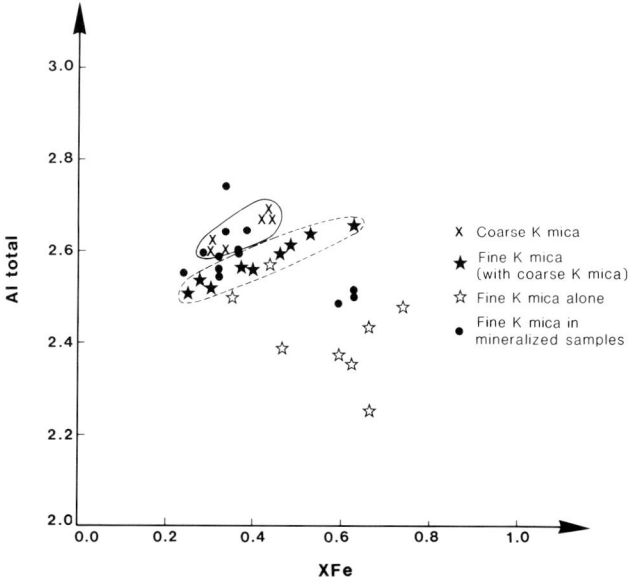

Figure 15. Graphic representation of K-micas from the Dominique-Peter area (X_{Fe} $\frac{Fe}{(Fe + Mg)}$ versus total Al).

TABLE VIII

AVERAGE CHEMICAL MICROPROBE ANALYSES OF WHITE TOURMALINE FROM DOMINIQUE-PETER AND OP. ANALYSES OF TOURMALINE FROM JABILUKA (DURAK, IN PREP.) AND RABBIT LAKE (RIMSAITE, 1979) ARE GIVEN FOR COMPARISON.

	DOMINIQUE-PETER		OP	RABBIT LAKE	JABILUKA
Number of Analyses	9	1	3	1	3
SiO_2	37.24	36.18	37.06	34.9	39.10
Al_2O_3	37.52	37.54	38.02	37.5	34.61
FeO	–	–	(0.20)	0.2	0.15
MnO	–	1.20	–	–	–
MgO	8.75	7.34	9.99	9.6	10.24
CaO	0.11	0.33	(0.07)	0.1	0.10
K_2O	(0.02)	–	(0.01)	–	–
Na_2O	0.65	0.47	0.58	0.8	0.66
TOTAL	84.28	83.06	84.93	83.1	84.86

CONCLUSIONS

The main conclusions of this petrographic and geochemical study are:

1) In the Carswell basement core, the rocks have been divided into the Earl River complex and the structurally overlying Peter River gneiss. Geochemically, the Earl River complex consists predominantly of arkose and sandstones whereas the Peter River gneiss have a typical shale composition. Basic volcanism of komatiitic affinity (Bell *et al.*, 1985) is evidenced especially in the Earl River complex. At the present time, no evaporitic sequence has been documented.

2) Metamorphism around 1800 Ma (Bell, 1985) reached the granulite facies in the presence of a fluid rich in carbon dioxide, methane, and nitrogen. The presence of garnet-cordierite-sillimanite-biotite gneisses, hypersthene-clinopyroxene-biotite gneisses, and fluid inclusion data indicate high temperature and medium pressure conditions of estimated 800°C and 5 to 7 kbars.

3) Anatexis was widespread throughout the structure, particularly in the Earl River complex. Anatectic granitoids and pegmatoids show a wide range in chemical compositions.

4) The average uranium content of the basement is higher than average for the crust, a fact which is not common in granulite facies terrains. The conservation of uranium during granulite facies is ascribed to the occurrence of graphite which maintains a low fO_2 during metamorphism and consequently a low solubility of uranium. Enrichment in uranium and thorium in anatectic material (pegmatoids) has been evidenced locally (Ruhlmann, 1985; Pagel and Ruhlmann, 1985).

5) The Peter River gneisses have high amounts of Cu, Pb, Zn, As, Mo, Co, and Ni; elements typically associated with uranium in mineralized zones. The increase of these elements is also related to an increase in the carbon and sulphur content. In the authors' opinion, the similarity between the trace element content of the Peter River gneisses and the element association in mineralized zones is not fortuitous. On the contrary, Hoeve *et al.* (1980) have invoked the Athabasca sandstones as the source of elements present in the ore deposit. Other areas must be studied to see if such an association is widespread, as it would be an important key to the understanding of the genesis of unconformity-related uranium deposits.

6) At least six alteration patterns have been recognized in the basement rocks of the Carswell structure, based on their spatial distribution, petrographic characteristics, paragenetic association, and mutual relationships: alteration related to retrograde metamorphism during the Hudsonian orogeny, alteration related to surface weathering before the deposition of the Athabasca Group, alteration subsequent to the deposition of the Athabasca Group and resulting from deep burial, alteration in Proterozoic mineralized zones, alteration related to the formation of the Carswell structure, and alteration related to remobilization of uranium deposits after 400 Ma.

7) Around the mineralized zone, there is a development of aluminous and magnesium chlorite, illite, and Mg-tourmaline. In the mineralized zone itself, hyperaluminous chlorite and mixed layer clays have been observed. Kaolinite, if present, always appears to be a later mineral phase.

8) Around the mineralized zones, there are also anomalies in various elements (Cu, Pb, Zn, As, Mo, Co, Au, Ni, V, and Y) which are sometimes erratically distributed. An increase in boron and fluorine during magnesium metasomatism is especially notable since these elements are related to clay formation. However, caution is necessary in the case of boron as it has been evidenced that Mg tourmaline is a late mineral phase and is not related to the main mineralizing process.

ACKNOWLEDGEMENTS

The authors are grateful to Amok Ltd. for providing geochemical data and financial support and to three anonymous reviewers who critically read and improved the manuscript. Thanks are also extended to D.M. Burton who improved the english text, to P. Lagrange who drafted the figures, and to L.L. McNally who patiently typed the many drafts.

REFERENCES

Adams, J.A.S., and Weaver, C.E., 1958, Thorium-to-Uranium Ratios as Indicators of Sedimentary Processes – An Example of Geochemical Facies: Bulletin of the American Association of Petroleum Geologists, v. 42, p. 387.

Audeoud, D., 1983, Les Minéralisations Uranifères et leur Environment à Kamoto, Kambove et Shinkolobwe (Shaba, Zaire); Pétrographie, Géochimie et Inclusions Fluides: Unpublished Thesis, Université de Lyon, France, 212 p.

Barbey, P., and Cuney, M., 1982, K, Rb, Sr, Ba, V, and Th Geochemistry of the Lapland Granulite (Fennoscandia), LILE Fractionation Controlling Factors: Contributions to Mineralogy and Petrology, v. 81, p. 304-316.

Bell, K., 1985, Geochronology of the Carswell Area, Northern Saskatchewan: *in* Lainé, R., Alonso, D., and Svab, M., eds., The Carswell Structure Uranium Deposits, Saskatchewan: Geological Association of Canada Special Paper 29.

Bell, K., Cacciotti, A.D., and Schnessl, J.H., 1985, Petrography and Geochemistry of the Earl River Complex, Carswell Structure, Saskatchewan — A Possible Proterozoic Komatiitic Succession: in Lainé, R., Alonso, D., and Svab, M., eds., The Carswell Structure Uranium Deposits, Saskatchewan: Geological Association of Canada Special Paper 29.

Breaks, F.W., 1982, Uraniferous Granitoid Rocks from the Superior Province of Northwestern Ontario: in Maurice, Y.T., ed., Uranium in Granites: Geological Survey of Canada Paper 81-23, p. 61-69.

Bruneton, P., 1981, Etude Pétrographique et Géochimique des Faciès Rencontrés dans la Structure de Carswell: Bordereau 6674, Amok Internal Report.

Cacciotti, A.D., 1983, The Geochemistry and Petrography of the Peter River Mafic Gneisses, Carswell Structure, Northern Saskatchewan: Unpublished B.Sc. Thesis, Carleton University, Ottawa, 44 p.

Cameron, E.M., and Garrels, R.M., 1980, Geochemical Compositions of some Precambrian Shales from the Canadian Shield: Chemical Geology, v. 28., p. 181-197.

Cuney, M., 1981, Comportement de l'Uranium et du Thorium au Cours du Métamorphisme, le Rôle de l'Anatéxie dans la Génèse des Magmas Riches en Radio-Eléments: Unpublished Thesis, Université de Nancy, France, 511 p.

Currie, K.L., 1969, Geological Notes on the Carswell Circular Structure, Saskatchewan (74K): Geological Survey of Canada Paper 67-32, 60 p.

Dunoyer de Segonzac, G., 1969, Les Minéraux Argileux dans la Diagénèse. Passage au Métamorphisme (Thèse, Université de Strasbourg): Mémoire du Service de la Carte Géologique d'Alsace et de Lorraine, v. 29, 320 p.

Ewers, G.R., and Ferguson, J., 1980, Mineralogy of the Jabiluka, Ranger, Koongarra, and Nabarlek Uranium Deposits: in Ferguson, J., and Goleby A., eds., Uranium in the Pine Creek Geosyncline: International Atomic Energy Agency, Vienna, p. 363-374.

Ey, F., Gauthier-LaFaye, F., Lillié, F., and Weber, F., 1985, A Uranium Unconformity Deposit: The Geological Setting of the D Orebody, (Saskatchewan — Canada): in Lainé, R., Alonso, D., and Svab, M., eds., The Carswell Structure Uranium Deposits, Saskatchewan: Geological Association of Canada Special Paper 29.

Fahrig, W.F., 1961, The Geology of the Athabasca Formation: Geological Survey of Canada Bulletin 68, 41 p.

Fahrig, W.F., and Eade, K.E., 1968, The Chemical Evolution of the Canadian Shield: Canadian Journal of Earth Sciences, v. 5, p. 1247-1252.

Harper, C.T., 1982, Geology of the Carswell Structure, Central Part (Parts of NTS Areas 74K-5, -6, -11, -12): Saskatchewan Mineral Resources Report 214, 6 p.

———, 1983, The Geology and Uranium Deposits of the Central Part of the Carswell Structure, Northern Saskatchewan, Canada: Unpublished Ph.D. Thesis, Colorado School of Mines, Golden, Colorado, 337 p.

Herring, B.G., 1976, The Metamorphism and Alteration of the Basement Rocks in the Carswell Circular Structure, Saskatchewan: Unpublished M.Sc. Thesis, University of British Columbia, Vancouver, 133 p.

Hoeve, J., Sibbald, T.I.I., Ramaekers, P., and Lewry, J.F., 1980, Athabasca Basin Unconformity-Type Uranium Deposits: A Special Class of Sandstone Type Deposits: in Ferguson, J., and Goleby, A., eds., Uranium in the Pine Creek Geosyncline: International Atomic Energy Agency, Vienna, p. 575-594.

Jarousse, J., Moine, B., and Sauvan, P., 1978, Contribution à l'Etude Géochimique des Séries Evaporitiques: Comptes Rendus de l'Académie des Sciences, Paris, t.286, Série D, p. 1057-1060.

Kish, L., and Cuney, M., 1982, Uraninite-Albite Veins from the Mistamisk Valley of the Labrador Trough, Quebec: Mineralogical Magazine, v. 44, p. 471-483.

Lambert, J.B., and Heier, K.S., 1968, Geochemical Investigations of Deep-Seated Rocks in the Australian Shield: Lithos 1, p. 30-53.

La Roche, H. de, 1968, Comportement Géochimique Différentiel de Na, K, et Al dans les Formations Volcaniques et Sédimentaires: Un Guide pour l'Etude des Formations Métamorphiques et Plutoniques: Comptes Rendus de l'Académie des Sciences t. 267, Série D, p. 39-42.

———, 1980, Granite Chemistry Through Multicationic Diagrams: Sciences de la Terre, Nancy, Série "Informatique Géologique", No. 13: Proceedings of GEPIC Meeting of 26-27 April, 1979, Nancy-Vandoevre (France) — IGCP Project 154, p. 65-88.

Leroy, J., 1978, The Margnac and Fanay Uranium Deposits of the La Crouzille District (Western Massif Central, France), Geology, Mineralogy, and Fluid Inclusion Studies: Economic Geology, v. 73, p. 1611-1634.

Leroy, J., and Cathelineau, M., 1982, Les minéraux Phylliteux dans les Gisements Hydrothermaux d'Uranium. I. Cristallochimie des Micas Hérités et Néoformés: Bulletin de Minéralogie, t. 105, p. 99-109.

Macdonald, C.C., 1980, Mineralogy and Geochemistry of a Precambrian Regolith in the Athabasca Basin: Unpublished M.Sc. Thesis, University of Saskatchewan, Saskatoon, 151 p.

Moine, B., 1974, Caractères de Sédimentation et de Métamorphisme des Séries Précambriennes Epizonales à Catazonales du Centre de Madagascar (Région d'Ambatofinandrahana). Approche Structurale, Pétrographique et Spécialement Géochimique: Sciences de la Terre, Nancy, Mémoire 34, 293 p.

Nguyen, C.T., amd Poty, B., 1977, Solubilité de UO_2 en Milieu Aqueux à 500° C et 1 Kilobar: 5é Réunion Annuelle des Sciences de la Terre, Rennes, p. 352.

Pagel, M., 1977, Microthermometry and Chemical Analysis of Fluid Inclusions from the Rabbit Lake Uranium Deposit, Saskatchewan, Canada: Institute of Mining and Metallurgy Transactions, Section B, v. 86 p., B157-B158.

Pagel, M., and Jaffrezic, H., 1977, Analyses Chimiques des Saumures des Inclusions du Quartz et de la Dolomite de Gisement d'Uranium de Rabbit Lake (Canada). Aspect Méthodologique et Importance Génètique: Comptes Rendus de l'Académie des Sciences, Paris, t.284, Série D, p. 113.

Pagel, M. and Ruhlmann, F., 1985, Chemistry of Uranium Minerals in Deposits and Showings of the Carswell Structure (Saskatchewan, Canada): in Lainé, R., Alonso, D., and Svab, M., eds., The Carswell Structure Uranium Deposits, Saskatchewan: Geological Association of Canada Special Paper 29.

Pettijohn, F.J., 1963, Chemical Compositions of Sandstones Excluding Carbonate Volcanic Sands: in Fleisher, ed., Data of Geochemistry: United States Geological Survey Professional Paper 440s.

Rimsaite, J., 1979, Petrology of the Basement Rocks at the Rabbit Lake Deposit and Progressive Alteration of Pitchblende in Saskatchewan: in Current Research, Part B: Geological Survey of Canada Paper 79-1B, p. 281-299.

Rogers, J.J.W. and Adams, J.A.S., 1969, Geochemistry of Uranium: in Wedepohl, K.H., ed., Handbook of Geochemistry: New York, Springer-Verlag, v. 2, ch. V.

Rosenberg, P.E. and Foit, F.F., 1975, Alkali-Free Tourmalines in the System $MgO-Al_2O_3-SiO_2-B_2O_3-H_2O$ (Abstract): Geological Society of America Abstracts and Programs 7, p. 1250.

Ruhlmann, F., 1985, Mineralogy and Metallogeny of Uraniferous Occurrences in the Carswell Structure: in Lainé, R., Alonso, D., and Svab, M., eds., The Carswell Structure Uranium Deposits, Saskatchewan: Geological Association of Canada Special Paper 29.

Schnessl, J., 1983, A Petrographic and Geochemical Study of the Leptynites in the Peter River Area, Carswell Structure: Unpublished B.Sc. Thesis, Carleton University, Ottawa, 40 p.

Shaw, D.M., Reilly, G.A., Muysson, J.R., Pattenden, G.E., and Campbell, F..A., 1967, An Estimate of the Chemical Composition of the Canadian Precambrian Shield: Canadian Journal of Earth Sciences, v. 4, p. 829-853.

Spry, A., 1969, Metamorphic Textures: Oxford, Pergamon Press Ltd., 271 p.

Taylor, S.R., 1964, Abundance of Chemical Elements in the Continental Crust: A New Table: Geochimica et Cosmochimica Acta, v. 28, p. 1273-1285.

Tona, F., Alonso, D., and Svab, M., 1985, Geology and Mineralization in the Carswell Structure – A General Approach: in Lainé, R., Alonso, D., and Svab, M., eds., The Carswell Structure Uranium Deposits, Saskatchewan: Geological Association of Canada Special Paper 29.

Tremblay, L.P., 1972, Geology of the Beaverlodge Mining Area, Saskatchewan: Geological Survey of Canada Memoir 367, 265 p.

_____, 1983, Some Chemical Aspects of the Regolithic and Hydrothermal Alterations Associated with the Uranium Mineralization in the Athabasca Basin, Saskatchewan: in Current Research, Part A: Geological Survey of Canada Paper 83-1A, p. 1-14.

Turekian, K.K., and Wedepohl, K.H., 1961, Distribution of the Elements in some Major Units of the Earth's Crust: Geological Society of America Bulletin, v. 72, no. 2, p. 175-191.

Vine, J.D., and Tourtelot, E.B., 1960, Geochemistry of Black Shale Deposits. A Summary Report: Economic Geology, v. 65, p. 253-272.

Ypma, P.J., and Fujikawa, K., 1980, Fluid Inclusion and Oxygen Isotope Studies of the Nabarlek and Jabiluka Uranium Deposits, Northern Territory, Australia: in Ferguson, J., and Goleby, A., eds., Uranium in the Pine Creek Geosyncline, International Atomic Energy Agency, Vienna, P. 375-395.

PETROGRAPHY AND GEOCHEMISTRY OF THE EARL RIVER COMPLEX, CARSWELL STRUCTURE, SASKATCHEWAN – A POSSIBLE PROTEROZOIC KOMATIITIC SUCCESSION

K. Bell, A.D. Cacciotti and J.H. Schnessl
Ottawa-Carleton Centre for Geoscience Studies, Department of Geology, Carleton University, Ottawa, Ontario K1S 5B6

ABSTRACT

The Earl River complex of the Carswell structure, Saskatchewan, consists of a Proterozoic series of intercalated quartzofeldspathic and mafic gneisses. The quartzofeldspathic pods and lenses, on the basis of both modal mineralogy and chemical composition, correspond to granites (sensu stricto) and have many of the criteria associated with S-type granites. The mafic gneisses (feldspar+pyroxene+biotite+hornblende) have MgO contents that range from 5 to 21 weight per cent. On a Jensen cation plot the mafic gneiss data define a magmatic trend that crosses the komatiite, tholeiite and calc-alkali fields. Although additional variation diagrams are consistent with a komatiitic affinity for the mafic gneisses their REE pattern is quite unlike the flat REE patterns normally associated with komatiitic suites. A similar light-REE enriched pattern for the quartzofeldspathic units, along with high K, Rb, La, and Ce contents for the mafic gneisses, suggest interaction of a volcanic-sedimentary pile with a fluid phase, perhaps during regional metamorphic activity.

RÉSUMÉ

Le complexe de Earl River, dans la zone de Peter River, est composé de séries Protérozoiques de gneiss interstratifiés quartzofeldspathiques et basiques. Les poches et lentilles quartzofeldspathiques, d'après leur composition modale chimique et minéralogique, correspondent à des granites. Leurs pourcentages de quartz-albite-orthoclase normatifs se rapprochent du minimum ternaire et un bon nombre de leurs caractéristiques géochimiques sont semblables à celles associées aux granites de Type-S. Les gneiss basiques, cartographiés par le passé comme amphibolites, renferment des feldspaths, pyroxènes, biotites, et hornblendes. Seule une de ces roches peut être considérée comme une vraie amphibolite. Les gneiss basiques contiennent de 5 à 21% en poids de MgO, et les données des gneiss, sur un diagramme de Jensen, soulignent une tendance magmatique qui recoupe les champs komatiitique, tholèiitique, et calc-alcalin. De fortes teneurs en Ni et Cr pour quelques-uns des gneiss les plus basiques, de même que les tendances sur diagrammes de variation, sont cohérentes avec une affinité komatiitique. Néanmoins, l'enrichissement en Terres-Rares légères des gneiss basiques est différent de l'allure plate généralement liée aux roches de la suite komatiitique. Un enrichissement semblable en Terres-Rares du matériel quartzofeldspathique, associé avec les fortes concentrations en K, Rb, La, et Ce dans les gneiss basiques, sont en harmonie avec un model qui impliquerait une intéraction de la pile volcano-sédimentaire avec une phase fluide, peut-être pendant l'activité métamorphique régionale.

INTRODUCTION

Basement rocks in the Carswell structure have been broadly divided into quartzofeldspathic rocks (including foliated and non-foliated varieties), mafic granulites + amphibolites and pelitic gneisses. Their complicated history includes metamorphism to granulite facies, retrogression to upper amphibolite facies and subsequent alteration by low-temperature hydration (Herring, 1976). Further details of the basement geology are given in the paper by Tona *et al.* (1985).

Near the southwestern margin of the Carswell structure, in the Peter River area, an interesting association of mafic gneisses intercalated with quartzofeldspathic units occurs. Locally the succession can be divided broadly into two units (see Fig. 1) (Tona *et al*, 1985). The lower unit or Earl River complex consists of quartzofeldspathic rocks, amphibolites, and pyroxene gneisses. Immediately above this lies a cataclastically-deformed zone marked by mylonites and ultramylonites wholly contained within a pale green alteration envelope characterized by chloritization, sericitization, and silicification. This zone dips to the southeast and parallels the regional foliation. A series of alumina-rich gneisses and bands of quartzofeldspathic material, the Peter River gneiss, occurs immediately above the cataclastically deformed zone.

The Earl River complex was originally thought to consist of quartzofeldspathic layers intercalated with pelitic gneisses and amphibolites. Thin bands of mafic material commonly occur within the quartzofeldspathic material and the reverse is also true. The contact between the two can be either sharp or diffuse and concentrations of garnet near the margins are common. In general, the units appear to be conformable and no crosscutting relationships are observed. The close association of both mafic and felsic units within the Earl River complex suggests a close genetic relationship between the two and an attempt has been made to describe petrographically the so-called amphibolites and quartzofeldspathic units, in addition to evaluating their petrogenesis on the basis of major and trace element geochemistry. It should be remembered, however, that this study is restricted to only a small part of the heterogeneous Earl River complex.

QUARTZOFELDSPATHIC ROCKS

Cores from eight drill holes from an E-W 600 m section in the Peter River area were examined. The quartzofeldspathic units are massive and medium-grained and in some cases pegmatitic. In general, the grain size increases with depth and the rocks contain less mafic material. At all depths the quartzofeldspathic layers are commonly brecciated and altered.

The mineral assemblage of the quartzofeldspathic rocks consists of: K-feldspar + quartz + plagioclase + garnet ± biotite ± sillimanite.

An average mode consists of 55% potash feldspar, 20% quartz, and 15% plagioclase feldspar with biotite, chlorite, garnet, muscovite, sillimanite, and opaques making up the remainder. The potassium feldspar is commonly microcline microperthite and, in most cases, plagioclase is so altered to a sericite-kaolinite mixture that identification is difficult. Garnet poikiloblasts are common and are invariably rimmed by fine-grained chlorite. Serrated biotite grains show brown to straw-yellow pleochroism. In some samples a subtle lineation is defined by quartz rodding. Most modal analyses of the samples fall in the granite field of the Streckeisen diagram.

Fifteen samples were analyzed for major oxides. Analyses were done by TSL laboratories, Saskatoon, using an inductively-coupled argon plasma method. Comparison of these results with international standards suggests an accuracy of ±1% at the 2 σ level for major elements. Results from the quartzofeldspathic units are given in Table I. Assessment of the geochemical data shows that 13 of 15 samples have similar compositions (two are highly altered). Most rocks have relatively restricted SiO_2 contents that range from 72% to 75% in addition to relatively high K_2O contents that aver-

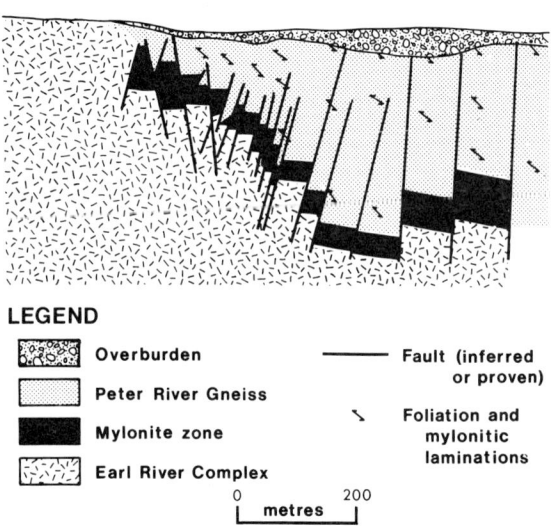

Figure 1. Schematic E-W cross section showing the distribution of the main rock types in the Peter River area, compiled from Amok bore-hole data.

TABLE I

MAJOR ELEMENT ANALYSES FOR QUARTZOFELDSPATHIC ROCKS

Sample #	SiO_2	Al_2O_3	Fe_2O_3*	CaO	MgO	Na_2O	K_2O	TiO_2	MnO	P_2O_5	L.O.I.	Total
JS 1637-571.5	74.23	12.90	1.58	0.38	0.95	2.34	5.72	0.11	0.04	0.23	0.64	99.1
JS 1637-900	72.62	13.93	0.52	0.65	0.28	2.57	7.11	0.07	0.01	0.57	0.19	98.5
JS 1641-321.5	75.17	13.74	0.84	0.46	0.30	3.01	4.25	0.03	0.03	0.31	0.61	98.8
JS 1641-748	73.19	15.03	0.57	0.95	0.72	3.18	5.60	0.05	0.01	0.26	1.21	100.8
JS 1687-464	71.94	13.80	0.48	0.48	0.39	2.62	7.42	0.02	0.01	0.16	0.66	98.0
JS 1687-900.5	67.36	15.20	0.66	0.29	3.73	0.24	6.71	0.04	0.01	0.10	3.46	97.8
JS 1689-409	73.48	12.98	0.70	0.48	0.81	2.59	5.96	0.06	0.01	0.16	1.00	98.2
JS 1689-942	72.78	13.50	0.30	0.37	0.18	2.48	7.86	0.03	0.01	0.10	0.47	98.1
JS 1693-239	73.22	13.00	1.58	0.43	0.92	2.13	6.59	0.10	0.03	0.08	0.88	99.0
JS 1693-1142	64.52	17.58	0.89	0.23	4.38	0.33	6.45	0.07	0.01	0.17	4.06	98.7
JS 1699-197	72.89	13.09	1.91	0.27	0.69	1.73	7.06	0.02	0.03	0.16	0.75	98.6
JS 1699-783	74.37	12.22	0.45	0.62	0.24	2.69	6.49	0.04	0.01	0.12	0.45	97.7
JS 1701-203	74.16	12.63	0.87	1.58	0.50	2.29	6.90	0.06	0.02	0.13	0.48	99.6
JS 1701-862	72.71	13.29	0.95	0.94	0.68	3.02	5.30	0.09	0.02	0.22	0.72	97.9
JS 1703-913.5	72.10	13.41	1.03	0.56	0.42	2.85	6.25	0.10	0.01	0.01	0.52	97.3

*total Fe as Fe_2O_3

age about 6.4%. An interesting feature to emerge from the data is that there is little or no chemical variation with depth. All of the rocks can be classified both geochemically and petrographically as granites. These observations are summarized in Figure 2.

Part of the difficulty of assessing the geochemical information is the complication brought about by the alteration processes associated with uranium mineralization. In spite of this, attempts have been made to evaluate the bulk chemistry. Figure 3 shows a de La Roche plot (1974), that emphasizes the contrast between magmatic and sedimentary source material and reflects the different geochemical behaviour of Al, K, and Na. Sedimentary rocks lie to the right of the origin, emphasizing the decrease in Na during weathering, and igneous rocks lie to the left indicating their high alkali, especially Na, contents. Data from the Carswell quartzofeldspathic rocks in Figure 3 plot in an intermediate field that lies between the fields of rhyolitic material and immature clastic sediments. The geochemical parameters, such as K/Na and Al/(Na+K+Ca/2), are also compatible with the criteria set for S-type granites (Chappell and White, 1974; Hine et al., 1978). Other than two samples (JS-1699-783, and JS-1701-203), these rocks also contain normative corundum, a feature consistent with an origin from a sedimentary source. The compositions of all of the analyzed quartzofeldspathic samples fall fairly close to the minimum melt composition in the quartz-albite-orthoclase system.

MAFIC GNEISSES

The mafic gneisses in the Carswell structure represent only a minor component of the basement, and form units of the order of a few metres thick. Earlier petrographic and geochemical work (Herring, 1976) produced a two-fold division of the mafic gneisses into: (i) amphibolites and (ii) two-pyroxene granulites.

Contacts between the mafic gneisses and the quartzofeldspathic and pelitic units in the Peter River area are complex. The contacts between the pelites and the mafic gneisses are usually gradational, and are commonly marked by a garnet-rich zone, up to 30 cm thick. The contacts between the mafic gneisses and quartzofeldspathic units can be either sharp or diffuse. Feldspathic material commonly occurs in the mafic gneisses at the gradational contacts and both types of contacts are associated with alteration and can be marked by garnet-rich zones. The sharp contacts are sub-parallel to the gneissic foliation and mafic rocks near the contacts generally become coarser in grain size.

Specimens from six of the eight drill holes were petrographically examined. The holes sample a line along the E-W profile, and extend to a depth of 300 m. The mafic gneisses vary greatly in colour, texture, mineralogy and degree of foliation, and can be divided into two groups on the basis of mafic mineral content. Type A rocks are greenish grey in colour with a colour index of less than 30. These grade vertically down or laterally into the Type B mafic gneisses which are dark green or black, massive units with a colour index greater than 60. All rocks are fine- to coarse-grained, although the grain size is seldom consistent throughout any part of the mafic sequence. In general, however, the Type A

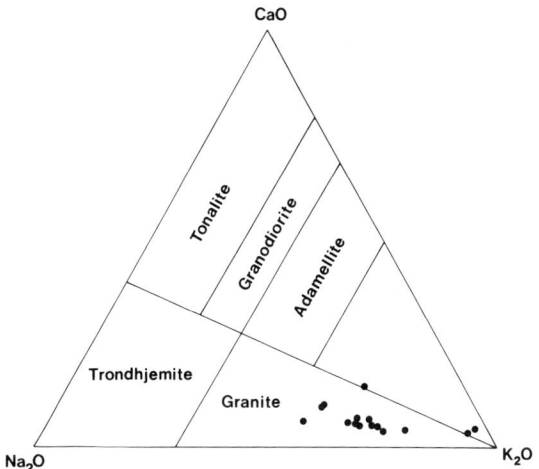

Figure 2. Triangular plot of CaO vs Na_2O vs K_2O (weight per cent) showing the data from the quartzofeldspathic units. The classification boundaries are from Glikson (1979).

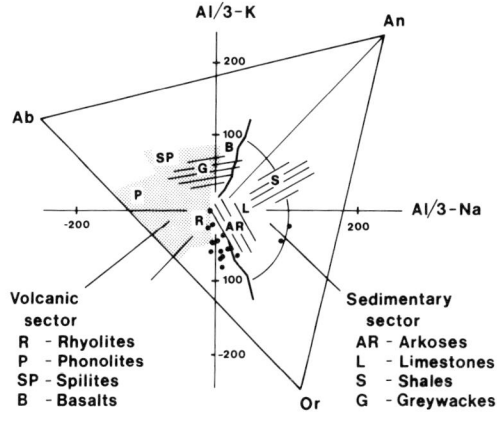

Figure 3. Al, K and Na diagram after de La Roche (1974). $Al/3 - Na$ is plotted on the abscissa and $Al/3 - K$ is plotted on the ordinate. Fields are shown for the main igneous associations on the left and mature sedimentary rocks on the right. Positions of Ab, An, and Or feldspars shown for comparison.

mafic gneisses are fine- to medium-grained, while the more mafic Type B rocks are generally coarser and show much less variation in grain size. In general, the Type B gneisses lie lower in the succession than the more felsic types.

The mineral assemblage in the mafic gneisses consists of: orthopyroxene + clinopyroxene + plagioclase + amphibole + biotite + K-feldspar.

In the more felsic units (Type A), plagioclase (An_{50-60}) makes up between 35 to 60%, potassium feldspar 5 to 20%, orthopyroxene and clinopyroxene 15 to 40%, and biotite 5 to 20% of the rock. Type B gneisses consist of 5 to 20% plagioclase (An_{50-60}), 25 to 60% orthopyroxene and clinopyroxene, 2 to 10% potassium feldspar, 5 to 75%

hornblende, and 5 to 35% biotite. Accessory minerals in both types include opaques, apatite, chlorite, carbonate, and sphene.

Pyroxenes in the Type A gneisses form small, lobate grains and larger anhedral aggregates (0.3 to 1.3 mm in size) interstitial to the relatively unaltered feldspar. Type B gneisses are characterized by large pyroxenes (1 to 10 mm) with subordinate altered feldspars. Pyroxenes in some of the Type B rocks enclose both hornblende and potassium feldspars in addition to apatite and biotite. All orthopyroxenes are variably altered to a green-brown alteration product along fractures and cleavage planes. Clinopyroxenes show exsolution lamellae and/or inclusions of opaque minerals.

Hornblende in the Type B gneisses is commonly green in colour and takes two forms. In the pyroxene-rich samples, with up to 60% pyroxene, hornblende occurs as small, rounded grains contained within feldspars and pyroxenes. In other Type B samples, hornblende occurs as large prisms up to 4 mm in size, enclosing pyroxene and feldspar. In the Type A gneisses, green-brown hornblende always occurs as small, rounded grains within pyroxene and feldspar.

Biotite, conspicuously fresh, takes two forms. It occurs as small (0.1 mm) flakes either interstitial to, or included within, pyroxene and feldspar; it is commonly kinked and defines a weak foliation. In the more mafic-rich gneisses, biotite forms small flakes and large porphyroblasts (up to 4 mm) overgrowing pyroxene and amphibole.

Probe data from three of the samples are shown in Table II. The analyses were obtained on a Cambridge Mark V electron probe at Carleton University using natural mineral standards. Clinopyroxenes are augite and diopside; the orthopyroxenes are hypersthene and bronzite. Temperature estimates using the method of Wood and Banno (1973) for co-existing orthopyroxene-clinopyroxene pairs indicate values of about 835°C, consistent with those obtained by Herring (1976). The method of Wells (1977) provided similar results.

The degree of alteration of the rocks varies from low to intense. Some samples are so altered that identification of some of the primary mafic minerals is virtually impossible.

Only one of the rocks examined from the Peter River area is an amphibolite. On the basis of the colour indices and overall mineralogy, these rocks are best described as mafic gneisses.

GEOCHEMISTRY

Chemical analyses for major elements from fourteen samples are given in Table III. The most interesting feature to emerge from the data is the fact that the Earl River mafic gneisses are rich in MgO. Eleven of the fourteen samples contain greater than 10 weight per cent MgO and, of these,

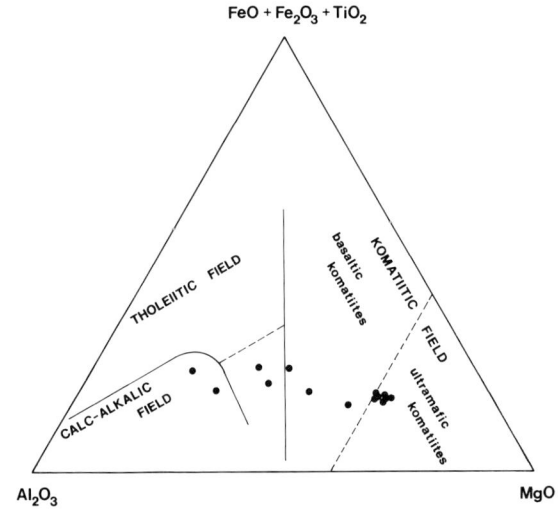

Figure 4. Jensen cation plot. Two fields are outlined for komatiites: ultramafic komatiites and basaltic komatiites. Tholeiitic and calc-alkalic fields are shown for comparison.

TABLE II

PROBE ANALYSES OF PYROXENES FROM MAFIC GNEISSES

	1	2	3	4	5	6	7	8	9	10	11
SiO_2	50.30	49.91	50.36	50.39	52.05	51.44	52.39	53.53	54.11	53.93	53.69
Al_2O_3	0.92	0.88	2.09	2.09	1.75	1.95	1.74	1.74	1.71	1.70	1.70
FeO*	28.46	28.37	13.08	12.17	5.82	6.39	16.73	6.31	6.63	5.39	2.88
MnO	0.62	0.64	0.31	0.31	0.25	0.23	0.51	0.01	0.00	0.03	0.06
MgO	18.17	18.20	12.37	12.22	14.60	15.55	26.88	16.25	16.65	16.55	16.20
CaO	0.91	0.95	22.18	21.50	23.62	23.28	0.69	22.48	22.56	22.64	22.35
TiO_2	0.12	0.15	0.30	0.27	0.06	0.06	0.02	0.29	0.28	0.29	0.27
Total	99.50	99.10	100.69	98.95	98.15	98.90	98.96	100.61	101.94	100.53	97.15

*total Fe as FeO

1-2	FC 1693-211	Hypersthene	2 spots same grain
3-4	FC 1693-211	Augite	2 separate grains
5-6	FC 1637-788	Diopside	2 separate grains
7	FC 1637-788	Bronzite	
8-11	FC 1699-814	Diopside	2 separate grains. Grain A (rim 8, core 9). Grain B (rim 10, core 11).

eight contain greater than 18 weight per cent MgO. High magnesian values such as these are unusual and may suggest an affinity with such rocks as komatiites, boninites, tholeiitic picrites or olivine-rich cumulates.

Brooks and Hart (1974) suggested that a komatiitic suite has $SiO_2<53$ weight per cent, MgO>9 weight per cent, $K_2O<0.9$ weight per cent and $CaO/Al_2O_3>1$. A recent proposal (Arndt and Nisbet, 1982) sets an arbitrary lower limit for komatiites of 18 weight per cent MgO on an anhydrous basis. Beswick (1982) also indicates that within komatiitic suites there is a wide range of MgO values from as low as 10 weight per cent to as high as 30 weight per cent and that the CaO/Al_2O_3 ratios can be 0.8 or less. Although several of the Earl River mafic gneisses have MgO contents greater than 18 weight per cent, the CaO/Al_2O_3 ratios for the group as a whole are fairly low and fall between 0.15 and 1.35 (average = 0.56). Similar CaO/Al_2O_3 ratios, however, have been recorded from komatiites and related rocks from southeastern Brazil (Jahn and Schrank, 1983).

A Jensen cation plot (Jensen, 1976), shown in Figure 4, divides subalkaline volcanic rocks on the basis of cation percentages of Al_2O_3, MgO, and $Fe_2O_3+FeO+TiO_2$. This plot is helpful in recognizing rocks of komatiitic affinity as well as differentiation trends related to komatiitic, tholeiitic, and calc-alkaline suites. Most of the Earl River gneisses plot in the komatiite fields. Seven of the Type B mafic gneisses plot within the ultramafic komatiite field and contain greater than 18 weight per cent MgO. The remaining samples plot within the basaltic komatiite, Mg-rich tholeiite and calc-alkaline fields and overall appear to define a trend, similar to that outlined for komatiites (e.g., Jensen, 1976; Jensen and Pyke, 1982). Two additional variation diagrams also indicate komatiitic affinities for the Earl River mafic gneisses. Figure 5, in which Al_2O_3 is plotted against $FeO/FeO+MgO$ (total Fe as FeO), separates komatiites from tholeiites (Arndt et al., 1977). Again the Earl River mafic gneisses fall well within the komatiitic field. The eight samples that cluster in the ultramafic komatiite field also plot in or near the ultramafic komatiite field of the Jensen plot. The remaining six samples lie in the pyroxenitic or basaltic komatiite fields. Figure 6 is a plot of weight per cent TiO_2 against MgO. The fields shown are based on chemical studies of Archean spinifex-textured peridotites and high-Mg and low-Mg tholeiites (Nesbitt and Sun, 1976). Figure 6 shows a general decrease in TiO_2 content with increase in MgO, and the eight Type B mafic gneisses lie within or near the komatiite field.

Although, on the basis of Fe (total), Mg, Al, and Ti contents, these rocks appear to have komatiitic affinities, some of the oxide abundances given in Table III are inconsistent with such an interpretation. In particular, the K_2O contents of 0.91 to 4.97% (average 2.7%) are much too high to give much credence to the interpretation that the mafic gneisses are part of a pristine high-MgO sequence. The trend shown in Figure 4, however, is so similar to many documented for

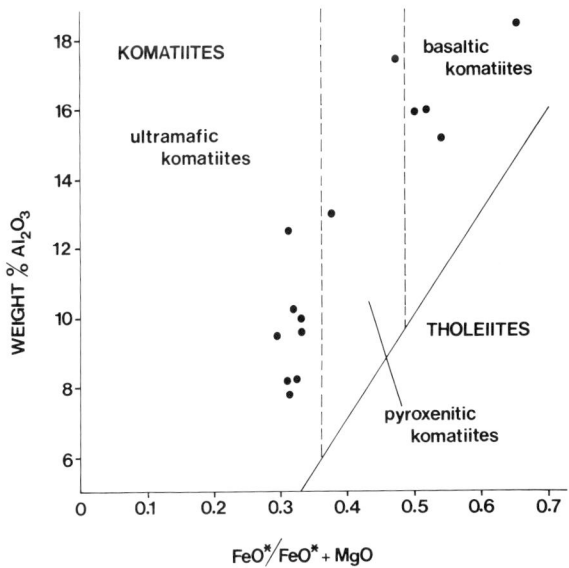

Figure 5. Plot of Al_2O_3 vs $FeO/FeO + MgO$ (weight per cent). FeO is total Fe. Komatiite fields based on information given by Arndt et al. (1977).

TABLE III

MAJOR ELEMENT ANALYSES FOR EARL RIVER MAFIC GNEISSES

Sample #	SiO_2	Al_2O_3	Fe_2O_3*	CaO	MgO	Na_2O	K_2O	TiO_2	MnO	P_2O_5	L.O.I.	Total
FC 1637-788	48.12	8.23	9.95	8.83	18.50	0.33	3.18	0.55	0.19	0.13	1.01	99.0
FC 1641-832	48.36	8.19	9.43	11.04	18.73	0.77	1.41	0.47	0.17	0.29	0.85	99.7
FC 1689-856.5	38.30	15.92	14.66	3.89	13.03	0.65	3.54	1.14	0.12	0.23	7.79	99.3
FC 1693-221	53.36	18.42	10.45	6.42	5.00	1.93	1.84	1.12	0.11	0.16	1.19	100.0
FC 1693-262	43.80	17.42	11.51	2.56	11.48	0.32	4.97	1.14	0.13	0.08	4.56	98.0
FC 1699-689	50.26	15.17	11.72	3.56	8.92	1.20	2.15	1.28	0.08	0.21	4.06	98.6
FC 1699-727.5	47.43	9.44	9.73	6.40	20.82	0.43	2.86	0.57	0.15	0.43	1.09	99.4
FC 1699-814	49.69	9.60	10.55	7.37	18.95	0.57	1.01	0.58	0.17	0.29	0.91	99.7
FC 1699-835	44.99	10.00	11.38	4.26	20.53	0.25	2.46	0.63	0.14	0.00	2.44	97.1
FC 1699-862	48.14	7.77	9.73	8.81	18.95	0.75	0.91	0.48	0.17	0.20	3.22	99.1
FC 1699-937	43.54	12.49	9.27	4.04	18.31	0.36	4.25	0.68	0.14	0.43	4.37	97.9
FC 1699-966	43.94	10.20	10.86	3.79	20.53	0.59	1.16	0.60	0.15	0.40	5.61	97.8
FC 1701-292	47.12	12.99	8.73	6.15	12.88	0.22	4.01	1.40	0.25	0.30	4.34	98.5
FC 1701-585	56.60	15.97	7.56	4.46	6.23	2.68	4.06	0.93	0.13	0.16	0.83	99.6

*total Fe as Fe_2O_3.

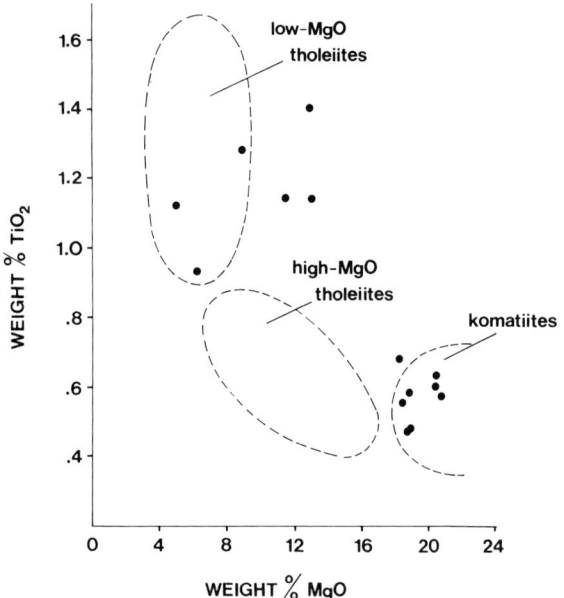

Figure 6. Plot of TiO₂ vs MgO (weight per cent). Fields based on data given in Nesbitt and Sun (1976).

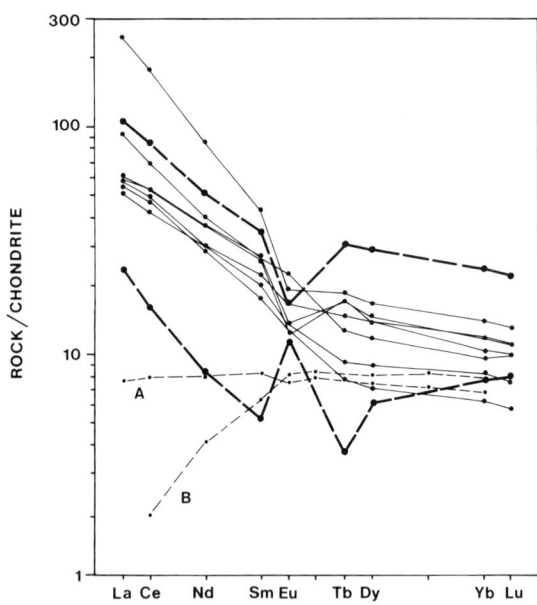

Figure 7. Plot of chondrite-normalized REE data for seven mafic gneisses. The REE data for the quartzofeldspathic rocks are shown by the heavy dashed lines. Line A corresponds to data from a typical Archean komatiite (Nesbitt et al., 1979) and line B to data from a rock of komatiitic affinity from Gorgona Island (Echeverría, 1980).

TABLE IV

TRACE ELEMENT DATA

Element (ppm)	FC-1641 −832	FC-1693 −221	FC-1699 −689	FC-1699 −727.5	FC-1699 −937	FC-1699 −966	FC-1701 −292	JS-1637* −571.5	JS-1641* −321.5
U	0.6	1.3	1.2	1.0	1.3	0.7	1.8	3.6	2.6
Th	0.6	2.7	2.5	2.3	7.1	1.5	6.7	16	0.4
Sc	25	27	42	32	35	38	45	11	2.3
Cr	2500	130	310	3200	2300	2800	850	200	200
Co	64	31	41	66	54	77	43	3.2	2.3
Ni	1500	50	350	1400	1300	1200	700	50	50
Zn	50	220	220	50	200	50	240	50	50
As	39	16	16	6	4	4	5	8	4
Se	1.1	0.5	0.9	1.7	0.6	0.6	0.5	0.9	0.5
Br	13	4.4	13	7.9	14	18	20	3.0	3.5
Mo	2	2	2	2	2	2	3	4	3
Sb	0.2	0.1	0.1	0.1	0.2	0.1	0.1	0.1	0.1
Cs	1.3	2.0	2.5	8.2	7.7	2.8	8.3	0.9	0.2
Ba	400	1000	730	310	900	800	1100	870	510
La	17.3	29.2	16.0	18.1	19.0	18.6	77.0	33.3	7.5
Hf	2.1	5.4	6.9	2.5	4.2	2.3	6.1	1.5	1.6
Ta	0.5	0.5	0.5	0.5	1.8	0.5	3.1	0.5	0.5
W	1	2	1	3	1	2	88	2	1
Ce	38	56	34	40	43	43	144	68	13
Nd	17	24	18	18	22	22	51	30	5
Sm	3.4	5.0	4.3	3.9	5.2	5.1	8.2	6.6	1.0
Eu	0.92	1.63	1.19	0.97	0.99	0.88	1.39	1.20	0.84
Tb	0.38	0.62	0.72	0.45	0.83	0.85	0.91	1.49	0.18
Dy	2.3	3.8	4.5	2.9	4.7	4.5	5.4	9.4	2.0
Yb	1.30	2.00	2.47	1.72	2.44	2.17	2.91	4.98	1.61
Lu	0.186	0.317	0.355	0.244	0.353	0.321	0.422	0.714	0.259
Sr	100	400	200	100	100	100	100	100	100
Rb	30	100	120	170	320	140	230	220	200

Uncertainties for most elements are between 2-10% of the quoted values at the 2σ level.

*Quartzofeldspathic material

komatiitic sequences from South Africa and elsewhere (Jensen, 1976; Jensen and Pyke, 1982; Nisbet *et al*., 1982 Viljoen *et al*., 1982) that these high-MgO rocks could represent a highly metamorphosed komatiitic succession.

Because of the incompatibility of high K_2O contents and the high MgO contents, it was hoped that trace element, including rare earth element (REE), data would help decipher at least part of the story. Rare earth elements are particularly useful in evaluating the sources of melts and magma modification by such processes as assimilation, differentiation, metasomatism, and fluid activity (Hanson, 1980). Most komatiites have characteristic flat REE patterns, not too dissimilar from MORB's, and values that are only slightly enriched (3 to 5 times) above chondritic values. Variations from these flat patterns include light-REE depletion or even significant heavy-REE depletion. The REE and trace element data from seven of the Earl River mafic gneisses and two of the quartzofeldspathic units are given in Table IV. The REE data are shown in Figure 7 on a chondrite normalized diagram. Analyses were done by neutron activation by Nuclear Activation Services Ltd. The rare earth patterns show an enrichment (20 to 300 times) in the light-REE relative to chondrites and are quite different to those of known komatiites. Also shown in Figure 7, for comparison, are the REE data from an Archean komatiite (line A; Nesbitt *et al.*, 1979) and a komatiite of Mesozoic or younger age (line B; Echeverría, 1980) from Gorgona Island that shows light-REE depletion. On the same diagram the REE data obtained from the intercalated quartzofeldspathic material are given. The REE patterns shown in Figure 7 are similar to patterns observed in basalts from oceanic islands (e.g., Zielinski, 1975) and continental granitoid rocks (e.g., Barker *et al.*, 1979).

Attempts have been made to compare geochemical ratios from komatiites with those of chondrites (see Ludden and Gélinas, 1982). Al_2O_3/TiO_2 ratios, involving two of the more "immobile" elements, from 13 of the Earl River mafic gneiss samples, average 16, a figure that is somewhat lower than the chondritic value of 20. The Ce and La abundances in the mafic gneisses are as much as ten times higher than those found in most komatiites. The normal range in komatiites for La is 0 to 6 ppm, and for Ce, 2 to 15 ppm. The Earl River mafic gneisses have La abundances between 16 to 77 ppm and Ce abundances between 34 to 144 ppm. The Ce data from two granitoid samples are 13 ppm and 68 ppm; the La data, 7.5 ppm and 33 ppm. A plot of Ce vs La for the mafic gneisses is shown in Figure 8; a chondritic trend (Ludden and Gélinas, 1982) is shown for comparison. The chondritic Ce/La ratio of approximately 2.5 is somewhat higher than the average ratio of 2.1 for the Earl River mafic gneisses.

Of particular interest are the Co, Cr and Ni abundances. For the mafic gneiss samples with high MgO contents (greater than 18 weight per cent) the Cr and Ni contents are similar to levels found in komatiites. The ranges found in the high-MgO gneisses are: Cr, 2300 to 3200 ppm and Ni, 1200 to 1500 ppm. Even for those samples with less than 18 weight per cent MgO, the Ni and Cr abundances overlap with those found in komatiitic suites. Figures 9 to 11 show plots of Ni versus MgO, Cr versus MgO and Co versus MgO. In all three cases there is a positive correlation, a feature that is also found in komatiitic suites (Beswick, 1982).

Plots of Sc vs MgO, Yb vs MgO, and Ce vs MgO show a scatter of points that suggests that some of these elements may have been fairly mobile. The high Rb content of the mafic gneisses is also consistent with enrichment in K.

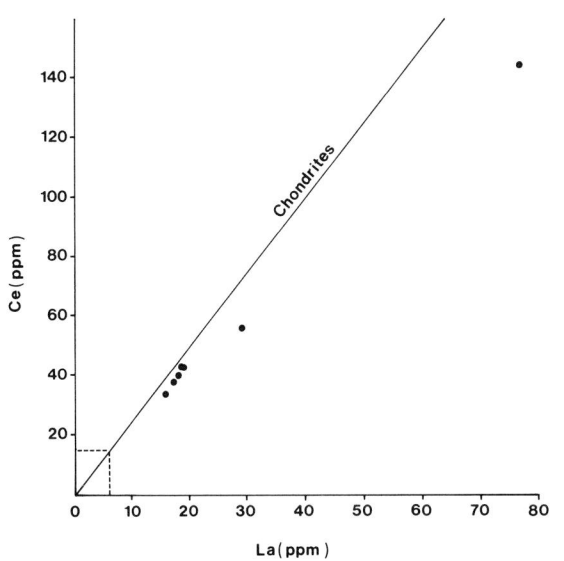

Figure 8. Plot of Ce vs La (ppm). The line shows the chondritic trend (see Ludden and Gélinas, 1982). Most komatiites have values that fall within the box shown near the origin.

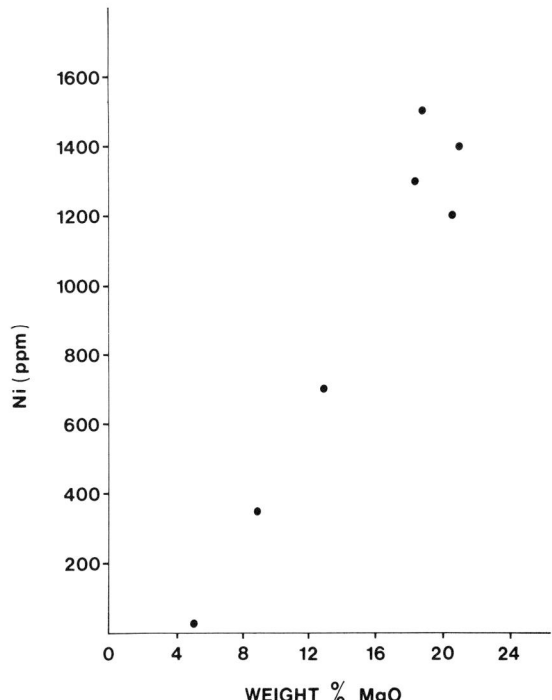

Figure 9. Plot of Ni (ppm) vs MgO (weight per cent).

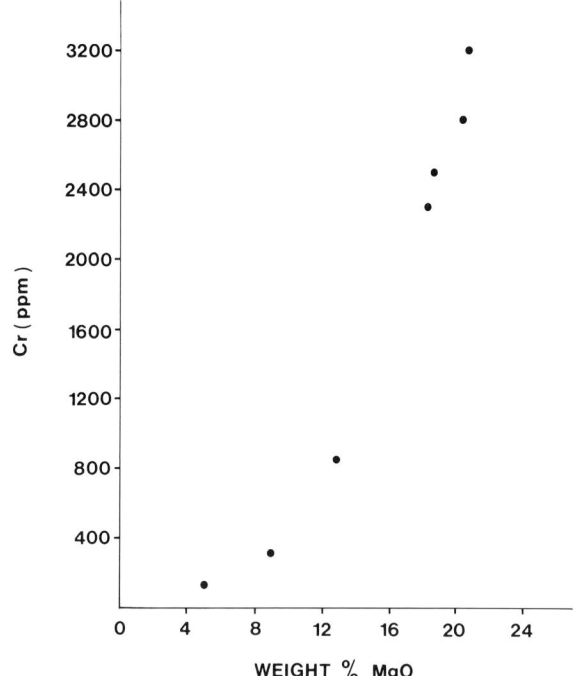

Figure 10. Plot of Cr (ppm) vs MgO (weight per cent).

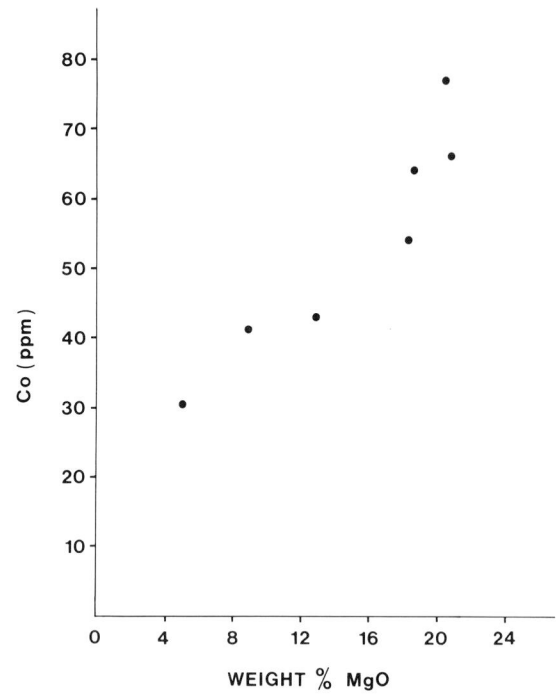

Figure 11. Plot of Co (ppm) vs MgO (weight per cent).

INTERPRETATION OF THE DATA

Any model for the genesis of the quartzofeldspathic material and mafic gneisses from the Peter River area must explain the following features: (i) the intimate association of both mafic and quartzofeldspathic material, (ii) the granitic nature of the quartzofeldspathic units, (iii) the MgO-rich nature of the mafic gneisses, (iv) the high K, Rb, Ce, and La contents of the mafic rocks, and (v) the light-REE enrichment in both quartzofeldspathic and mafic units. The question surrounding the origin of the Peter River area association centres on how much of the geochemical information reflects original composition and how much reflects open-system behaviour during metamorphism and/or alteration. The high K, Rb, Ce and La concentrations contrast with the high MgO contents of the mafic gneisses and seemingly rule out any model that claims that these rocks reflect original compositions. Few rocks, as far as we are aware, with similar SiO_2 contents have such a chemistry.

There are three possible models to explain the overall geochemical trends observed in the mafic gneisses: (i) metamorphism of a layered igneous intrusion, (ii) metamorphic differentiation, and (iii) metamorphism of a high-Mg (komatiitic?) succession, coupled with metasomatism.

The possibility of an extremely fractionated sequence with olivine and pyroxene accumulation at the base of a tholeiitic pile could explain some of the geochemical features observed in the Earl River mafic gneisses. This model, however, can be dismissed on the grounds that the mafic gneisses are intercalated with both quartzofeldspathic and metapelitic units, suggesting an originally heterogeneous succession. There is also little evidence to suggest that these units are tectonic slivers.

An alternative explanation involves metamorphic differentiation, a process that could create heterogeneous layers and pods from initially homogeneous units. This normally involves cation exchange reactions in the presence of intergranular fluids (e.g., Vidale, 1974). Orville (1969) argues that for amphibolites, a trend similar to that seen in igneous rocks is not incontrovertible proof that the rocks are themselves of igneous origin. Such trends can be mimicked from any mixture of plagioclase and hornblende because a tie-line between these two minerals on an ACF diagram passes through the field of basalt compositions. Generation of thin-layered amphibolites in Orville's model involves migration of H_2O, CO_2, and K_2O as well as the reciprocal exchange of Ca with Mg and Fe in carbonate-rich pelites. In spite of the fact that the basement succession has been affected by high-grade regional metamorphic activity, it seems unlikely that metamorphic differentiation alone can produce the chemistry, the required thicknesses and the spatial distribution of the mafic units within the Earl River complex. The restriction of most of the high-Mg (Type-B) gneisses to the lower parts of the succession argues strongly against such a model.

We feel that the simplest interpretation of the geochemical, petrographic, and field evidence is one that involves high-grade metamorphism of a volcanic succession, coupled with regional metasomatism and perhaps anatexis. The most interesting feature to emerge from the geochemistry of the Earl River succession is the contrast between the uniform major element chemistry and petrography of the quartzofeldspathic material and the systematic chemical variation (particularly Fe, Mg and Al) of the mafic gneisses. High-grade metamorphism and metasomatism of an original bimodal sequence of mafic and felsic volcanics perhaps associated with volcaniclastic material could, under appropriate P-T conditions, have produced compositions similar to those of the quartzofeldspathic and metasomatized mafic units.

In regions of transition from amphibolite to granulite facies metamorphism, incipient granulite facies metamorphism can be associated with partial melting and metasomatism. Several authors have emphasized a close association between charnockite and migmatite development. Janardhan *et al.* (1982) documented anatexis and K-metasomatism prior to granulite facies metamorphism, while Sen (1974) interpreted a large spread in Mg/Fe ratios with variable alkalis as evidence for migmatization of charnockites from Madras. Weaver (1980) came to a similar conclusion on the basis of variable K, Rb and REE abundances and the lack of coherence of these elements with such ratios as Fe/Mg. Extensive metasomatism and palingenesis of the country rocks were proposed either prior to or during granulite metamorphism.

The systematic trends seen in many of the variation diagrams for data from the Earl River mafic gneisses, coupled with the unusual K, Rb, Ce, and La concentrations for rocks with such high MgO contents, are consistent with a regional metasomatism but one that involved mainly the LIL elements. The similarity in light-REE enrichment in both the mafic and quartzofeldspathic units suggests pervasive fluid activity, consistent with enrichment in alkalis proposed by Herring (1976) for the Carswell basement and by Burwash *et al.* (1973) for other parts of the Churchill Province.

CONCLUSIONS

Although the mafic gneisses have had a complicated history that involved both prograde and retrograde metamorphism and later alteration, their Mg, Al, and Fe contents still record a differentiation trend similar to that shown by many komatiitic sequences. The de-coupling of the LIL elements from Fe, Mg, and Al, and the similar light-REE enriched patterns in both mafic and quartzofeldspathic units is attributed to regional metasomatism, perhaps related to high-grade metamorphic activity sometime during the Hudsonian orogeny. Partial melting of either felsic volcanic or sedimentary units intercalated with the precursors to the mafic gneisses, most easily explains the uniform chemical and mineralogical composition of the quartzofeldspathic units along with their possession of many of the criteria associated with S-type granites.

Aphebian Mg-rich extrusives and related subvolcanic rocks are not unknown in Saskatchewan. A semi-continuous belt of high-Mg extrusives has been delineated over a distance of 150 km in the central La Ronge domain, approximately 500 km southeast of the Carswell structure (Fox and Johnston, 1981). Confined mainly to the basal section of the La Ronge/Wasekwan metavolcanic sequences, the volcanic rocks were extruded onto sialic crust. Similar rocks can be traced further to the east at Lynn Lake, Manitoba. A more recent finding by Sibbald *et al.* (1983) documents ultramafic rocks of possible komatiitic affinity in the Nicholson Bay area, east of the Beaverlodge area and only 100 km from the Peter River area. Although these rocks have low CaO/Al_2O_3 ratios, their Co, Cr, Mg, and Ni abundances are somewhat higher than those of the Earl River complex and their K_2O contents extremely low. These differences, however, could be attributed to variation in alteration and perhaps to metamorphic grade. The existence of such rocks within the Churchill Province adds additional support to our model.

ACKNOWLEDGEMENTS

We would like to thank Amok Ltd. for their generous support throughout the course of this work. This study was initially suggested by F. Tona and later discussions with D. Alonso, E. Koning, R.T. Lainé and M. Svab added greatly to our understanding of the geology of the Peter River area. We thank P.C. Jones for his help with the probe analyses, and J.W. Card's significant contribution to the final version of this paper is gratefully acknowledged.

REFERENCES

Arndt, N.T., and Nisbet, E.G., 1982, What is a Komatiite?: *in* Arndt, N.T., and Nisbet, E.G., eds., Komatiites: London, George Allen and Unwin Ltd., p. 19-27.

Arndt, N.T., Naldrett, A.J., and Pyke, D.R., 1977, Komatiitic and Iron-Rich Tholeiitic Lavas of Munro Township, Northeast Ontario: Journal of Petrology, v. 18, p. 319-369.

Barker, F., Arth, J.G., and Millard, H.T., Jr., 1979, Archean Trondhjemites of the Southwestern Big Horn Mountains, Wyoming: A Preliminary Report: *in* Barker, F., ed., Trondhjemites, Dacites, and Related Rocks (Developments in Petrology 6): Amsterdam, Elsevier Scientific Publishing Co., p. 401-414.

Beswick, A.E., 1982, Some Geochemical Aspects of Alteration, and Genetic Relations in Komatiitic Suites: *in* Arndt, N.T., and Nisbet, E.G, eds., Komatiites: London, George Allen and Unwin Ltd., p. 283-308.

Brooks, C., and Hart, S.R., 1974, On the Significance of Komatiite: Geology, v. 2, p. 107-110.

Burwash, R.A., Krupivcka, J., and Culbert, R.R., 1973, Cratonic Reactivation in the Precambrian Basement of Western Canada, III, Crustal Evolution: Canadian Journal of Earth Sciences, v. 10, p. 283-291.

Chappell, B.W., and White, A.J.R., 1974, Two Contrasting Granite Types: Pacific Geology, v. 8, p. 173-174.

Echeverría, L.M., 1980, Tertiary or Mesozoic Komatiites from Gorgona Island, Colombia: Field Relations and Geochemistry: Contributions to Mineralogy and Petrology, v. 73, p. 253-266.

Fox, J.S., and Johnston, W.G.Q., 1981, Komatiites, "Boninites" and Tholeiitic Picrites in the Central La Ronge Metavolcanic Belt, Saskatchewan and Manitoba, and their Possible Economic Significance: Canadian Mining and Metallurgical Bulletin, v. 74, no. 831, p. 73-82.

Glikson, A.Y., 1979, Early Precambrian Tonalite-Trondhjemite Sialic Nucleii: Earth-Science Reviews, v. 15, p. 1-73.

Hanson, G.N., 1980, Rare Earth Elements in Petrogenetic Studies of Igneous Systems: Annual Review of Earth and Planetary Sciences, v. 8, p. 371-406.

Herring, B.G., 1976, The Metamorphism and Alteration of the Basement Rocks in the Carswell Circular Structure, Saskatchewan: Unpublished M.Sc. Thesis, University of British Columbia, Vancouver, 134 p.

Hine, R., Williams, I.S., Chappell, B.W., and White, A.J.R., 1978, Contrasts Between I- and S-type Granitoids of the Kosciusko Batholith: Journal of the Geological Society of Australia, v. 25, no. 4, p. 219-234.

Jahn, B.M., and Schrank, A., 1983, REE Geochemistry of Komatiites and Associated Rocks from Piumhi, Southeastern Brazil: Precambrian Research, v. 21, p. 1-20.

Janardhan, A.S., Newton, R.C., and Hansen, E.C., 1982, The Transformation of Amphibolite Facies Gneiss to Charnockite in Southern Karnataka and Northern Tamil Nadu, India: Contributions to Mineralogy and Petrology, v. 79, p. 130-149.

Jensen, L.S., 1976, A New Cation Plot for Classifying Subalkalic Volcanic Rocks: Ontario Division of Mines, Ministry of Natural Resources Miscellaneous Paper 66, p. 1-22.

Jensen, L.S., and Pyke, D.R., 1982, Komatiites in the Ontario Portion of the Abitibi Belt: in Arndt, N.T., and Nisbet, E.G., eds., Komatiites: London, George Allen and Unwin Ltd., p. 147-159.

La Roche, H. de, 1974, Geochemical Characters of the Metamorphic Domains: Survival and Testimony of their Premetamorphic History: Sciences de la Terre, Nancy, v. 19, no. 2, p. 101-117.

Ludden, J.N., and Gélinas, L., 1982, Trace Element Characteristics of Komatiites and Komatiitic Basalts from the Abitibi Metavolcanic Belt of Quebec: in Arndt, N.T., and Nisbet, E.G., eds., Komatiites: London, George Allen and Unwin Ltd., p. 331-346.

Nesbitt, R.W., and Sun, S-S., 1976, Geochemistry of Archaean Spinifex-Textured Peridotites and Magnesian and Low-Magnesian Tholeiites: Earth and Planetary Science Letters, v. 31, p. 433-453.

Nesbitt, R.W., Sun, S-S., and Purvis, A.C., 1979, Komatiites: Geochemistry and Genesis: Canadian Mineralogist, v. 17, p. 165-186.

Nisbet, E.G., Bickle, M.J., Martin, A., Orpen, J.L., and Wilson, J.F., 1982, Komatiites in Zimbabwe: in Arndt, N.T., and Nisbet, E.G., eds., Komatiites: London, George Allen and Unwin Ltd., p. 97-104.

Orville, P.M., 1969, A Model for Metamorphic Differentiation Origin of Thin-Layered Amphibolites: American Journal of Science, v. 267, p. 64-86.

Sen, S.K., 1974, A Review of some Geochemical Characters of the Type Area (Pallavaram, India) Charnockites: Journal of the Geological Society of India, v. 15, p. 413-420.

Sibbald, T.I.I., Schwann, P.L., and Dunn, C.E., 1983, Uranium-Gold Metallogenic Studies: Nicholson Bay Ultramafic Complex: in Summary of Investigations 1983 Saskatchewan Geological Survey Miscellaneous Report 83-4, p. 75-79.

Tona, F., Alonso, D., and Svab, M., 1985, Geology and Mineralization in the Carswell Structure – A General Approach: in Lainé, R., Alonso, D., and Svab, M., eds., The Carswell Structure Uranium Deposits, Saskatchewan: Geological Association of Canada Special Paper 29.

Vidale, R., 1974, Metamorphic Differentiation Layering in Pelitic Rocks of Dutchess Country, New York: in Hofmann, A.W., Giletti, B.J., Yoder, H.S., Jr., and Yund, R.A., eds., Geochemical Transport and Kinetics: Carnegie Institution of Washington Publication 634, p. 273-286.

Viljoen, M.J., Viljoen, R.P., and Pearton, T.N., 1982, The Nature and Distribution of Archaean Komatiite Volcanics in South Africa: in Arndt, N.T., and Nisbet, E.G., eds., Komatiites: London, George Allen and Unwin Ltd., p. 53-79.

Wells, P.R.A., 1977, Pyroxene Thermometry in Simple and Complex Systems: Contributions to Mineralogy and Petrology, v. 62, p. 129-139.

Weaver, B.L., 1980, Rare-Earth Element Geochemistry of Madras Granulites: Contributions to Mineralogy and Petrology, v. 71, p. 271-279.

Wood, B.J., and Banno, S., 1973, Garnet-Orthopyroxene and Orthopyroxene-Clinopyroxene Relationships in Simple and Complex Systems: Contributions to Mineralogy and Petrology, v. 42, p. 109-124.

Zielinski, R.A., 1975, Trace Element Evaluation of a Suite of Rocks from Reunion Island, Indian Ocean: Geochimica et Cosmochimica Acta, v. 39, p. 713-734.

The Carswell Structure Uranium Deposits, Saskatchewan,
edited by R. Lainé, D. Alonso and M. Svab,
Geological Association of Canada Special Paper 29, 1985

THE STUDY OF THE BASAL ATHABASCA SUCCESSION IN THE D, E, L, F, AND S AREAS OF THE CARSWELL STRUCTURE

A. Pacquet
Service de Minéralogie, COGEMA, 1, Avenue Einstein, B.P. 103, 78191 TRAPPES Cedex, France

S. McNamara
Amok Ltd., 817-825 45th St. W., P.O. Box 9204, Saskatoon, Saskatchewan S7K 3X5

ABSTRACT

The base of the Athabasca Group was studied in drill hole CAR 114, located 2.5 km east of the Carswell structure, to obtain a reference section that could be compared with the Athabasca succession observed within the structure. The basal succession exposed within the Carswell structure is very different from that in CAR 114. It was studied in drill holes and surface exposures in the D, E, L, F, and S areas along the southern contact between the sub-Athabasca basement and the inner sandstone ring. The succession in the D, E, L, F, and S areas is composed of five units in three sequences up to 78 m thick.

The presence of Athabasca Group sandstone clasts within the Athabasca Group indicates that, initially, the Carswell area was part of the topographic high, while sediments were deposited in the lower parts of the Athabasca Basin. Renewed tectonic activity uplifted and reworked this material in the first two units of the Carswell succession. Altered volcanic shards in the mudstone and siltstone interbeds in Units 1 and 2 of Sequence I are evidence of acid volcanism at the time of sedimentation.

RÉSUMÉ

La partie inférieure des grès d'Athabasca a été étudiée dans la carrière du gisement D, quelques tranchées et plusieurs sondages carottés des zones D, E, L, et S dans la partie Sud de la structure de Carswell. Le sondage CAR 114, foré à l'est de la structure de Carswell, a fourni une colonne stratigraphique de référence pour la base du groupe Athabasca. On peut y distinguer une unité basale, d'une centaine de mètres d'épaisseur, constituée d'une succession monotone de grès moyens à grossiers et moyens, à bandes de grains grossiers épars ou groupés. Un léger affinement des grès marque la partie supérieure de l'unité.

Dans la carrière de D, la succession gréseuse apparaît très différente. A sa base, l'unité du conglomérat basal de 0 à 18 mètres d'épaisseur remplit les irregularités de la paléotopographie du socle. Son dépôt est localement contrôlé par des failles synsédimentaires. Cette unité, caractérisée par des variations latérales de faciès importantes et par un matériel détritique polygénique, très mal classé et en partie peu transporté, semble représenter un dépôt de cône alluvial proximal de type "debris-flow". La première mégaséquence se termine par l'unité inférieure de grès à galets, silt et pélite dont l'épaisseur, une dizaine de mètres, est assez constante. Elle contient des chenaux, bien visibles, dans le parement nord-est de la carrière de D et dans certains sondages carottés tel CLU 2301. Ils sont d'épaisseur inférieure à un mètre et d'extension réduite (quelques mètres), et contiennent de nombreux fragments de silt et d'argile, arrachés aux niveaux fins voisins. L'unité 2 correspond à un dépôt de cône alluvial distal.

La deuxième mégaséquence débute par l'unité de grès grossier et grès à graviers de 0 à 33 mètres d'épaisseur, dépôt fluviatile en tresse proximal, proche d'un type Donjek distal, constitué de grès moyens à grossiers dominants, à minces bandes de grès à graviers et de conglomérats. L'unité supérieure des grès, silt et pélite de 0 a 11.5 mètres d'épaisseur termine la mégaséquence.

L'énergie de transport des sédiments diminue ensuite dans l'unité des grès moyens à fins et fins à moyens, dépôt de rivière en tresse, de type South Saskatchewan.

La présence, dans le conglomérat basal et l'unité surincombante de blocs généralement anguleux à caractéristiques pétrographiques de grès Athabasca, témoigne de l'existence d'une sédimentation gréseuse, antérieure à la succession de Carswell et déposée dans des parties plus basses du Bassin Athabasca. L'étude pétrographique d'un bloc montre qu'il a subi une silicification antérieure à son dépôt dans le conglomérat basal. Celle-ci a pu intervenir au cours d'une période d'arrêt de sédimentation. L'érosion de tels blocs et leur dépôt sur la région de Carswell a nécessité une remontée des zones de sédimentation, jusque-là en dépression, grâce à une reprise de l'activité tectonique.

Les niveaux fins, dans l'unité du conglomérat basal et l'unité inférieure des grès à galets, silt et pélite contiennent de minuscules fragments de quartz, à formes caractéristiques d'échardes volcaniques (Figs. 6, 7, et 8). De tels vestiges témoignent de l'existence d'un volcanisme de type probablement ignimbritique, contemporain du dépôt de ces unités.

INTRODUCTION

The lower portion of the Athabasca Group was studied in the southern portion (D, E, L, F, and S areas) of the Carswell structure in order to construct a stratigraphic succession for the basal sediments that could be utilized in diamond drill core interpretation. Seven diamond drill holes (217, 823, 965, 1805, 2289, 2301, and 2303) in the D, E, L, F, and S areas, exposures in the D open pit, and diamond drill hole CAR 114, drilled slightly east of the Carswell structure, were examined (Fig. 1).

THE BASAL ATHABASCA SUCCESSION IN CAR 114

The lowermost part of the Athabasca Group in CAR 114 may correlate with the Lazenby Lake Formation of Hoeve *et al.* (1985), or may represent facies within the Manitou Falls Formation (J. Wilson, pers. commun., 1983). The succession was studied to obtain a reference section that could be compared with the Athabasca Group observed in diamond drill holes within the Carswell structure.

In CAR 114 a 700 m thick section of the Lazenby Lake Formation is composed of units similar to the basal unit. These units are approximately 100 m thick and show only subtle variations in grain size with the exception of the uppermost part which is composed of thin fine-grained beds.

A basal unit 103 m thick can be differentiated from the slightly coarser overlying unit. From the base upwards, the basal unit is composed of: 1) a few decimetres of coarse-grained sandstone with a 10 cm thick bed of well-rounded quartz pebble conglomerate; 2) a 62 m thick trough cross-bedded (10 to 20 cm thick sets), poorly sorted, medium- to coarse-grained sandstone and medium grained sandstone with a 2 cm thick bed of closely packed pebbles. Heavy mineral laminae are common throughout the basal unit; 3) a 31 m thick upper part, composed predominantly of trough cross-bedded (10 cm thick), medium- to fine-grained sandstones with scattered coarse grains.

Slight differences in the distribution of the coarser grains within the basal unit allow for the differentiation of elementary sequences (approximately 8 m thick). These sequences generally have a sharp lower contact, are composed predominantly of medium-grained sandstone mixed with an upwardly decreasing number of coarse grains. A thin (3.5 cm thick) fine-grained to silty layer completes the sequence.

The unit above the basal unit has a sharp, possibly disconformable, lower contact which is overlain by 0.2 m of coarse-grained to granule-sized sandstone.

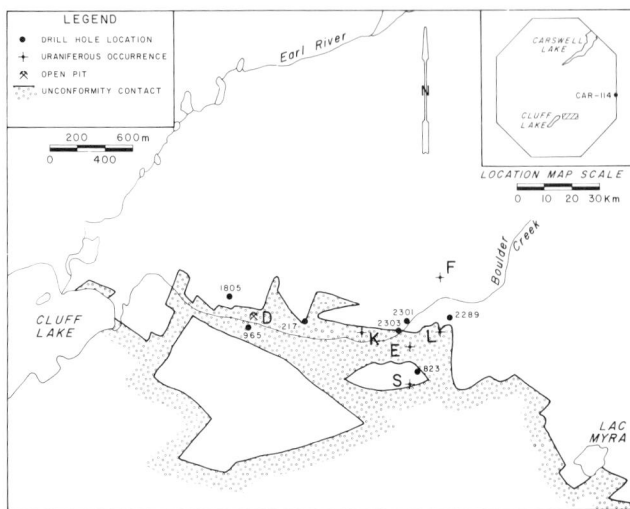

Figure 1. Location of the study area along the southern basement-sandstone contact of the Carswell structure.

The basal succession of CAR 114 is not comparable to the lower portion of the Athabasca Group in the southern portion of the Carswell structure. Exposures in the D open pit, and drill holes in the D, E, L, F, and S areas were therefore examined to provide a stratigraphic sequence recognizable in the diamond drill core from that area.

THE BASAL ATHABASCA SUCCESSION IN THE D, E, L, F, AND S AREAS

Only seven drill holes with thick sandstone intersections could be used in this study. Others had to be eliminated because of stratigraphic repetition by faulting or frequent dip changes in the siltstones and mudstones. Sedimentary units were defined by dominant lithology and sequential evolution.

The stratigraphic section shown in Figure 2 is based on data obtained from the D open pit and the cores of the seven diamond drill holes in the D, E, L, F, and S areas. A stratigraphic section composed of five units grouped in three sequences is proposed for the basal Athabasca succession in these areas (Table I, Fig. 2). They are presented here in their order of deposition.

TABLE I
STRATIGRAPHY IN THE D, E, L, F, AND S AREAS

Sequence	Unit	Thickness of Unit
I	Unit 1: Basal Conglomerate Unit	0 - 18 m
	Unit 2: Lower Pebbly Sandstone, Siltstone, and Mudstone Unit	0 - 13.5 m
II	Unit 3: Coarse-Grained and Granule-Sized Sandstone Unit	0 - 33 m
	Unit 4: Upper Sandstone, Siltstone, and Mudstone Unit	0 - 11.5 m
III	Unit 5: Medium- to Fine-Grained Sandstone Unit	Over 14 m

Unit 1: Basal Conglomerate Unit

The basal conglomerate is variable in thickness (0 to 18 m) and lithology. In the D orebody, the thickness varies from 0 to 1.8 m over a distance of less than 10 m. The variability in thickness is due to the uneven topography on the underlying basement and to the action of minor synsedimentary faulting.

From the west to the east of the D orebody it changes from a very poorly sorted, matrix-supported, polymictic conglomerate to mudstone and siltstone. The types of clasts include white quartz pebbles and cobbles, unaltered or altered hematized basement boulders, deformed mudstone and siltstone boulders, and trough cross-bedded medium- to coarse-grained and granule-sized sandstone boulders. Clasts are either sub-rounded to rounded (quartz) or very angular (sandstone). Also, they are randomly oriented within a muddy sandstone or mudstone matrix.

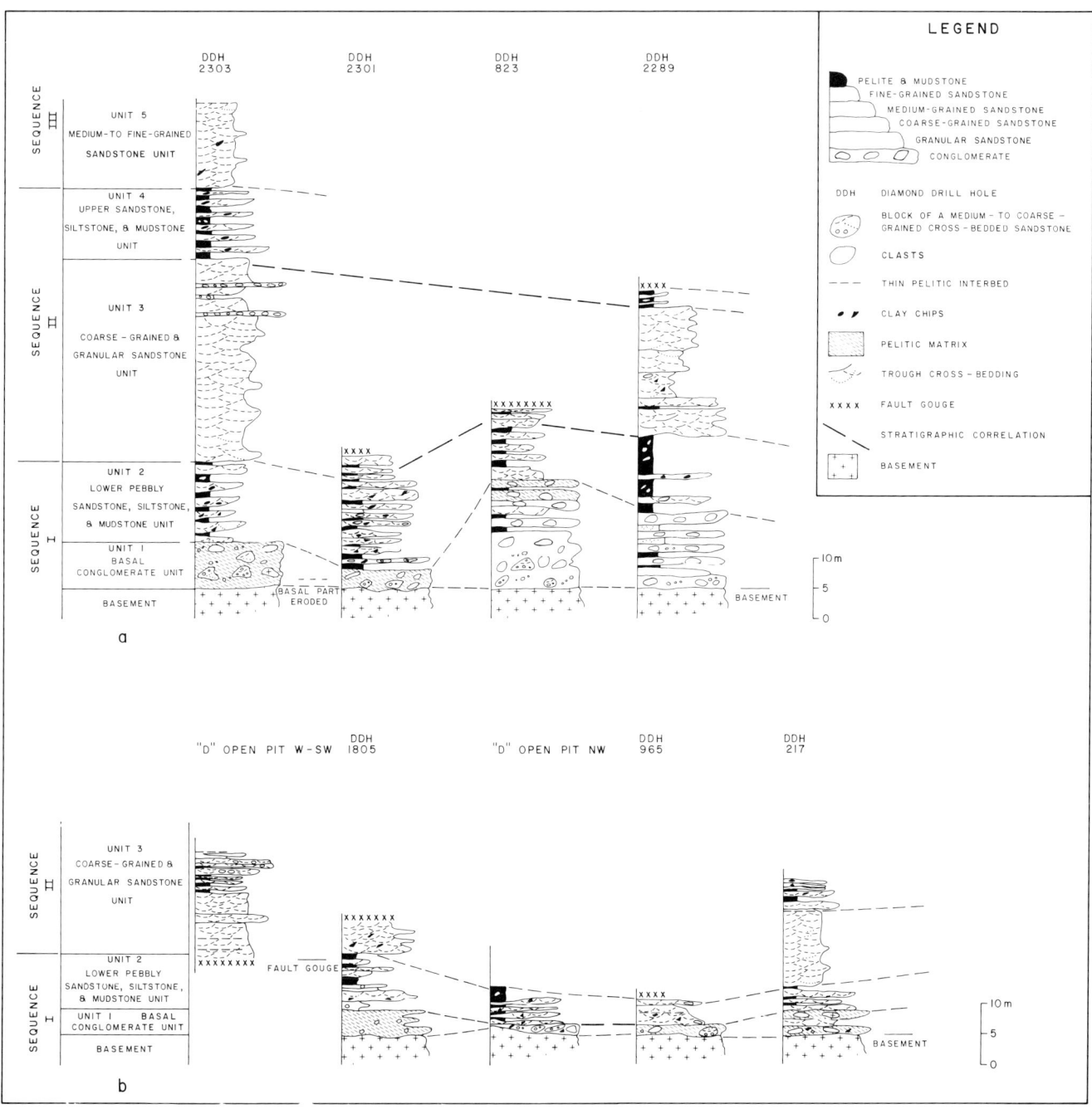

Figure 2. Stratigraphic sections based on diamond drill core, and exposures in the D open pit along the southern margin of the Carswell structure. (a) E, L, F, and S areas. (b) the D area.

Unit 2: Lower Pebbly Sandstone, Siltstone, and Mudstone Unit

The lower pebbly sandstone, siltstone, and mudstone unit varies from 0 to 13.5 m in thickness and is composed of a variable number of small sandstone lenses which are interfingered and interbedded with siltstone and mudstone.

The siltstone and mudstone beds are laminated or massive with scattered coarser grains, pebbles, and cobbles, and occasional thin conglomerate lenses. The sandstone lenses vary from 0.2 m to 1 m in thickness and from 2 m to 5 m in apparent width. They are generally composed of poorly sorted, medium- to coarse-grained and granule-sized sandstone with scattered pebbles and cobbles (quartz and sandstone clasts), conglomerate lenses, and fairly abundant mudstone and siltstone intraclasts. The average grain size may decrease upwards. Primary sedimentary structures include trough cross-bedding (10 cm thick sets), current-rippled surfaces with mud drapes on top of the sandstone lenses, and reactivation surfaces with pebble lags. The sandstone lenses usually have an uneven and erosional base lined by a pebble lag. Heavy minerals are common along laminae.

Unit 3: Coarse-Grained and Granule-Sized Sandstone Unit

The coarse-grained and granule-sized sandstone unit varies from 0 to 33 m in thickness and is composed of a monotonous succession of medium- to coarse-grained sandstone with a variable amount of granule-sized sandstone, conglomerate, and minor siltstone. The clasts within this unit are generally finer grained than those occurring in Unit 1. These variably rounded clasts generally are centimetric pebbles with some cobbles. Mudstone and siltstone intraclasts are common. Some of the beds within this unit show an upward fining from a conglomerate to a coarse-grained and granule-sized sandstone. Trough cross-bedded sets of decimetric or greater thicknesses and heavy mineral concentrations along laminae are common throughout this unit. Some of the sandstone beds show an erosional lower contact.

Unit 4: Upper Sandstone, Siltstone, and Mudstone Unit

The upper sandstone, siltstone, and mudstone unit varies from 0 to 11.5 m in thickness. It is composed of alternating metre thick beds of siltstone and mudstone containing scattered pebbles as well as beds of poorly sorted, trough cross-bedded, medium- to coarse-grained and granule-sized sandstone containing mudstone intraclasts.

Unit 5: Medium- to Fine-Grained Sandstone Unit

The medium- to fine-grained sandstone unit is greater than 14 m thick and is composed predominantly of trough cross-bedded, medium- to fine-grained sandstones with layers and bands of medium-to coarse-grained sandstone. Large clay intraclasts are present throughout the sandstone. Heavy mineral beds, where present, are almost totally bleached.

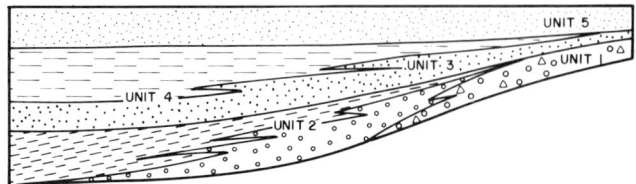

Figure 3. Schematic diagram showing a possible relationship between units, with reference to an alluvial fan model (Unit 1: proximal fan; Unit 2: distal fan) and a braided river model (Unit 3: proximal facies; Unit 4: distal alluvial plain). Unit 5 is a more widespread braided river deposit over a more level topography.

Environment of Deposition

The basal Athabasca succession in the D, E, L, F, and S areas is composed of five depositional units representing three sequences of an alluvial fan (Fig. 3).

Sequence I represents the first stage of Athabasca sedimentation and is composed of Units 1 and 2.

The basal conglomerate unit, Unit 1, represents a proximal alluvial fan deposit. The very poor sorting, matrix support, lack of imbrication, and internal stratification in the basal conglomerate unit are consistent with debris-flow deposition. Its variability in thickness indicates deposition over an uneven topography. Poor sorting and angularity of some clasts indicates a short transport distance. Unit 1 is similar to the lower member of the Cannes de Roche Formation of Gaspe, Quebec interpreted as a debris flow deposit (Rust, 1979). In both localities the clast orientation is random, bedding is absent, and imbrication has not been observed.

The lower pebbly sandstone, siltstone, and mudstone unit, Unit 2, shows some similarity to the middle member of the Cannes de Roche Formation (Rust, 1979), interpreted as a distal fan/valley flat deposit. Both are composed of laminated to apparently massive mudstone and siltstone with minor interbeds of fine conglomerate. The characteristics, relationships, and contacts of the sandstone bodies with the adjacent mudstones and siltstones indicate a distal alluvial fan environment with numerous stream channels of the type described by Bull (1964) in modern alluvial fans in California.

Sequence II represents the second stage of Athabasca sedimentation and is composed of Units 3 and 4.

The coarse-grained and granule-sized sandstone unit, Unit 3, exhibits trough cross-bedding and erosional surfaces overlain by mudstone and siltstone intraclasts and contains thin lenses of pebbles and conglomerate. This particular unit is similar to those described by Rust (1979) in the Malbaie Formation which is interpreted as a proximal braided river deposit in a sand-dominant braided system. Unit 3 is also similar to a distal equivalent of the Donjek-type model of Miall (1977).

The upper sandstone, siltstone, and mudstone unit, Unit 4, is similar to Unit 2 but has slightly finer grained sandstone in the channels. Unit 4 possibly represents a distal alluvial plain deposit.

Sequence III represents the third stage of Athabasca sedimentation and is composed of Unit 5.

The medium- to fine-grained sandstone unit, Unit 5, is thought to be distal, braided and fluviatile, of possible South Saskatchewan River type where trough cross-bedded sand is the dominant facies (Miall, 1977).

DISCUSSION

Sandstone Clasts within Units 1 and 2

The basal conglomerates of Unit 1 and the channels in Unit 2 contain angular clasts of variable shape and size that are composed of medium- to coarse-grained and coarse-grained to granule-sized sandstone (Figs. 4 and 5). These sandstone clasts are commonly trough cross-bedded and primarily contain quartz grains with minor altered biotite set in a clayey matrix. These macroscopic clast characteristics are similar to those of the sandstone matrix within the basal conglomerate unit.

A thin section study of a sandstone boulder from Unit 1 (drill hole 2303) confirmed the similarity of detrital grain morphology and composition between the boulder and sandstones in the basal conglomerate matrix and in Unit 2. Subrounded to angular quartz grains are definitely detrital grains and not regolith fragments. The heavy mineral suite (tourmaline, zircon, rutile, and altered iron-titanium oxide) is identical to those found throughout the Athabasca Group sandstones in the Carswell structure. Secondary quartz overgrowths around quartz grains and an illitic cement, mainly derived from the biotite alteration, are both common to these matrix sandstones and the studied boulder.

Part of the cement in the sandstone boulder consists of cryptocrystalline silica, stained by finely disseminated iron oxides. This silicification is restricted to the clast and is absent in the surrounding conglomerate matrix, and therefore is older than the conglomerate deposition. Before erosion and transportation, the clast of sandstone was silicified and iron-stained, possibly during a period of weathering.

The sandstone clasts in Unit 1 and Unit 2 do not show any soft sediment deformation and were well consolidated before erosion and sedimentation. Their type of cross-bedding and lithology, identical to the Athabasca Group sandstones, indicates that they were derived from Athabasca Group sediments previously deposited in parts of the basin lower than the Carswell area topographic high. A hiatus in sedimentation then occurred prior to the Athabasca deposition in the Carswell area, during which older Athabasca sediments were weathered and silicified. A renewal of tectonic activity must have occurred to uplift the Athabasca succession already deposited and subsequently caused its erosion and sedimentation in the Carswell area.

Evidence of Acid Volcanism

Thin section examination of a few mudstone and siltstone interbeds in Units 1 and 2 from the D, and E, L, and F areas shows the presence of quartz splinters and microfragments. The shape of these fragments is characteristic of volcanic shards (Figs. 6, 7, and 8) and is different from acicular and highly angular fragments originating from conchoidally

Figure 4. Sandstone clast in Unit 1, the basal conglomerate unit (overturned succession) in the northwestern section of the D open pit. Note the deformation of the overlying beds around the boulder.

Figure 5. Angular block of sandstone in Unit 1, the basal conglomerate unit, in the northwestern section of the D open pit.

fractured quartz pebbles. The smooth contours of these shards differentiates them from ragged quartz grains corroded by clay.

Such delicate fragments are likely to come from acid volcanism emitted during the early stages of the Athabasca sedimentation in the Carswell area. Tuffs have been reported by Ramaekers (1980) in the Wolverine Point Formation where the quiet environment of deposition may explain their preservation.

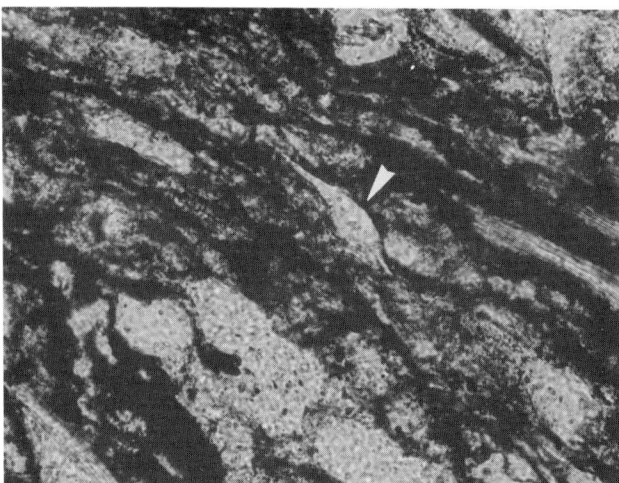

Figure 6. The arrow indicates a quartz devitrified volcanic shard within Unit 1, the basal conglomerate unit, at 19.9 m in diamond drill hole 2302. Plane light × 250.

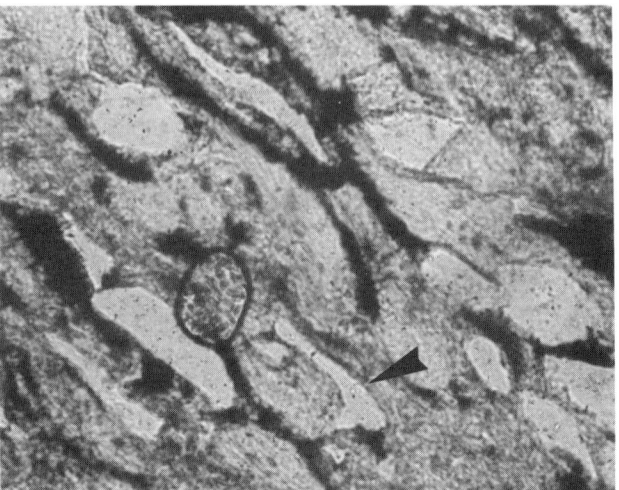

Figure 7. The arrow indicates a hook-shaped quartz devitrified volcanic shard near the D orebody. Plane light × 160.

Figure 8. The arrow outlines a quartz devitrified volcanic shard from the E, L, F area. Plane light × 250.

CONCLUSIONS

Five depositional units representing three sequences of an alluvial fan compose the lowermost 78 m of the Athabasca Group intersected in the D, E, L, F, and S areas.

Within Sequence I (Units 1 and 2), the first stage of Athabasca sedimentation in the D, E, L, F, and S areas is a debris flow proximal fan deposit which fills the irregularities of the uneven topography of the weathered basement. This unit grades laterally and upward into a more regular distal fan deposit with a variable degree of stream channelling.

Sequence II (Units 3 and 4) includes an episode of proximal braided river sedimentation (a possible distal equivalent of the Donjek model) which is widespread due to the levelling of the topography by the first sequence, and is followed by a deposit similar to Unit 2.

Sequence III (Unit 5) is a second episode of braided river deposition. This episode is of the South Saskatchewan River type and shows an overall decrease in transport energy.

The presence of Athabasca sandstone clasts within Sequence I indicates that sedimentation had taken place in low areas of the Athabasca Basin prior to sedimentation in the Carswell area. As a result of tectonic activity, these sediments were subsequently uplifted, eroded, and then redeposited over the Carswell structure weathered basement. Prior to erosion, a portion of these sandstones were silicified by weathering during a possible episode of nondeposition.

A contemporaneous, possibly ignimbritic volcanism supplied a portion of the detrital material present in Sequence I. The detrital material is primarily preserved in the fine-grained mudstone and siltstone horizons.

ACKNOWLEDGEMENTS

The authors are grateful to Paul Ramaekers of the Saskatchewan Mining and Development Corporation and John Wilson of the Alberta Research Council for their critical evaluation of the manuscript. In addition, we would like to thank Amok Ltd. geologists for their assistance.

REFERENCES

Bull, W.B., 1964, Alluvial Fans and Near-Surface Subsidence in Western Fresno County, California: United States Geological Survey Professional Paper 437-A.

Hoeve, J., Quirt, D., and Alonso, D., 1985, Clay Mineral Stratigraphy of the Athabasca Group: Correlation Inside and Outside the Carswell Structure: *in* Lainé, R., Alonso, D., and Svab, M., eds., The Carswell Structure Uranium Deposits, Saskatchewan: Geological Association of Canada Special Paper 29.

Miall, A.D., 1977, A Review of the Braided River Depositional Environment: Earth Sciences Review, v. 13, p. 1-62.

Ramaekers, P., 1980, Stratigraphy and Tectonic History of the Athabasca Group (Helikian) of Northern Saskatchewan: *in* Summary of Investigations 1980, Saskatchewan Geological Survey Miscellaneous Report 80-4, p. 99-106.

Rust, B.R., 1979, Depositional Models for Braided Alluvium: *in* Miall, A.D., ed., Fluvial Sedimentology: Canadian Society of Petroleum Geologists Memoir 5, p. 605-625.

THE CARSWELL FORMATION, NORTHERN SASKATCHEWAN: STRATIGRAPHY, SEDIMENTOLOGY, AND STRUCTURE

H. E. Hendry
Department of Geological Sciences, University of Saskatchewan, Saskatoon, Saskatchewan S7N 0W0

K. L. Wheatley
Amok Ltd., 817-825 45th Street West, P.O. Box 9204, Saskatoon, Saskatchewan S7K 3X5

ABSTRACT

The dolostones of the Carswell Formation, the youngest formation of the Helikian Athabasca Group, crop out in a distinctive annular pattern in northwestern Saskatchewan. The dolostone ring encloses Aphebian metamorphic rocks, and sandstones of the Athabasca Group. The rocks of the Carswell Formation were intensely folded and faulted about 480 Ma ago during the so-called Carswell event. The dolostones now are overturned in many places and cut by so many faults that it is difficult to establish the original thickness of the formation; the estimated minimum thickness is 400 to 500 m. The Carswell Formation has been divided into a lower and upper member at the top of a marker bed of conical stromatolites. The lower member is composed of algal laminites and hemispheroidal stromatolites of variable shape, interbedded with oncolitic and oolitic beds, and rudites. The upper member consists of algal laminites interbedded with dolomicrites, and hemispheroidal and irregularly-shaped stromatolites; the top of the upper member and the limit of exposure is marked by a structureless dolostone. Both lower and upper members contain evidence of the former presence of evaporites. Throughout the Carswell Formation there are beds of dolostone breccia and fracture-filling breccias which were produced partly by tectonic movements and partly by collapse following dissolution of evaporites at depth. Some of the larger faults which cut the formation are believed to have been active before and during deposition of the Carswell sediments. The sediments were deposited on a shallow tidal marine carbonate platform in environments ranging from subtidal to supratidal.

Structural interpretations of sections through the formation confirm the view that in places, particularly in the north and east of the ring, the Carswell Formation has been deformed into a synclinal form with a circular axial trace and overturning of beds towards the outside of the ring.

RÉSUMÉ

Les dolostones de la Formation Carswell, les roches les plus jeunes du Groupe Hélikien de l'Athabasca, affleurent selon un anneau circulaire discontinu au sud du Lac Athabasca dans le nord Saskatchewan (Fig. 1). Les datations des argiles des grès du Groupe Athabasca et des brèches de Cluff, qui se sont mises en place en même temps que la structure circulaire de Carswell, indiquent un âge supérieur de 1200 à 1300 Ma et un age inférieur de 480 Ma pour la formation Carswell. Pendant la formation de la structure circulaire les dolostones furent plissées, retournées en bien des endroits, et recoupées par de nombreuses failles (Figs. 2 and 4). Ni le contact supérieur ni le contact inférieur de la Formation Carswell ne sont visibles à l'affleurement, l'anneau est en contact faillé avec la Formation Douglas, sous-jacente, à la fois à l'extérieur et à l'intérieur. L'épaisseur stratigraphique originale des dolostones de Carswell est estimée à un minimum de 400 à 500 m.

Les dolostones sont lithologiquement diversifiés mais consistent principalement de stromatolites (Figs. 5, 6 et 7; Tableau II), qui comprennent les types côniquehémisphéroidaux, en plaquettes, irréguliers, et de laminites algaires (Fig. 8), avec des lits à grains enduits d'algues (Fig. 10), des conglomérats intraformationnels (Fig. 11), et des brèches de dissolution et tectoniques (Figs. 11 et 12). De nombreux dolostones contiennent des pseudomorphes dolomitique de gypse. La Formation comprend un marqueur très distinct de larges stromatolites côniques surmontés par un lit de dolomicrite blanche (Figs. 5 and 6). Ce marqueur est visible sur la quasitotalité de la partie nord de l'anneau et la Formation a été ainsi divisée en un membre supérieur et un membre inférieur le long de la limite entre le lit de stromatolites côniques et la dolomicrite (Fig. 3; Tableau IV). Les membres ainsi définis ne correspondent pas à ceux proposés par Currie (1969).

La plupart des sédiments se dont déposés dans un en-

vironement intertidal, sujet à des dessications fréquentes et des érosions par courants locaux. Durand les périodes d'extrêmes évaporation le gypse était précipité, et durand les orages des oolithes et de la boue en provenance du large du rivage s'échouaient sur la zone de battement des marées. L'horizon marqueur de stromatolites côniques du sommet du membre inférieur peut être tracé sur 15 km et s'est formé dans un environnement subtidal. Les brèches abondantes dans la formation, constituent des corps tabulaires à la fois concordants et discordants sur la stratification. Elles sont particulièrement abondantes dans le membre supérieur et se sont formées quand les dolostones furent fracturées pendant et après le dépôt de la séquence, en partie à la suite de la dissolution des évaporites en profondeur. La présence de quelques brèches particulièrement épaisses à côté de failles majeures semble indiquer que l'activité tectonique régionale ait pu contribué à leur formation.

La structure de Carswell est complexe. Au nord et à l'est de l'anneau, la structure est synclinale avec la stratification retournée en direction de l'exterieur de l'anneau, alors qu'à l'ouest et au sud la stratification est typiquement normale et moins déformée, et l'épaisseur totale de dolostone préservée est moindre (Fig. 4). Les lits ont été déplacés à la fois le long de failles courbes, qui sont parallèles à la direction locale de la stratification, et le long de failles qui coupent au travers de la structure circulaire (Figs. 2 and 4). Il y avait probablement des mouvements le long de la plupart des failles pendant la formation de la structure circulaire mais les failles de direction nord-est et nord-ouest étaient actives à la fois avant et après son développement.

INTRODUCTION

The dolostones of the Carswell Formation are the youngest rocks of the Proterozoic Athabasca Group in northern Saskatchewan (Fig. 1, Table I). They crop out as a series of ridges in a ring approximately 39 km in diameter and 1 to 4 km wide, and encircle Helikian sandstones and Aphebian metamorphic rocks (Figs. 2 and 3). The Carswell Formation may be as much as 500 m thick. It consists mostly of stromatolites, algal laminites and dolomicrites with some ooid and rudite lenses. Penecontemporaneous and post-depositional breccias are common at many stratigraphic levels in the formation. Selective silicification, detrital quartz grains, and clays are common in the lower parts of the dolostone close to the contact with the Douglas Formation.

The sandstones of the Athabasca Group have been dated radiometrically at 1350 ± 50 Ma (Ramaekers and Dunn, 1977), 1428 ± 30 Ma (Ramaekers, 1981), and at 1513 ± 24 Ma (Bell, 1981), all by the Rb-Sr method. The Douglas Formation has been dated by the K-Ar method at 1293 ± 26 Ma on samples of illite in a red pelite, and at 1220 ± 43 Ma using an illite-chlorite mixture in a sandstone (Clauer et al., 1985). The ages for clay minerals of the Douglas Formation may have been reset by hydrothermal activity because the same ages have been derived from samples in the D uranium orebody and the surrounding sandstones at Cluff Lake. Syndepositional deformation in the dolostones may have been associated with this hydrothermal event. This places an upper limit on the age of the Carswell Formation of 1200 to 1300 Ma. The lower limit for the age is determined by tectonism that produced the ring-structure and the Cluff breccias. This deformation, during the so-called "Carswell event", has been dated at 480 Ma by Bell (1985).

The first description of the Carswell dolomite was published by Blake (1956) who named the formation the "Trout Lake Limestone," and commented on its complex structure. Fahrig (1961) was the first to draw attention to the ring-like distribution of the formation and to note the presence of stromatolites; he inferred that breccias in the formation

TABLE I

ROCKS OF THE ATHABASCA GROUP (THICKNESS OF TERRIGENOUS CLASTIC FORMATIONS MODIFIED FROM RAMAEKERS, 1981)

Carswell Formation	? 500 m	dolostone
Douglas Formation	? 300 m	sandstone, siltstone, mudstone
Tuma Lake Formation	100 m	pebbly sandstone
Otherside Formation	200 m	sandstone, siltstone
Locker Lake Formation	75-200 m	pebbly sandstone
Wolverine Point Formation	200-500 m	siltstone, clayey sandstone phosporite, tuff
Lazenby Lake Formation	130 m	pebbly sandstone
Manitou Falls Formation	1300 m	pebbly sandstone
Fair Point Formation	300 m	pebbly and cobbly sandstone

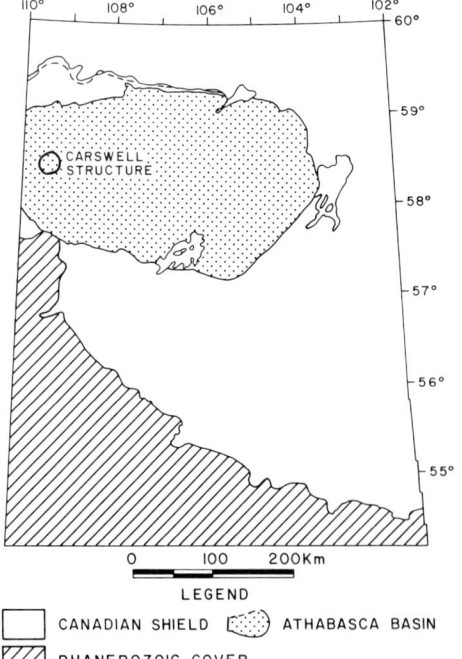

Figure 1. Location of the Carswell ring structure (Modified from Lewry et al., 1978).

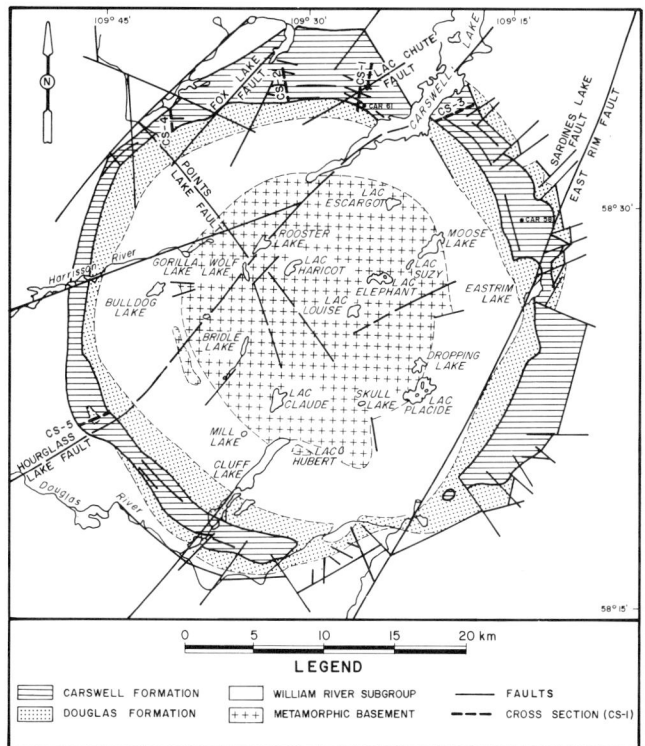

Figure 2. Geological map of the Carswell ring structure showing locations of cross sections.

formed contemporaneously with deposition. Currie (1969) divided the Carswell Formation into two members: a lower member of thinly bedded and stromatolitic dolomite, and an upper member of thickly bedded, structureless dolomite. Study of the stromatolites allowed Currie to determine facing directions in some of the steeply dipping beds. He noted the

Figure 3. Dolostone ridges of the Carswell Formation west of Carswell Lake. View looking north.

TABLE II

CHARACTERISTIQUES OF STROMATOLITES IN THE CARSWELL FORMATION

Characteristics	Conical	Hemispheroidal	Platy	Irregular
Size and shape	Two main sizes. The larger ones are circular or lensoid in plan-view and have a zig-zag pattern in cross section; 10 to 200 cm long, 8 to 50 cm wide, 8 to 40 cm synoptic height. The smaller ones ("egg carton") are oval in plan view and vary in cross section; 1.5 to 5 cm in diameter, 0.75 to 2.5 cm synoptic height.	Wide range of sizes from 2 cm to 200 cm diameter and up to 50 cm in synoptic height.	Up to 30 cm across and up to 5 cm in synoptic height. Very irregular in plan view but individual columns are equant. Outlines in cross section resemble stacked inverted plates.	Generally up to 30 cm across and up to 25 cm in synoptic height; shape extremely variable.
Branching	Common in large stromatolites; not seen in "egg cartons".	Common	Uncommon and poorly developed.	Common
Intercolumnar Area	In large conical types may contain dolomicrite, coated grains, or columnar stromatolites. In "egg carton" types there is no intercolumnar debris.	In some beds they contain flakestone conglomerate, laminated dolarenite, dolomicrite, or breccias.	Contain mud.	Contain crumpled stromatilitic layers or lumps of stromatolite in dolomicrite.
Continuity of laminae	High; almost all stromatolites laterally linked; internal erosion surfaces are rare.	High in laterally-linked types, but internal truncation of laminae is common in some beds. Solitary hemispheroids are common also.	Poor continuity of laminae	Common in some thin beds; absent in others.
Vertical persistence	Large: 4 m Egg carton: 30 cm	10 m	30 cm	1 m
Lateral extent	15 km	Variable	15 km	Variable
Preferred orientation	Long dimension of large types oriented north-south.	None	None	None

large areas of brecciated rock in the Carswell Formation and suggested that the breccias were produced during folding before the development of the Cluff breccias, but he did not make a clear distinction between breccias composed of dolomite and those breccias produced by formation of the Carswell circular structure. Both Fahrig (1961) and Currie (1969) interpreted the Carswell Formation as having a synclinal structure with the trace of the synclinal axis following a circular trend, and the axial plane dipping towards the centre of the ring.

The Carswell Formation crops out only in the ring structure and the intense deformation with overturning of the beds is local and restricted to the ring structure. In the rest of the Athabasca Basin rocks of the Athabasca Group are flat-lying or gently dipping. Speculations about the origin of the ring structure and the nature of the "Carswell event" are many, with Innes (1964), Pagel (1975) and Harper (1982) favouring meteorite impact, and Currie (1969) interpreting the structure as cryptovolcanic.

In the course of field work for this study all outcrops of dolostone were visited and five stratigraphic cross sections were measured in the north and southwestern parts of the ring (Fig. 2). There is no continuous stratigraphic section through the Carswell Formation, and measured sections across the ring normally are broken by steeply dipping faults (Figs. 2 and 4). Distinctive stratigraphic units and sequences facilitate correlation between fault blocks and establishment of the stratigraphic sequence. A unit containing conical stromatolites overlain by light coloured dolomicrite is the most distinctive marker in the lower part of the formation and can be traced along strike for 15 km. This unit provides evidence of repetition of the stratigraphic sequence by faults in sections 1 and 2 (Fig. 4).

STRATIGRAPHY
Lithological Descriptions

Stromatolites. Stromatolites in the Carswell Formation include a variety of conical types and hemispheroids of various sizes as well as asymmetrical, platy, and irregular forms (Table II). The most striking shapes are laterally-linked conical forms, mainly conical lensoid. They have sharp arcuate crests, V-shaped troughs and steep sides (Fig. 5a). Individual cones are up to 50 cm wide with synoptic heights of up to 40 cm; lengths measured parallel to the crests generally are less than 1 m but may reach 2 m. Most of the stromatolites are laterally-linked by new growth between columns or by branching from the sides of columns (Fig. 6). In vertical cross section successive crests follow near vertical sinuous traces which persist for up to 4 m; in places the crests are broken or brecciated. In plan view the long dimensions of conical lensoid stromatolites normally are aligned north-south. Laminations generally are slightly thicker at crests than on flanks or in troughs, and on the sides they commonly are crenulated. Conical columnar stromatolites are present in the troughs between some conical lensoid forms. Generally there is no debris in the intercolumnar spaces, but between some columns there are deposits of dolomicrite, oncolites, and breccias; the breccias contain pieces of broken stromatolite in a matrix of dolomicrite. The conical

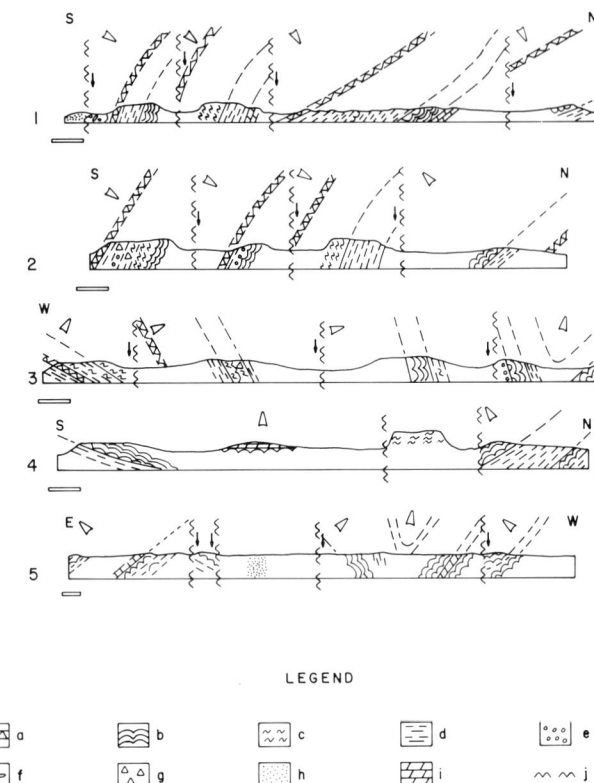

Figure 4. Geological cross section, across the Carswell Formation. All cross sections have been oriented with the inside of the Carswell ring structure to the left of the diagram. Legend: a = marker bed of large conical stromatolites, thickness not to scale; b = hemispheroidal stromatolites; c = irregular stromatolites; d = algal laminites; e = oolites; f = intraformational conglomerates (dolorudites): g = breccias; h = quartz arenites of the Douglas Formation; j = faults; open arrows point to stratigraphic top; closed arrows indicate direction of throw on faults. Where the conical stromatolite marker bed is shown only above the topographic profile it has not been observed and its presence has been inferred. There is no vertical exaggeration, the scale bar at the bottom left of each cross section represents 100 m.

stromatolites are the principal constituents of a 3 to 5 m bed that is traceable along strike for over 15 km in the northern part of the Carswell ring structure. The base of the bed is gradational over about 30 cm to platy stromatolites, and the top of the bed is very irregular and overlain sharply by white dolomicrite (Figs. 5b and 6). In the main part of the bed the columns are of uniform shape and size, but near the top they are relatively small with many branches, and individual columns lean to one side or the other (Figs. 5b and 6). Conical stromatolites of this size occur in abundance in only the one bed.

A second type of conical stromatolite is here designated as "egg carton" because the stromatolites are the same size and shape as egg cartons (Fig. 5c). They have diameters up to 5 cm and synoptic heights up to 2.5 cm. They are laterally-linked, but the lateral extent of individual laminae is limited to a few tens of centimetres. In places they are asymmetrical and resemble ripple drift laminations in cross section.

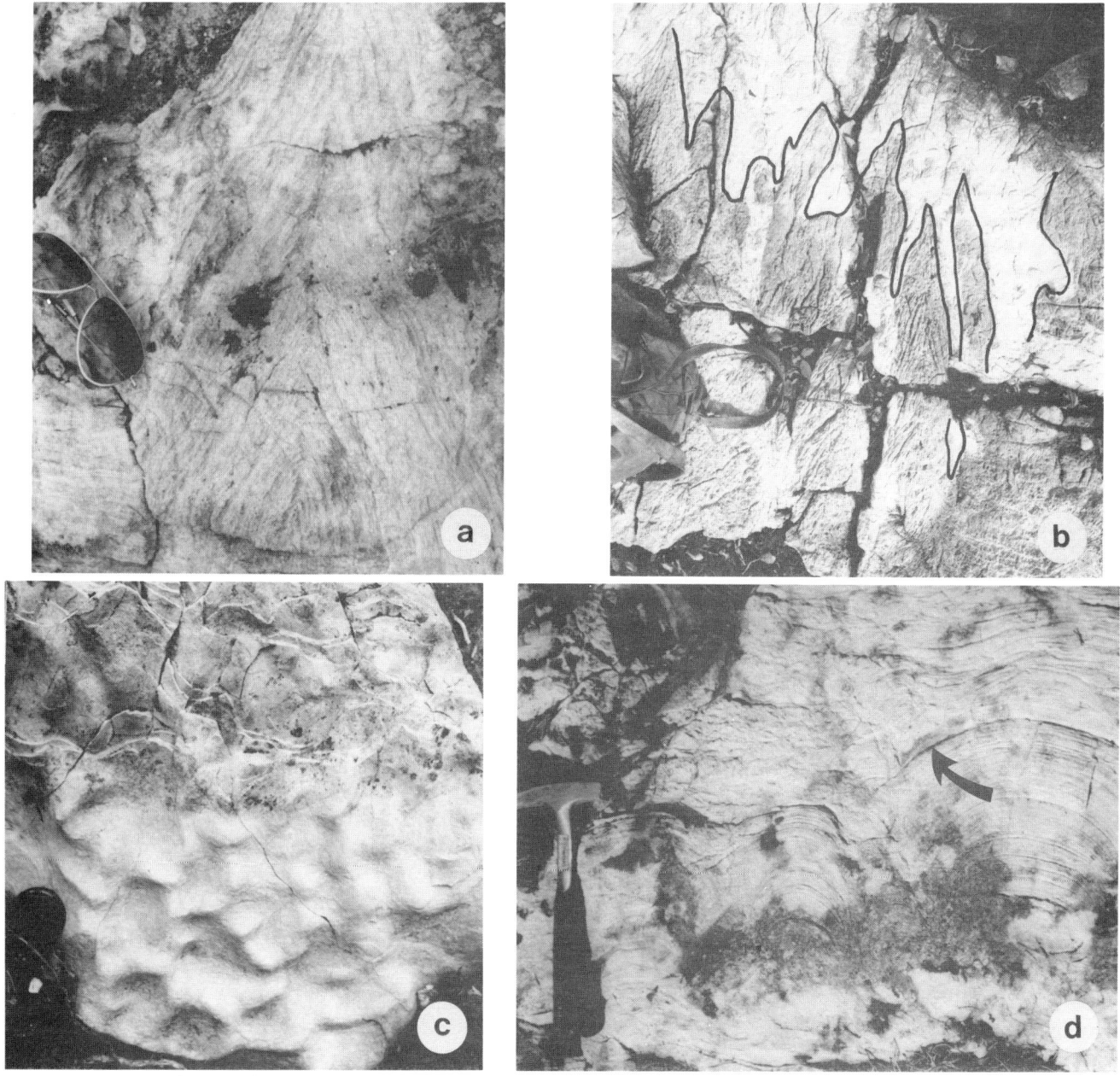

Figure 5. Stromatolites. (a) Large-scale conical stromatolites with sharp crests and wrinkled laminations on steep flanks, lower member, cross section 1. (b) Large-scale conical stromatolites at top of marker bed overlain by white dolomicrite; note very irregular sharp contact; backpack for scale; lower member, cross section 1. (c) Small-scale conical stromatolites, view of upper surface of bedding plane; lower member, cross section 1. (d) Hemispheroidal stromatolites showing branching (arrowed). Upper member, cross section 2.

Hemispheroidal stromatolites up to 50 cm in diameter are abundant as both laterally-linked and solitary forms. Solitary hemispheroidal stromatolites occur mainly in beds of algal laminite where they have developed on top of irregularities or disrupted layers in an algal mat, over local deposits of breccia, or over mud clasts. The solitary forms generally are small, typically 10 to 20 cm in diameter and 2 to 5 cm in synoptic height, though the largest solitary hemispheroid was 50 cm across and 25 cm high. Vertical persistence of such isolated forms is very limited and normally they are obliterated by smoothing out of the mat.

Laterally-linked hemispheroids (Fig. 5d) are of many different sizes but commonly up to 40 to 50 cm in diameter and 10 to 20 cm in synoptic height. All are circular in plan view and the main variation in shape is in the ratio of diameter to synoptic height. There are several different types of branching, and through bed thicknesses of several centimetres there may be a change from small to large, and back to small, stromatolites. It is also common for there to be stromatolites of several different sizes on one bedding plane. In some beds individual stromatolites are asymmetrical or recumbent and between the laminae there are thin layers of flakestone con-

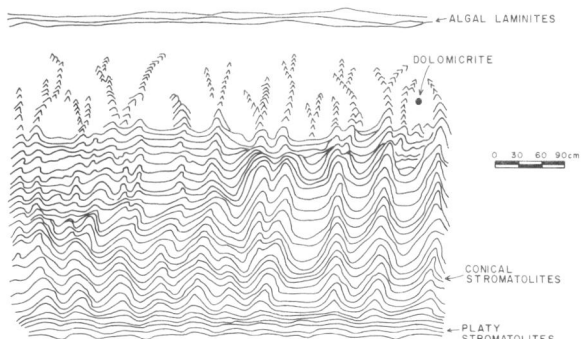

Figure 6. Sketch of the bed of large conical stromatolites at the top of the lower member and the overlying dolomicrite. Note the gradual development of the conical forms from platy forms at the base and the change in shape at the top of the bed.

glomerate, cross-laminated dolarenite, lenses of dolomicrite, or intraformational breccias. Though the hemispheroidal stromatolites vary in size and distribution, there are some thick, well developed beds. One 25 m thick bed of interbedded hemispheroidal stromatolites and algal laminites can be traced for 1.5 km along strike south of Carswell Lake.

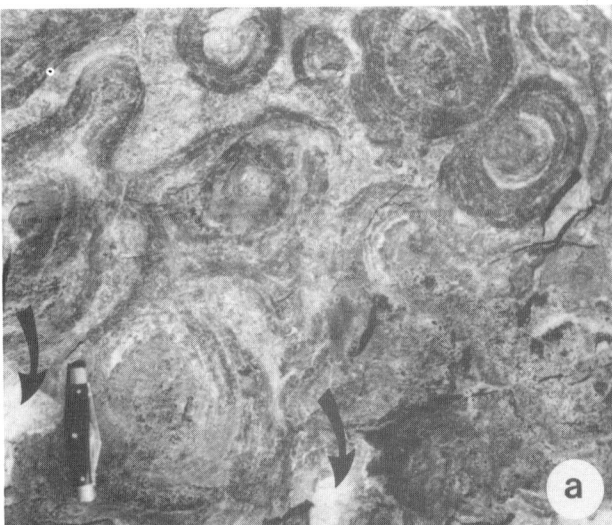

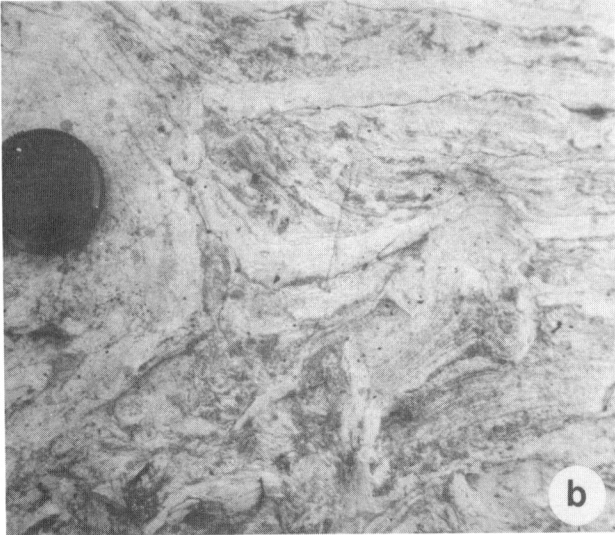

Figure 7. (a) Platy stromatolites viewed in section parallel to bedding plane. Note white dolomicrite (arrowed) in intercolumnar areas; lower member, cross section 1; knife is 8 cm long. (b) Irregular stromatolites in vertical cross section; note continuous layers separated by crumpled and brecciated laminae; upper member, cross section 2. Lens cap is 5 cm across.

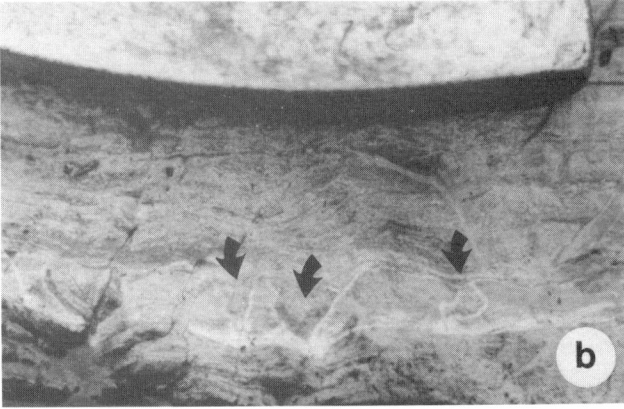

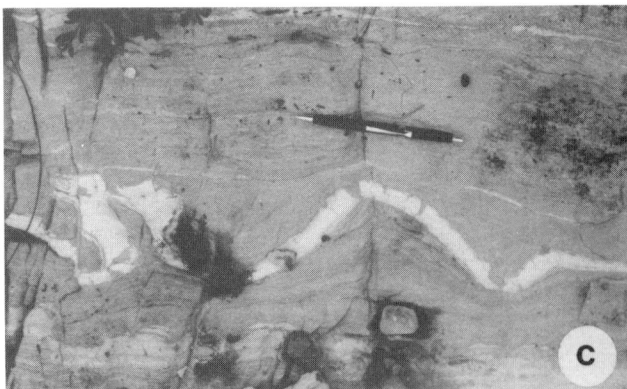

Figure 8. (a) Algal laminites with interbedded dolomicrite. (b) Mud cracks in dolomicrite layer (hammer head for scale). (c) Folded dolomicrite layer in sequence of algal laminites and small stromatolites. All in upper member, cross section 1.

Figure 9. (a) Large circular, bowl-shaped depressions in algal laminites, upper member cross section 1. View of underside of bedding plane. (b) Moulds of symmetrical ripple marks on bottom of bowl-shaped depression.

Platy stromatolites have low relief with a maximum diameter of 30 cm and a synoptic height of up to 5 cm. They are irregular in plan view. Lateral linkage is poor, and interstromatolite areas are filled with dolomicrite (Fig. 7a).

The irregular stromatolites have great variations in shape. They are relatively small and have forms that are conical, columnar, rectangular, triangular, ellipsoidal, or extremely irregular. They have a maximum vertical persistence of about a metre and exhibit variable degrees of lateral linkage. It is quite common for intercolumnar troughs to be replaced by columns within 30 cm of bedding thickness. Many forms have branches but their shapes commonly changed where a branch developed. Lateral linkage is common but took place intermittently during deposition. Layers in which lateral linkage is present alternate with layers in which stromatolites are separated by masses of crumpled and broken algal laminae (Fig. 7b) or, less commonly, by mud with algal clasts. Deposits of broken algal mats generally are confined to individual layers and occur mainly where the algal laminae are steep. The deposition of the irregular stromatolites seems to have involved repeated smoothing out of the relief on a very uneven surface.

Algal Laminites. Beds of algal laminite are common throughout the Carswell Formation. Typically they consist of wavy or wrinkled layers of dolomicrite or silt-grade dolostone (Fig. 8a). The layers are continuous with the layering in solitary stromatolites and pass laterally into beds of conical stromatolites and laterally-linked hemispheroids. Interbedded algal laminites and stromatolites have identical laminae. Units of algal laminites contain layers of flakestone conglomerate, intraformational breccias, pebble-grade polymictic conglomerate, ooids, dolarenite, and dolomicrite. Dolomicrites within the algal laminites are common as thin discontinuous lenses but in places they are up to 1 m thick; they contain mud cracks and folds (Figs. 8b and 8c). The algal laminates contain gypsum pseudomorphs. Units up to several tens of metres thick composed mainly of algal laminite with interbedded stromatolites and other rock types can be traced along strike through most of the northern rim of the ring structure. At one locality algal laminites coat symmetrical ripples in the bottom of large, shallow, bowl-shaped depressions (Figs. 9a and 9b).

Coated Grains. Beds of ooids are common in the lower part of the formation. There are also a few oncolite beds, but oncolites occur mainly as grains in oolitic and rudaceous beds. Oolitic beds are up to 50 cm thick, lensoid, and can seldom be traced for more than a few tens of metres. In some beds the lower surface clearly is erosional but in others the ooids were deposited on an undisturbed surface. Upper surfaces of ooid lenses may be gradational to mud, or overlain by algal laminations or stromatolites. The ooids are well-sorted, but it is quite common to have ooids of markedly different sizes in adjacent beds (Fig. 10a). Both oosparites and oomicrites are present. Oolitic layers contain other clasts, including dolomicrite flakes and oncolites. Platy clasts of dolomicrite may be scattered in an ooid bed or may be concentrated in lenses, either with or without any preferred orientation (Fig. 10b). The ooids in some of the beds are relatively large (3 mm or more) attesting to continuous agitation in very strong currents. Small ooids generally have a radial texture, whereas larger ooids have a radial inner cortex and an outer radial-concentric cortex similar to those described from the Cambrian of Pennsylvania by Heller *et al.* (1980) (Fig. 10c). Oncolites are most abundant near the base of the formation. Oncolites up to 5 mm across occur in well-laminated grainstones, packstones, and wackestones (Fig. 10d). Individual laminae are less than 1 cm thick, and oncolites are most abundant in the coarser laminae; other grains include platy clasts of dolomite mud up to 3 cm long. Some of the laminae are cross-bedded in sets up to 20 cm thick.

Rudites. The rudites include flakestone conglomerate, polymictic dolostone conglomerate and various types of breccia (Table III). Flakestone conglomerates form beds up to 30 cm thick. Generally the bases of the beds are erosional and the tops are mantled by algal laminites. They contain

TABLE III

TYPES OF RUDITES IN THE CARSWELL FORMATION

Type of Rudite	Description	Origin
Flakestone Conglomerate	Monomictic; rounded platy clasts; variable clast orientation; lenticular beds with erosional or non-erosional bases.	Clasts produced by desiccation then transported by wave and tidal currents.
Polymictic conglomerate	Granule pebble grade rounded clasts of dolomicrite, algal laminite, stromatolite, ooids, in lenticular beds or in intercolumnar areas of stromatolites.	Clasts produced by desiccation or intertidal erosion then rounded and mixed by tidal currents.
Crackle Breccia	Monomictic: angular clasts in irregular masses, especially near fractures.	Deformation as a result of tectonic movements or solution collapse during and after deposition of the Carswell Formation.
Mosaic Breccia	Mainly in or near cracks.	Formed by tectonic movements or solution collapse during and after deposition of the Carswell Formation.
Matrix-poor Rubble Breccia	Monomictic and polymictic; angular clasts; no preferred orientation; in cracks and beds.	Tectonic; solution collapse; sedimentary; formed at sediment surface and within sediment during and after deposition of the Carswell Formation.
Matrix-rich Rubble Breccia	Polymictic; subangular clasts; no preferred orientation; sand-grade matrix; as beds, dykes, and sills.	Formed within sediment by injection along fractures from time of deposition until time of the "Carswell Events".
	Polymictic with mud matrix; found in intercolumnar areas of stromatolites.	Sedimentary; washed between stromatolites by waves and tides.

mainly flat pebbles of dolomicrite which may be imbricated, packed vertically, or have an apparently random orientation (Fig. 11a); in some beds the elongate clasts have been bent during compaction. The source of the clasts generally was in an underlying layer or a laterally adjacent undisturbed layer. Thin (2 to 3 cm) layers of flakestone conglomerate are common in some stromatolite beds. Dolostone conglomerates are up to 30 cm thick and may overlie an erosional surface. They contain ooids and well-rounded, equant or elongate grains of algal laminate and dolomicrite up to 5 mm long in a dolomicrite matrix. There are both grain-supported and matrix-supported beds. Some beds are cross-bedded, and elongate clasts commonly are oriented parallel to one another (Fig. 11b). The conglomerates are interbedded mainly with oolitic beds, dolomicrites, and poorly developed stromatolites.

Breccias. The matrix-poor breccias contain blocky clasts up to about 50 cm across, in close contact with each other (Fig. 11a). The interstices generally are filled by granule-size or sand-size grains of the same material. The breccias occur as beds, sills, and inclined dykes. Beds of breccia are present in sequences of algal laminites and hemispheroidal stromatolites and are recognizable where their upper surfaces have been buried by younger beds (Fig. 11d). They are rarely thicker than about 1 m and may be as thin as a few centimetres, and commonly consist mainly of clasts derived from the underlying beds. Some contain more than one clast-type (Fig. 11d). The clasts are not generally rounded, and presumably were not transported far; they lack any preferred orientation. Some of the stratabound, matrix-poor breccias do not have clearly depositional contacts at their upper surfaces, and they can be traced along strike into compositionally and texturally similar breccias that fill cracks above and below the bed and at an angle to the bedding in a manner similar to breccias described by Baldry (1938). The crack-fill breccias in the lower member show a wide variation in both composition and texture but generally contain clasts of a composition similar to the wall-rock, and in some breccias, the wall-rock can be seen to have settled down into the crack (Fig. 12a); in others, clasts must have been derived from beds several metres higher in the section. The breccia-filled cracks vary in thickness from 1 to 100 cm. Breccias about 40 to 50 cm thick can be traced across the bedding for as much as 10 m, and their full extent may be greater. Thicknesses of crack-fill breccias are variable along their lengths; very short, thin breccias, which tend also to be monomictic, are lenticular in cross section with thickness decreasing upwards and downwards. Mantling of the upper ends of some breccias by algal laminations demonstrates that the formation of the breccias took place during deposition of the sequence.

Mosaic breccias (Morrow, 1982) are most common in narrow cracks, whereas rubble breccias are common in wider cracks. At the margins of some cracks, mosaic breccias merge with rubble breccias or with faulted rock (Fig. 12b); folds are uncommon but are present in some breccias. Stratabound breccias are mainly rubble breccias. The breccias are distributed unevenly and tend to be localized within the different rock types. In the hemispheroidal stromatolites, for example, the breccias occur in abundance in certain outcrops, but are separated from each other laterally by many metres of unbrecciated stromatolites. It is not possible to demonstrate displacement of rocks on either side of the breccia-filled cracks. No systematic study was made of the orientation of the cracks, but near Carswell Lake they have apparent dips that seem to be consistently to the west.

The matrix-rich breccias generally are polymictic and consist of angular and sub-angular clasts of most rock types of the Carswell Formation in a sand-grade dolomite matrix

(Fig. 12c) which may make up as much as 75% of the rock. In this type of breccia, it is not generally possible to trace clasts to a source in the local wall rock, and there is a wide variety of clasts in cracks where the wall rock is uniform. Matrix-rich breccias have been injected along bedding planes and up and down cracks (Fig. 12d). There is no flow banding or alignment of clasts in the breccias or sorting of clasts from margins towards the centres of the breccia layers; nor are the breccias bedded. Matrix-rich breccias form some of the largest breccia outcrops in the Carswell Formation and are best exposed near faults. Large masses of breccia are present on either side of a NNE-trending fault at Points Lake. Near the Lac Chute Fault there is a breccia layer that has a minimum thickness of 5 m and cuts across 50 m of stratigraphic section. Currie (1969) described similar breccias in the vicinity of East Rim Lake. In contrast to the breccias described by Currie (1969), the breccias near Points Lake and Carswell Lake contain clasts from only the Carswell Formation. The breccia near the Lac Chute Fault was examined in detail to establish stratigraphic relationships. It is monomictic in some places, polymictic in others, and passes into a crackle breccia in the wall rocks. It is overlain and underlain by wall rock, but there is no evidence that it was buried by dolomites of the Carswell Formation. The breccia does not contain clasts of dolomite.

The matrix-poor and the matrix-rich breccias do not normally occur in contact, but at least one sill of matrix-rich breccia was formed adjacent to a matrix-poor crack-fill breccia. It is difficult to estimate the abundance of the breccias. In some outcrops they are virtually absent, in others they make up between 50% and 100% of the exposed rock. They are found in all rock types except in the large conical stromatolites.

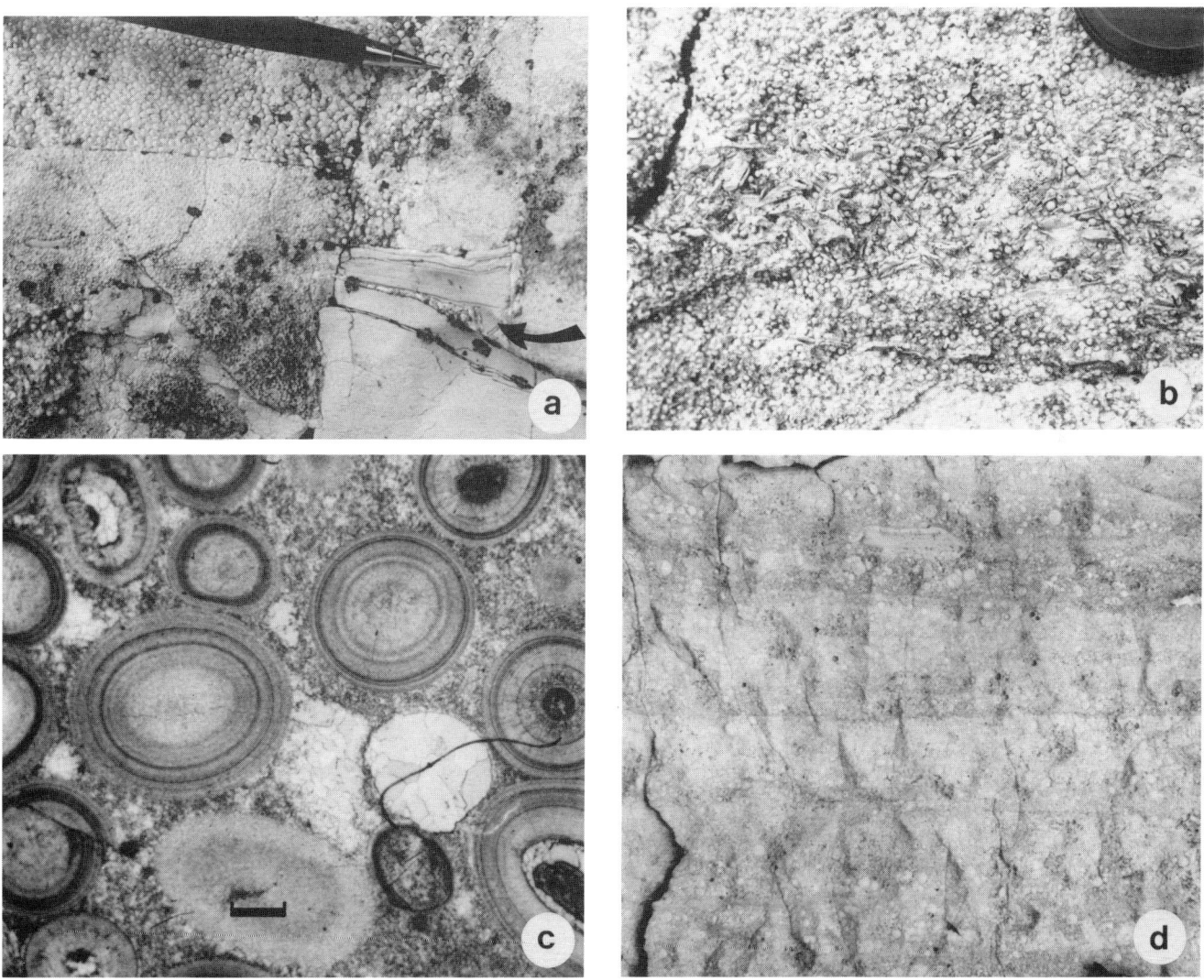

Figure 10. (a) Oolite beds with grains of distinctly different sizes. Note ooids deposited in crack in dolomicrite (arrowed). (b) Oolite bed with elongate dolomicrite intraclasts; lens cap is 5 cm across. (c) Photomicrograph showing ooids with both concentric and radial cortices; scale bar is 0.5 mm. (d) Oncolitic grainstones; height of photograph is 25 cm. All outcrops in lower member, cross section 1.

TABLE IV

INFERRED STRATIGRAPHIC SEQUENCE IN THE CARSWELL FORMATION.

Carswell Formation	Thickness	
		Laterally-linked hemispheroidal stromatolites with interbedded algal laminites. Desiccation cracks and crystal-lined cavities (presumed to be up-faulted from the lower part of the upper member)
		——— Fault ———
Upper Member	? 30-100 m	Massive dolostone with hemispheroidal stromatolites in places.
	65 m	Algal laminites and dolomicrite with desiccation cracks and intraformational folds; interbedded dolorudite and dolarenite; penetrated by crack-fill breccias in places.
	? 60 m	Irregular stromatolites
		——— Fault ———
	? 60 m	Laterally-linked hemispheroidal stromatolites passing laterally to irregular stromatolites with intercolumnar breccias, interbedded flakestone conglomerate and algal laminite.
	100 m	Hemispheroidal stromatolites, laterally-linked in upper part, solitary in lower part. Solitary stromatolites occur in algal laminites with thin layers of flakestone conglomerate and pebble-grade polymictic conglomerates and breccias; shallow channels. Dolomicrite at base.
		——— Sharp irregular contact ———
	5 m	Large conical stromatolites; transition to underlying beds through a thin bed of platy stromatolites.
Lower Member	110 m	Hemispheroidal stromatolites with algal laminites and oolites; crack-fill breccias. Algal laminites and small conical stromatolites near base; flakestone conglomerates; polymictic conglomerates.
	25 m	Interbedded dolomicrites, dolorudites, algal laminites, and oolites. Detrital quartz grains near base.
		——— Fault ———

Amount of overlap between faults or thickness cut out by faults is generally unknown.

Other Rock Types. Other common rock types in the formation are dolomicrite and massive dolostone. White or pale dolomicrite is particularly abundant in the upper part of the upper member where it occurs as lenses only a few millimetres thick or in more extensive beds up to several centimetres thick. It is very commonly interbedded with algal laminites and small stromatolites. Discontinuous layers of it typically have filled depressions in the underlying surface and it is commonly found between stromatolite columns. One of the thickest layers blankets the conical stromatolites at the top of the lower member. In some places it contains mud cracks or syndepositional folds (Figs. 4a and 4b). The massive dolostone is equivalent to the upper member of the Carswell Formation defined by Currie (1969). In the field it forms dark brownish-grey cliffs and bluffs of unbedded granular dolomite. It bears no distinctive features in the field other than a pitted weathered surface. In thin-section it appears to be composed mainly of fractured and partly recrystallized algal laminites and stromatolites. It appears to overlie the algal laminites conformably; its upper limit is not exposed.

Stratigraphic Sequence

Stratigraphic sections were measured at four localities in the northern part of the Carswell structure, at one locality in the southwest, and in drill-hole CAR 58 (Figs. 2 and 13). There is no continuous section through the Carswell Formation. All sections are interrupted by covered areas in valleys between the dolostone ridges (Fig. 3) and broken by steeply dipping faults (Fig. 4). Establishment of the stratigraphic sequence and correlation between fault blocks was made possible because of the presence of distinctive stratigraphic units. The unit containing conical stromatolites overlain by dolomicrite persists for over 15 km along strike in the lower part of the formation, and the Carswell Formation has been divided into an upper and lower member along the contact between the conical stromatolites and the overlying dolomicrite (Fig. 5b). The definition of the members proposed is not the same as was suggested by Currie (1969). The lower member defined by Currie (1969) includes all of the lower member and most of the upper member proposed here; the upper member of Currie comprises only the "massive

Figure 11. (a) Flakestone conglomerate overlain by algal laminites. Note vertically oriented clast; lower member, cross section 1. (b) Dolorudite overlain by algal laminites; upper member, cross section 2. (c) Matrix-poor breccia with angular clasts of algal laminite; note pebble-grade clasts and white dolomicrite (arrowed) in matrix; upper member, cross section 2. (d) Matrix-poor breccia; clasts are angular and closely packed; there is one clast of dolomicrite (arrowed); note the sharp contact with the overlying algal laminites; upper member, cross section 2.

dolostone'' of this report (Table IV). The basal contact of the Carswell Formation with the Douglas Formation is faulted so the minimum thickness of the lower member is about 140 m; the upper member is at least 300 or 400 m thick depending on the amount of repetition of the sequence by faults. Neither the top nor bottom of the formation is exposed, but detrital quartz grains and clays in the dolostones near the faulted contact with the Douglas Formation indicate that most of the lower part of the formation has been penetrated in drill holes.

The lowest part of the formation generally is exposed on the inside of the ring structure, and in sections where the beds are steeply dipping rocks are younger towards the outside of the ring. The outcrops of hemispheroidal stromatolites with algal laminites at the northern ends of sections 1 and 2 cannot be assigned unequivocally to any part of the sequence, but for purposes of reconstruction of the depositional history and the geologic structure they have been considered correlative with the hemispheroidal stromatolites in the lower part of the upper member (Table IV).

Figure 12. (a) Crack-fill breccia with large lumps of locally-derived rock (arrowed); lower member, cross section 1. (b) Faulted rock adjacent to crack-fill breccia; lower member, cross section 1. (c) Matrix-rich breccia; upper member, cross section 3. (d) Matrix-rich breccia which has been injected along crack parallel to local bedding; upper member, cross section 1.

SEDIMENTOLOGY
Lower Member

Throughout the lower member there is abundant evidence of transport of sediment up to gravel-grade. Cross-bedded oncolitic and oolitic grainstones indicate periods of strong current action, but beds generally are thin and lenticular, suggesting that no large shoals were developed. The non-erosional base beneath many of the oolitic beds and the common occurrence of mud matrix in only the upper portions of oolitic beds may be explained by deposition as a result of storm wash-over into areas where currents were relatively weak. The presence of oncolites, and particularly of platy clasts of dolomicrite in oolitic beds, is evidence that the ooids were mixed with other sediments and further supports the suggestion that they were transported during storms. Scattered ooids in stromatolitic rocks are most easily explained as washed-in sediment. Erosional surfaces beneath some oolitic beds testify to scour before deposition, and the incorporation of ooids in intraformational conglomerates at the bases of some oolitic beds indicates that the erosional currents introduced the ooids. Very good sorting of ooids within beds that now contain other types and sizes of clasts indicates that they were deposited somewhere other than the place in which they first formed. The preservation of both radial and radial-concentric textures in the same ooids and the restriction of purely radial textures to smaller ooids suggest that early growth occurred while the ooids were carried in suspension (Heller et al., 1980).

The intraformational conglomerates indicate contemporaneous erosion of muds and algal laminites. The most distinctive beds are flakestone conglomerates composed entirely of flat pebbles of dolomicrite. Relatively thin beds with no evidence that they fill deep channels favour an origin by wave or tidal transport after formation of the clasts by desiccation.

Though the clasts in all of the flakestones are similar, the mode of deposition seems to have varied. The variability in clast orientation can be explained if it is assumed that both wave and tidal currents played a role. Those beds in which flat clasts are packed vertically may have formed as a result of wave activity (Sanderson and Donovan, 1974), whereas the beds in which the clasts are imbricated more likely were produced by unidirectional currents, perhaps during tidal ebb. The variability in clast orientation and, in particular, the very irregular upper surface of some of the beds suggests that periods of transport were short and that the clasts were not reworked into stable positions. Short periods of relatively strong currents, which disrupted beds and deposited flakestone conglomerates, are best explained by storms, though the lack of graded bedding and the intercalation with algal laminites favours an intertidal setting (Akhtar, 1976; Eriksson and Truswell, 1974; Siedlecka, 1978), rather than deposition below storm-wave base as inferred for the Precambrian of Norway by Tucker (1982). The polymictic conglomerates typically show evidence of more transport than the flakestones; clasts are well-rounded and equant or elongate, suggesting considerable transport and abrasion. Lenses of oolites, flakestones, and polymictic conglomerates with erosional bases may be the fills of shallow channels. They are similar in thickness and lateral extent to intraclast beds in the tidal channels of Shark Bay (Davies, 1970; Logan et al., 1974), and the channels are of similar scale to those of the modern tidal flats of southwest Andros Island (Gebelein, 1974).

Detrital beds predominate in the lower part of the member but the relative proportion of stromatolites increases up-section. The two main types of stromatolite in the lower part of the unit, hemispheroidal and small conical (egg carton) types, show little or no evidence of current activity, although the ripple-like profile of some of them may have been produced by currents. Near the top of the unit, the transition from platy stromatolites to well-aligned large conical stromatolites indicates strong N-S currents. The widespread distribution of the conical stromatolite layer together with the continuity of laminae, a lack of internal erosion surfaces, and only small amounts of intercolumnar debris, suggest a subtidal setting.

Breccias are found mainly near the top of the lower member and fill cracks up to several metres deep in the algal laminites. The cracks do not have the typical polygonal pattern of desiccation cracks and are presumed to have formed as a result of contemporaneous deformation.

Upper Member

Deposition of the upper member was marked by extensive intraformational reworking of many sediments and the development of abundant breccias. Contemporaneous erosion of the sediments was very common and currents were sufficiently strong to transport large clasts. The channels indicate an intertidal setting. Disruption and brecciation in the algal laminites resulted from both desiccation and structural disturbances. The clasts in the very coarse breccias likely were produced mainly by fracturing.

The lateral changes suggest that depositional conditions were variable. The algal laminites contain several different kinds of breccia and conglomerate, some of them filling

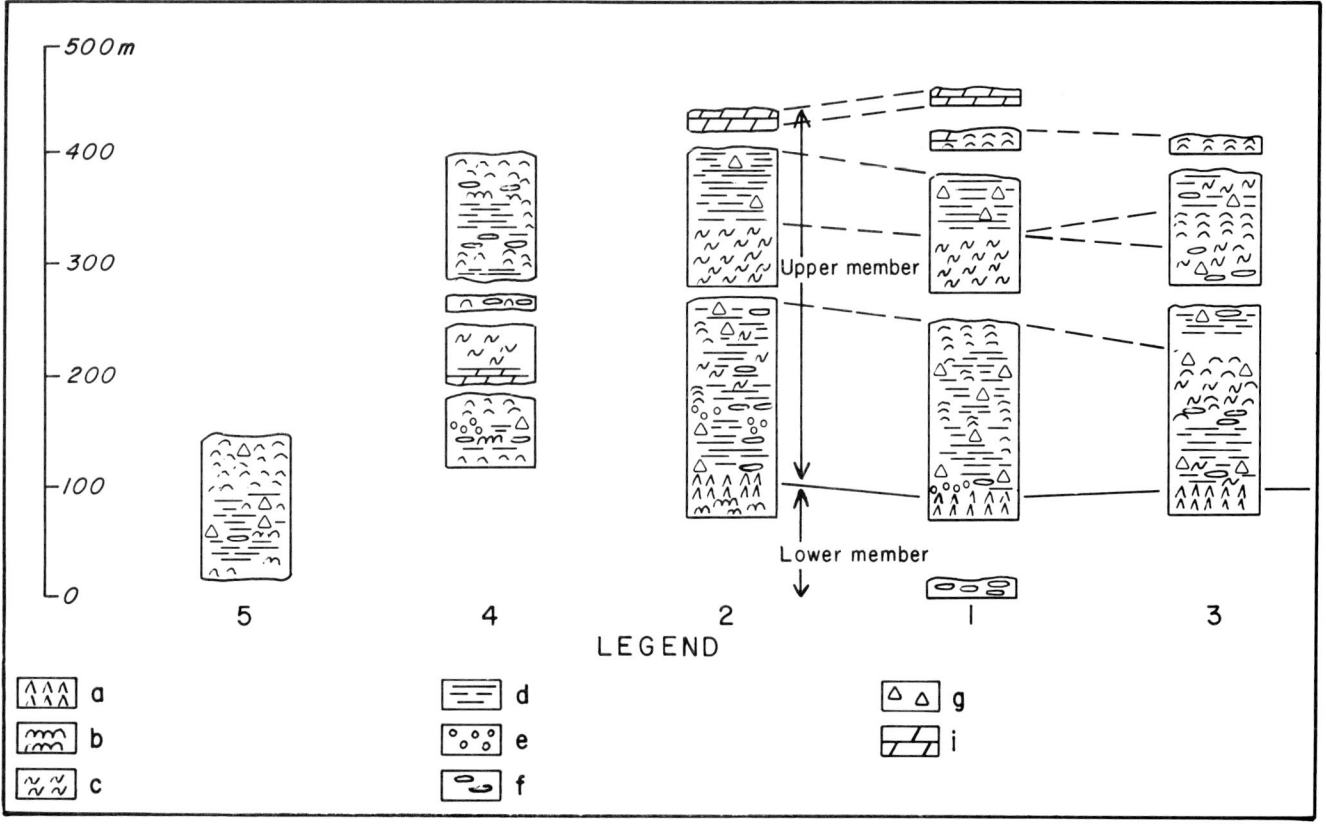

Figure 13. Stratigraphic sequence in the Carswell Formation interpreted from measured cross sections. Legend: a = marker bed of large conical stromatolites; b = hemispheroidal stromatolites; c = irregular stromatolites; d = algal laminites; e = oolites; f = intraformational conglomerates (dolorudites); g = breccias; h = massive dolostones.

shallow scours, and oolitic beds. The capping of detrital beds, and even of individual clasts by laminae of apparently algal origin, suggests that active currents repeatedly interrupted conditions in which the algal mats flourished. Disruption of the sediment by contemporaneous erosion increased through time because the upper parts of the algal laminites contain abundant beds of intraformational conglomerate, oolites, and flakestones. The stromatolites above the algal laminites developed gradually. They can be traced laterally into algal laminites with intraformational conglomerates and breccias. Local development of large hemispheroidal stromatolites probably was controlled by currents, with the development of stromatolites in areas where currents had the least influence. Limited vertical persistence of individual stromatolites indicates that shallow channels may have changed position frequently, and internal erosion surfaces indicate current scour. Where the hemispheroidal stromatolites are best developed they do not contain intercolumnar breccias; where they have more irregular shapes, and lean or are recumbent, breccias are common between columns indicating that scouring and sediment deposition was localized.

The laterally-linked hemispheroidal stromatolites exposed in the outermost part of the ring (sections 1 and 2) developed in a protected area where desiccation caused shrinkage and cracking of algal layers, and periodic flooding washed debris from the stromatolites into the intercolumnar areas. The crystal-lined vugs in the stromatolites are fairly large, up to several centimetres, and probably developed around gas bubbles. A high intertidal setting, perhaps in a sheltered pond, would be a likely depositional environment.

In some cross sections the layers of large hemispheroidal stromatolites are laterally equivalent to beds of irregular stromatolites. The irregular ones seem to have developed in an area where currents were very weak so that mats of relatively high relief could develop. The repeated wrinkling and brecciation of the layers, particularly where very steep layers appear to have collapsed, and the many changes in mat morphology throughout the bed could be a result of growth in a sheltered intertidal zone.

Deposition continued with accumulation of more algal laminites, laminated muds, and intraformational conglomerates. Oolitic and oncolitic beds are uncommon, and there is no evidence of channels, but desiccation cracks indicate periodic exposure. The depositional setting probably was high intertidal. The uppermost part of the upper member appears structureless in the field but contains irregularly shaped masses of algal-laminated material in a dolomicrite matrix. Original bedding may have been destroyed by brecciation, which is visible only in thin section. Lack of primary structures make an interpretation of the depositional environment difficult.

The upper member contains the most abundant breccias in the Carswell Formation. Breccias in carbonate sequences may form in a variety of ways: by development of karst topography beneath unconformities, soil formation, mass flow, meteorite impact, dissolution of evaporites, or by tectonic deformation. The Carswell breccias are not karst deposits. Karst breccias typically are bedded (Brain, 1958) and contain speleothems and large cavities (Jennings, 1971). The Carswell breccias have none of the features of caliche breccias, which contain acicular carbonate cement and pisolites (Blount and Moore, 1969).

Though some breccias occur in beds, they lack grading, clast-alignment, and the other features common in mass-flow deposits (Cook, 1979; Richter and Fuchtbauer, 1981). It has been proposed that the ring-like pattern of the Carswell Formation was produced by meteorite impact (Innes, 1964; Pagel, 1975; Harper, 1982; Pagel et al., 1985) and breccias in large masses, beds, and dykes occur in many circular structures that are considered to have formed in this way (Horz et al., 1977; Lambert, 1981; Offield and Pohn, 1977; Pohl et al., 1977; Reiff, 1978; Wilshire et al., 1972a, 1972b). The Carswell breccias have several features in common with impact breccias, particularly with those in carbonate rocks, but lack flow foliation, mylonitic matrix, broken shatter cones, evidence of intense granulation and fracturing of clasts, and other features that occur in impact breccias. The evidence for formation of many of the crack-fill breccias during deposition of the Carswell Formation also provides a strong argument against relating their development to that of the circular structure. The most reasonable explanation for the origin of most of the discordant breccia masses in the Carswell Formation is that they formed as a result of deformation during and after deposition of the sediments. Fahrig (1961) recognized that some of the breccias were depositional, but Currie (1969) demonstrated that breccias cut across bedding with a "zig-zag" pattern and pointed out the location of breccia masses close to faults and to synclinal axes. All of the discordant breccia bodies appear to fill tensional cracks. Mosaic breccias in cracks appear to have formed because clasts separated as cracks opened, and the common presence of only a small proportion of clasts different from the adjacent wall-rock indicates that clasts generally fell from above into the space created as cracks opened. The lack of cavities, or sparry calcite cement, in most of the breccias is surprising and most of the matrix must have filtered in between the clasts. The breccias may also have been compacted if the rock settled into position after tension had created the cracks. The development of large numbers of breccia clasts at the depositional surface probably occurred infrequently because clearly demonstrable beds of breccia are much less common than sills. The horizontal upper surfaces of breccia beds may have been produced by wave action. The absence from most of the sediments of clasts larger than small pebbles indicates that even when breccias formed at the surface, the particles could not be transported far. Crystalline basement clasts such as those discovered by Currie (1969) were not seen in the breccias so the upward emplacement of clasts from below the Carswell Formation was not common. Evidence of upward injection of breccias with only dolostone clasts is very scarce and is limited to a few small dykes (Fig. 12d). Many of the large bodies of matrix-rich breccias are located close to faults that were active after deposition of the Carswell sediments. It is reasonable to conclude that movements on these faults and others may have caused deformation of the sediment during deposition and contributed to the formation of the crack-fill breccias. The crackle breccias in many of the dolomites could have developed either at the time of deposition of the Carswell Formation, or during later deformation.

Breccias in the Carswell Formation may have formed also as a result of solution-collapse. The dolostones do contain evidence of former gypsum crystals and dissolution of evaporites at depth may have caused foundering of beds and the penetration of some cracks to the depositional surface. The step-like form of some of the breccia layers would be expected if large masses of rock had collapsed. Collapse as a result of salt dissolution would tend also to occur by fracturing along existing fault zones.

Depositional History

Most of the lower member was deposited in the intertidal zone where algal mats and small stromatolites were developing. Flooding of the area during storms left beds of ooids and intraclasts on algal flats and in shallow tidal channels, but the surfaces of the detrital deposits commonly were covered by algal mats. Periodic evaporation caused precipitation of gypsum. Deepening of the water was accompanied by formation of hemispheroidal stromatolites followed by widespread development of conical stromatolites. The lack of development of thick oolitic grainstones during the deepening phase may be explained by stagnation of oolite shoals as the water deepened (Gebelein, 1974; Harris, 1979). Strong currents, perhaps tidal, caused the crests of conical lensoid stromatolites to be aligned parallel to one another.

The massive, white dolomicrite is common as thick beds in the upper part of the upper member, as thin, lenticular, mud drapes between stromatolites, and as discontinuous layers in the algal laminites throughout the formation. The dearth of stromatolites and absence of algal laminites from the thick mud layer at the base of the upper member suggests that deposition was at such a high rate that algal growth was inhibited, but the presence of conical stromatolites rules out sudden catastropic burial during a storm; the bed is also considerably thicker than deposits produced by hurricanes in modern environments (Ball *et al.*, 1967), and does not contain debris that would have been washed in from the intertidal zone during the ebb (Perkins and Enos, 1968). Termination of stromatolite growth by detrital carbonate has been observed by Hoffman (1976) in the Pethei Group of the Slave Craton.

The upper member records tidal flat progradation and shallowing of the basin. After the layer of conical stromatolites had been buried, a decrease in the sedimentation rate permitted re-establishment of algal mats and a return to the conditions that had prevailed during deposition of the lower part of the lower member. Development of a network of shallow tidal channels resulted in the formation of stromatolite "meadows" in the interchannel areas. Curling and crenulation of algal mats is common in modern ponded tidal flats (Gebelein, 1976), and the irregular stromatolites with variability of form, thick layering, and breccias of wrinkled stromatolites may have accumulated in such a setting. They may be comparable to the "aberrant stromatolites" described by Pratt and James (1982). Subsequent progradation of the high intertidal and supratidal deposits led to increased exposure and more frequent desiccation. Dolomicrite layers probably formed during storms.

Throughout deposition of the upper member fractures developed as a result of tectonic activity and solution collapse, and breccias formed from the breakdown of the relatively brittle sediment. Most of the breccias were confined to cracks, some of which extended to the sediment surface, but some of the breccias were laid down on the surface, presumably after being transported by wave or tidal currents. In the algal laminites above the sequence of irregular stromatolites, the white dolomicrites were folded, perhaps also as a result of contemporaneous deformation.

The sediments were deposited on a shallow marine platform affected by tides, currents, and tectonic activity, but were virtually unaffected by clastic terrigenous contamination.

Geochemical analysis shows a considerable decrease in SiO_2, K_2O, Al_2O_3, and Na_2O above the Douglas Formation contact. There are also interbedded marls and pelites in the upper parts of the Douglas Formation. The gradual decrease in the rate of supply of terrigenous detritus likely came about because of peneplanation of the source area. The sediments of the Carswell Formation are entirely shallow water so subsidence of several hundred metres must have taken place during their deposition.

STRUCTURE

The dolostones are well exposed in the north, east, and south to southwest parts of the ring but are poorly exposed in the west and southeast. Diamond drill holes in the north and east have intersected the dolostones to a depth of 275 m and in most places did not reach the base of the formation. However, three of the four holes drilled in the southern part of the ring, just south of Cluff Lake, intersected sandstones of the Douglas Formation beneath relatively flat lying dolostones with faulted contacts at depths ranging from 35 to 45 m. Silicification of the dolostone, along with pyrite and carbonaceous material in fractures, is common near the Douglas Formation contact. The lack of dolostone exposure in the west may be because glacial deposits have covered flat-lying beds, but in the southeast the dolostones appear to have been completely eroded.

The striking ring-like outcrop pattern of the Carswell Formation and the interior rocks has provoked much debate over the history of deformation of the area (Fahrig, 1961; Currie, 1969; Pagel, 1975; Harper, 1982, Pagel *et al.*, 1985). Detailed structural studies are hampered by limited exposures and lack of good stratigraphic markers. Of all the formations in the Athabasca Basin, the Carswell Formation is lithologically the most diverse, and there is a distinct marker bed containing large conical stromatolites at the top of the lower member. The marker bed of conical stromatolites is present in all of the cross sections in the northern part of the ring and has been used to demonstrate direction of throw on several of the faults. It is absent in the southern part of the ring where correlation between fault blocks is very difficult in spite of the less complex structure. Other beds in the formation are much less distinctive than the conical stromatolite marker bed, and much less persistent laterally, but they can be used for correlation over short distances, and in some cross sections the position of the conical stromatolite marker bed has been inferred to be about 100 m below the base of the irregular stromatolites.

The new stratigraphic information has allowed the construction of structural cross sections across five traverses of the ring structure (Fig. 4).

The structure of the Carswell Formation is essentially that proposed by Fahrig (1961) and Currie (1969). The dolostone is deformed into a series of broad folds with trends roughly parallel to the circumference of the ring. Deformation has been most intense in the north where beds dip steeply or are overturned. Rocks of the Douglas Formation have been detected adjacent to the Carswell Formation both inside and outside the ring, but the contacts are faulted and the folds in the Carswell are not simply synclinal (Fig. 4). In cross section 1 the structure is synclinal and all of the beds are overturned towards the outside of the ring. In cross section 2 there is a syncline overturned towards the outside of the ring. In cross section 3 the beds are right way up and form a syncline with the hinge-line close to the outer part of the ring. Sections 4 and 5 contain gently dipping beds and broad folds. All of the folds have been cut by arcuate faults that follow the trend of the ring, and by straight faults that cut across the circular structure. The direction of dip of the fault planes is not known so they have been portrayed as vertical in the figure. The Carswell Formation has been faulted out in the southern part of the structure, and the circular outcrop has been offset by faults in several places, most noticeably on either side of Carswell Lake. Small folds with steeply plunging hinge-lines along the north shore of Carswell Lake may have been produced by drag along the Carswell Lake Fault (see Map in pocket).

Most of the structural features in the Carswell Formation were produced during the so-called "Carswell event" when the beds of dolostone were folded and faulted and overturned in many places to produce a ring-shaped structure in which the folds and faults followed a circular trend. Disruption and offsetting of the annular pattern may have taken place during later stages of the same event or as a result of movement along northwest- and northeast-trending faults at a later date. Many of the longer faults are part of a system that extends far beyond the development of the Carswell structure and the presence of intraformational breccias close to faults such as the East Rim Fault and the Points Lake Fault indicates that they probably were active during deposition of the Carswell Formation. Pacquet and McNamara (1985) have described evidence of tectonic activity in the Carswell area prior to deposition of the Carswell Formation, so the area was tectonically active, at least intermittently, before formation of the ring structure.

ORIGINAL EXTENT OF THE CARSWELL FORMATION

The Carswell Formation is exposed only in the ring structure. The presence of the Douglas Formation adjacent to the dolostones, both inside and outside the ring, indicates that more extensive dolostones probably were removed by erosion. It is difficult to judge just how extensive the depositional area of the dolostone might have been because the present outcrops are in an area where there has been the least amount of erosion of the Athabasca Group. Ramaekers (1981) has pointed out that the western part of the basin received marine sediments throughout most of the depositional history of the Athabasca Group. This suggests that the depositional basin, during carbonate sedimentation, was also open to the ocean to the west.

If the full thickness of both Carswell and Douglas Formations was present at the time of the "Carswell event", then there must have been about 800 m of erosion immediately outside of the Carswell structure since Ordovician time.

CONCLUSIONS

1) The sediments of the Carswell Formation were deposited between 1200 and 1300 Ma ago in a shallow sea to which there was no terrigenous clastic contribution.

2) During deposition, sediments and rocks of the formation were fractured by both tectonic movements and by collapse of parts of the sequence following dissolution of evaporites at depth. All of the post-depositional breccias in the Carswell Formation can be explained satisfactorily by processes other than meteorite impact.

3) Most deformation of the Carswell dolostone took place at the time of the "Carswell event", 480 Ma ago. The rocks were intensely folded and in many places overturned towards the outside of the ring structure. The nature of the "Carswell event" is still unclear.

ACKNOWLEDGEMENTS

Amok Ltd. provided logistic support in the field and access to samples and core, and D. Koning and G. Rees provided invaluable assistance in the field. The authors appreciate the comments of W. G. E. Caldwell, J. A. Donaldson, and M. R. Stauffer on an early version of the manuscript.

REFERENCES

Akhtar, K., 1976, Facies Analysis and Depositional Environments of the Bhauder Limestone (Precambrian), Southeastern Rajasthan and Adjoining Madhy Pradesh, India: Sedimentary Geology, v. 16, p. 299-318.

Baldry, R. A., 1938, Slip-planes and Breccia Zones in Peru: Quarterly Journal of the Geological Society, v. 94, p. 347-358.

Ball, M. M., Shinn, E. A., and Stockman, K. W., 1967, The Geologic Effects of Hurricane Donna in South Florida: Journal of Geology, v. 75, p. 583-597.

Bell, K., 1981, A Review of the Geochronology of the Precambrian of Saskatchewan — Some Clues to Uranium Mineralization: Mineralogical Magazine, v. 44, p. 371-378.

_____, 1985, Geochronology of the Carswell Area, Northern Saskatchewan: in Lainé, R., Alonso, D., and Svab, M., eds., The Carswell Structure Uranium Deposits, Saskatchewan: Geological Association of Canada Special Paper 29.

Blake, D. A. W., 1956, Geological Notes on the Region South of Lake Athabasca and Black Lake, Saskatchewan and Alberta: Geological Survey of Canada Paper 55-33, 12 p.

Blount, D. N., and Moore, C. H., Jr., 1969, Depositional and Non-Depositional Carbonate Breccias, Chiantla Quadrangle, Guatemala: Geological Society of America Bulletin, v. 80, p. 429-442.

Brain, C. K., 1958, The Transvaal Ape Man-Bearing Cave Deposits: Transvaal Museum Memoir 11, 131 p.

Clauer, N., Ey, F., and Gauthier-Lafaye, F., 1985, K-Ar Dating of Different Rock Types From Cluff Lake Uranium Ore Deposit: in Lainé, R., Alonso, D., and Svab, M., eds., The Carswell Structure Uranium Deposits, Saskatchewan: Geological Association of Canada Special Paper 29.

Cook, H. E., 1979, Ancient Continental Slope Sequences and Their Value in Understanding Modern Slope Development: in Doyle, L. H., and Pilkey, O. H., eds., Geology of Continental Slopes: Society of Economic Paleontologists and Mineralogists Special Publication No. 27, p. 287-305.

Currie, K. L., 1969, Geological Notes on the Carswell Circular Structure, Saskatchewan (74K): Geological Survey of Canada, Paper 67-32, 60 p.

Davies, G. R., 1970, Algal-Laminated Sediments, Gladstone Embayment, Shark Bay, Western Australia: in Logan, B. W., Davies, G. R., Read, J. F., and Cebulski, D. E., eds., Carbonate Sedimentation and Environments, Shark Bay, Western Australia: American Association Petroleum Geologists Memoir 13, p. 169-205.

Eriksson, K. A., and Truswell, J. F., 1974, Tidal Flat Associations from a Lower Proterozoic Carbonate Sequence in South Africa: Sedimentology, v. 21, p. 293-309.

Fahrig, W. F., 1961, The Geology of the Athabasca Formation: Geological Survey of Canada Bulletin 68, 41 p.

Gebelein, C. D., 1974, Modern Bahaman Platform Environments: Field Trip Guidebook Annual Meeting, Geological Society of America, 96 p.

_____, 1976, Open Marine Subtidal and Intertidal Stromatolites (Florida, the Bahamas, and Bermuda): in Walter, M. R., ed., Stromatolites: Developments in Sedimentology, v. 20, Amsterdam, Elsevier, p. 381-388.

Harper, C. T., 1982, Geology of the Carswell Structure, Central Part (Parts of NTS Areas 74K-5, -6, -11, -12): Saskatchewan Mineral Resources, Report 214, 6 p.

Harris, P. M., 1979, Facies Anatomy and Diagenesis of a Bahamian Ooid Shoal: Sediments VII, Comparative Sedimentology Laboratory, Rosentiel School of Marine and Atmospheric Science, University of Miami, Florida, 163 p.

Heller, P. L., Komar, P. D., and Pevear, D. R., 1980, Transport Processes in Ooid Genesis: Journal of Sedimentary Petrology, v. 50, no. 3, p. 943-952.

Hoffman, P., 1976, Environmental Diversity of Middle Precambrian Stromatolites: in Walter, M. R., ed., Stromatolites: Developments in Sedimentology, v. 20, Amsterdam, Elsevier, p. 599-611

Horz, F., Gall, H., Huttern, R., and Oberbeck, V. R., 1977, Shallow Drilling in the "Bunte Breccia" Impact Deposits, Ries Crater, Germany: in Roddy, D. J., Pepin, R. O., and Merrill, P. B., eds., Impact and Explosion Cratering: New York, Pergamon Press, p. 425-448.

Innes, J. J. S., 1964, Recent Advances in Meteorite Crater Research of the Dominion Observatory: Meteoritics, v. 2, p. 219-249.

Jennings, J. N., 1971, Karst: Cambridge, Massachusetts, Michigan Institute of Technology Press, 252 p.

Lewry, J. F., Sibbald, T. I. I., and Rees, C. J., 1978, Metamorphic Patterns and their Relation to Tectonism and Plutonism in the Churchill Province in Northern Saskatchewan: in Metamorphism in the Canadian Shield: Geological Survey of Canada Paper 78-10, p. 139-154.

Lambert, P., 1981, Breccia Dykes: Geological Constraints on the Formation of Complex Craters: in Schultz, P. H., and Merrill, R. B., eds., Multi-Ring Basins: Proceedings of Lunar and Planetary Sciences, 12A, p. 59-78.

Logan, B. W., Hoffman, P., and Gebelein, C. D., 1974, Algal Mats, Cryptalgal Fabrics, and Structures, Hamelin Pool, Western Australia: in Logan, B. W., Read, J. F., Hagan, G. M., Hoffman, P., Brown, R. G., Woods, P. J., and Gebelein, C. D., eds., Evolution and Diagenesis of Quaternary Carbonate Sequences, Shark Bay, Western Australia: American Association of Petroleum Geologists Memoir 22, p. 140-194.

Morrow, D. W., 1982, Descriptive Field Classification of Sedimentary and Diagenetic Breccia Fabrics in Carbonate Rocks: Bulletin of Canadian Petroleum Geology, v. 30, p. 227-229.

Offield, T. W., and Pohn, H. A., 1977, Deformation at the Decaturville Impact Structure, Missouri: in Roddy, D. J., Pepin, R. O., and Merrill, R. B., eds., Impact and Explosion Cratering: New York, Pergamon Press, p. 321-341.

Pagel, M., 1975, Cadre Géologique des Gisements d'Uranium dans la Structure Carswell "Etude des Phases Fluides": Thèse d'Etat, Université de Nancy, France, 157 p.

Pagel, M., Wheatley, K., and Ey, F., 1985, Origin of the Carswell Circular Structure: in Lainé, R., Alonso, D., and Svab, M., eds., The Carswell Structure Uranium Deposits, Saskatchewan: Geological Association of Canada Special Paper 29.

Pacquet, A., and McNamara, S., 1985, A study of the Basal Athabasca Succession in the D, E, L, F, and S Areas of the Carswell Structure: in Lainé, R., Alonso, D., and Svab, M., eds., The Carswell Structure Uranium Deposits, Saskatchewan: Geological Association of Canada Special Paper 29.

Perkins, R. D. and Enos, P., 1968, Hurricane Betsy in the Florida-Bahama Area – Geologic Effects and Comparison with Hurricane Donna: Journal of Geology, v. 76, p. 710-717.

Pohl, J., Stoffler, D., Gall, H., and Ernston, K., 1977, The Ries Impact Crater: in Roddy, D. J., Pepin, R. O., and Merrill, R. B., eds., Impact and Explosion Cratering: New York, Pergamon Press, p. 343-404.

Pratt, B. R., and James, N. P., 1982, Cryptalgal-Metazoan Bioherms of Early Ordovician Age in the St. George Group, Western Newfoundland: Sedimentology, v. 29, p. 543-569.

Ramaekers, P., 1981, Hudsonian and Helikian Basins of the Athabasca Region, Northern Saskatchewan: in Campbell, F. H. A., ed., Proterozoic Basins of Canada: Geological Survey of Canada Paper 81-10, p. 219-233.

Ramaekers, P., and Dunn, C., 1977, Geology and Geochemistry of the Eastern Margin of the Athabasca Basin: in Dunn, C. E., ed., Uranium in Saskatchewan: Saskatchewan Geological Survey Special Publication No. 3, p. 297-322.

Reiff, W., 1978, Monomict Movement Breccias: An Indicator of Meteoritic Impact: Meteoritics, v. 13, p. 605-609.

Richter, D. K. and Fuchtbauer, H., 1981, Merkmale und Genese von Breccien und ihre Bedeutung im Mesozoikum von Hydra (Griechenland): Zeitschrift der Deutschen Geologischen Gesellschaft Band 132, p. 451-501.

Sanderson, D. J., and Donovan, R. N., 1974, The Vertical Packing of Shells and Stones on some Recent Beaches: Journal of Sedimentary Petrology, v. 44, p. 680-688.

Siedlecka, A., 1978, Late Precambrian Tidal-Flat Deposits and Algal Stromatolites in the Batsfjord Formation, East Finnmark, North Norway: Sedimentary Geology, v. 21, p. 277-310.

Tucker, M., 1982, Precambrian Dolomites: Petrographic and Isotopic Evidence that they Differ from Phanerozoic Dolomites: Geology, v. 10, p. 7-12.

Wilshire, H. G., Howard, K. A., and Offield, T. W., 1972a, Impact Breccias in Carbonate Rocks, Sierra Madera, Texas: Geological Society of America Bulletin, v. 82, p. 1009-1018.

Wilshire, H. G., Offield, T. W., Howard, K. A., and Cummings, D., 1972b, Geology of the Sierra Madera Cryptoexplosion Structure, Pecos County, Texas: United States Geological Survey Professional Paper 599-11, 42 p.

MINERALOGY AND METALLOGENY OF URANIFEROUS OCCURRENCES IN THE CARSWELL STRUCTURE

F. Ruhlmann
Cogema, BU-DRM, 78140, VELIZY, France

ABSTRACT

A study of the mineralogy and metallogeny of uraniferous occurrences was undertaken by Amok Ltd. to characterize both economic and non-economic uranium mineralization in the Carswell structure. It is possible to define multiple parageneses as a result of polyphase hydrothermal metallogenic activity which affected both the metamorphic basement and the overlying Athabasca sandstones. The descriptions of the principal occurrences are based on the host rocks and have been divided into sandstone-hosted and gneiss- or granite-hosted mineralization.

Sandstone-hosted mineralization was studied in detail in the D orebody. It is an excellent example of an unconformity-related uranium deposit resulting from an episode of hydrothermal mineralization around 1050 Ma. Other sandstone-hosted examples, particularly the Donna boulder train and the E, S, and U showings contain different mineral associations and belong to a different mineralizing episode.

Studies on the basement-hosted deposits at Claude, Dominique-Peter, and to a lesser extent the N and OP areas, indicate a major mineralizing episode of the same age and nature as the D mineralization. The element association originating from this episode is U-Mo-Bi-Se-S, with minor Te-Ni-Co. Other mineralizing episodes succeeded the major event and each contain distinct assemblages. They appear to be episodes of lesser significance which are superimposed over the earlier mineralization and occur within the same structures. Mineralization in the Carswell structure ranges in age from 1800 to 200 Ma, and fits very well with the regional mineralization described for the entire Athabasca Basin.

Two occurrences within the Carswell structure, in the Sophie and Numac areas, contain unusual uraniferous assemblages. Both areas are examples of mineralization that is much older than the economic deposits in the southern part of the structure.

RÉSUMÉ

L'étude minéralogique et métallogénique des occurrences uranifères de l'anneau de Carswell s'inscrit dans le vaste programme de caractérisation des minéralisations économiques ou non, entrepris par Amok.

A partir des caractéristiques propres à chacune des occurrences, il a été possible de définir plusieurs paragénèses metallifères qui sont les reflets d'une activité métallogénique polyphasée à caractère hydrothermal. Cette activité qui s'étale dans le temps, de 1800 Ma pour la plus ancienne à 200 Ma pour la plus récente, affecte aussi bien les formations métamorphiques du socle que la couverture gréseuse hélikienne constituée par les grès Athabasca.

Nous décrivons donc les principales occurrences encaissées d'une part dans la couverture, et, d'autre part celles localisées dans le socle.

Les minéralisations intragrès ont été particulièrement étudiées dans le corps minéralisé "D". Ce dernier, situé à la discordance grès-socle est un excellent exemple de cet épisode hydrothermal dont la mise en place se situe aux alentours de 1050 Ma. Cet épisode présente une paragénèse extrêmement complexe où aux uraninites et brannérites précoces succède un dépôt polymétallique à séléniures, téllurures et sulfures. L'association caractéristique est à U-Mo-Bi-Te-Se-S-Au.

D'autres minéralisations ont été définies dans les occurrences de Donna, E, S et U; elles diffèrent de la précédente par la nature de la paragénèse et par l'époque de mise en place. Parmi ces paragénèses, la plus importante est celle à pechblende-hématite de Donna; l'âge de la mise en place de cette paragénèse se situe vers 380 Ma.

Les minéralisations intrasocle métamorphique ont été étudiées dans les occurrences de Claude, Dominique-Peter et OP. Les différences paragénétiques sont peu marquées d'une occurrence à l'autre; par contre on note de grandes différences texturales qui nous ont permis de différencier les minéralisations de Claude de celles de OP.

Les minéralisations de Claude ainsi que de Dominique-Peter sont encaissées dans une "zone mylonitique" intra-socle; l'existence au sein de celle-ci de deux systèmes de fracturation permet d'une part l'individualisation des "boules" et, d'autre part la mise en place de la minéralisation. Cette dernière associée à une gangue chloriteuse présente les mêmes caractéristiques que celles de la minéralisation du corps "D". Il s'agit donc d'un seul et même épisode minéralisateur qui affecte à la fois le socle et la couverture.

Les minéralisations de OP se distinguent de celles de Claude par la présence d'une paragénèse exclusivement urano-sulfurée en remplissage de fractures obliques sur la foliation des gneiss. Cette paragénèse est associée à une gangue de quartz et sa mise en place se situerait vers 850 Ma.

L'occurrence de Dominique-Peter nous confirme que ces deux types de minéralisations, issus de deux épisodes minéralisateurs séparés dans le temps, peuvent coexister dans la même structure.

Dans la zone des gisements, deux autres épisodes minéralisateurs de moindre importance se superposent aux épisodes précédents; il s'agit de la paragénèse à pechblende-carbonate et celle à coffinite-sulfures qui constituent des remaniements des minéralisations précédentes.

En dehors des zones de gisement, il a été mis en évidence deux paragénèses particulières dans les occurrences de Sophie et de Numac. Ces deux paragénèses sont les témoins de l'existence des épisodes métallogéniques antérieurs à la mise en place des gisements dans la partie sud de la structure Carswell. L'occurrence de Sophie caractérisée par une paragénèse uranothorifère (monazite-uraninite) est liée à un pegmatoïde à grenat. Il s'agit sans aucun doute d'une minéralisation issue d'un épisode minéralisateur nécessitant des conditions physicochimiques différentes de celles des épisodes hydrothermaux définis dans les gisements.

L'occurrence de Numac semble, de par son association avec une roche feldspathique, beaucoup plus proche des minéralisations hudsoniennes de type Beaverlodge.

Les minéralisations métallifères de l'anneau de Carswell sont issues d'une succession de plusieurs épisodes minéralisateurs hydrothermaux. Ces épisodes caractérisés chacun par une paragénèse uranifère ne sont pas propres à la structure Carswell. L'exemple des minéralisations de l'Est Athabasca nous confirme qu'il s'agit d'épisodes régionaux qui affectent l'ensemble du plateau Athabasca.

INTRODUCTION

Mineralogical studies on the metalliferous occurrences in the Carswell structure began after the discovery in 1968 of a mineralized boulder train in the Cluff Lake area. Boulders in glacial overburden comprised the first samples of massive mineralization to be studied (Geffroy, 1975). Detailed studies from numerous drill holes and mining operations in the OP, Claude, and D areas allowed the uranium mineralization to be placed in a petrographic and structural context. Uranium was observed to have originated from multiple hydrothermal metallogenic episodes affecting the same mineralized structure.

Mineralizing episodes are characterized by important alteration of the host minerals and by successive deposition of Fe-Ti-U oxides, tellurides, selenides, and sulphides. This depositional sequence indicates the complexity of mineralizing solutions and the nature of the mineralizing environment. The sequence of deposition is based on metallographic examinations and is confirmed by a crystallochemical study of uranium oxides as well as the remaining metallic suite of minerals (Pagel and Ruhlmann, 1985). The sandstone-hosted and basement-hosted mineralization discussed in the following pages indicates that metallogenic processes equally affected the basement and the younger sandstones in the Carswell structure.

SANDSTONE-HOSTED MINERALIZATION

Sandstone-hosted mineralization has been studied in two main occurrences, the first of which corresponds to a group of boulders sampled in the Donna, S, and E boulder train, and the second to samples collected in the D orebody along the different units in the open pit, and also from drill holes. Several episodes of mineralization, differentiated more by their parageneses than by their texture, have been defined. The four main mineralogical assemblages: pitchblende-hematite, pitchblende-sulphide, uraninite-selenide-telluride, and uraninite-sulphide shall be described.

Pitchblende-Hematite Assemblage

Uranium mineralization appears in fine- to medium-grained, poorly sorted sandstones containing 70% detrital material. Dominant quartz is slightly stained by iron hydroxides and is associated with a few tourmalines and zoned zircons. In the sandstones, there are commonly thin, heavy mineral layers where various mineralogical forms of titanium oxide are predominant. As well rounded grains, they contain titanium grids of anatase or rutile.

All the detrital minerals are set in a matrix (30%) of variable composition, either clayey, or composed of tetravalent or hexavalent uranium, hematite, or limonite. The abundance of one phase or the other depends on the degree of supergene alteration, which is well developed in the boulders sampled in the glacial overburden.

All of the transitional stages between an iron oxide such as fibrous hematite and an iron hydroxide such as limonite are present in the sandstone matrix. This fibrous hematite, generally associated with limonite, is different from the hypogene acicular hematite that is associated with the uranium mineralization.

The hydrothermal metallogenic process needs open fractures to penetrate the sandstone host. This results in massive mineralization from which mineralizing solutions can diffuse into the fractured walls and then into the non-tectonized sandstones. Due to the microfractures and the sandstone microporosity, this results in three textures of mineralization: massive, microfracture filling, and matrix impregnation.

This mineralizing process is associated with hypogene alteration in the sandstone. Quartz grains are corroded as shown in Figure 1. The corrosion, slight in the walls of the mineralized zone, increases strongly in the inner zone and may result in a complete dissolution of the quartz grains and a subsequent filling of the quartz grain network with a mixture of iron hydroxides and clay minerals. Some weakly mineralized samples show earlier diagenetic quartz overgrowths which decreased the volume of the sandstone matrix and subsequently inhibited the circulation of mineralizing solutions.

Uranium mineralization, in the primary paragenesis, is exclusively composed of two U(IV) oxides: pitchblende and uraninite which each have a distinctive habit. Pitchblende seems to be dominant and is associated with hematite. Textural relationships (alternation of hematite and pitchblende) between the two minerals (Fig. 2) confirm that they belong to a unique depositional event.

An iron oxide-uranium oxide association with a different mineralogy has been observed in one boulder (i.e., a magnetite-uraninite association) which may reflect less oxidizing conditions than that for hematite and pitchblende. Magnetite is martitized with hematite forming along its cleavages. Some native gold may be associated with the iron and uranium oxides as xenomorphic, micrometre-size inclusions in the massive pitchblende. This paragenesis does not contain sulphides, selenides, or tellurides which are usually abundant and associated with the regional uranium metallogenic process. As a matter of fact, in most mineralization an important reducing phase has been observed; its absence seems to be one of the many characteristics which differentiates this iron and uranium oxide paragenesis from the other types of mineralization in the Athabasca Basin.

Supergene alteration, which is well developed in these samples, is marked in the metalliferous mineralization by a secondary paragenesis and a leaching of uranium. This leaching is made easier by the microporosity and microfracturing of mineralized sandstones. The secondary paragenesis consists of uraniferous, yellow weathering products, fibrous hematite, goethite, limonite, and psilomelane-type manganese hydroxides. These minerals are closely imbricated on the sandstone matrix and also appear as microfracture and vug fillings.

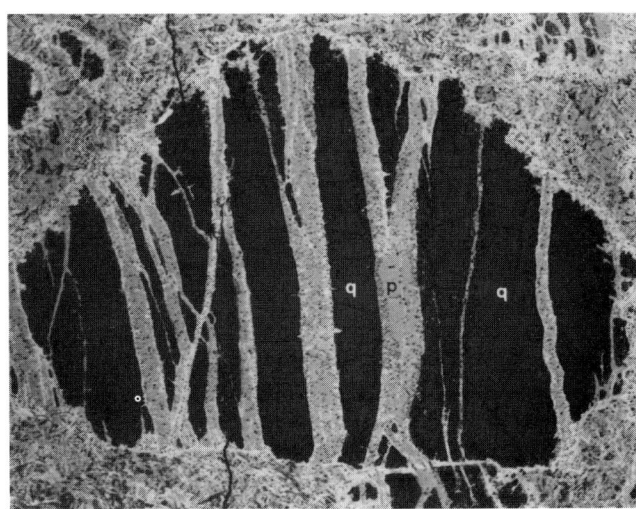

Figure 2. Polished section photomicrograph of the pitchblende (P)-hematite assemblage as a fracture filling in quartz (q). S boulder, × 125, plane light.

Pitchblende-Sulphide Assemblage

Pitchblende-sulphide mineralization is characterized by simple sulphides crystallized after the uranium oxide phase. It is well illustrated in a sample taken from a N140° fracture in the sandstones at the D orebody. It consists of intergranular pitchblende, which is associated with a green clay (possibly a chlorite), and rims detrital quartz grain boundaries. Corrosion of quartz grains, which is especially well developed if there is a pitchblende-hematite association, was not observed in this sample.

Pitchblende, with its usual botryoidal habit, is associated with a few sulphides, of which galena is predominant, along with chalcopyrite, pyrite, and a few patches of marcasite. The sulphides form cross-cutting veinlets in the pitchblende.

Uraninite-Selenide-Telluride Assemblage

Mineralization as replacement of host rock minerals has been observed in the open pit and in drill hole CLU 131 from the D orebody, and consists of massive mineralization in sandstones. At the base of the metalliferous mineralization, a heavy mineral bed containing 2% TiO_2 in total rock analysis consists of a several centimetres thick layer of zircons and titanium grid minerals in well-rounded grains. The titanium grids may come from the transformation of primary minerals, such as titanomagnetites, with ilmenite or rutile exsolutions. This alteration, common in hydrothermal and supergene processes, consists of a dissolution of the magnetite to iron and a transformation of ilmenite exsolutions into anatase. The presence of rutile exsolutions in the titanomagnetites and titanium grids as products of their alteration emphasizes the refractory character of rutile. This refractory character seems to be confirmed by the presence of detrital rutile in the heavy mineral beds. Titanomagnetite

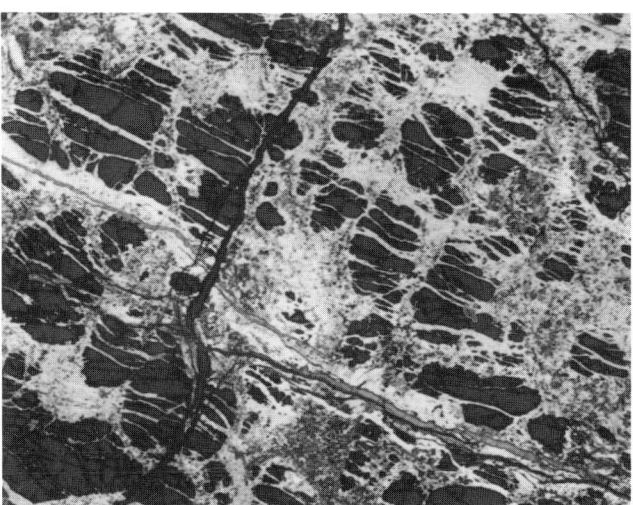

Figure 1. Polished section photomicrograph of oriented microfracturing at the edge of mineralization. The fractured quartz grains are encased in a hematite and pitchblende matrix. Donna area, × 65, plane light.

exsolutions show a different behaviour in alteration depending upon whether they are derived from rutile or from ilmenite.

Redox processes which destabilize iron-titanium minerals were studied by Goldhaber *et al.* (1978) in a roll-type orebody in Texas. These minerals, very sensitive to alteration, were used by Adams and Saucier (1981) in their study of the Grants orebodies in New Mexico. The authors show that mineralization is surrounded by a zone where ilmenite and magnetite alteration is well developed. In the sandstones directly in contact with mineralization, these two minerals are completely removed by dissolution of their components (Fe and Ti). The titanium minerals also act as nucleii for uranium precipitation, as shown in the many grids found in the middle of U(IV) oxides (Fig. 3) and also the many uranium-rich leucoxene pods. These titanium-uranium compounds (Fig. 4) are formed by precipitating uranium in the mineralizing solutions on a titanium bearer which originated from the alteration of a detrital titanium mineral.

In hydrothermal conditions, the detrital titanium mineral alteration is generally followed by a reducing episode with intense sulphur, selenium, or telluride activities depending on the complexity of the hydrothermal phenomenon. Its mineralogical expressions are exhibited in the D orebody: bismuth and lead selenides and tellurides, the texture of which is stongly influenced by the titanomagnetite grain texture. These minerals have the rounded shape of detrital grains (Fig. 5). The mineralogy of the titanium lamellae (rutile, leucoxene, and anatase) enclosed in the selenides reflects the Fe-Ti mineral alteration prior to the selenides deposition. This process is close to the one more frequently observed in formations (volcanics, sandstones) which have undergone sulphidic hydrothermal alteration. Titanomagnetites are completely transformed into iron sulphides either by sulphurization of the magnetite iron, or by pyrite crystallization in the loose structures of the titanium grid. This latter process requires prior alteration of magnetite to release iron into the solution.

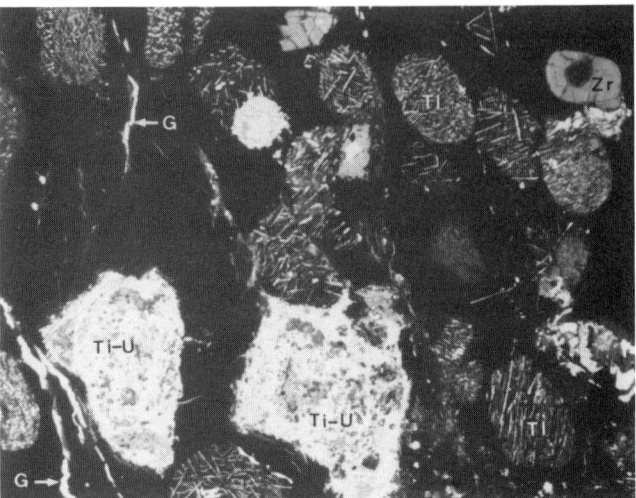

Figure 4. Polished section photomicrograph of a heavy mineral bed containing: zircon (Zr), titanium grids (Ti), uranium – titanium aggregates (Ti-U), and galena (G). CLU 429-51.2, × 125, oil immersion, plane light.

Mineralization within the clayey matrix has two distinct mineralogical expressions: uranium titanates and U(IV) oxides. Uranium titanates fringe the mineralized zones. The study of their habits and of their chemical components distribution by scanning electron microscope (SEM) shows the differentiation of two types. The first type is composed of globular patches which are disseminated in the heavy mineral beds. The SEM shows a very random distribution of elements like titanium, uranium, and accessory iron. These compounds originated from absorption of uranium on a titanium bearer of detrital origin. The second type differs from the former by a subidiomorphic crystallization tendancy and a more homogeneous distribution of Ti, U, and Fe (Fig. 6). These minerals probably belong to the brannerite group and are an integral part of the uranium phase (Table I).

Reflected light microscopy shows variable reflectance in the brannerite group minerals. This phenomenon is a function of their different degrees of alteration, and alteration may progress to a total destruction of the mineral, transforming it into a mixture of titanium oxide (anatase-leucoxene) and uranium oxide.

The uranium titanates of the brannerite group form only a small part of the uranium mineralization which is mainly composed of uraninite and not of pitchblende. Uraninites form perfectly idiomorphic patches, enclosed in selenium-bismuth minerals (Fig. 7). These minerals, which crystallize after the uranium minerals, are part of a metallic paragenesis composed of many mineral species: clausthalite-paraguanajuatite-guanajuatite, trogtalite-calaverite-altaite, freboldite-gersdorffite-gold-bismuth-copper, nickeline-skutterudite-jordisite, and galena-chalcopyrite-pyrite-sphalerite. Although many samples occur in massive mineralization, others show a very rib-boned texture, where chronological relations between mineral species can be observed.

This hydrothermal mineralizing process starts with uranium deposition first as brannerite and then as uranium

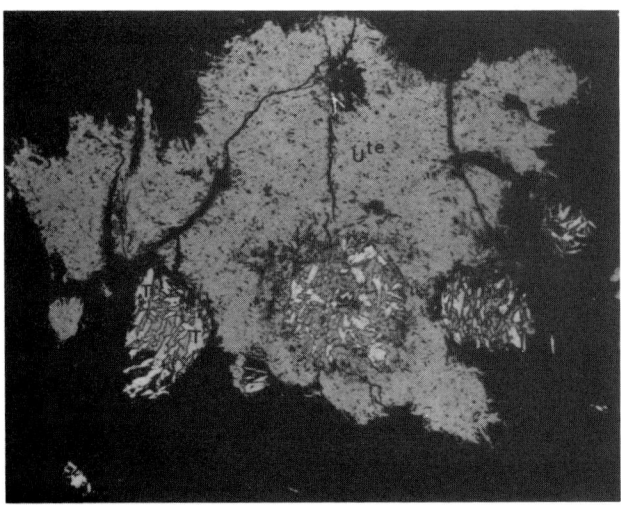

Figure 3. Polished section photomicrograph of a titanium grid acting as nucleii for precipitation of uraninite (U^{te}). CLU 909, × 125, oil immersion, plane light.

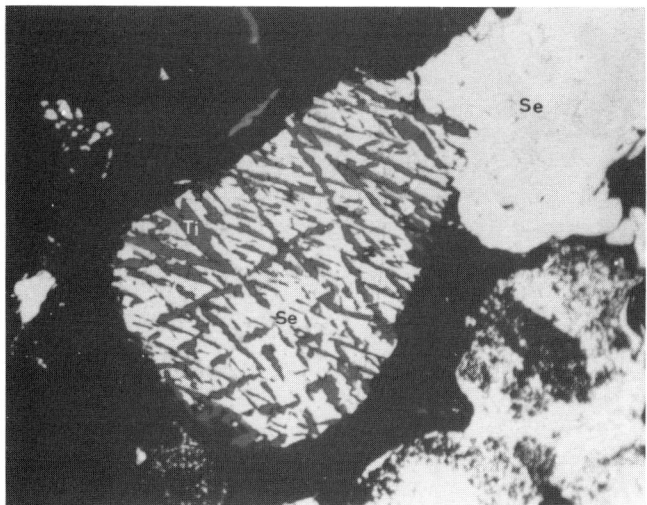

Figure 5. Polished section photomicrograph of a titanomagnetite grain which is being replaced by bismuth selenides (Se). Ti = rutile exsolutions. CLU 131, × 220, oil immersion, plane light.

Figure 7. Polished section photomicrograph of uraninite (U^{te}) inclusions in a mixture of clausthalite and paraguanajuatite. CLU 131, × 125, oil immersion, plane light.

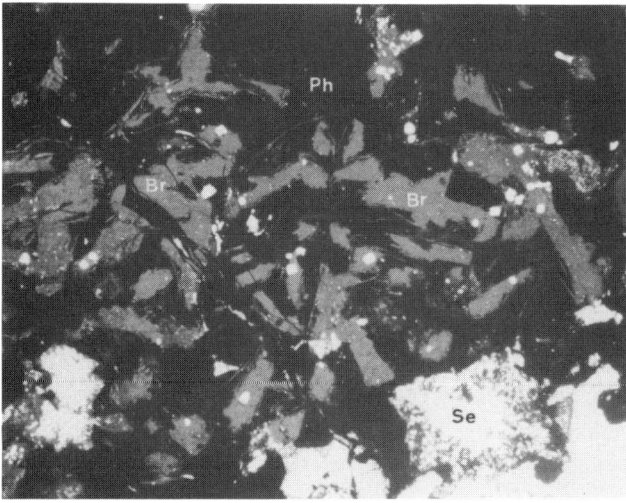

Figure 6. Polished section photomicrograph of disseminated brannerite (Br) in a phyllitic (Ph) matrix. Se = bismuth selenides. CLU 131-624, × 125, oil immersion, plane light.

TABLE I

ANALYSES OF URANIUM-TITANIUM MINERALS FROM SANDSTONE-HOSTED MINERALIZATION IN THE CARSWELL STRUCTURE (ANALYSES BY CAMEBAX ELECTRON MICROPROBE, NANCY, FRANCE).

FeO	4.09	0.62	1.25
P_2O_5	0.03	–	0.23
CaO	1.17	1.29	1.35
SiO_2	3.26	4.04	2.98
K_2O	0.20	–	0.12
TiO_2	36.57	33.92	34.05
Al_2O_3	0.91	0.83	0.52
UO_2	48.20	49.96	49.14
ThO_2	–	–	–
PbO	1.20	0.93	1.80
ZrO_2	–	–	–
TOTAL	95.63	91.59	91.44

oxide (uraninite). After the oxides deposition (brannerite and uraninite), there is a second phase, always a reducing one, with an early crystallization of tellurides, followed by various forms of selenides which are more abundant. The nickel and cobalt arsenides and sulphoarsenides chronological relationship is not clearly defined, and the same is true for native elements such as gold and bismuth. Sulphides seem to come later in the mineralization sequence. They fill the inner part of mineralized structures or form veinlets crosscutting the uranium-selenide mineralization.

The D orebody perfectly illustrates the different stages needed for metallic mineralization (Pagel and Svab, 1985; Ey et al., 1985): an episode of hydrothermal alteration of the host rock followed by an important polymetallic deposition of successive tellurides, selenides, and sulphides.

Uraninite-Sulphide Assemblage

There are uranium occurrences in sandstones with very simple metalliferous assemblages, contrasting with the polymetallic assemblages previously described in hole CLU 131. The uraninite-sulphide assemblage has been intersected at different levels in drill holes CLU 429 and 433 in the D orebody. They lie between two heavy mineral layers similar to the mineralization in CLU 131 containing zircons, rutiles, tourmalines and many titanium grid minerals.

The uranium mineralization develops in a clayey matrix

where chlorite seems especially abundant, and is composed of a uraninite-type U(IV) oxide of various textural habits. A more or less fibrous fabric is due to the clayey matrix impregnation by the U(IV) oxide. A second type forms as thin ribbons of the uranium oxides, which may or may not be affected by a late tectonic phase. A few rare, idiomorphic uraninites appear in the central part of the mineralized structure.

The frequent occurrence of all three fabrics of U(IV) oxide in the same mineralized zone is one of the characteristics of U(IV) oxides in mineralization localized in the basement metamorphic rocks and related to a clayey matrix.

The non-uranium metalliferous mineralization is exclusively composed of sulphides. The minerals appear in small quantities in cracks across the ribboned uraninites, or in the middle of veins. The paragenesis is as follows: chalcopyrite-galena and pyrite-marcasite. Native gold appears in one micrometre blebs disseminated in the ribboned uraninites.

Identification of the Main Uranium Deposition Events

Standard metallogenic techniques have been used to identify the various mineralogical associations, and microanalysis has been performed on the U(IV) oxide which is common to all these associations. Detailed results and the use of this study are given in another paper of this volume (Pagel and Ruhlmann, 1985). Consideration of elements such as Ca, Si, and Pb allows for the identification of two uranium depositional events. As the whole lead content is assumed to be radiogenic in origin, common lead not being considered, estimates cannot be used in the same way as in classical geochronology. One can, however, within a certain limit, differentiate the gap of deposition for several generations of U(IV) oxides present in one mineralized zone.

The uranium oxides observed in the four mineralogical associations discussed previously were systematically analyzed by microanalysis. Results for three mineralogical associations are given in Table II. Distinction and the relative timing of events have been confirmed by U-Pb isotopic analyses on the same samples by Bell (1985).

From the results in Table II and U-Pb analyses, at least two hydrothermal mineralizing episodes can be differentiated. The first, at 380 Ma, corresponds to the pitchblende-hematite paragenesis and perhaps also to the pitchblende-sulphide assemblage, and the second one, at about 1100 Ma, sets the time of the uraninite-sulphides and selenides-tellurides associations.

Metallogenic studies of metalliferous mineralization in the Athabasca sandstones has facilitated the definition of four mineralogical assemblages, each with their own characteristics (see Table III). At least two mineralizing episodes can be observed. They are strictly hydrothermal episodes which correspond to fairly distinct parageneses and are separated by an important period of time.

The oldest mineralizing episode, producing uraninite-selenide-telluride and uraninite-sulphide assemblages may be linked to a unique metallogenic episode. Mineralogical variations observed in this older assemblage are marked by the presence or absence of tellurides and selenides that are typically associated with a U(IV) oxide. Such variations compared with the basement-hosted mineralization can be related to several factors: the influence of geochemical characteristics of the source environment and of the environment likely to be mineralized, and the influence of fractional crystallization leading to horizontal and vertical variations in mineralized structures.

A pitchblende-sulphide assemblage occurs in the vicinity of the D orebody and probably originated from a remobilizing of older mineralization which composed the main D deposit. Such remobilizations of small magnitude are likely to be spatially restricted.

Characteristics of the pitchblende-hematite paragenesis are quite different. This very simple paragenesis, characterized by an Fe-U association, corresponds to a different metallogenic process. Its age, about 380 Ma, corresponds to

TABLE II

MICROPROBE ANALYSES FOR URANIUM OXIDES IN SANDSTONES FROM THE CARSWELL STRUCTURE, CLASSIFIED ACCORDING TO MINERALOGICAL ASSOCIATIONS.

	PITCHBLENDE - HEMATITE PARAGENESIS				PITCHBLENDE - SULPHIDE PARAGENESIS						URANINITE - SULPHIDE PARAGENESIS				
SAMPLE IDENTIFICATION	DONNA SANDSTONE BOULDERS				MAIN OREBODY OF "D", CLU 369, FT 81						MAIN OREBODY OF "D", MP 74-10				
UO_2	77.96	83.32	81.20	82.35	83.66	87.73	87.66	86.12	86.90	87.96	80.63	79.50	80.19	82.08	85.18
PbO	3.90	3.43	3.28	2.81	5.73	3.14	2.82	2.72	2.70	2.28	14.73	14.25	13.07	13.07	9.80
CaO	8.65	8.27	7.70	8.94	2.95	3.07	3.00	3.03	3.41	2.71	0.99	1.00	1.18	1.15	1.11
FeO	–	–	1.56	–	0.52	0.46	0.85	0.72	0.74	0.85	0.88	0.07	0.05	0.07	0.24
MnO	–	–	–	–	0.87	1.10	0.36	0.18	0.50	0.44	–	–	0.16	0.16	0.08
TiO_2	–	0.13	–	–	–	0.28	0.45	0.42	0.30	0.49	–	–	–	–	0.90
SiO_2	0.14	0.25	–	–	0.84	1.10	0.41	1.09	0.54	0.66	–	–	–	–	–
P_2O_5	0.19	0.20	–	–	0.35	0.25	0.24	–	0.10	0.20	0.22	0.03	0.22	0.20	0.16
TOTAL	90.84	95.60	94.96	94.52	94.92	96.79	95.79	94.41	97.19	95.65	95.65	94.85	95.70	96.73	97.47
APPARENT AGE (Ma)	330	270	260	220	450	230	210	210	200	170	1190	1170	1130	1040	750

TABLE III

SUMMARY OF THE SANDSTONE-HOSTED MINERALIZATION IN THE CARSWELL STRUCTURE.

	PITCHBLENDE – HEMATITE	PITCHBLENDE – SULPHIDE	URANINITE – Te – Se – Bi	URANINITE SULPHIDE
TEXTURE	VEIN	VEIN – IMPREGNATIONS	VEIN	VEIN
URANIFEROUS EPISODE	PITCHBLENDE	PITCHBLENDE	URANINITE & BRANNERITE	URANINITE
NON-URANIFEROUS EPISODE	HEMATITE (MAGNETITE) LIMONITE – GOETHITE – NATIVE GOLD	GALENA – PYRITE CHALCOPYRITE – MARCASITE	PARAGUANAJUATITE – GOLD CLAUSTHALITE – ALTAITE CALAVERITE – FREBOLDITE TROGTALITE – GERSDORFFITE NICKELINE – SKUTTERUDITE GALENA – CHALCOPYRITE PYRITE – MARCASITE JORDISITE	GALENA (CLAUSTHALITE ?) CHALCOPYRITE – PYRITE MARCASITE
MINERALIZATION ASSOCIATIONS	?	PHYLLITES (CHLORITES ?) ± CARBONATE	PHYLLITES (CHLORITES)	PHYLLITES (CHLORITES)
CHARACTERISTIC ASSOCIATIONS	U – Fe	U ± Pb ± Cu (S)	U – Te – Se – Bi – (Ag – Au) – Mo	U ± Pb (– S) (Se ?)
AGE U-Pb	380 BELL (1985)		1050 (BELL, 1985) 1050 (GANCARZ, 1979)	1150 (BELLON et al., 1976)
TYPES OF OCCURRENCES	DONNA – S – E – U BOULDERS	MAIN OREBODY OF D (FT 81) D – PIT	CLU 131 MAIN OREBODY OF D	CLU 429 & 433 MAIN OREBODY OF D

a mineralizing episode known in many uranium occurrences, for example, the Beaverlodge district and Rabbit Lake deposits. Similar to Carswell, it occurs as hematite and pitchblende fillings in fractures.

BASEMENT-HOSTED MINERALIZATION

The description of mineralization hosted by basement gneisses and granitoids includes problems occurring in the study of the Dominique-Peter and the Claude orebodies. At the time of this research, these orebodies, along with the N orebody, constituted the principal reserves in the Carswell structure. The Donna, Numac, and OP occurrences are of some significance. Although of limited economic value, their examination contributes information regarding the regional metallogenic events.

Basement mineralization differs markedly between deposits due to their textural relationships with the host gneisses and differences in ore paragenesis. The following descriptions are more the result of textural considerations rather than paragenesis. The textural differences are particularly clear if we take into account mineralization such as at Claude and OP. In the first case, mineralization is associated with a phyllitic gangue and subjected to plastic deformation. This mineralization, which also occurs subparallel to the gneissic foliation, is very different from that in OP, which occurs as open fracture fillings in quartz gangue.

These two types of mineralization can occur very close together within the same orebody, as in the case of Dominique-Peter. It is only with the determination of apparent ages, confirmed by geochronological methods, that the ore mineral associations could be defined.

Petrographic Context

The diverse nature of the rocks has a certain effect on the nature of the mineralization which it hosts. Studies by Herring (1976) and Bruneton (1981) give the petrographic and geochemical characteristics of the different facies observed in the basement. The rocks can be divided into three major divisions. The basic gneisses are made up of garnetites, amphibolites, biotite-amphibole gneisses, garnet-rich gneisses, calcic gneisses, gneisses rich in garnets and aluminosilicates and granulitic amphibole-pyroxene gneisses. These gneisses are generally dark coloured and usually well-foliated. They may consist of thin units or metre-thick bands which contacts with other rock types can be diffuse. These basic rocks are cut by pegmatoids and other associated rocks.

The acid assemblage consists of biotite-rich, biotite-aluminous-rich, and granulitic pyroxene-garnet gneisses. They are light-coloured and very well foliated. The more evolved gneisses look like granitic gneisses, where foliation is partially destroyed. There are local zones of granitization in this assemblage.

The third division is a remobilized assemblage composed of a leptynitic, garnet-bearing gneiss, granite gneiss, and pegmatoids. They have been injected into the two preceeding assemblages from which they were probably derived by anatexis.

The youngest rocks in the petrographic assemblage are the Cluff breccias, which have been geochemically and petrographically described by von Einseidel (1981).

Mineralization in the Orebodies

Claude Type Mineralization. The Claude orebody contains mineralization similar to that observed in the mylonite zones in the Dominique-Peter orebody, and at the present time, it yields the most complete metallogenic paragenesis in a phyllitic gangue. This type of mineralization was studied using samples that came from the first reconnaissance mining which began in 1979. Samples came from diamond drill holes CLU 749, 785, 1107, and 1113 as well as from the walls of the experimental pit that was dug in the summer of 1982.

Uraniferous mineralization does not show true vein-type textures and fracture filling, but rather it is associated with a phyllitic gangue which is commonly coated with iron oxides and hydroxides. The gangue mineral is generally a magnesian chlorite, which spatially defines the uranium oxide facies but creates a problem regarding the deep hydrothermal alteration which affected the host rock.

Observations in the Claude orebody reveal the existence of a "zone à boules" similar to that described in the D orebody (Ey et al., 1985). This type of faulting results from a double fracture system. Hydrothermal solutions circulating through the fractures caused alteration of the host gneiss and ultimately the deposition of uranium. Ore minerals were deposited as a coating on the fault blocks which are frequently tens of metres in size. Depending upon the degree of alteration, the gneissic core of the fault blocks may be totally destroyed and contain nothing but a mineralized argillaceous material. The major metalliferous minerals observed in these structures include: brannerite-uraninite, paraguanajuatite-clausthalite, molybdenum sulphide-galena, and pyrite-chalcopyrite. Uranium minerals which appear early in the core of this assemblage include uraninite and uranium titanates similar to the brannerite group.

Uranium titanates can be divided into two types. The first type consists of poikilitic globular aggregates which coat the main uraniferous zones. The semi-quantitative study of these optically heterogeneous minerals defines the partitioning of the constituent elements. Qualitative analyses obtained from the scanning electron microscope yield a Si-Ti-U assemblage with minor Fe. Although the titanium and silica are present in a regular array, the other elements are not as well ordered. These silica-titanium-uranium assemblages do not correspond to strictly defined minerals. They do not show crystallographic diffraction patterns, and possibly represent a disordered state that was created by metamictization which increases its effect with time. This phenomenon is apparent in the case of brannerite, where the primary array is disrupted by metamictization, however, this is not the case where the primary array was never well defined. The problem with these uranium-titanium minerals was noted in the Blind River deposit at Elliot Lake, Canada (Theis, 1979), where uranium present in mineralizing solutions appeared to have become trapped in titanium-related minerals. These titanium assemblages may correspond to relic ilmenites or titanium-magnetites in metamorphic gneisses, but may also be titanium minerals derived from the alteration of biotite. As a result, the presence of these assemblages in association with relics of biotites should be noted where in direct association with uranium oxides.

The association of uranium and titanium implies the contemporaneous liberation during hydrothermal alteration of titanium from the host rocks at the time of uranium deposition. Titanium oxides are an excellent trap for uranium (Ruhlmann, 1980).

The second type of titanium-uranium association is characterized by sub-to-euhedral, homogeneous crystallization of individual grains. These crystals probably correspond to relic brannerites and are generally strongly altered (Fig. 8). There is frequently a fresh core composed of Ti-U-Fe-Ca, which increases in variation toward the edge of the grain, where alteration is more intense. There appears to be an increase in the leaching of uranium as the titanium recrystallizes to anatase. This type of alteration, which may be either hypogene or supergene, is facilitated by the metamictization of brannerite. The form and habit of uranium and titanium minerals are dependant upon physio-chemical conditions at the time of transformation. For example, in the Claude orebody, it is not out of the question that uranium is reprecipitated as coffinite within the same mineralized zones. Anatase is present as numerous crystals in the coffinite aggregates, which supports the hypothesis of this transformation.

Uranium oxides deposited after brannerite are uraninites, although the textures are extremely variable depending upon whether they border the mineralizing zone or occur in the

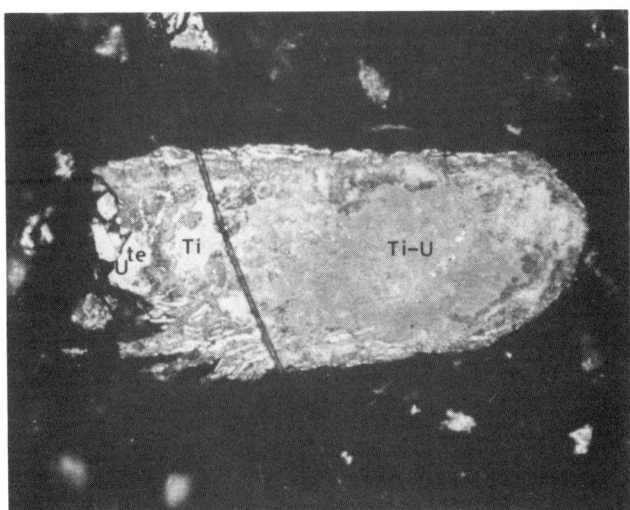

Figure 8. Polished section photomicrograph of brannerite (Ti-U) destabilizing to form uraninite (U^{te}) and anatase (Ti). Claude Mine, × 220, oil immersion, plane light.

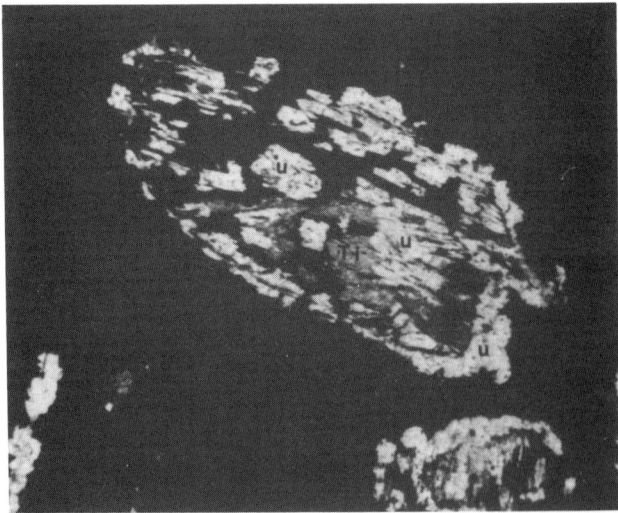

Figure 9. Polished section photomicrograph of uranium mineralization (U), outlining the contours and cleavage planes of a mica. Dominique-Peter, × 220, oil immersion, plane light.

centre. There are three major textures: fibrous, ribboned, and euhedral. Fibrous textures (Fig. 9) develop within strongly altered relic biotites from the metamorphic host rocks. These biotites, with good cleavage and associated titanium oxides, are excellent traps for uranium. This type of mineralization only occurs at the edges of a mineralized zone where there are relics of the host gneiss.

Ribboned textures, which are part of the massive uraniferous mineralization, are illustrated in Figure 10. The distribution of minerals in this zone is often disturbed by later tectonic events. In a few basement samples from the Donna boulder train, strong alteration has not occurred to destroy the mineralogy and textures, and the paragenesis is relatively undeformed.

This ribboning corresponds to radial, wispy fibres described by Geffroy (1975) and called "roué dentée" or cogwheel texture. Many hypotheses have been developed over the years to explain the pseudospherulitic texture which is reminiscent of pitchblende, coinciding with the spearshaped fibers which resemble uraninite. Studies of the lead, calcium, and silica distribution in this uranium oxide show no variation between the centre of the fibre and the euhedral edge. This homogeneity implies a single mineralizing episode that resulted in crystallization of ribboned uraninites with a cogwheel configuration. This type of texture has also been observed by the author in a few samples from James Bay, Quebec. Mineralization associated with hydrothermal alteration appears to have originated from syn-and post-metallogenic activity associated with the evolution of the basement.

The euhedral texture (Fig. 11) generally occurs in the central part of the mineralized zone. Uranium oxides may be euhedral and finely disseminated in a phyllitic matrix. The progressive change from a ribboned to euhedral texture is rarely observed in these samples.

The development of different textures, illustrated by the uranium oxide variation from the walls to the centre axis, yields a clear image of the mineralized structure. The entire sequence is only observed in zones of a certain thickness. In zones which are only centimetres wide, some textures may be missing entirely, particularly the euhedral and the fibrous textures. A sulphide deposition accompanied by minor amounts of selenides occurs after uranium deposition. Molybdenum sulphide and galena are the two main sulphides associated with this mineraliztion.

Molybdenum sulphides are very interesting considering their heterogeneous composition and the observed optical properties. A qualitative study on the mineral, generally referred to as jordisite, shows a heterogeneity developed from the disordering of constituent elements, which is evidenced by the heterogeneity of the reflectance and optical anisotropy noted within the grains. Although molybdenum, sulphur, and iron are major elements, there are always minor amounts of nickel, cobalt, and accessory yttrium and vanadium. The composition is crystallographically poorly defined and does not match with other known formulae. These are probably colloidal phases that are evolving towards a more ordered state. The various stages in this evolution are characterized by variable reflectance and optical anisotropy.

One other minor mineral associated with this mineralization is euhedral galena, which contains inclusions of bismuth and lead selenides. This Se-Bi-Mo association is very similar to that previously described in the sandstones of the D orebody.

Organic material is abundant in the Claude orebody. It is not only graphitic material observed frequently in the walls of the mineralized zone, but also bitumen that has migrated into the central portion of the mineralized zone. The metalliferous assemblage described above is found as inclusions within the organic material. The nature of the inclusions enables a reconstruction of the entire mineralized zone.

The organic material shows variable textures which contain different reflectance properties. This is clearly illustrated in Figure 12 where three different reflectances are observed in the centre of the same grain. A possible explanation for the phenomena may be the consideration of diffe-

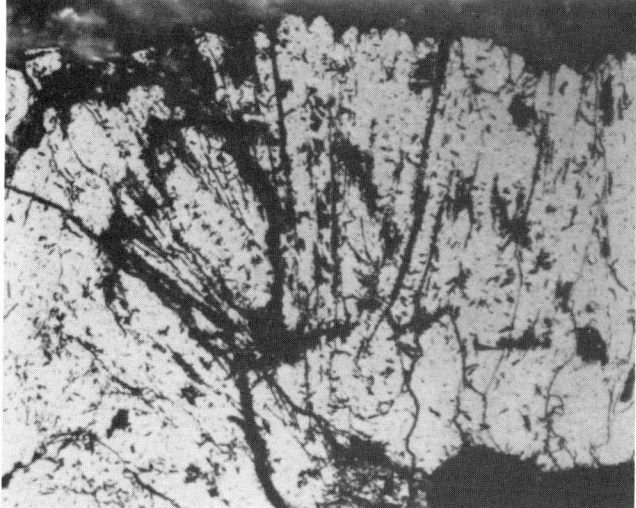

Figure 10. Polished section photomicrograph of ribbons of uraninite with lance-like grain terminations. Donna basement, × 280, oil immersion, plane light.

Figure 11. Polished section photomicrograph of euhedral uranium oxides. "D" orebody, × 200, plane light.

rent degrees of evolution of the organic material. The optical variations appear to be most common in grains containing fine, uraniferous inclusions. It should be noted, however, that there is a fine layer, on the order of a micron, of isotropic organic material between a uraniferous inclusion and other organic material. There was probably a transformation of organic matter by emission of radioactive particles from the uranium, although in some cases this aureole was not observed.

The evolution of the organic matter toward an isotropic state may occur during alteration. In some highly oxidized samples, there is a decrease in the reflectance of organic material, accompanied by a total isotropy along the edges of the grains. This phenomenon progresses inwards from the edges, also illustrated in Figure 12. Another notable observation is the presence of numerous galena crystals, whose lead is probably originally radiogenic, which occur as inclusions in the isotropic aureoles.

From all observations, it appears that the organic matter originates from an independant and later event than the uranium-sulphide deposition. This does not exclude the possibility of non-mechanical uranium transport as a metal-organic complex by organic material in the course of its migration from the sedimentary cover to the traps which constitute the mineralized zone (Bonnamy, 1981).

In the Claude orebody, as in the Dominique-Peter deposit, there are two other mineralogical associations that occurred after the development of the primary ore association. Figure 13 illustrates the over-printing of these associations. The two assemblages are a pitchblende-carbonate and coffinite-sulphide assemblage.

The pitchblende-carbonate assemblage occurs as fracture filling associated with calcite, the carbonate gangue mineral (Fig. 14), or as an aureole around uraninite. The uranium oxides are pseudospherulitic and totally different from the fibrous, euhedral, or ribboned textures previously described. Galena-pyrite, and chalcopyrite-marcasite also occur in this assemblage. It is a simple paragenesis of small, disseminated sulphides in uranium oxides or as veinlets in the oxides. Galena is the most abundant phase, where a large amount of the lead may be originally radiogenic.

In the coffinite-sulphide assemblage, uranium occurs as silicates (Fig. 15) closely associated with the preceding assemblage of pitchblende-carbonate. The silicate is derived from a remobilization of primary uraniferous minerals and is found as stringers in the mineralized zone, rimming sulphides, or as impregnations in a phyllitic matrix. It is particularly well developed in Cluff breccia. Numerous stringers of Cluff breccia which are mineralized contain only coffinite as the uraniferous mineral.

In the Claude orebody, a slight variation in the mineralogy is reflected by the appearance of tellurides, predominantly altaite, and nickel and cobalt arsenides. In other parts of the orebody there is an abundance of selenides.

Donna Type Mineralization. Donna boulder samples illustrate the variations that the Claude type mineralization has gone through. The samples in this study originate from a boulder train situated southeast of the D and OP mineralized zones. Within this train there is evidence of two petrographic assemblages: basement and sandstone. The mineralization associated with sedimentary cover has been previously described. Within the basement boulders, there appears to be a metalliferous assemblage associated with a totally deformed phyllitic gangue. Uranium oxide, as uraninite, follows the deposition of galena, chalcopyrite, and native gold. There is no trace of selenides or molybdenum sulphides in this mineralization.

The variation in assemblages is probably related to the geochemical characteristics of favourable depositional environments. In the case of Donna basement boulders, for example, the host gneiss is a hematized quartzofeldspathic rock, rather than the garnet- and aluminosilicate-rich gneisses typical of the Claude type host gneisses.

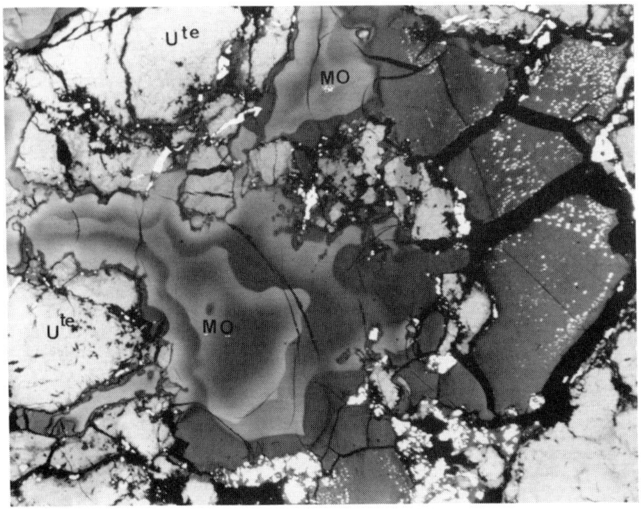

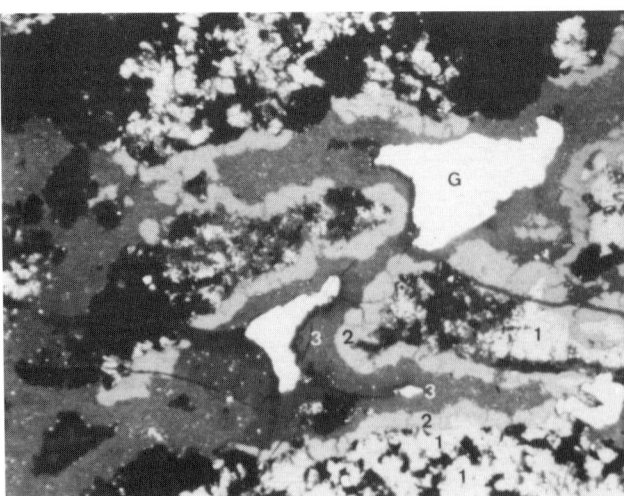

Figure 12. Polished section photomicrograph of organic matter (MO) in the center of a mineralized zone. Claude, × 65, oil immersion, plane light.

Figure 13. Polished photomicrograph of the deposition of uraninite (1), pitchblende (2), and coffinite (3). G = Galena. Dominique-Peter, CLU 1571, × 200, oil immersion, plane light.

Figure 14. Polished section photomicrograph of the pitchblende (p) – calcite (c) association as a fracture filling. U= a uraninite fragment from the previous mineralizing phase. Dominique-Peter CLU 1607, × 65, oil immersion, plane light.

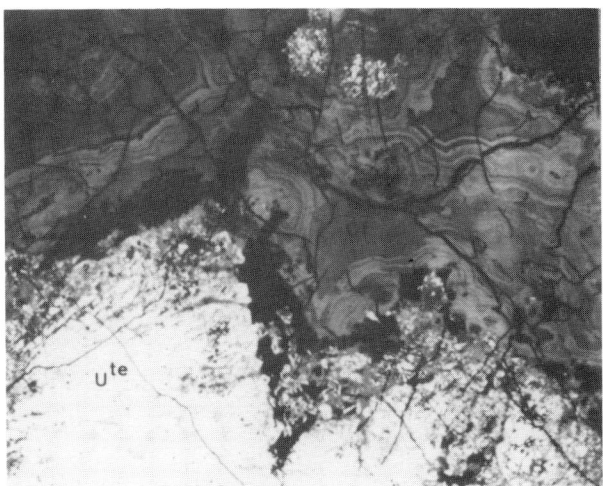

Figure 15. Polished section photomicrograph of the development of coffinite around uraninite. Dominique-Peter, CLU 1571, × 125, oil immersion, plane light.

OP Type Mineralization. The OP type mineralization was first studied in the three mineralized structures intersected by mining in the OP area, approximately 500 m north of the D orebody. It is fracture-associated and totally different from the Claude type, which was not observed in any of the structures. The textural relationship between mineralization and host rock shows, on outcrop scale, the possibility of the existence of vein-type and breccia-associated uranium. Metallogenic studies show that mineralization is essentially vein-type, oblique or subparallel to the gneissosity, and ultimately brecciated.

The most typical vein-type mineralization is found in mineralized structure 1. It appears to be a true vein stockwork of beige quartz which cross-cuts the gneissosity. Regular filling of the open fractures creates a banded texture where the sequence of deposition from the wall to the centre is clearly defined. From the walls, palisade quartz is well crystallized (Fig. 16). This quartz has typical hydrothermal characteristics, being colorless and without inclusions. There is no hematization within the grains. The zoned mineralization is made up of ribboned uraninite, similar to that previously described in the Claude type assemblage. On the edge closest to the vein axis, there are a few euhedral uraninites. Following the uranium, a sulphide assemblage containing chalcopyrite, galena, sphalerite, pyrite, and tennantite was deposited. These minerals generally occur as small anhedral grains and form fracture fillings in the uraniferous bands, or at the base of the quartz crystals. The latter represents an important corrosion phase by phyllitic material which has been analysed as pennine. Very fine disseminated sulphides have been observed in this material.

The sequential deposition that follows the quartz-uraninite can also produce a totally different assemblage. In another sequence, in the centre of the vein, there is a second deposition of beige quartz and at the base there are crystals of colourless tourmaline(Fig. 17). This tourmaline is close to the composition of dravite, and a magnesian phase such as tourmaline is not unique to the mineralization in the Carswell structure. The association with beige quartz is one of its characteristics. Quartz-tourmaline assemblages have been observed outside of mineralized zones, and suggest a paragenesis completely different from the quartz-uraninite-Cu and Pb sulphide assemblages.

The banded texture defined by the two paragenesis is disrupted in structures 2 and 3 in the OP decline. There is intense brecciation due to brittle tectonics that affected mineralization and host rock, and overturned sandstone and basement sequences. This tectonic activity created new textural relationships not originally present at deposition. Mineralization shows deformation and brecciation, and becomes boudinaged in places.

Organic material similar to that described for the Claude

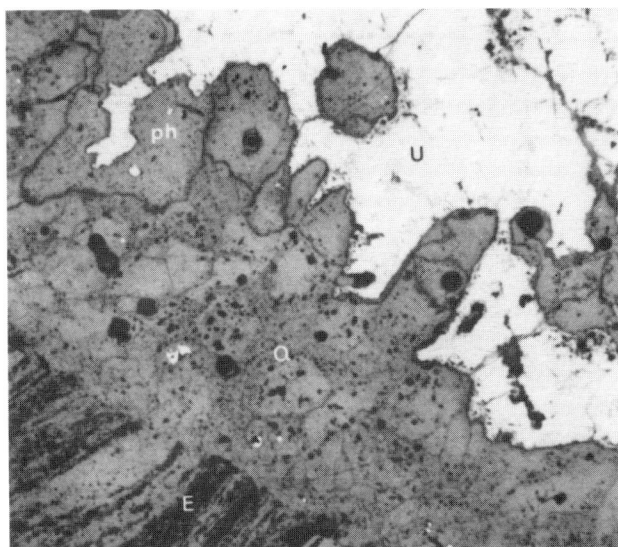

Figure 16. Polished section photomicrograph of ribboned texture, perpendicular to the host gneiss (E). Q = quartz at the wall, U = uranium oxide in the centre of the vein. Dominique-Peter, CLU 1607, × 15, plane light.

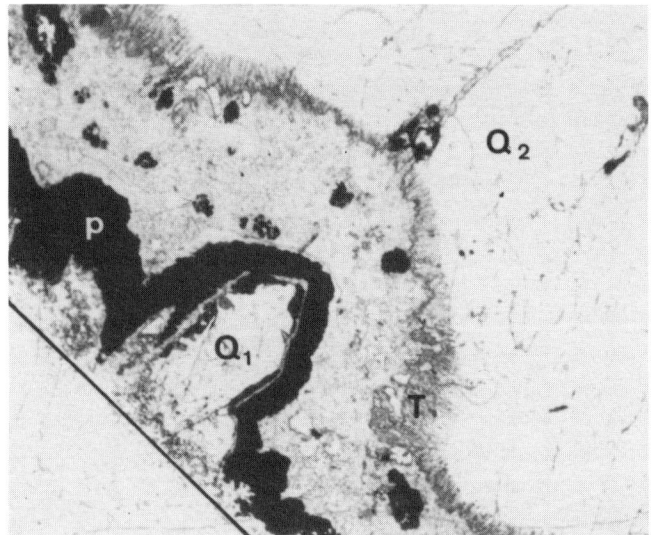

Figure 17. Polished section photomicrograph of ribboned texture in a vein where the quartz at the wall (Q1) crystallized first, followed by uranium oxide (P), tourmaline (T), and finally beige quartz (Q2). OP, × 125, plane light.

type is also observed in this deposit.

In summary, OP type mineralization is the result of two mineralizing episodes, one uraniferous and the other non-metalliferous. The uraniferous episode consists of a very simple assemblage which is very different from the polymetallic associations observed in Claude type deposits.

Dominique-Peter Type Mineralization. Uranium mineralization in Dominique-Peter was studied using samples from about twenty drill hole intersections of mineralization. The occurrence contains mineralization of both the Claude and the OP type. The Claude type mineralization is economic, and all characteristics are found in Dominique-Peter. A few mineralogical variations were observed. For example, if molybdenum sulphide corresponds to the most abundant sulphide phase, there is an overall increase in the abundance of selenides, tellurides, and nickel and cobalt arsenides.

Mineralization within fractures is identical to that described in OP. The banded texture, a uranium-sulphide assemblage, and a quartz-tourmaline phase are observed. There are frequent variations of this type of mineralization between drill holes. There may be a brecciated quartz-uraninite phase with no quartz-tourmaline episode, or quartz may be absent at the base of the uraniferous mineralization. In rare cases, both quartz and the quartz-tourmaline episodes are absent, and there remains only banding by uranium oxides, with or without the sulphide phase. Other variations include mineralogical variations within the uranium-sulphide assemblage. Drill hole CLU 1607 is a good example of such variations.

Two mineralized zones in CLU 1607 occur at 127 m (480') and 160.4 m (515'). Uranium oxides are deposited on quartz similar to that in OP. The only notable variations occur in the sulphide phase associated with uranium. The upper mineralized zone is hosted by an aluminous feldspathic rock and contains predominantly galena. Pyrite, chalcopyrite, and sphalerite are associated with the galena. The lower mineralized zone cuts a biotite-rich mesocratic gneiss. The dominant sulphide is molybdenum sulphide. Minor galena and chalcopyrite also occur.

There are, therefore, two different mineral associations within the same mineralizing episode, one with galena and the other with molybdenum sulphides. The first may be directly comparable with that in OP, however, the second is more closely related to that found in fault zones in Claude. It is different from Claude, however, due to the absence of selenides, tellurides, and arsenides. Apparent ages confirm the differences between the mineralization contained within the mylonite zones and those along fractures.

Cluff Breccia Type Mineralization. Cluff breccias are associated with all principal uranium occurrences in the Carswell structure. Recent studies by von Einseidel (1981) defined three major types of Cluff breccia: Type I, which is a volcanic-like breccia (VLB), Type II, which corresponds to a classic polymictic breccia (CCB), and a tectonic breccia of Type III, which may be associated with pseudotachylites (TLB).

Observations that can be made concerning uraniferous mineralization show that it can be spatially related to Types II and III when they are developed in the mineralized zones. The principal occurrences all contain variable quantities of mineralized breccia, indicating that there was brecciation, or a mechanical remobilization of pre-existing mineralization. It is not surprising, therefore, to find fragments of mineralization similar to the principal occurrences within these mineralized breccias. Breccia zones are excellent traps for the circulation of alteration solutions, and often contain a secondary paragenesis where coffinite is the principal uranium mineral.

Classification of the Parageneses

Certain mineralogical assemblages discussed above occur spatially very close to one another. It is usually very difficult to place these assemblages into one or more parageneses, due to the heterogeneity of deposition as well as variations in the host gneiss composition. As in the case of sandstone-hosted deposits, mineralization has been classified on the basis of the crystal chemistry of uranium oxides present in each assemblage. The analytical measurements on parageneses of lesser importance indicate the limits of this method.

Table IV contains the main characteristics of uranium oxides in the Dominique-Peter, Donna, and OP occurrences. From the table, it is apparent that there have been at least two major mineralizing events. The earlier episode corresponds to Dominique-Peter and Donna basement mineralization, and the later one corresponds to mineralization in OP. Each episode illustrates different textures which emphasize the difference in apparent ages. Subsequently an age of 1050 Ma (Bell, 1985) for Dominique-Peter and an age around 820 to 890 Ma have been determined for OP by U-Pb isotopic analyses.

In light of these results, there appears to be two mineral parageneses which are separated by space and time, both of which result from polyphase hydrothermal activity. The first paragenesis is uraninite-selenides-sulphides of the Claude and Dominique-Peter type. This corresponds to an intra-mylonite host zone and associated phyllitic material. The mineralogical variations are reflected by the relative abundance of selenides, tellurides and molybdenum sulphides. An extreme example is Donna basement, where mineralization is strictly uranium-sulphides. The second major paragenesis is uraninite-sulphides of the OP type. This paragenesis occurs as fracture filling with quartz as a gangue mineral. Variations in these two parageneses are very clear in the Dominique-Peter zone. It was possible to define two subgroups within this mineralization: the first with abundant molybdenum sulphides, and the second with Pb-Cu-Zn sulphides.

Two other parageneses of lesser importance are superimposed on the major two. There is a pitchblende-carbonate phase with minor sulphides, and a coffinite-sulphide phase. This is probably associated with the process of coffinite development on the uranium oxides.

Table V summarizes the main characteristics of four basement-hosted metallogenic assemblages observed in the Carswell area.

Mineralization Outside of the Main Mineralized Zone

There are two occurrences which are different from those previously described, and they occur outside of the main mineralized zone. This mineralization, in the Numac and Sophie occurrences, is situated in the northwest part of the Carswell structure and could not be chronologically linked with the main mineralized zone.

Numac Occurrence. Samples originated from four trenches in the Numac area: GM1, TR14, TR16, and TR18. Mineralization is associated with a garnet-cordierite-sillimanite-biotite gneiss containing accessory sphene, ilmenite, magnetite, rutile, pyrite, and arsenopyrite. There is also mineralization in a red coarse-grained feldspathic rock which contains some ferromagnesians. The typical mineralogy of the feldspathic rock is plagioclase, potassium feldspar, chlorite, albite, and carbonate.

It is interesting to note the presence of albite in these rocks. The grains are clear, but show a deep colouration by iron hydroxides, and occur as either rimming of earlier feldspars or in small veinlets associated with chlorite and carbonate. It is perhaps an indication of the sodium metasomatism which has often been observed in feldspathic rocks elsewhere (Tremblay, 1972). This albitization

TABLE IV

MICROPROBE ANALYSES AND CHEMICAL AGES OF URANIUM OXIDES FROM BASEMENT-HOSTED MINERALIZATION IN THE DOMINIQUE-PETER AND OP AREAS, AND THE DONNA BASEMENT BOULDERS.

	DOMINIQUE - PETER									O-P						DONNA BASEMENT BOULDERS				
UO_2	80.54	79.76	82.89	81.31	80.62	81.02	80.03	80.10	79.68	83.64	83.62	84.05	80.93	82.20	82.37	79.96	77.31	80.91	79.36	78.33
PbO	17.29	16.10	13.30	12.75	12.52	11.26	15.23	15.16	14.33	9.80	9.32	9.18	8.90	8.77	8.56	15.34	14.04	11.92	11.01	10.32
CaO	0.72	0.64	0.91	0.80	0.93	1.01	0.62	0.47	0.59	1.21	1.33	1.44	1.20	1.44	1.80	0.89	0.86	0.83	1.32	1.35
FeO	–	0.05	0.12	–	0.11	–	–	–	–	0.37	0.35	0.19	0.49	0.77	0.37	0.27	0.07	0.01	0.25	0.14
MnO	–	0.11	0.03	0.12	0.09	–	–	0.24	0.06	–	0.06	0.35	0.09	0.22	0.11	0.32	–	–	–	–
TiO_2	–	0.34	0.36	–	–	1.44	–	–	0.49	0.08	0.46	0.18	0.31	0.35	–	0.07	0.16	–	–	0.39
SiO_2	0.33	0.10	0.31	0.31	0.44	1.02	–	–	–	0.17	0.59	0.49	0.46	0.57	0.60	–	–	0.08	0.03	0.16
P_2O_5	–	0.19	–	–	–	–	–	–	–	–	–	–	–	–	–	0.05	0.12	–	0.11	–
TOTAL	98.88	97.29	97.92	95.29	94.71	97.19	95.88	95.97	95.79	95.27	95.73	96.01	92.38	94.32	93.81	93.90	92.56	93.75	92.08	90.69
CHEMICAL AGE (Ma)	1400	1320	1050	1020	1010	900	1240	1240	1170	770	730	710	720	700	680	1300	1190	960	910	860

TABLE V

SUMMARY OF THE BASEMENT-HOSTED MINERALIZATION IN THE CARSWELL STRUCTURE.

	URANINITE – SELENIDE – SULPHIDE		URANINITE – SULPHIDE		PITCHBLENDE – CARBONATE	COFFINITE – SULPHIDE
URANIFEROUS PHASE	BRANNERITE URANINITE		URANINITE		PITCHBLENDE	COFFINITE
SELENIDE	ALTAITE – PARAGUANAJUATITE CLAUSTHALITE – ARSENIDES (Ni – Co)					
SULPHIDE PHASE	MOLYBDENUM SULPHIDES GALENA – CHALCOPYRITE PYRITE – GOLD		MOLYBDENUM SULPHIDES GALENA – CHALCOPYRITE PYRITE – SPHALERITE – TENNANTITE		CHALCOPYRITE PYRITE – GALENA	PYRITE – GALENA
GANGUE	Mg CHLORITE Al CHLORITE		QUARTZ		CALCITE	
CHARACTERISTIC ASSOCIATION	U – Mo – Bi – Se – S – (Te – Ni – Co)		U – Mo – Pb – Zn – Cu		U – Ca – Pb – Cu	U – Si
HOST ROCK	BIOTITE GNEISS	QUARTZOFELDSPATHIC GNEISS	BIOTITE GNEISS	QUARTZOFELDSPATHIC GNEISS	ALL	ALL
TYPES OF OCCURRENCES	U – Mo – Bi – Se – S (Te) DOMINIQUE – PETER CLAUDE	U – Pb – Zn – Cu DONNA	U – Mo ± (Pb – Cu) DOMINIQUE – PETER	U – Pb – Zn – Cu OP	ALL	ALL
AGE (Ma) BELL (1985)	1050		ANALYSED ON SAMPLES FROM OP 820 – 890			

phenomemon has been accurately described by Beletsev (1972) following an episode of desilicification comparable with episyenitization phenomena. There is a third petrographic facies in Numac trenches: albitized pegmatoids rich in garnets and orthopyroxenes.

Uranium occurs as uranium oxide, sometimes pseudospherulitic pitchblende (Fig. 18) or euhedral uraninite. They occupy microfractures and are associated with chalcopyrite, pyrite, and galena. Supergene oxidation processes produced a secondary digenite-covellite-chalcocite assemblage. A few hexavalent uranium oxides, forming as yellow weathering products, are disseminated in fractures.

There is no evidence to allow this occurrence to be categorized with those in the main ore deposits. It is possible that the association of feldspathic rocks with mineralization is comparable with the Beaverlodge type deposits. Some samples from our laboratory will be the object of a geochronology study by K. Bell. The results of this study should confirm or refute this hypothesis.

Sophie Occurrence. The host rock for mineralization in the Sophie area is largely garnetite with a chloritized quartzo-feldspathic matrix. Uranium occurs as euhedral inclusions within the garnets. Monazite is frequently associated with the uraninite. Figure 19 illustrates the uraninite-monazite-sulphide assemblage. Semi-quantitative SEM studies revealed thorium in the monazite.

The sulphides fill fractures in uraninites and are accessory in this sample. They include molybdenite, galena, and pyrite. The molybdenum sulphide is not the jordisite that is found in Claude and Dominique-Peter orebodies, although it is not the only occurrence in the Carswell structure. Molybdenite has also been observed in a mineralized zone intersected by drill hole CLU 749 in the Claude orebody.

The Sophie area contains the most unusual type of mineralization observed to date. The thorian monazite-uraninite and sulphide paragenesis represents a high temperature mineralizing episode. This mineralization is comparable to a thoriferous anomaly to the north of the ELF zone. Sample EK180 from the anomaly is a monazite-bearing garnetite; the mineralogy is similar to that in the Sophie area: garnet, biotite, chlorite, epidote, sphene, sericitized plagioclase, and monazite. The metalliferous assemblage is a thorian monazite-graphite association with accessory sulphides. Whole rock U and Th values are 147 ppm and 7880 ppm, respectively.

Uranium-thorium associations create problems in defining the paragenesis of high temperature mineralization in the Carswell structure. This metallogenic process is different from the hydrothermal processes which created the economic orebodies. The higher temperature deposition is probably related to the metamorphic history of the pre-Athabasca basement gneisses. The yttrium, thorium, and lead contents of uraninite confirm the uniqueness of this metallogenic episode. The results are listed in Table VI.

The Sophie type mineralization, related to a metamorphic event in the basement, demonstrates that earlier mobilization of uranium in Aphebian metasediments has occurred. As these concentrations are expressed as uraninite when they are not contained within refractory minerals, they are

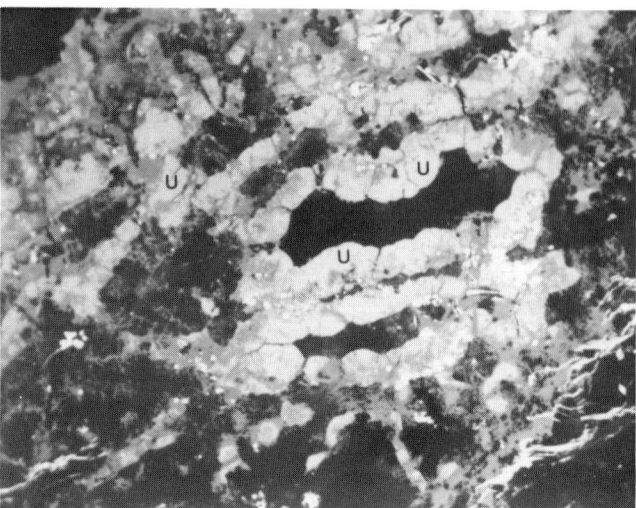

Figure 18. Polished section photomicrograph of the pseudospherulitic texture formed by uranium oxides (U). Numac TR18-5, × 280, oil immersion, plane light.

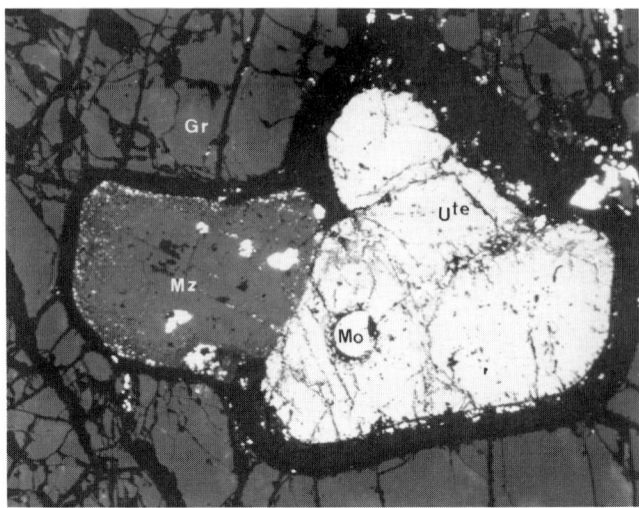

Figure 19. Polished section photomicrograph of uraninite inclusions (U^{te}) and monazite (Mz) in a garnet (Gr). Mo = molybdenite. Sophie PC-164, × 125, plane light.

easily leached and can be remobilized by hydrothermal processes if physical and chemical conditions are favourable.

CONCLUSION

The principal metallogenic characteristics of the mineralization in both sandstone and basement indicate the existence of multiple episodes of mineralization. Some of the uranium remained in situ, as in the case of Sophie and Numac, whereas other uranium migrated through the basement and the sandstone.

These episodes (Table VII) correspond with regional metallogenic events outside of the Carswell structure. The most important event that caused the emplacement of economic mineralization in the Claude, Dominique-Peter, and D areas occurred around 1050 Ma. Although this event and those that followed, are becoming well defined in the

Carswell structure, they are not as yet related to the earliest event evidenced by Sophie mineralization.

The uranium-thorium occurrence in Sophie raises questions regarding uranium metallogenesis prior to the economic mineralizing event. The Sophie mineralization is symptomatic of high temperature mineralization. Garnet, thorian monazite, uraninite, and molybdenite are incompatible with the hydrothermal processes that occurred in the orebodies. The high temperature processes, with the proper physical and chemical conditions, must allow remobilization of uranium, mobilize thorium, and introduce both into the monazite and uraninite structure. These processes are the same as those found in all post-metamorphic, catazonal metallogenic, or post-magmatic processes.

The second type of mineralization is probably later than the first episode and is represented by the Numac occurrence. Thorium is absent, and the nature of the uranium oxide is different from that in Sophie. The feldspathization or albitization phenomena is well developed in Numac. This type of mineralization is comparable with that of the Beaverlodge district, and thought to be Hudsonian age deposition.

The two above-mentioned mineralizing episodes coincide with the metallogenic scheme proposed by Beletsev (1972) for uranium deposits in Precambrian shield terrains. If the mineralizing episodes created economic deposits such as Beaverlodge and the uranium district in Kirovograd, Ukraine, they remain limited in the Carswell structure to poorly known zones.

The uranium in the Carswell area is a product of polyphase metallogenic events which formed three main episodes, all of which are typically hydrothermal. The main mineralizing event at 1050 Ma is comparable with ages obtained on many occurrences in the east part of the Athabasca Basin (Hoeve et al., 1981). This major episode affects both the metamorphic basement and the sandstone cover, as is suggested by the occurrences at Claude, D, and Dominique-Peter. The mineralizing solutions were tectonically controlled by mylonite zones which formed in both the sandstone and the basement. Mineralization is associated with phyllitic gangue, and coats the fault blocks which were formed by two directions of tectonic movement. The most complex mineralogy in the fault zone is found in the D orebody. After uranium-titanium oxide deposition, titanium oxides, tellurides, selenides, and sulphides formed. An oxidizing phase, then a reducing phase is the pattern during hydrothermal uraniferous processes.

The mineral associations in Claude, Dominique-Peter, and Donna only locally vary in principal paragenesis. The variations may be qualitative, as is the case for the sulphide-rich Donna basement, or quantitative, as is the case of Dominique-Peter orebody. There is only one paragenesis for these occurrences and it corresponds to regional metallogenesis in the Athabasca Basin.

The succeeding episode is evidenced in OP and Dominique-Peter occurrences, dated at about 850 Ma (Bell, 1985). Uranium and sulphides were deposited in an important fracture network. Mineralization is associated with quartz gangue in a typical hydrothermal phase. Banded textures observed in the two occurrences imply the coexistence

TABLE VI

MICROPROBE ANALYSIS OF SOPHIE AREA MINERALIZATION.

FeO	0.24
Y_2O_3	1.50
CaO	0.66
UO_2	69.96
ThO_2	1.86
Pb	19.59
TOTAL	93.81
APPARENT AGE	1800 Ma

TABLE VII

CHARACTERISTICS OF ALL URANIUM-METALLIFEROUS ASSEMBLAGES IN THE CARSWELL STRUCTURE.

	MONAZITE – URANINITE	PITCHBLENDE	URANINITE – SELENIDE – SULPHIDE	URANINITE – SULPHIDE	PITCHBLENDE – HEMATITE	PITCHBLENDE – CARBONATE	COFFINITE – SULPHIDE
URANIFEROUS PHASE	MONAZITE (th) URANINITE	PITCHBLENDE	BRANNERITE URANINITE	URANINITE	PITCHBLENDE	PITCHBLENDE	COFFINITE
NON-URANIFEROUS PHASE	MOLYBDENITE PYRITE	PYRITE	ALTAITE – PARAGUANAJUATITE GUANJUATITE – CLAUSTHALITE CALAVERITE – TROGTALITE FREBOLDITE – GERSDORFFITE NICKELINE – SKUTTERUDITE GALENA – CHALCOPYRITE MOLYBDENUM SULPHIDES GOLD – BISMUTH	MOLYBDENUM – SULPHIDE GALENA – PYRITE CHALCOPYRITE SPHALERITE TENNANTITE	HEMATITE MAGNETITE LIMONITE GOETHITE GOLD	CHALCOPYRITE GALENA – PYRITE	PYRITE GALENA
GANGUE	GARNET	ALBITE – CHLORITE	CHLORITE (Mg)	QUARTZ	---	CALCITE	---
HOST ROCK	GARNET – RICH PEGMATOID	FELDSPAR – ROCK	BASEMENT & SANDSTONE	BASEMENT	SANDSTONE	BASEMENT	SANDSTONE BASEMENT
TEXTURAL CHARACTERISTICS	DISSEMINATED	FRACTURES	COATINGS AROUND THE "ZONE A BOULES"	FRACTURES	FRACTURES	FRACTURES	FRACTURES
CHARACTERISTIC ASSOCIATIONS	U – Th	U	U – Mo – Bi – Se – S – (Te – Ni – Co)	U – Mo – Pb – Zn – Cu – S	U – Fe	U – Ca – Pb – Cu	U – Si
AGE (Ma)	1800	?	1050	820 – 890	380	---	---
TYPES OF OCCURRENCES	SOPHIE	NUMAC	CLAUDE – D – DOMINIQUE – PETER	OP	DONNA	ALL TYPES	ALL TYPES

of two episodes: a uraniferous mineralizing episode, and a non-uraniferous episode containing tourmaline and quartz. The host gneiss composition controls the nature of the sulphide phase that follows uranium deposition.

The hydrothermal episode probably corresponds to a regional metallogenic reactivation which allowed remobilization of uranium and selenides from outside the main ore zones. This episode is evidenced elsewhere in the Athabasca Basin. A remobilizing episode has been noted at Key Lake and Maurice Bay. Ages obtained for Key Lake are 918 Ma and 760 Ma for Maurice Bay (Hohndorf et al., 1981a and 1981b). These ages are close to those obtained by K. Bell (1985) on samples from OP.

The event which followed the 850 Ma hydrothermal episode is the formation of Cluff breccia. This event, whose origin is still in dispute, was accompanied by mechanical remobilization of pre-existing minerals. There is no hydrothermal episode directly linked with the formation of the structure that is reflected in the main uraniferous occurrences. A young hematite-pitchblende episode is recorded in Carswell around 380 Ma. This age is similar to others noted in the Athabasca Basin (e.g., Maurice Bay, Hohndorf et al., 1981b). It appears to be a paragenesis where the sulphide phase is almost non-existent.

Two parageneses reflect the latest hydrothermal processes that occurred in the Carswell structure: pitchblende-carbonate and coffinite-sulphide. Yellow uraniferous weathering products are particularly abundant in boulders found in glacial till.

The mineralizing episodes described in the occurrences within the Carswell structure reflect metallogenic activity in this part of the Athabasca Basin. Except for the Sophie area, it appears to parallel hydrothermal events similar to those described in the eastern part of the basin. Although the episodes may be the same, the prevailing elements in the paragenesis are very different. U-Te-Se-Mo-Bi-Au associations are found in the Carswell structure, as opposed to the U-As-Ni-Co-Mo association found in the eastern basin deposits. This difference is probably due to geochemical characteristics prevalent in each area, which may in turn, correspond to varying structural domains in the Churchill province.

REFERENCES

Adams, S.S., and Saucier, A.E., 1981, Geology and Recognition Criteria for Uraniferous Humate Deposits, Grants Uranium Region, New Mexico, Grand Junction, Colorado: Bendix Field Engineering Corporation, GIBX-2 (81), 225 p.

Beletsev, Ya, N., 1972, The Modern Problems of the Endogenic Ore Formation: The Academy of Science of the Ukranian SSR.

Bell, K., 1985, Geochronology of the Carswell Area, Northern Saskatchewan: in Lainé, R., Alonso, D., and Svab, M., eds., The Carswell Structure Uranium Deposits, Saskatchewan: Geological Association of Canada Special Paper 29.

Bonnamy, S., 1981, Relations Matières Organiques-Métaux (U-V-Ni): Etude en Microscopie et Diffraction Electronique – Thèse 3ᵉ Cycle, Université d'Orléans, 82 p.

Bruneton, P., 1981, Etude Pétrographique et Géochimique des Faciès Rencontrés dans la Structure de Carswell: Bordereau 6674, Amok Internal Report.

von Einsiedel, C.A., 1981, Petrography and Geochemistry of the Cluff Lake Breccias, Carswell Structure, Northern Saskatchewan: Unpublished B.Sc. Thesis, Carleton University, Ottawa, 44 p.

Ey, F., Gauthier-Lafaye, F., Lillié, F., and Weber, F., 1985, A Uranium Unconformity Deposit: The Geological Setting of the D Orebody: in Lainé, R., Alonso, D., and Svab, M., eds., The Carswell structure Uranium Deposits, Saskatchewan: Geological Association of Canada Special Paper 29.

Geffroy, J., 1975, Données Préliminaires sur le District Uranifère du Lac Cluff: Cadre Géologique, Pétrographie et Paragénèses: Commissariat à l'Energie Atomique Internal Report.

Goldhaber, M.B., Reynolds, R.L., and Rye, R.O., 1978, Origin of a South Texas Roll-Type Uranium Deposit II. Sulphide Petrology and Sulfur Isotope Studies: Economic Geology v. 73, p. 1690-1705.

Herring, B.G., 1976, The Metamorphism and Alteration of the Basement Rocks in the Carswell Circular Structure: Unpublished M.Sc. Thesis, University of British Columbia, Vancouver, 134 p.

Hoeve, J., Cumming, G.L., Baadsgaard, H., and Morton, R.D., 1981, Geochronology of Uranium Metallogenesis in Saskatchewan: Canadian Institute of Mining Uranium Symposium, September 8-13, 1981, Saskatoon, Technical Program, p. 11-12.

Hohndorf, A., Lenz, H., Wendt, I., von Pechmann, E., and Voultsidis, V., 1981a, Radiometric Age Determination on Samples of the Key Lake Uranium Deposit: Canadian Institute of Mining Uranium Symposium, September 8-13, 1981, Saskatoon, Technical Program, p. 24.

Hohndorf, A., Voultsidis, V., and von Pechmann, E., 1981b, U-Pb Isotope Investigations of the Maurice Bay Uranium Deposit, Lake Athabasca (Preliminary Results): Canadian Institute of Mining Uranium Symposium, September 8-13, 1981, Saskatoon, Technical Program, p. 23-24.

Pagel, M., and Ruhlmann, F., 1985, Chemistry of Uranium Minerals in Deposits and Showings of the Carswell Structure (Saskatchewan-Canada): in Lainé, R., Alonso, D., and Svab, M., eds., The Carswell Structure Uranium Deposits, Saskatchewan: Geological Association of Canada Special Paper 29.

Pagel, M., and Svab, M., 1985, Petrographic and Geochemical Variations within the Carswell Structure Metamorphic Core and their Implications with Respect to Uranium Mineralization: in Lainé, R., Alonso, D., and Svab, M., eds., The Carswell Structure Uranium Deposits, Saskatchewan: Geological Association of Canada Special Paper 29.

Ruhlmann, F., 1980, Quelques Exemples de Relation Uranium-Titane: Bulletin de la Société de France, Minéralogie et Cristallographie, v. 103, p. 240-244.

Theis, N.J., 1979, Uranium-Bearing and Associated Minerals in their Geochemical and Sedimentological Context, Elliot Lake Ontario: Geological Survey of Canada Bulletin 304, 50 p.

Tremblay, L.P. 1972, Geology of the Beaverlodge Mining area, Saskatchewan: Geological Survey of Canada Memoir 367, 265 p.

A URANIUM UNCONFORMITY DEPOSIT: THE GEOLOGICAL SETTING OF THE D OREBODY (SASKATCHEWAN — CANADA)

F. Ey
Institut de Géologie — 1, Rue Blessig STRASBOURG 67084 Cedex, France

F. Gauthier-Lafaye
Centre de Sédimentologie et Géochimie de la Surface-1, Rue Blessig STRASBOURG 67084 Cedex, France

F. Lillié
TOTAL Compagnie Minière — 218-228 Av. du Haut Lévêque 33600 PESSAC, France

F. Weber
Centre de Sédimentologie et Géochimie de la Surface-1, Rue Blessig STRASBOURG 67084 Cedex, France

ABSTRACT

The D deposit, located at the southern edge of the uplifted basement core of the Carswell structure, lies parallel to the overturned unconformity between Aphebian metamorphic rocks and Athabasca sandstones. The deposit consists of two major units: the basement to the north and the sedimentary cover to the south, separated from each other by the zone of uranium mineralization.

Structural analysis of the deposit reveals a zonation of intensity of deformation. The major tectonic structure is a mylonite zone, which mainly affects the basement near to its contact with the sandstones. North of the mylonites, deformation is less intense and is expressed as a zone of lenticulation. Further away from the mylonites, deformation has induced extensive microcracks. Deformation in the sandstones at the unconformity is expressed by the "zone à boules" where the bulk of uranium mineralization occurs.

Development of these tectonic structures must have been accompanied by considerable hydrothermal fluid flow, as evidenced by intensive host rock alteration displaying a pattern of clay mineral zonation parallel to that of the deformation zonation. The mineralized zone, characterized by Mg-chlorite, is surrounded by an envelope rich in Al-chlorite, which is developed in grey sandstone and bleached basement. Within the sandstone, immediately adjacent to the mineralization, a 3T-polytype illite is found, suggesting anomalously high fluid pressures. Away from mineralization in both hematized basement and white sandstone, a 1M-polytype illite prevails, indicating crystallization temperatures less than 250°C.

K-Ar isotopic ages indicate that uranium mineralization, deformation, and alteration correspond to a single event. These ages are not modified by the subsequent Carswell event, to which overturning of the D slab and the present disposition of the basement lying on inverted Athabasca sandstones, is attributable.

RÉSUMÉ

Le premier indice reconnu dans la structure de Carswell a conduit à la découverte du gisement D. Situé dans la zone indicielle, sur la bordure sud de la remontée centrale du socle, le gisement a été exploité en mine à ciel ouvert, sur la discordance entre la série sédimentaire des grès de l'Athabasca et les roches métamorphiques aphébiennes.

La carrière, allongée parallèlement au contact socle couverture, peut être divisée en deux grandes unités lithostructurales: le socle au Nord, la couverture au Sud. Ces deux unités appartiennent à une même écaille qui a été retournée au cours de la formation de la structure circulaire de Carswell. Le retournement explique la dispositon actuelle où le socle repose sur la couverture. Le contact des grès de l'Athabasca avec les roches métamorphiques aphébiennes est cependant un contact stratigraphique.

La minéralisation uranifère se situe dans un chapelet de lentilles allongées le long du contact socle-couverture selon une direction voisine de E-W, sur une longueur de 120 m. Le minerai de ce gisement a la caractéristique d'être à très forte teneur (4,25% UO_2, en moyenne).

La structure tectonique principale mise en évidence dans le gisement de D est une zone mylonitique (métrique à décamétrique) qui affecte exclusivement le socle, au niveau du contact avec la couverture sédimentaire. Toutes les autres structures tectoniques, zone à boules dans la couverture, zone de cisaillement et fentes d'ouverture dans le socle, *apparaissent comme des structures tectoniques associées à cette mylonite.*

La mylonite se présente comme une roche entièrement recristallisée et orientée. Elle résulte d'un épisode tectonique majeur en compression faisant intervenir des circulations importantes de fluides qui sont à l'origine de la transformation totale de la roche. La zone mylonitique se cantonne dans un couloir limité par les accidents F1 et F7 qui contrôle l'extension latérale de la zone de déformation maximale.

Le toit de cette mylonite est constitué d'une zone de déformation moins intense représentée par la zone de cisaillement. Cette dernière peut s'interpréter comme constituée d'un réseau de couloirs mylonitiques décimétriques à métriques qui découpent la roche en lentilles.

Enfin, cette tectonique de compression se marque également dans le socle par le développement de microfentes d'ouvertures où la circulation des fluides induit la cristallisation de séricites.

Dans la couverture, les déformations associées à la mylonite ne s'expriment que par la zone à boules située à son mur. Elle peut s'interpréter comme étant l'équivalent structural de la zone de cisaillement du socle.

Au niveau d'érosion actuel, l'enracinement de la mylonite dans les grès peut être interprété comme étant son amortissement dans la couverture et sa parallélisation dans les plans de stratification.

Les faibles rejeux associés à cette mylonite ne permettent pas de l'interpréter comme un plan de décollement ou d'écaillage important. Par contre, l'intense transformation de la roche encaissante souligne l'importance des circulations de fluides au sein de cette zone tectonique qui a sans doute été le siège de pressions fluides anormales. Les zones à boules de la couverture distantes de la zone mylonitique du socle doivent alors s'interpréter comme des zones de surpression fluides locales contrôlées par les plans de cisaillement qui se prolongent dans la couverture ou par les plans d'anisotropie tel que le contact socle-couverture ou grès-pélites.

L'étude des altérations de D a montré l'importance des phénomènes hydrothermaux qui sont guidés par les structures tectoniques et surtout par les zones mylonitisées. Ces circulations hydrothermales ont pour effets majeurs un départ massif de silice et une néoformation d'argile, aussi bien dans le socle que dans la couverture.

La minéralisation uranifère se cantonne dans la zone à boules plus ou moins mylonitisée de la couverture. Elle s'accompagne d'une zonéographie de la nature des minéraux argileux. La zone minéralisée, formée de grès et d'argiles noirs est en effet riche en chlorite exclusivement magnésienne. A ses épontes, dans le socle blanchi (mylonite et zone de cisaillements) et les grès gris, la proportion de chlorite magnésienne diminue au profit d'une chlorite alumineuse et l'illite devient de plus en plus importante jusqu'à devenir exclusive loin de la minéralisation, dans les grès blancs et dans le socle hématisé. Au toit de la zone minéralisée et dans les zones très tectonisées de la couverture l'illite est de type 3T, traduisant des pressions fluides anormales.

Les données géochronologiques ont montré que l'essentiel des minéralisations (1150 Ma; Bellon et al., 1975) et les altérations associées aux structures tectoniques (âge K-Ar de 1293 Ma sur argiles des zones mylonitisées: Clauer et al., 1985) correspondent à une seule et même phase et donnent des âges qui n'apparaîssent pas être perturbés par l'événement Carswell (478 Ma; Wanless et al., 1968). Ceci est confirmé par les observations structurales qui ne mettent pas en évidence des structures majeures pouvant être associées avec certitude à cet événement. Ce dernier responsable du retournement de la série entière, socle collé à la couverture, ne doit provoquer qu'un écaillage à large maille.

Dans le détail, à l'intérieur de l'écaille de D, les perturbations que l'on rattache à l'événement ne s'expriment que par la rotation et la réorganisation des boules selon F1 et par des rejeux (effondrements) du rubannement du socle et des fractures qui provoquent un glissement vers le Sud de tout le bâti sans pour autant perturber la configuration générale des grands ensembles structuraux.

INTRODUCTION

One of the first radiometric anomalies found in the Carswell structure led to the discovery of the D deposit, which was mined out by open pit during 1980 and 1981.

The open pit, straddling the unconformity, can be divided according to two major lithostructural units: the metamorphic basement to the north and the sedimentary cover to the south. These two units belong to one single block, which was overturned during the formation of the Carswell circular structure. Nevertheless, the contact between the basement and the sandstones is of a stratigraphic rather than a tectonic nature. The bulk of the mineralization occurs in elongate lenses, along the unconformity, resulting in an ellipsoidal orebody about 140 m long, 25 m wide, and 7 m to 8 m thick (Tapaninen, 1975).

Various exploratory methods showed a relationship between the D orebody, the unconformity, and the tectonic structures. The purpose of this work was primarily to define the nature, the effects, and the chronology of the different tectonic structures in the control of the uranium mineralization and to differentiate the regional tectonic patterns from those created by the formation of the Carswell circular structure, and to petrographically characterize the alteration zones and their possible relationships with the tectonic structures and the uranium mineralization.

LITHOLOGIC DESCRIPTION

A summary of the different rock types constituting the D deposit is given here. The facies are described in more detail in the discussion of the alterations.

The Metamorphic Basement

The basement, as exposed in the north wall of the open pit, consists of garnet-rich aluminous gneisses which display two types of alteration: hematization and bleaching (Fig. 1).

The hematized basement is primarily composed of quartz and remnants of hematized garnets surrounded by a red and green argillaceous matrix. Clay mineral zones, such as described for the sub-Athabasca paleoweathering profile elsewhere in the Athabasca Basin (Macdonald, 1980), were not identified.

The bleached basement is associated with tectonic structures and is restricted to the vicinity of the basement-sandstone contact. It forms a halo around the mineralization in the central part of the open pit.

The Athabasca Group

The western wall of the open pit offers an inverted stratigraphic section through the basal Athabasca sediments. Here, the contact with the basement was scoured prior to deposition of a conglomeratic layer, 10 cm to 3.50 m thick, which is composed of well-rounded quartz clasts and angular sandstone blocks set in a red and green clay-rich matrix. The size of the quartz clasts ranges from a few centimetres to 30 cm, whereas the sandstone blocks are 0.3 m to 1.0 m in diameter. The clast size decreases away from the contact with the basement.

The basal conglomerate is structurally underlain by a 7 m thick, purplish to maroon coloured sandy pelite, which is locally interbedded with thin sandstone lenses as well as a monomictic conglomerate layer containing quartz clasts up to 5 cm across. The sandstone lenses display sedimentary features such as oblique stratification and cross-bedding which are often enhanced by detrital magnetite grains and show a reversed grading.

The sandy pelite is succeeded by a white, massive, medium- to coarse-grained sandstone unit containing a few pelitic interlayers. The bedding planes are oblique and separate metre-thick beds. The configuration of channel features indicates reversed grading. These white sandstones are exposed in the southern wall of the pit.

The contact between the white sandstone and the sandy pelite forms the locus of extensive tectonic deformation. Within a 10 m wide zone, the sandstone is broken up into ball-shaped blocks that are enclosed by a clay-rich matrix. This particular tectonic facies, which has also been found elsewhere in the Carswell structure, is denoted here as a "zone à boules" (Ey *et al.*, 1981) (Fig. 2).

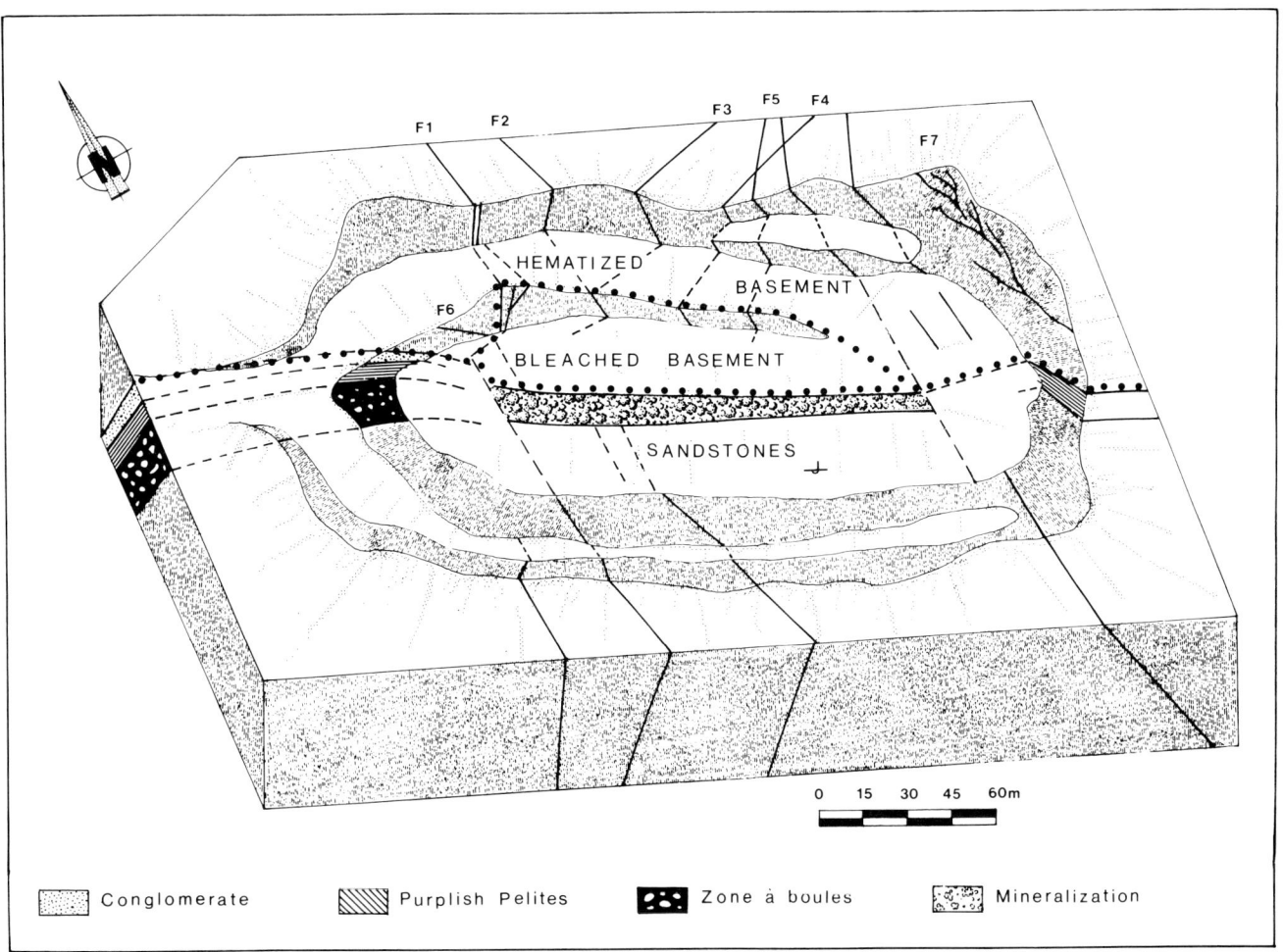

Figure 1. Major lithostructural units in the D open pit at the 323 m mining level (walls are shaded, horizontal surfaces are in white).

Figure 2. "Zone à boules" and basal sedimentary units on the western wall of the D pit. The maroon sandy pelite overlies the argillized "zone à boules". The sandstone balls and the argillized envelopes are subparallel to the overlying stratigraphic planes.

Figure 3. Basement-sandstone contact on the eastern wall of the D pit. The basal sedimentary series is composed of a purplish to maroon coloured sandy pelite lying on top of white Athabasca sandstones. The basement and the sandstones are in a reverse position. The basement slabs outlined by white paint on the pit face are isolated in the maroon pelites.

The Basement-Sandstone Contact

The basement-sandstone contact is highly variable within the open pit. On the western wall, the contact is of a stratigraphic nature. Channel structures account for a significant reduction in the thickness of the basal conglomerates, decreasing from 3.50 m at the 330 m mining level to only 10 cm at the 308 m level. The disturbed zone is located between the purplish sandy pelite and the white sandstone 9 m from the contact.

On the east wall, neither the basal conglomerate nor the "zone à boules" is present. The contact is made up of quartz clasts and basement fragments enclosed in a hematitic clay matrix. On this wall, the purplish sandy pelite undergoes a significant reduction in thickness from 15 m at the 330 m mining level to 5 m at the 308 m level. This is due to a combination of sedimentary channel features and tectonic shearing, which isolate a few metre-sized basement slices in the sandy pelite (Fig. 3).

In the central part of the open pit, at the lowest mining level (308 m), the contact is bleached on either side. Bleached basement overlies a cream coloured pelite which is locally intercalated with conglomeratic layers up to 10 cm thick.

STRUCTURAL SETTING OF THE BASEMENT

The most important structure within the open pit is a mylonite zone which follows the sandstone-basement contact. Away from the contact, towards the north, the mylonite gives way successively to a zone of lenticulation and a zone of microfracturing, signifying a decrease in the intensity of deformation. The mylonite zone and the zone of lenticulation together constitute an inner deformation envelope which coincides with the earlier mentioned halo of bleached basement surrounding the mineralization. The zone of microfracturing, corresponding to the zone of hematized basement, forms an outer envelope. Thus, the basement displays a zonation in the intensity of deformation (Fig. 4).

Figure 4. Contact between the hematized and bleached basement on the northern wall of the D pit. The dark zone is the shadow on the F3 fault. Note the trending of the joints and foliation which decreases away from F3 into the lenticulation zone.

Zone of Microfracturing

Within the zone of microfracturing, an eastern and a western block are distinguished on the basis of different structural characteristics. The former comprises most of the northern wall of the open pit and is characterized by jointing and multiple faults, denoted F1 to F6, whereas the latter is dominated by a transverse fault zone, denoted F7, and associated lenticular bodies (Fig. 5).

Fracturing is characterized by the combination of different tectonic features such as metamorphic foliation, joints, and tension gashes and produces in the basement a prismatic structure plunging south-southwest (Fig. 6), which is a

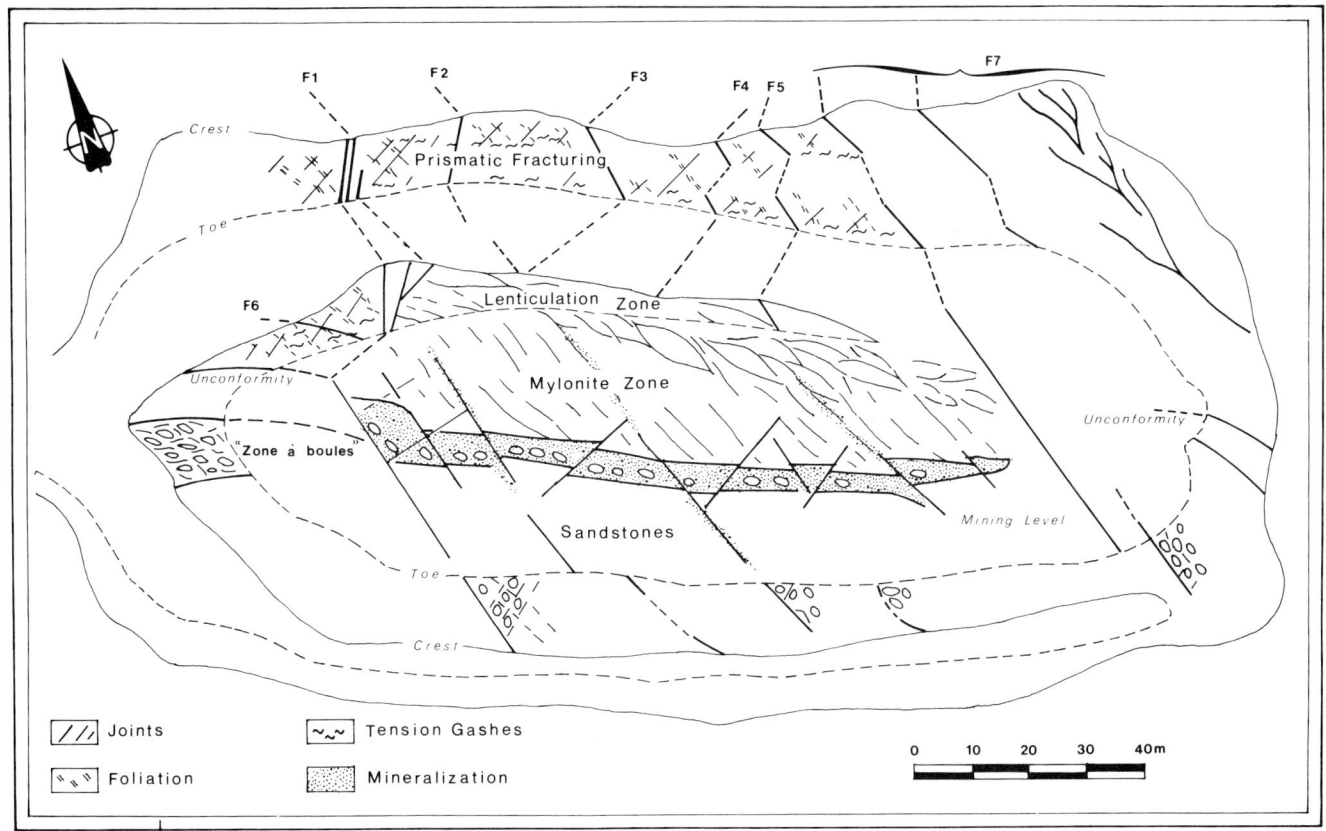

Figure 5. Synoptic map of the basement structure (dotted lines indicate direction, full line apparent dip).

characteristic feature of the western block.

Metamorphic foliation is characterized, on the hand specimen scale, by stretched, hematized garnet remnants in an argillaceous matrix. Variable in orientation, foliation planes are parallel to a zone axis plunging 45° SE. In the western block, foliation ranges from N25° to N60°, with dips about 45° SE, whereas in the eastern block, foliation is subvertical (N160°, 70°E) and may become inverted (N120°, 60° SW) near the contact with the sandstones.

Fractures dissect the north wall of the open pit into units 10 cm to 1 m in size. A single fracture set is well developed, especially in the western block where its average orientation is N170°, 55°W. Striae indicate a normal movement.

Tension gashes trend N100° to N120°, but dips range from 60° in the western block to 25° in the eastern block. In the western block, they appear as veins with openings of a few millimetres. Microscopic studies showed that they are filled with cryptocrystalline, argillaceous material and that they cut the quartz and garnet remnants. They were rejuvenated during later displacements, as expressed by the F6 fault, and reverse movements due to late overthrusting of the basement onto the Athabasca sediments are frequently observed (Fig. 7).

The western block is characterized by two dominant fault trends, N160° (F1, F2) and N40° (F3-F5, Fig. 1). The N130° to N160° trending, 60° SW dipping, F1 fault is the most important brittle structure in this block. It has dilated up to 1 m and is filled with clays, slightly mineralized Cluff breccias

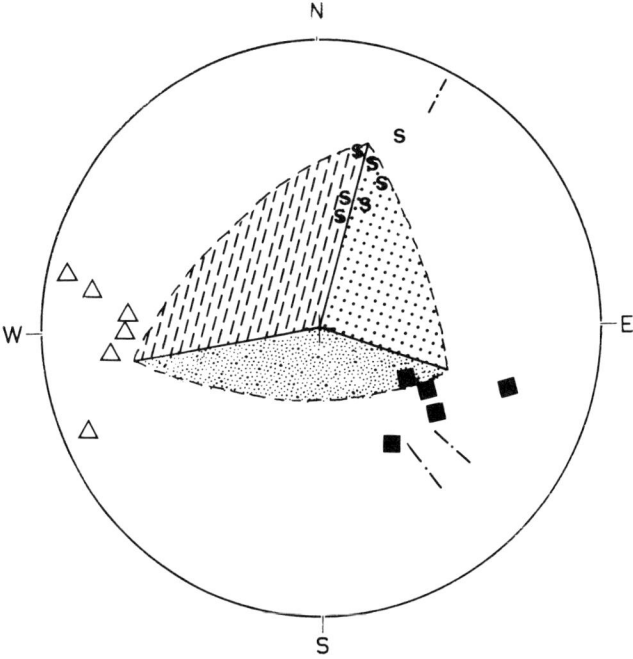

Figure 6. Synoptic stereogram illustrating prismatic fracturing in the western block by combination of the foliation (■), joints (△), and tension gashes (s) (upper hemisphere).

Figure 7. Alteration of the hematized basement. Tension gashes filled with pale green clay minerals offset the reddish hematized garnet remnants.

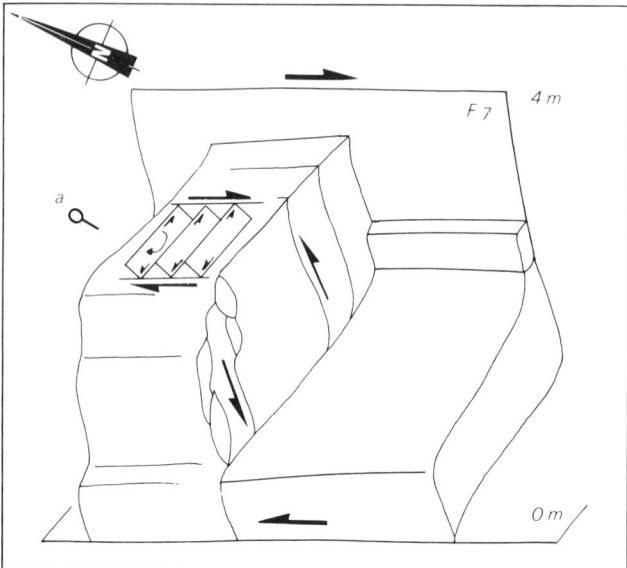

Figure 8. Tectonic lenses in the eastern block formed by F7 faults and N160° fractures. a = lens axis N160° plunging 40°S.

and gersdorffite. It has induced flexuring of the basement-sandstone contact and limits the westward extent of the basement bleaching and of multiple minor shears parallel to F1 which repeatedly offset the contact (see Fig. 5). Net displacement on the F1 fault has both sinistral strike-slip and normal components downthrowing the western hanging wall.

The N40° to N60° trending, 50° SE dipping faults parallel the metamorphic foliation. Locally they have opened up a few centimetres and have been filled with specular hematite. Slickensides indicate both normal and dextral strike-slip components to the net displacements.

The eastern block is characterized by a subvertical, N160° trending, strike-slip zone of anastomosing faults (F7 in Fig. 5) more or less parallel to the metamorphic foliation. The net

Figure 9. Sliding structure formed by joints (N170°, 50°W) and foliation (N40°, 60°SE) and used during the downthrow towards the south.

effect of the combination of the faults and the N160° trending fracture set is to dissect the eastern block into a series of lenses, elongate along the N160°, 40° SE plunging intersection. Near the contact with basal Athabasca sediments, F7 shear planes curve towards parallelism with the tension gashes and slickensides, indicating that they acted as contraction faults (thrusts). This suggests that the displacement sense of the F7 strike-slip zone was dextral (Fig. 8).

The combined effect of movements on all planar structures in the external envelope (fractures, faults, metamorphic foliation) was to produce metre-wide, subsident zones (Fig. 9), parallel with the overall N160°, 40° SE displacement direction, which is a radial direction from the center of the Carswell structure through the D deposit. The movement of the different blocks (Fig. 10) always has a component of downthrowing towards the south and is a late movement due to isostatic readjustment after overthrusting the basement onto Athabasca sediments.

Zone of Lenticulation

Although the structures described for the external envelope are more or less represented, the central part of the inner envelope is characterized by the presence of mylonites near the contact with the sedimentary cover. Between this mylonite zone and the hematized basement to the north, there is a zone of subparallel shears (lenticulation zone).

The lenticulation zone corresponds to a larger scale de-

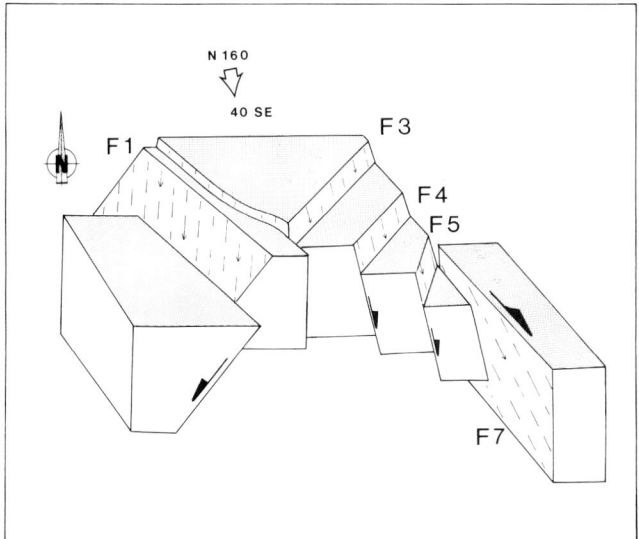

Figure 10. Illustration of the resultant movement indicating a generalized downthrow of the different blocks toward the south.

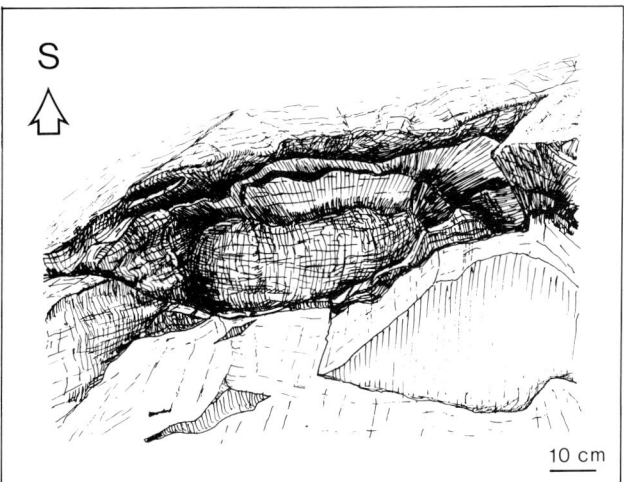

Figure 11. Shears forming metre-sized lenses in the bleached basement.

velopment of the anastomosing shear planes, trending N100° to N140° and dipping 40° to 50° NNE, which dissect the basement into metre-size lenses (Fig.11). Local recrystallization of micas is seen along the shear planes. Deformation is most intense near the mylonite zone.

Mylonite Zone

The mylonite zone is the dominant tectonic feature seen in the D deposit. Between the F1 and F7 faults, the extent of the zone corresponds to the extent of the bleached basement (Ey et al., 1983).

Trending N130° to N150° and dipping 35° to 70°NE, the mylonitic foliation is oblique to bedding in the sandstones which parallels the contact. The exposed thickness of the mylonite zone varies from several metres to several tens of metres according to position in the open pit. Where affected by the mylonitization, the basement appears as a white, fri-

Figure 12. A mylonitic argillized sample from the central part of the D pit in a mylonite zone near the sandstone contact (note the mylonitic foliation (m) intersected by the extensive microcracks (E)).

able, finely laminated and highly altered rock. In hand specimen, the original basement structures are completely overprinted (Fig. 12).

Microscopic observation of a mylonitic sample shows that although primary minerals have disappeared, their outlines remain and are represented by a brownish colouration. Anastomosing shear planes, dissecting the rock into lenses, are locally the sites of muscovite recrystallization. Within the lenses, microcracks are extensively developed and filled with cryptocrystalline clay minerals. The original foliation is microfolded and transposed subparallel to the shear planes. The altered remnants of primary minerals are extensively microfractured at the intersection of shear planes. The presence of microcracks filled with randomly oriented clay minerals within the lenses (Fig. 13) gives the rocks the appearance of hydraulically fractured microbreccias, comparable to the cataclasites or ultracataclasites of Sibson (1977). The mylonitization was thus achieved in the presence of hydrothermal fluids.

Locally, both mylonite and lenticulation zones are seen to be slightly offset by late movements along the F1 and F3 faults.

STRUCTURAL SETTING OF THE ATHABASCA SANDSTONES

The "Zone à Boules"

The most important tectonic structure in the Athabasca sandstones is the development of the "zone à boules". The cross-section exposed on the west wall of the open pit (Fig. 14) is taken as a reference section. Here the "zone à boules" occurs in the stratigraphically lower part of the Athabasca sediments and separates the massive, white sandstones and the purplish sandy pelite unit. The "zone à boules" is a 4 m to 5 m thick zone in which the sandstone is seen to be altered into three different forms:

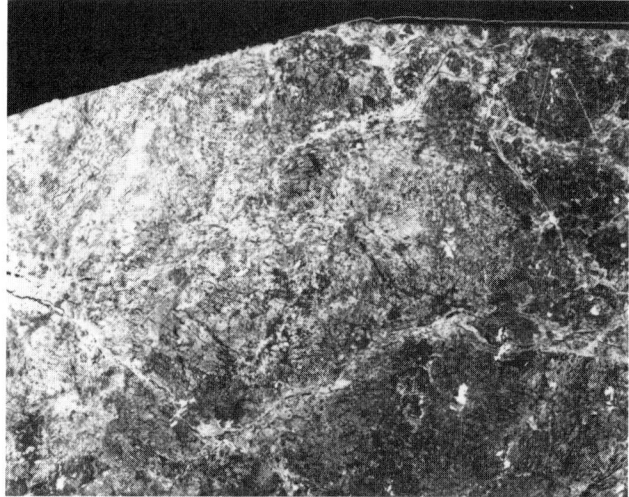

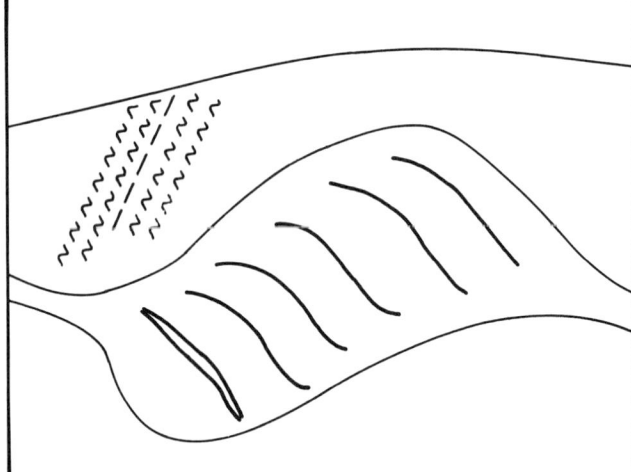

Figure 13. Photomicrograph of a mylonite from the D pit. The structure of the rock is represented by shear lenses containing extensive microcracks. The primary foliation is transposed, crenulated, and microfolded. No minerals are observable, and the argillization is homogeneous and ubiquitous in the entire rock.

1) Completely altered into greenish-white clay minerals, in which some bedding planes are still discernable;
2) 10 cm to 1 m diameter, striated bowl-shaped bodies or lenses, wrapped by the same clayey material; and,
3) Regular beds of coarse-grained, soft to powdery sandstones, where bedding planes are underlined by a small clayey layer.

The formation of such zones, which exist elsewhere in the Carswell structure, requires several processes:
1) Displacements along joints, bedding, and shear planes to form the basic "ball" or lens geometry;
2) Hydrothermal alteration to form the abundant clay minerals and large scale dissolution of quartz; and,
3) Deformation of both rounded blocks and matrix, inducing rotation of the sandstone balls and formation of striae on their margins (Fig. 15).

In the central part of the pit, at the lowest mining level, where the bleached creamy pelite unit is close to the bleached basement, the contact is laminated by N100° to N140° trending mylonitic shear planes. Adjacent to the F1 fault, displacements with both normal and sinistral strike-slip components, combined with displacements along shear and bedding planes, have produced balls of creamy pelite wrapped in argillized envelopes.

The Purplish Sandy Pelite Unit

The structure of the purplish sandy pelite unit on the eastern wall of the pit differs from that of the white sandstones only in the more intense development of fractures and in the tectonic thinning of the unit due to displacements along E-W trending shears. These shears also affect the adjacent basement rocks, give the contact a serrated form, and isolate basement slabs in the pelitic unit.

The White Sandstones

The white Athabasca sandstones forming the dip slope of the open pit, limit the southern extent of the D deposit. Bed-

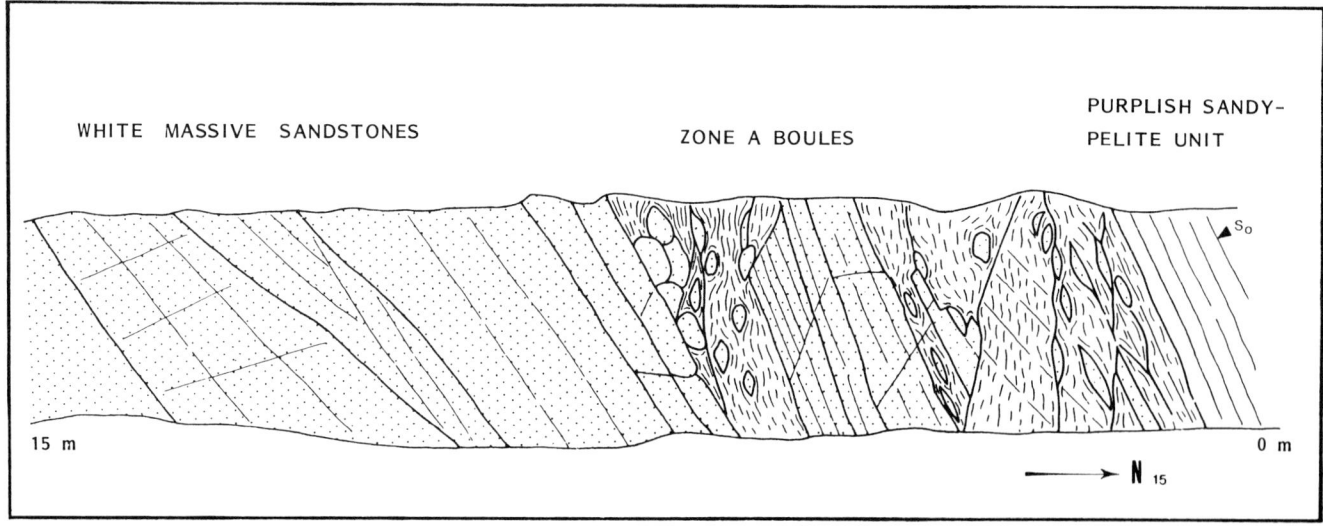

Figure 14. Schematic cross section of the "zone à boules" on the western wall (modified from Ey et al., 1983).

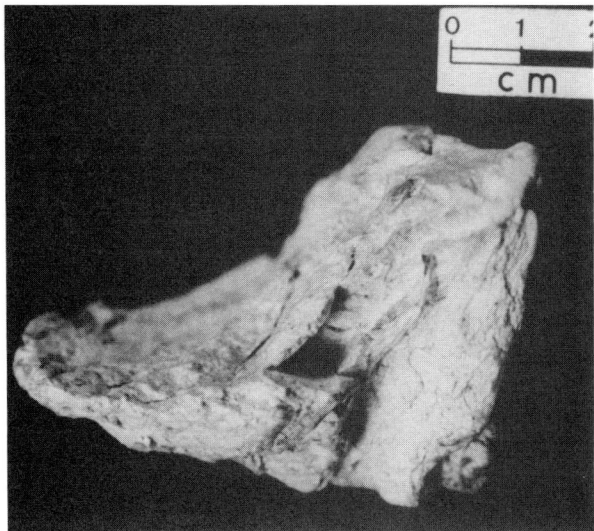

Figure 15. A curved illite-rich envelope of a faulted sandstone ball, "zone à boules", west wall.

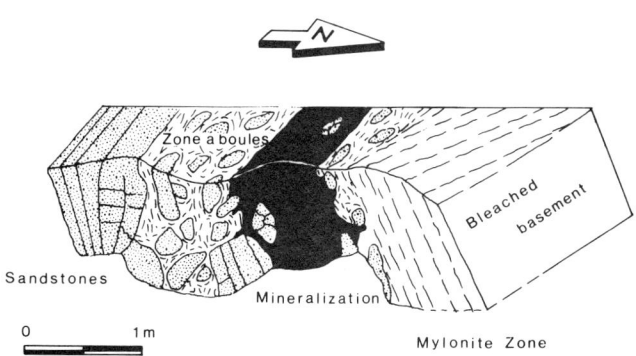

Figure 16. Cross section in a mineralized "zone à boules" (modified from Ey et al., 1983).

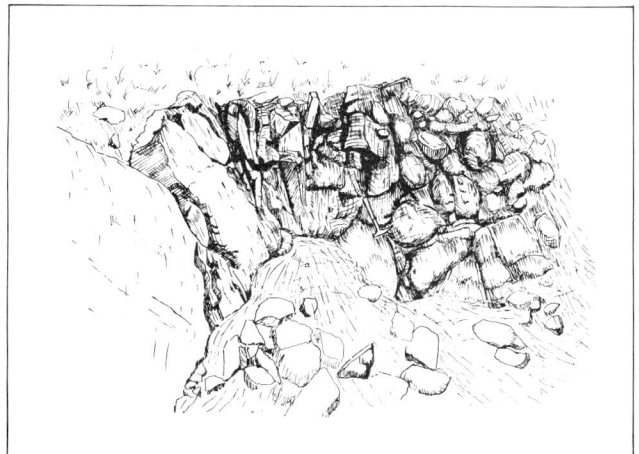

Figure 17. "Zone à boules" with metre-sized sandstone balls enveloped by the uranium mineralization.

ding is oriented N90° to N120°, dipping 50°N, subparallel to the basement-sandstone contact. Only brittle structures (faults of the N160° and N40° sets) affect the sandstones by producing local lenticulation of the sandstones at intersections with bedding planes.

STRUCTURAL SETTING OF THE MINERALIZATION

The bulk of the uranium mineralization occurs as discontinuous lenses in the "zone à boules" of the sedimentary cover, though minor disseminated mineralization does occur in the bleached basement clay, the N160°, and N40° fault sets.

The Unconformity Mineralization

The greater part of the mineralization forms an elongated ellipsoidal body trending WNW and dipping 50°N, subparallel to bedding in the sandstones. Sheared chloritic clays form the principle host for the mineralization, rather than organic-rich pelites as has been previously stated (Tapaninen, 1975; Harper, 1978).

The mineralized sequences begin at the footwall with a succession of grey sandstones, which are overlain structurally by black sheared clays and mineralization. The upper structural limit of the mineralization is generally the contact with mylonitic, bleached basement, though locally a few grey, fine-grained sandstone and pelitic layers, less than 50 cm thick, occur between the basement and the zone of mineralization. In places it is difficult to exactly locate the basement-sandstone contact, due to intense shearing and alteration in both units. A cross section through the central part of the pit (Fig. 16) shows the characteristic position of the mineralized zone. South of this cross section the grey, uncemented sandstones are followed by black, red, and then green silicified sandstones. The uranium mineralization both envelops the rolled sandstone balls (Fig. 17) and also forms bulky, metre-sized masses of pitchblende, wrapped in black chloritic clays and associated with pockets of organic matter (Landais and Dereppe, 1985). Beyond the orebody a layer of sandstone lenses and sheared clays is in contact with the bleached basement (Fig. 18a, 18b, and 18c).

The linearity of the orebody is disturbed by the large shear planes of the mylonitized zone, which refract into the bedding planes in the sandstones. The intersection of F1 faults and the basement-sandstone contact creates intensification of the "zone à boules", which then extends parallel to the faults, forming small appendages in the sandstones.

The Basement Mineralization

Mineralization in the basement occurs as deformed veinlets and disseminations of pitchblende on shear planes at the footwall of the F1 fault, or as infillings of remobilized pitchblende (Pagel and Ruhlmann, 1985) and alteration products, up to 10 cm thick, in the N160° and N40° fault sets where they offset the orebody (Fig. 19). These mineralized basement structures then appear as oxidized, hematized zones, tens of metres long, in the bleached basement.

Figure 18. a) Mineralized "zone à boules" from the D pit. The sandstone balls enveloped by pitchblende and chloritic sheared claystones overlie the grey sandstone unit. b) Mineralized "zone à boules" from the D pit. c) Pitchblende mineralization from the "zone à boules". The mineralization (P) and sheared claystone (c) envelope a sandstone ball (S). The center of the vein is composed of galena (G) and the borders, a more whitish colour, of Pb and Bi selenides.

ALTERATIONS AROUND THE D OREBODY

The Hematized Basement

The hematized basement differs from the unaltered basement by its depletion in mineral phases with only quartz, clay minerals, and hematite left (see Fig 7).

Quartz makes up 50% to 70% of the rock according to its degree of alteration. Quartz grains are often fractured and elongated, defining the foliation and showing an undulose extinction. Their boundaries are often corroded by the argillaceous matrix. Detrital quartz has, for the most part, a gneissic origin. Some secondary quartz occurs in altered minerals, such as in garnets.

Clay mineral analyses by X-ray diffraction show dominant illite (Table I) with minor chlorite. Fine-grained clay forms the whole matrix around corroded quartz grains. All minerals of the unaltered basement (feldspar, biotite, sillimanite, garnet, etc.), quartz excepted, have undergone the illitic alteration.

Illite, mainly of the 1M-polytype, with minor 2M-polytype, shows a crystallinity index of 3.5 to 4 following the Dunoyer de Segonzac index (1969) or of 2.5 to 3 on the Kubler index. This corresponds to the crystallization of illites in the anchizone and indicates a crystallization temperature range of 200° to 250°C. In addition, the $I(002)/I(001)$ peak ratio of about 0.4 to 0.5 shows that the illite has an aluminous tendency, which is confirmed by the microprobe analyses. A 0.8 calculated tetrahedral substitution ratio is common for these illites.

Most of the red staining observed in the pit is due to hematite development along cracks in altered garnets. Hematite also appears in the illitic matrix around the corroded quartz, or in microfractures.

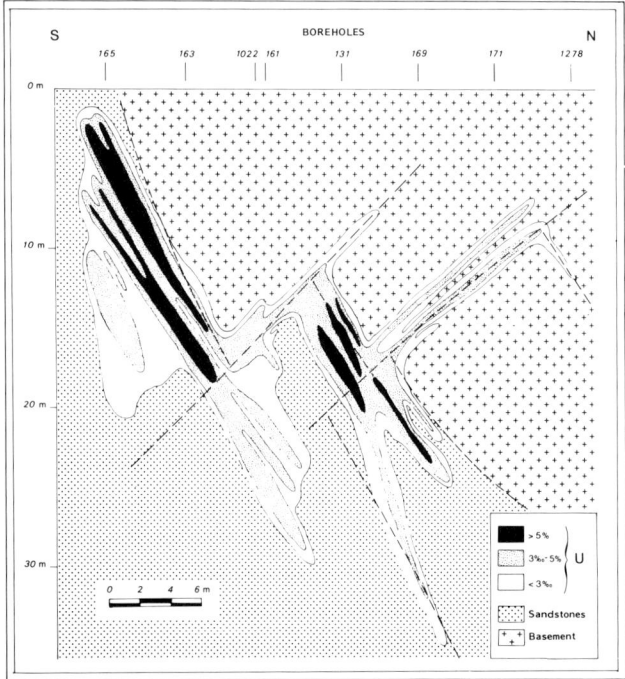

Figure 19. Configuration of the D orebody in a N-S cross section. Note the F1 fault offsets (modified from Amok — unpublished data).

The Bleached Basement

Along the unconformity, the bleached basement is characterized by a disappearance of hematite which is progressive toward the mineralization and is accompanied by argillization related to the mylonitization of the basement (see Fig. 4). Next to the orebody, the argillization is almost complete. Only a few dislocated and microfractured quartz are set in a pale-green fine-grained clay matrix which can comprise up to 90% of the rock.

Some minor uranium and nickel mineralization occurs in the bleached basement next to tectonic structures and is described by Ruhlmann (1985). The argillization results in a sharp increase of the Al, K, Mg, and Ti contents, up to twice the "original" amount, whereas the Fe content is constant. These chemical variations can be interpreted by the disappearance of quartz and hematite as shown by comparison of the chemical analyses of the hematized and bleached basement (Table II).

In fact, the hematized basement contains 75% silica of which two-thirds (50% reported in the total rock) represents quartz minerals and one third (25% reported in the total rock) is contained in clays. The bleached basement composition can be derived from that of the hematized basement by quartz elimination, which doubles the content of Al, K, Mg, and Ti, and the silica quantity, which considering the total rock, increases from 25% to 50%.

A clay zonation is noticeable in the bleached basement. The illitic phase is dominant near the hematized basement whereas the chloritic phase becomes dominant near the unconformity.

Illite. Illite is mainly of the 1M-polytype, with accessory amounts of the 2M-polytype. The crystallinity index (3.5 to 4), the I(002)/I(001) peak ratio (0.5 to 0.6), and the tetrahedral substitution ratios are similar to those of illites in the hematized basement and therefore indicate similar crystallinity conditions (Table III).

Chlorite. The XRD analysis showed one principal type of chlorite: an aluminous chlorite (Ey et al., 1983). The powder data diagram (Fig. 20) is characterized by the high intensity of the (003) peak and the position of the (060) peak at 1.50 angström which indicate a di-trioctahedral structure (Eggleton and Bailey, 1967). The microprobe analyses (Table III) confirm the aluminous character of these chlorites and give an overall composition:

$(Ca_{0.01} Na_{0.05} K_{0.12} Fe_{0.12} Mg_{1.68} Al_{2.96}) (Al_{0.67} Si_{3.33}) O_{10} (OH)_8$.

The small octahedral occupancy, the small tetrahedral substitution ratio (0.7), and the amount of alkalis are in accord with the di-trioctahedral nature of these minerals. These crystal chemical patterns are more comparable to the hyperaluminous chlorite described by Hayashi and Oinuma (1964) indicating hydrothermal conditions, than to a theoretical sudoite.

TABLE I
MICROPROBE ANALYSES OF THE HEMATIZED BASEMENT CLAY MINERALS

	HEMATIZED BASEMENT	
	DIOCTAHEDRAL MINERALS ILLITE	
	m(8)	σ
SiO_2	48.11	1.69
Al_2O_3	33.93	1.71
MgO	0.98	0.51
CaO	0.03	0.04
FeO	1.61	1.45
MnO	0.04	0.05
TiO_2	0.04	0.09
Na_2O	0.04	0.04
K_2O	8.23	0.30
Total	93.01	
	STRUCTURAL FORMULA	
Si^{4+}	3.21	
Al^{IV}	0.79	
Al^{VI}	1.87	
Mg^{2+}	0.10	
Fe^{2+}	0.09	
Mn^{2+}	–	
Ti^{2+}	–	
Total O	2.06	
Ca	–	
Na	0.01	
K	0.70	

TABLE II
CHEMICAL ANALYSES OF A BLEACHED AND A HEMATIZED BASEMENT SAMPLE

	LOSS ON IGN. 1000°C	SiO_2	Al_2O_3	MgO	CaO	Fe_2O_3	MnO	TiO_2	BaO	SrO	NaO	K_2O
HEMATIZED BASEMENT	3.20	75.2	15.1	2.79	<0.2	1.2	0.010	0.40	<0.01	0.03	0.10	2.95
BLEACHED BASEMENT	5.97	51.3	27.4	4.35	<0.2	1.0	0.014	1.07	0.06	0.02	<0.05	5.95

The Athabasca Sediments

The white sandstones of the southern wall are composed of up to 90% quartz grains enclosed in an argillaceous matrix. Quartz grains are always fairly rounded, with well developed quartz overgrowths in optical continuity with the detrital quartz grains.

Matrix filling in the sandstone pores is primarily illitic. Illites have a crystallinity index of 3.5 to 4 and are mainly of the 1M-polytype. According to the microprobe analyses (Table IV), they have a chemical composition identical to those in the hematized basement.

Detrital tourmaline, zircon, ilmenite, and magnetite are the principal accessory minerals which have been identified. Some important concentrations of heavy minerals, especially ilmenite, occur near a fault offset, with chemical analyses indicating a 3% Ti content.

The basal conglomerate differs from the white sandstone by the presence of rounded pebbles of different composition (quartz pebbles or clasts of sandstone or pelite). The conglomerate and the purplish sandy pelite unit matrix have a hematized clayey cement. The clay fraction is almost exclusively illitic, with some traces of chlorite and smectite; the latter is probably supergene in origin.

In the western "zone à boules" where intensive argillization affects the microcataclastic sandstones, the clay minerals are mainly 1M-polytype illite with accessory 3T-polytype illite, (identical to those which will be described further in the mineralized zone) and a small amount of Al-chlorite of the type seen in the bleached basement (Table IV).

A 10 cm thick layer with abundant magnesium tourmaline occurs in the hanging wall within the "zone à boules". This

TABLE III
MICROPROBE ANALYSES OF THE BLEACHED BASEMENT CLAY MINERALS

	BLEACHED BASEMENT			
	DIOCTAHEDRAL MINERALS		DI-TRIOCTAHEDRAL MINERALS	
	ILLITE		AL-CHLORITE	
	m(29)	σ	m(25)	σ
SiO_2	48.43	1.78	36.58	1.58
Al_2O_3	33.49	1.65	33.84	1.57
MgO	1.20	0.60	12.27	1.13
CaO	0.042	0.05	0.08	0.07
FeO	0.85	0.34	1.64	0.85
MnO	0.018	0.03	0.024	0.04
TiO_2	0.065	0.15	0.06	0.18
Na_2O	0.11	0.08	0.26	1.20
K_2O	8.61	0.83	1.01	0.54
Total	92.82		85.76	
	STRUCTURAL FORMULA			
Si^{4+}	3.24		3.33	
Al^{IV}	0.76		0.67	
Al^{VI}	1.87		2.96	
Mg^{2+}	0.12		1.68	
Fe^{2+}	0.05		0.12	
Mn^{2+}	–		–	
Ti^{2+}	–		–	
Total O	2.04		4.76	
Ca	–		0.01	
Na	0.01		0.05	
K	0.73		0.12	

TABLE IV
MICROPROBE ANALYSES OF THE SANDSTONE CLAY MINERALS

	SANDSTONES			
	DIOCTAHEDRAL MINERALS		DI-TRIOCTAHEDRAL MINERALS	
	ILLITE		AL-CHLORITE	
	m(11)	σ	m(9)	σ
SiO_2	47.49	1.37	35.94	2.33
Al_2O_3	34.36	0.68	34.10	1.70
MgO	1.13	0.61	10.59	3.09
CaO	0.03	0.04	0.17	0.17
FeO	1.42	1.06	3.25	3.15
MnO	0.04	0.04	0.02	0.03
TiO_2	0.11	0.14	0.12	0.28
Na_2O	0.19	0.25	0.34	0.95
K_2O	9.54	1.16	1.01	0.95
Total	94.31		85.53	
	STRUCTURAL FORMULA			
Si^{4+}	3.15		3.31	
Al^{IV}	0.85		0.69	
Al^{VI}	1.84		3.01	
Mg^{2+}	0.11		1.46	
Fe^{2+}	0.08		0.25	
Mn^{2+}	–		–	
Ti^{2+}	0.01		0.01	
Total O	2.04		4.73	
Ca	–		–	
Na	0.02		0.06	
K	0.81		0.12	

layer is composed of acicular dravite set in an illitic matrix. These hydrothermal tourmalines occur in different locations but always near the contact in the "zone à boules", or on the eastern wall between the purplish sandy pelite unit and the first white sandstone bed. This layer does not seem to be a well defined stratigraphic bed, but is always associated with the shear zones affecting the contact, and could explain the high content of boron observed near the unconformity (Mellinger, 1983).

The Mineralized Zone

The uranium mineralization has a complex mineral assemblage consisting of sulphides, selenides, tellurides, and gold associated with uraninite, pitchblende, and organic matter. This paragenesis is described by Ruhlmann (1985), and only the mineralization in the alteration aureoles around the orebody is presented in this paper.

The Sandstones. Near the mineralization different facies of sandstones can be distinguished:

Grey crumbly sandstones occur in the "zone à boules". They are generally sandstone balls wrapped in a mass of pitchblende. These sandstones are unconsolidated, have no clay matrix, and are generally transformed into a loose sand.

Greenish black sandstones comprise the majority of the sandstone balls and the first stratified beds in the footwall of the mineralization. Pyrite veins up to 3 cm thick occur in the bedding and fracture planes. The sandstone balls are speckled with hydrocarbon globules up to 3 mm in diameter (Landais and Dereppe, 1985). These sandstones contain a small amount of hematite which show different habits and are evidence of a late remobilization. The first hematite appears at the boundaries of the corroded detrital quartz grains, in the pressure dissolution joints, at the quartz grains contacts, and in the overgrowth aureoles. The remobilized hematite appears in nearly all the fractures and microcracks which affect the quartz minerals, the quartz overgrowths and the clay matrix. Carbonates are observed in fractures associated with the hematite but do not exist in the sandstone matrix.

Red silicified sandstones appear locally, a few metres from the footwall of the mineralization, but do not form a continuous envelope. The degree of silicification is variable. In the less silicified units, the quartz grains show pressure dissolution joints at their contacts with a matrix of chlorite and hematite. In the more silicified units, the quartz overgrowths fill up all the pores. Pressure dissolution joints do not affect any of the quartz grains but only the external part of the overgrowths. The chloritic matrix is not abundant and hematite is always present in the overgrowths.

Light green sandstones constitute a transitional unit between the chloritic sandstones and the white sandstones of the southern wall. The sandstone matrix is a mixture of illite and chlorite. The illite proportion increases away from the mineralization.

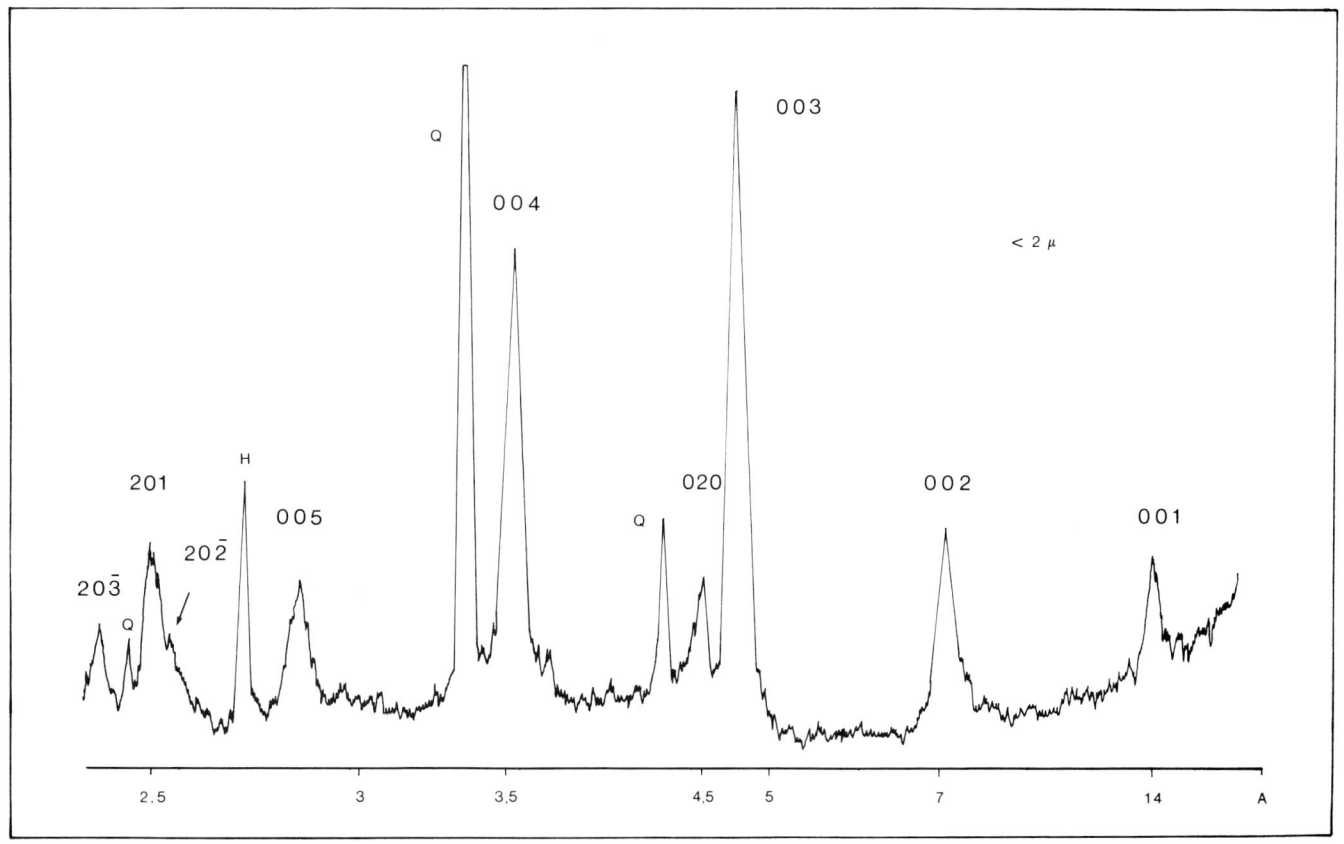

Figure 20. XRD diagram on $< 2\mu$m fraction of an aluminous chlorite (H = hematite).

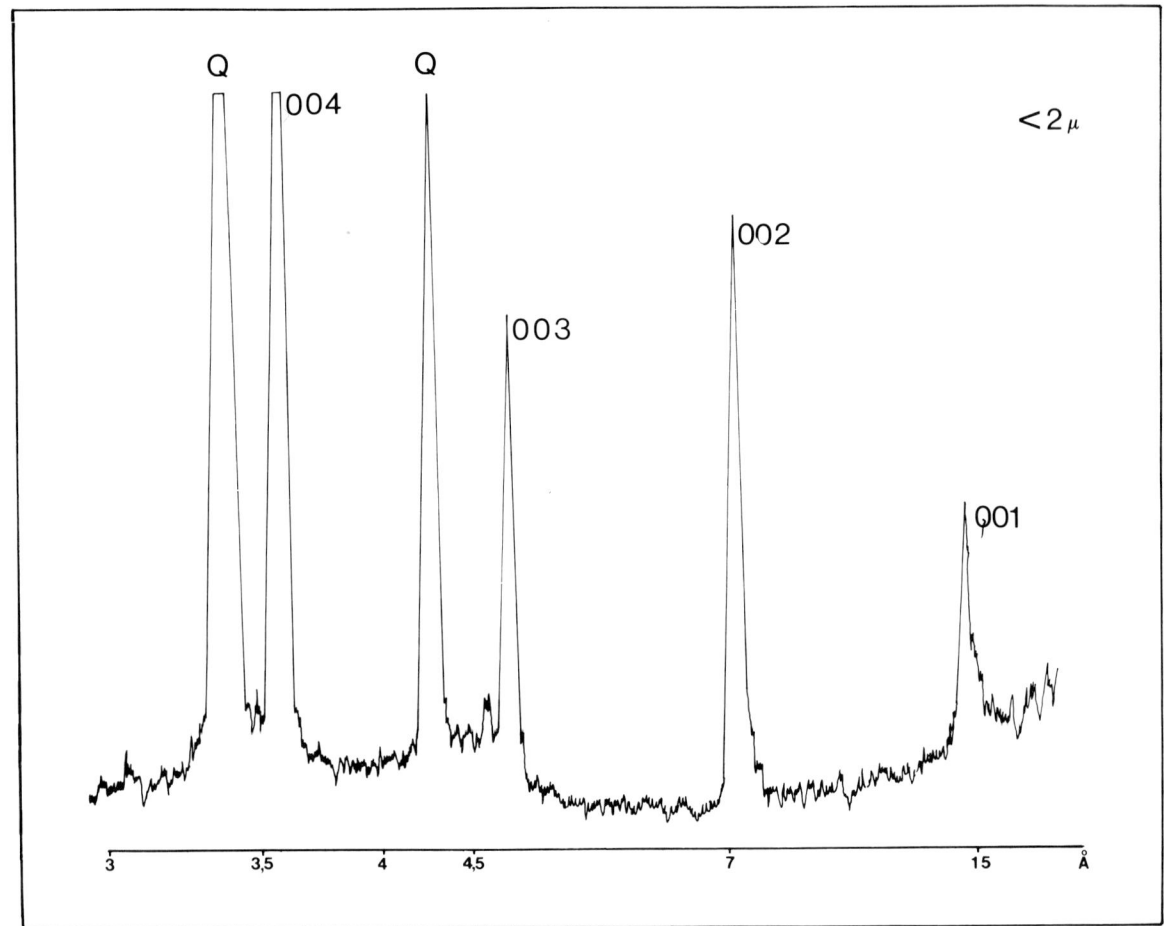

Figure 21. XRD diagram on < 2μm fraction of a magnesium chlorite (Q = quartz).

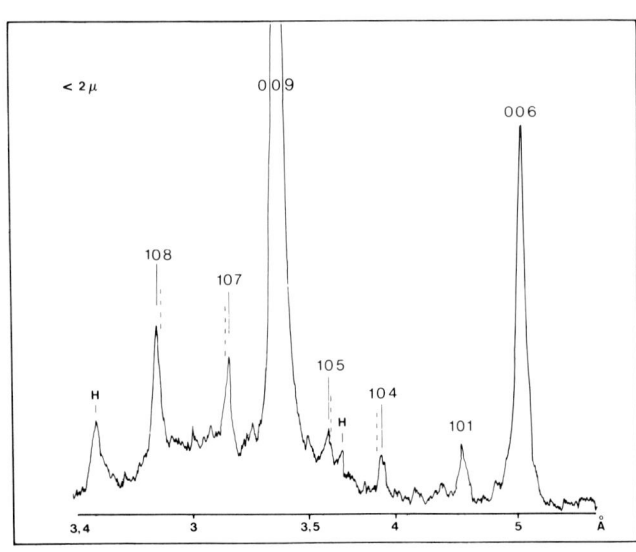

Figure 22. XRD diagram on < 2μm fraction of a 3T illite polytype. (H = hematite, the dotted lines correspond to the theoretical position as described by Yoder and Eugster, 1955).

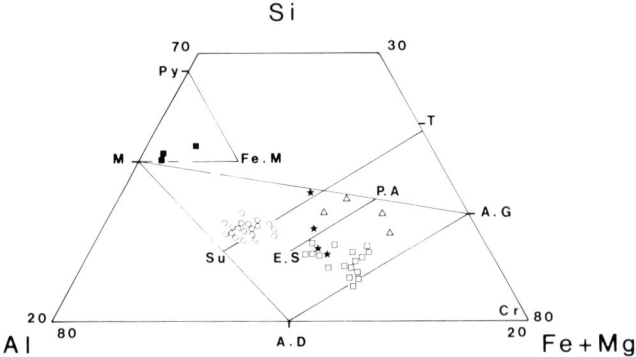

Figure 23. Partitioning in a Si-Al-Fe+Mg ternary diagram of the different clay minerals in the mineralized zone. The illites (■) are near the muscovite pole (M) with a tendancy toward the ferrimuscovite pole (F.M.). The magnesian (□)chlorites close to the uranium mineralization are, for the most part, near the eastonite-siderophyllite pole (E.S.). The aluminous (o) chlorites of the mineralization envelope are situated near the sudoite pole (Su). The other Fe (△) and Fe-Mg (*) chlorites are more dispersed (Py: pyrophyllite, T: talc, A.G.: antigorite-greenalite, A.D.: amesite-daphnite, P.A.: phlogopite-annite).

The Claystones. The black claystone, which forms the gangue of the mineralization and the envelope of the sandstone balls, cannot be attributed as it was before (Geffroy, 1975) to the purplish sandy pelite unit, known on each side of the mineralization toward the basal sedimentary sequence. The structure of these claystones is mylonitic, according to the phyllite crystallization along the shear planes in which veinlets of calcite and hematite occur. These claystones are, as are the sandstones, composed of chlorite, and small amounts of illite are present locally. The matrix often contains acicular dravite and locally neoformed apatite.

The abundance of heavy minerals in some samples suggests that the claystones are derived from the sandstone transformation. Locally, some small pelite clasts occur, set in the chloritic matrix. The mylonitic clays wrapping the balls seem to originate from the sandstone argillization and from the original pelitic material.

Clay Minerals. The dominant clay phase in the mineralized zone is made up of trioctahedral minerals. The most common chlorite is magnesian (Fig. 21). The microprobe analyses show that nearly two thirds of the octahedral and interlayer sheets are occupied by Mg cations and show a tetrahedral substitution ratio close to 1 (Table V).

More iron-rich, ferrous chlorite occurs in small fractures cutting the mineralization. In these chlorites, the amount of Fe cations is equal to or even greater than that of Mg cations and shows a very low tetrahedral substitution ratio. They also contain a small amount of potassium.

Going away from the mineralization, the illite proportion increases while the magnesian chlorite is replaced by an aluminous chlorite (di-trioctahedral mineral) identical to that described in the bleached basement.

Illite forms a proportion of the clay mineralization too small to be characterized with precision. However, a few microprobe analyses indicate a tetrahedral substitution

TABLE V
MICROPROBE ANALYSES OF THE MINERALIZED ZONE CLAY MINERALS

MINERALIZED ZONE					
	DIOCTAHEDRAL MINERALS	TRIOCTAHEDRAL MINERALS			DI-TRIOCTAHEDRAL MINERALS
	ILLITE m(3) σ	MG-CHLORITE m(25) σ	FE-CHLORITE m(4) σ	FE, MG-CHLORITE m(4) σ	AL-CHLORITE m(25) σ
SiO_2	44.58 2.89	29.22 2.22	34.06 1.31	33.21 4.01	34.32 1.53
Al_2O_3	32.05 3.05	20.79 2.56	12.90 3.40	20.11 2.27	30.07 2.71
MgO	1.19 0.96	22.45 2.70	10.99 3.47	13.94 2.00	11.56 1.44
CaO	0.007 0.006	0.086 0.10	0.44 0.19	0.36 0.14	0.17 0.88
FeO	3.00 1.05	9.44 3.26	22.53 3.95	14.92 1.45	5.10 3.18
MnO	– –	0.03 0.05	0.11 0.13	0.06 0.06	0.05 0.05
TiO_2	0.04 0.36	0.16 0.42	0.01 2.02	0.005 0.01	0.034 0.06
Na_2O	0.47 0.03	0.06 0.08	0.07 0.05	0.11 0.08	0.18 0.10
K_2O	9.83 0.48	0.50 0.75	1.10 1.40	0.56 0.82	1.26 0.47
Total	91.53	82.74	82.21	82.38	82.74
STRUCTURAL FORMULA					
Si^{4+}	3.09	2.98	3.74	3.43	3.32
Al^{IV}	0.91	1.02	0.26	0.57	0.68
Al^{VI}	1.71	1.47	1.41	1.88	2.75
Mg^{2+}	0.12	3.43	1.81	2.16	1.68
Fe^{2+}	0.17	0.80	2.06	1.21	0.41
Mn^{2+}	–	–	0.01	–	–
Ti^{2+}	0.02	0.01	–	–	–
Total O	2.03	5.72	5.29	5.25	4.85
Ca	–	0.01	0.05	0.04	0.02
Na	0.06	0.01	0.02	0.02	0.03
K	0.88	0.07	0.15	0.07	0.16

slightly higher than for the illites of the barren zones. But at the footwall of the mineralization, a particular species of illite occurs; its XRD diagram is close to the 3T muscovite structural form (Fig. 22) as described by Yoder and Eugster (1955). The presence of these illites could indicate crystallization under high pressure conditions (Baronnet, 1976). Further away from the mineralization, the common 1M-polytype illite with some 2M-polytype becomes the main clay mineral. These different clay minerals are shown in the Si-Al-Fe+Mg ternary diagram (Fig. 23).

INTERPRETATION AND CONCLUSION

The north-south cross section (Fig. 24) of the deposit synthesizes the spatial relationship between the different tectonic structures observed. The main tectonic structure displayed in the D deposit is a mylonitic structure affecting exclusively the basement near its contact with the Athabasca sandstones. All other tectonic structures of the basement, such as lenticulation zones and extensive microcracks, and of the sedimentary cover, such as the "zone à boules", appear to be structural features related to the mylonite zone.

The mylonite is a totally recrystallized and re-oriented rock. It results from a major compressive tectonic event with important fluid circulations which cause the argillization of the rock. North of, and parallel to the mylonite zone, the less intense deformation is marked by a lenticulation zone. This can be interpreted as small mylonitic shears cutting the rock into lenses. Finally, far away from the mylonite zone, this tectonic event induces a development of extensive microcracks where the fluid circulations control cryptocrystalline clay deposition.

At the unconformity, the deformation in the sandstones associated with the mylonite is expressed at the intersection of the mylonite zone and the sedimentary cover by the "zone à boules" host of the bulk of mineralization. The mylonite zone and the associated structures are limited by the F1 fault set which controls the lateral extension of the maximum deformation. The intense transformation of the surrounding rocks emphasizes the importance of fluid circulation under anomalously high fluid pressure. The "zone à boules" away from the mylonite zone must be interpreted as a local fluid super pressure zone controlled by shear planes which affected the sedimentary cover, or by anisotropic planes such as basement-sandstone or sandstone-pelite contacts.

The alteration study shows the importance of the hydrothermal phenomenon guided by the tectonic structures and especially the mylonite zone. This intense alteration is marked, in the basement as well as in the Athabasca sediments, by massive dissolution of quartz, neoformation of clay, leaching of the basement hematite, and temperature conditions below 250°C.

The alteration is accompanied by a zonation of the clay minerals (Fig. 25) which is conformable with the deformation zonation. The mineralization zone, composed of sandstone and black clays, is rich in Mg-chlorites. At its walls, in the bleached basement and in the grey sandstones of the "zone à boules", the proportion of the Mg-chlorites decreases with a concomitant increase in Al-chlorite and illite; away from the

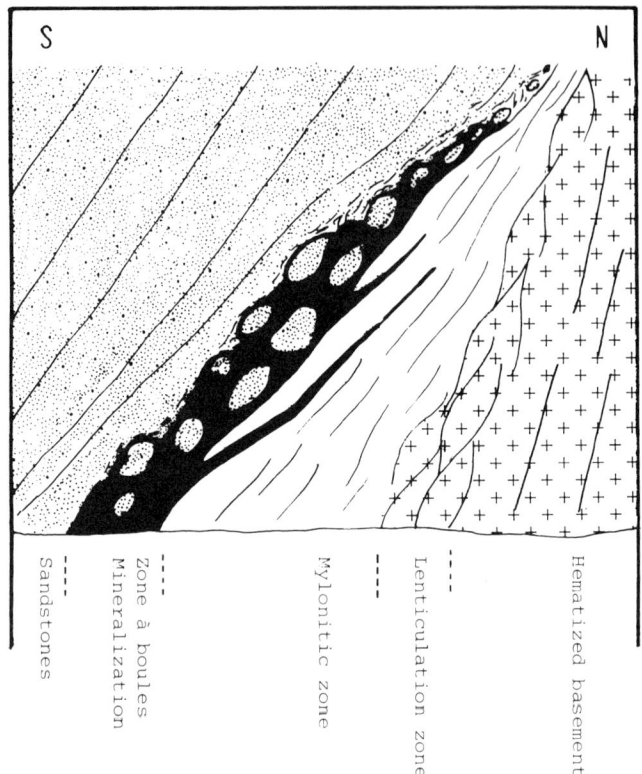

Figure 24. Relationships between the major tectonic structures of the D deposit (modified from Ey et al., 1983).

mineralization it becomes exclusively illite in the white sandstones and the hematized basement. Near the mineralization and the tectonized zones of the sedimentary cover, the 3T-polytype illite appears, indicating anomalous fluid pressures.

Uranium mineralization (1150 Ma, Bellon et al., 1976) and the alterations associated with the tectonic structures (1260 Ma on mylonitic clays, Clauer et al., 1985) have ages which show that the deformation and alteration correspond to the same event and are not disturbed by the Carswell event (478 Ma: Wanless et al., 1968). This seems to be confirmed by structural analyses which do not show any major structures related to the Carswell event. This event in the D deposit overturned an entire block of basement and sedimentary cover (Fig. 26).

At the present level of erosion, the rooting of the mylonite zone in the sandstones can be interpreted as its damping in the sedimentary cover and its parallelism to the bedding planes. In detail, inside the D slab, the perturbations due to the Carswell event are only expressed by the rotation and the reorganization of the sandstone balls by the F1 fault and by the downthrows on the major faults which induced a sliding towards the south and offset the orebody. None of these later movements disturb the general organization of the larger structural units.

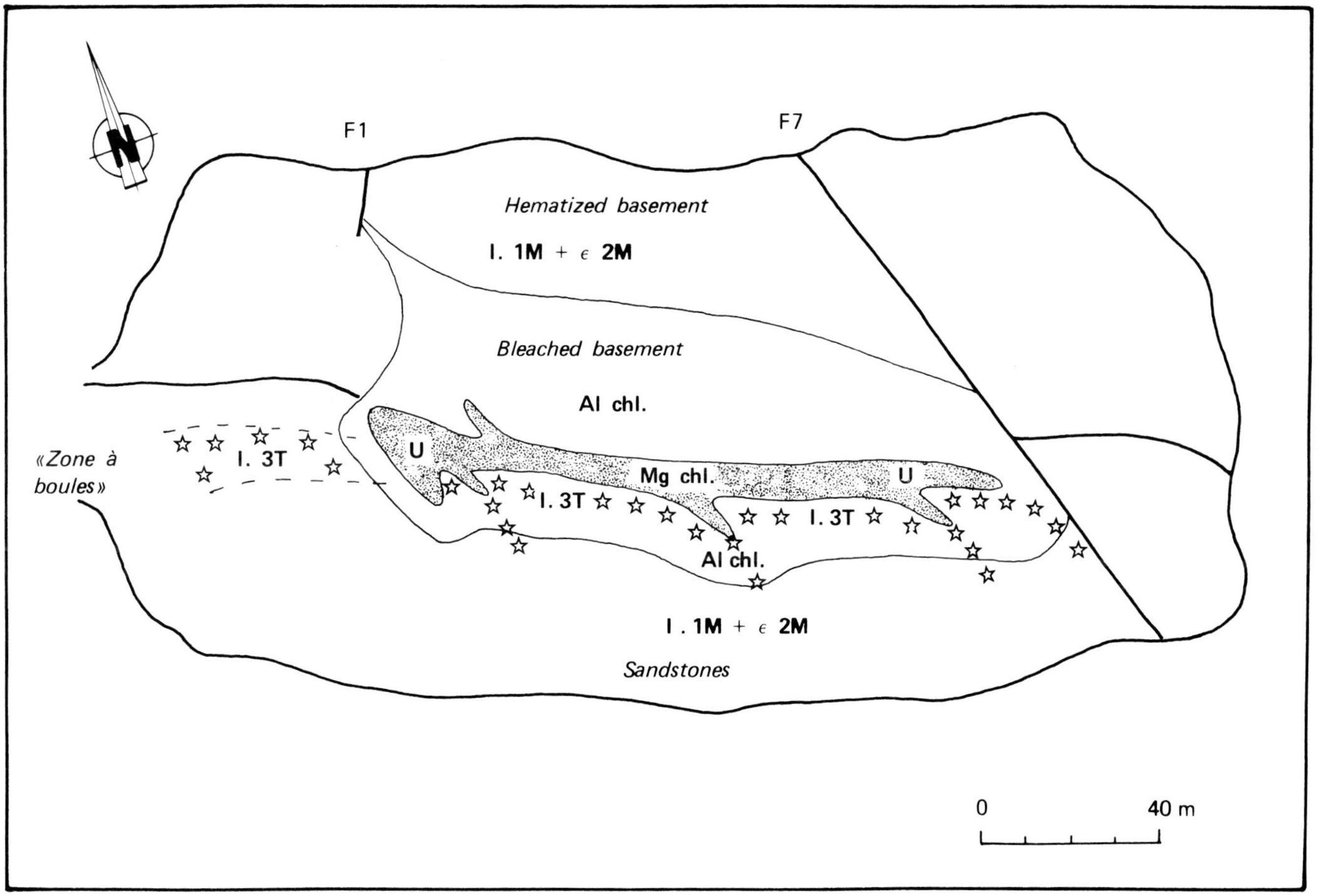

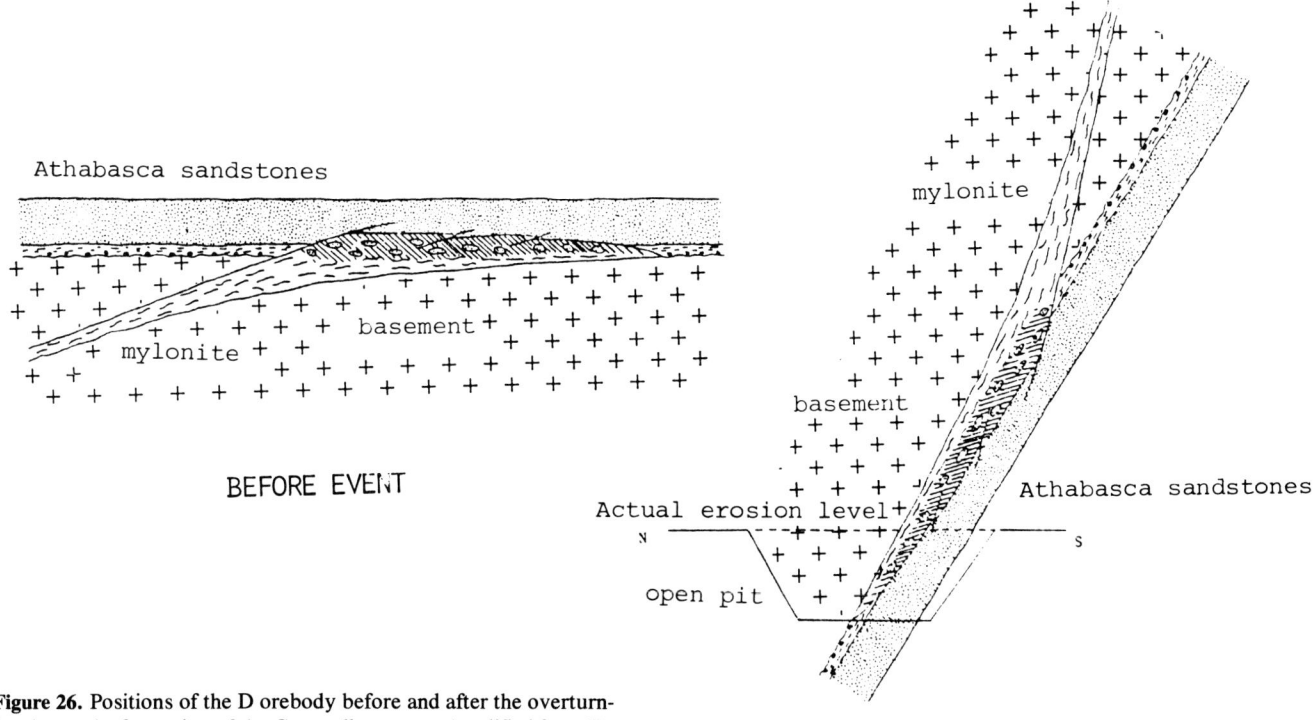

Figure 25. Clay mineral zonation in the D deposit (I = illite, chl = chlorite).

Figure 26. Positions of the D orebody before and after the overturning due to the formation of the Carswell structure (modified from Ey et al., 1983).

ACKNOWLEDGEMENTS

We acknowledge Amok Ltd. for their support, cooperation, field, and financial assistance for this work. Dr. A.W. Siddans is acknowledged for his extensive rewriting along with Dr. D. Jeannette for his helpful comments. Thanks are extended to Mrs. C. Erb and C. Ballot for typing the manuscript and to Mr. D. Baudemont for drafting.

REFERENCES

Baronnet, A., 1976, Polytypisme et Polymorphisme dans les Micas. Contribution à l'Etude du Rôle de la Croissance Cristalline: Thèse d'Etat, Université de Aix-Marseille, France.

Bellon, H., Devillers, C., Hagemann, R., and Touray, J.C., 1976, Dater les Minéralisations: Mémoire Hors Série de la Société Géologique de France, no. 7, p. 265-268.

Clauer, N., Ey, F., and Gauthier-Lafaye, F., 1985, K-Ar Dating of Different Rock Types from the Cluff Lake Uranium Ore Deposits (Saskatchewan-Canada): in Lainé, R., Alonso, D., and Svab, M., eds., The Carswell Structure Uranium Deposits, Saskatchewan: Geological Association of Canada Special Paper 29.

Dunoyer de Segonzac, G., 1969, Les Minéraux Argileux dans la Diagénèse. Passage au Métamorphisme: Mémoire du Service de la Carte Géologique d'Alsace et de Lorraine, v. 29, 317 p.

Eggleton, R.A., and Bailey, S.W., 1967, Structural Aspects of Dioctahedral Chlorite: American Mineralogist, v. 52, p. 673-689.

Ey, F., and Gauthier-Lafaye, F., 1982, Données Préliminaires sur les Altérations du Gisements D: Amok Internal Report.

Ey, F., Gauthier-Lafaye, F., Lillié, F., and Schumacher, F., 1981, Analyse Tectonique du Gisement D, Cluff Lake Saskatchewan: Amok Internal Report.

Ey, F., Gauthier-Lafaye, F., Lillié, F., Schumacher, F., and Weber, F., 1983, Le Gisement d'Uranium de Cluff D (Saskatchewan, Canada). Etudes Structurales et Pétrographiques: in Pagel, M., ed., Les Gisements d'Uranium Liés Spatialement aux Discordances: Géologie et Géochimie de l'Uranium, Mémoire 1, p. 97-113.

Geffroy, J., 1975, Données Préliminaires sur le District Uranifère du Lac Cluff (Saskatchewan): Cadre Géologique, Pétrographique et Paragénèses: Amok Internal Report.

Harper, C.T., 1978, Geology of the Cluff Lake Uranium Deposits: Canadian Institute of Mining Bulletin, v. 81, no. 800, p. 67-78.

Hayashi, H. and Oinuma, K., 1964, Aluminium Chlorite from Kamitika Mine, Japan: Clay Science, v. 2, p. 22-30.

Landais, P. and Dereppe, J.M., 1985, A Chemical Study of the Carbonaceous Material from the Carswell Structure: in Lainé, R., Alonso, D., and Svab, M., eds., The Carswell Structure Uranium Deposits, Saskatchewan: Geological Association of Canada Special Paper 29.

Macdonald, C.C., 1980, Mineralogy and Geochemistry of a Precambrian Regolith in the Athabasca Basin: Unpublished M.Sc. Thesis, University of Saskatchewan, Saskatoon, 151 p.

Mellinger, M., 1983, Investigation of Mineralogical and Geochemical Data from the Cluff Lake Area: Saskatchewan Research Council, Publication no. G-740-1-C-83, 29 p.

Pagel, M., and Ruhlmann, F., 1985, Chemistry of Uranium Minerals in Deposits and Showings of the Carswell Structure (Saskatchewan-Canada): in Lainé, R., Alonso, D., and Svab, M., eds., The Carswell Structure Uranium Deposits, Saskatchewan: Geological Association of Canada Special Paper 29.

Ruhlmann, F., 1985, Mineralogy and Metallogeny of the Uraniferous Occurences in the Carswell Structure: in Lainé, R., Alonso, D., and Svab, M., eds., The Carswell Structure Uranium Deposits, Saskatchewan: Geological Association of Canada Special Paper 29.

Sibson, R.H., 1977, Fault Rocks and Fault Mechanisms: Journal of the Geological Society, v. 133, p. 191-213.

Tapaninen, K., 1975, Geology and Metallogenesis of the Carswell Area Uranium Deposits: Canadian Institute of Mining Annual Western Meeting, Edmonton.

Wanless, R.K., Stevens, R.D., Lachance, G.R., and Edmonds, C.M., 1968, Age Determinations and Geological Studies, K-Ar Isotopic Ages, Report 8: Geological Survey of Canada Paper 67-2, Part A, 141 p.

Yoder, H.S., and Eugster, H.P., 1955, Synthetic and Natural Muscovites: Geochimica et Cosmochimica Acta, v. 6, p. 157-185.

The Carswell Structure Uranium Deposits, Saskatchewan,
edited by R. Lainé, D. Alonso and M. Svab,
Geological Association of Canada Special Paper 29, 1985

MINERALOGICAL AND STRUCTURAL ASPECTS OF THE DOMINIQUE-PETER URANIUM DEPOSIT

J.R. Blaise, and E. Koning
Amok Ltd., 817-825 45th St. W., P.O. Box 9204, Saskatoon, Saskatchewan S7K 3X5

ABSTRACT

The Dominique-Peter uranium deposit is situated 1 km to the north-northeast of Cluff Lake, Saskatchewan. The orebody occurs entirely within Aphebian age basement rocks metamorphosed to upper amphibolite to granulite facies. The basement is divided into two major lithologic units: a sequence of garnet ± cordierite ± sillimanite gneisses, the Peter River gneiss, dated at around 1760 Ma, and the underlying Earl River complex, composed of feldspathic and mafic gneisses which were dated at around 1870 Ma. A mylonite zone is developed along the transitional boundary between the two units. It affects the Peter River gneiss particularly but penetrates locally into the Earl River complex. Both units are cut by the later Cluff breccia, dated at around 400 to 500 Ma. In the northwest portion of the area, an overturned block of paleoweathered basement overlies Helikian age basal Athabasca sandstone, suggesting that a genetic connection exists between the deposit and the sub-Athabasca unconformity.

Uranium mineralization is found within the mylonite zone in northerly and east-northeast trending fracture networks and breccia zones dipping 45° to 75° to the west and northwest. Mineralization is vein-type, occurring as disseminations and stringers within breccia zones, single veins, and vein stockworks. Beyond the mylonite zone, the mineralization is discontinuous and generally uneconomic. Late north-northeast and northwest faults displace the mineralized structures. Three uraniferous assemblages exist: uraninite-sulphide-selenide accompanied by molybdenum and titanium, dated at around 1050 Ma, uraninite-copper and lead sulphides associated with quartz veins and secondary dravite, dated at around 900 Ma, and pitchblende-carbonate dated between 300 and 200 Ma. Alteration minerals include magnesium chlorite, sericite, and illite. Kaolinite only occurs in narrow zones around faults. Calculated reserves for the Dominique-Peter deposit are 1,761,000 tonnes of ore at a grade of 0.66% U.

RÉSUMÉ

Situé à 1 km au NNE du lac Cluff, Saskatchewan (Fig. 1), le gisement de Dominique-Peter est entièrement situé dans le socle métamorphique. Le contexte géologique du gisement peut être schématisé par la superposition à cet endroit de trois grandes entités pétrographiques: une série de gneiss alumineux au sommet, une série gneissique mixte à la base, et une zone mylonitique à l'interface des deux ensembles précédents (Figs. 3 et 12). Les gneiss alumineux mésocrates, gneiss Peter River, datés à 1760 Ma, présentent un assemblage à grenat, cordiérite, sillimanite, quartz et feldspaths. Cette série atteint 120 à 200 m d'épaisseur dans la zone minéralisée. Dans la série gneissique mixte, complexe Earl River, on distingue des faciès gneissiques quartzofeldspathiques francs à biotite, amphiboles, des faciès granitoïdiques et des niveaux moins importants d'amphibolites. Le passage entre les gneiss alumineux et la série mixte s'éffectue par une alternance de gneiss alumineux et de granitoïdes. Une zone mylonitique, de 40 à 60 m, s'est développée dans la zone de transition; elle affecte les différents faciès, mais plus particulièrement les gneiss alumineux et s'étend localement dans les gneiss inférieurs. Injectées dans les séries précédentes, sans direction tectonique privilègiée, on observe les brèches de Cluff, brèches qui se sont mises en place tardivement entre 500 et 400 Ma. Au nord de Dominique on observe en position renversée un lambeau de grès surmonté de socle régolithisé, grès caractéristique de la base des grès Athabasca. D'autres lambeaux, de plus faible puissance ont été recoupés en sondage le long de structures N45° à N60° au SO de Dominique. Ces lambeaux de grès suggèrent qu'il y a un lien génétique entre le gisement et la discordance Athabasca et que le gisement de Dominique-Peter représenterait la racine d'un gisement de type discordance classique.

Dans le gisement de Dominique-Peter on observe une relation étroite entre les zones d'altération et les minéralisations économiques. L'altération liée à des phénomènes hy-

drothermaux est surtout développée dans les faciès mylonitisés et avec un degré moindre dans la série alumineuse. Elle se caractérise par l'apparition d'illite et de chlorite magnésienne. Plus l'altération est proche de la minéralisation, plus la chlorite magnésienne est abondante. La kaolinite est parfois présente au niveau des failles. Les datations absolues (K-Ar) sur les zones d'altération indiquent plusieurs épisodes de circulations hydrothermales entre 1450 et 1000 Ma.

La minéralisation est liée directement aux mylonites, mais est limitée à des failles Nord-Sud et Est Nord-Est. Au delà de la mylonite, dans ces mêmes failles, les minéralisations existent mais sont plus discontinues et rarement économiques. Une fracturation tardive Nord-Est et Nord-Ouest recoupe les minéralisations provoquant des déplacements mètriques à décamètriques (Fig. 2). Dans la partie SE du gisement six structures minéralisées N-S à pendage ouest compris entre 45° et 70° ont été individualisées. La minéralisation s'y présente sous forme d'imprégnations ou de veinules à l'intérieur d'une brèche tectonique. Ces structures sont séparées de la partie NO du gisement par trois failles minéralisées ENE à pendage NO, dans lesquelles la minéralization est franchement filonnienne. Au NO du gisement trois structures N-S à pendage Ouest présentent des minéralisations semblables à celles du SE du gisement. Au SO existent des minéralisations peu profondes orientées ENE à pendage NO.

L'étude métallogénique montre trois paragénèses essentielles: uraninites-sulfures-séléniures accompagnée de molybdène et de titanates datée à 1050 Ma; uraninite-sulfures de cuivre et plomb dans des veines de quartz à dravite tardive, datée à 900 Ma; pechblende-carbonate, datée entre 300 et 200 Ma.

Les réserves du gisement Dominique-Peter, calculées avec la méthode de projection frontale, se montent à 1,761,000t de minerai à une teneur de 0.66% U soit 11,587 tonnes d'U.

INTRODUCTION

The Dominique-Peter uranium deposit is situated approximately 1 km north-northeast of Cluff Lake, about midway between the Claude and OP deposits, and near the southwest margin of the Carswell structure basement core (Fig. 1 and map in pocket). Initial interest in the Dominique area began in 1974 with the discovery of mineralization in the overburden.

Between 1974 and 1979, numerous small showings of mineralization were found in Cluff breccias which intrude the basement rocks. During 1979, geological mapping within an exploratory adit in the Claude deposit revealed that mineralized Cluff breccias occur near highly mineralized zones, and this led to a re-evaluation of the Dominique area.

In 1980, reconnaissance geological drilling was completed in the area located between two airborne magnetic anomalies (the OP and Dominique zones) which both contain uraniferous occurrences and overturned and fault-bounded blocks of Athabasca sandstone. Drilling was carried out to investigate the geological and structural relationships between these two mineralized zones. Economic mineralization (4.25 m of 2.5% U) was intersected below Peter River. Subsequent drilling to 1982 outlined the Dominique-Peter deposit. The deposit covers an area approximately 800 m long and 600 m wide, with mineralization located at depths varying from 120 m to 300 m below surface.

GEOLOGY OF THE DOMINIQUE-PETER URANIUM DEPOSIT

The Dominique-Peter area exhibits a consistent lithologic sequence of metamorphosed Aphebian age basement rocks. Based primarily on drill core interpretation, the basement succession is divided into two broad lithologic units: the pyritic and graphitic Peter River gneiss composed of garnet ± cordierite ± sillimanite gneisses which have been Rb-Sr dated at 1760 ± 80 Ma (Bell, 1985) and minor amounts (less than one metre thick) of interfoliated granitoid, and the underlying Earl River complex comprised of feldspathic and mafic gneisses with numerous conformable lenses and layers of granitoid (one to ten metres in thickness) Rb-Sr dated at 1870 ±75 Ma (Bell, 1985). The transition between these two units is commonly characterized by a thick sequence of quartzofeldspathic rock which contains numerous thin bands and discontinuous lenses, up to a few metres thick, of Peter River gneiss and local Earl River complex. All units strike northwest or north-northwest, although the dip of the compositional layering which parallels gneissosity varies between 30° and 50° to the east and east-northeast.

Peter River Gneiss

The Peter River gneiss is a unit composed of well banded semi-pelitic to pelitic gneisses. Generally they are grey to greyish-black or blackish-green, coarse-grained, well-foliated, and compositionally banded on a scale of centimetres to tens of centimetres.

The aluminous-rich bands contain xenoblastic porphyroblasts of almandine garnet, commonly less than 1 cm in diameter, typically augen-shaped, flattened, and concentrated in planes parallel to the metamorphic foliation. Pseudomorphic replacement by chlorite is common. Locally, high concentrations of equant garnets form metre-thick zones which are found near contacts with granitoid rocks, suggesting a concentration of refractory minerals during anatectic processes (Harper, 1983; Pagel and Svab, 1985). Porphyroblastic cordierite which is almost always altered and flattened, is distinguished by its blue-green colour and waxy appearance. Sillimanite is whitish to yellow, occurs as felted masses of fine prismatic crystals, and commonly wraps or curves around garnet and cordierite porphyroblasts. Biotite is present but is normally a minor constituent.

The leucocratic bands are typically quartz-rich but also contain both potassium and plagioclase feldspar. On the average, the leucocratic bands comprise 25% to 40% of the rock by volume. Pyrite occurs as disseminations and fracture fillings. Disseminated graphite is present, but more commonly accompanies chlorite along shear and fracture surfaces. Within these gneisses there exist numerous conformable bands of massive garnetiferous pegmatite, tens to hundreds of centimetres thick. Apart from retrograde metamorphism, alteration of this unit is weak. The vertical extent of the Peter River gneiss is 150 m to 200 m in the deposit area.

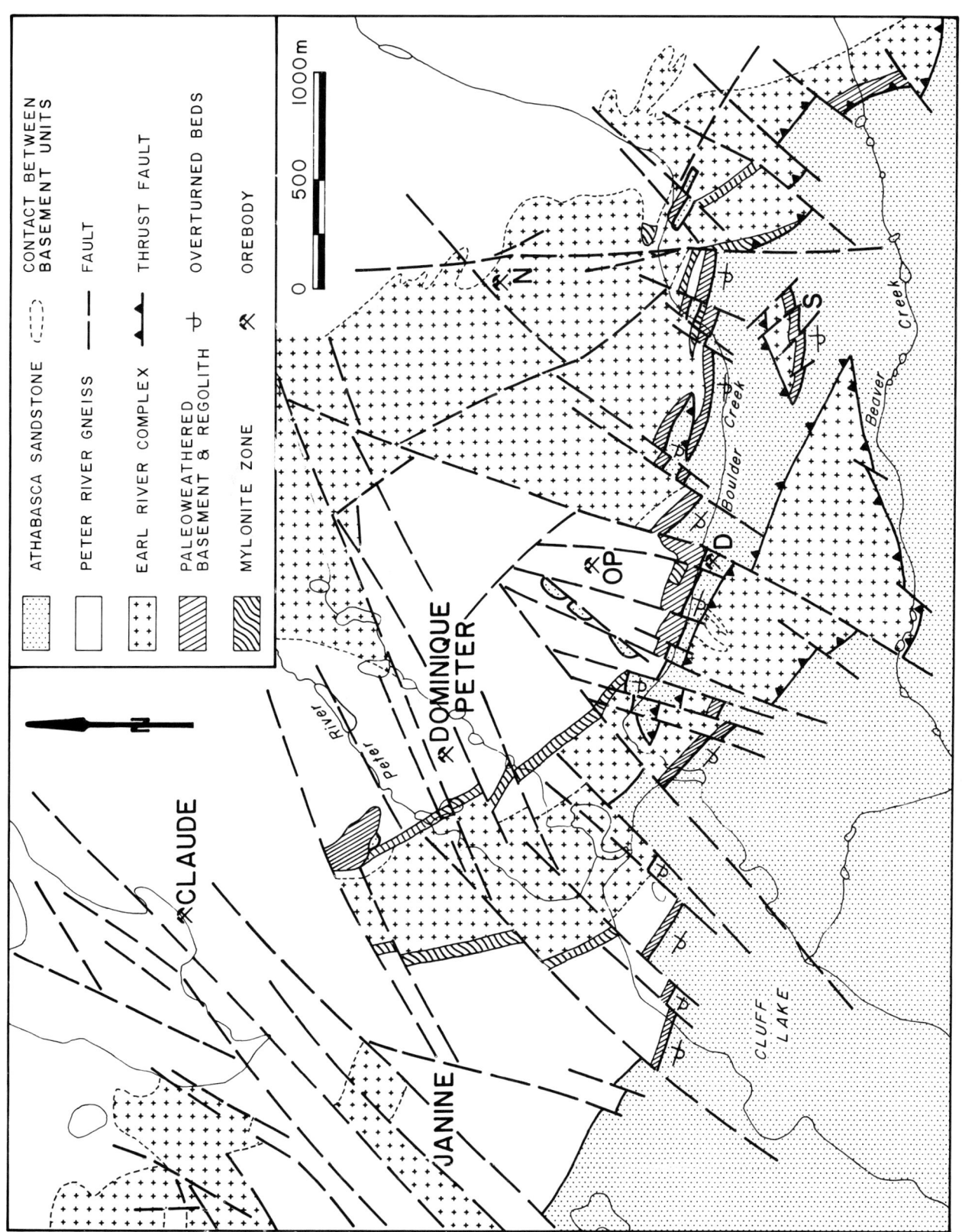

Figure 1. Generalized basement geology of the area surrounding the Cluff Lake uranium occurrences and deposits (Tona et al., 1985).

Earl River Complex

Underlying the Peter River gneiss is the Earl River complex, composed of feldspathic and mafic gneisses. Within the feldspathic gneiss, two main sub-types are recognized: biotite-feldspathic and mafic-feldspathic gneiss. Feldspathic gneisses are light to dark grey and pinkish-grey, fine-to medium-grained, poorly to well foliated, and composed mostly of quartz, potassium and plagioclase feldspar with biotite, pyroxene ± amphibole in amounts up to 20%. Compositional banding, developed on a 1 cm to 5 cm scale, consists of biotite-rich layers or pyroxene, biotite ± amphibole layers, and coarse-grained pinkish to greyish quartzo-feldspathic layers. Locally, augen textures are observed in the feldspathic layers. Disseminated pyrite and small equant garnets occur in the mafic-rich layers. Feldspathic gneisses found near the Peter River gneiss typically contain garnet, cordierite, and sillimanite (Pagel and Svab, 1985).

Mafic gneisses occur as discontinuous lenses or layers, a few metres thick, within the feldspathic gneisses. Mafic gneisses are usually greenish-grey, dark green or black, fine-to coarse-grained, poorly to well foliated, and composed essentially of pyroxene, amphibole, biotite, and plagioclase feldspar. Small quantities of quartz and trace amounts of garnet and sulphides, generally pyrite, are also present. Two major sub-types are identified: coarse-grained amphibolite, in which the predominant mafic phase is amphibole, and a well-foliated granulite, in which two pyroxenes, plagioclase, and small quantities of quartz are present (Herring, 1976). Small equant garnets, when observed, appear to be restricted to the two pyroxene granulite gneiss. The coexistence of hypersthene and clinopyroxene implies that these rocks have undergone prograde granulite facies metamorphism (Herring, 1976; Harper, 1983; Pagel and Svab, 1985). Later they underwent retrograde metamorphism to upper amphibolite facies.

Rocks of the Earl River complex crop out to the west and produce the large positive ovoid aeromagnetic anomaly observed in the Dominique area (Fig. 1). Throughout the Cluff Lake area the Earl River complex forms a series of domal structures (Powell et al., 1985) overlain and draped by the Peter River metasedimentary sequence. Separating the domes and along the flanks of these structures are synformal troughs filled with thick sequences of Peter River gneiss underlain by rocks of the Earl River complex. Various authors (e.g., Lewry and Sibbald, 1977, 1979; Ray, 1977; de Carle, 1980) have noted similar domal structures in the Wollaston Domain along the eastern rim of the Athabasca Basin and have attributed their formation to thermotectonic events during the Hudsonian orogeny.

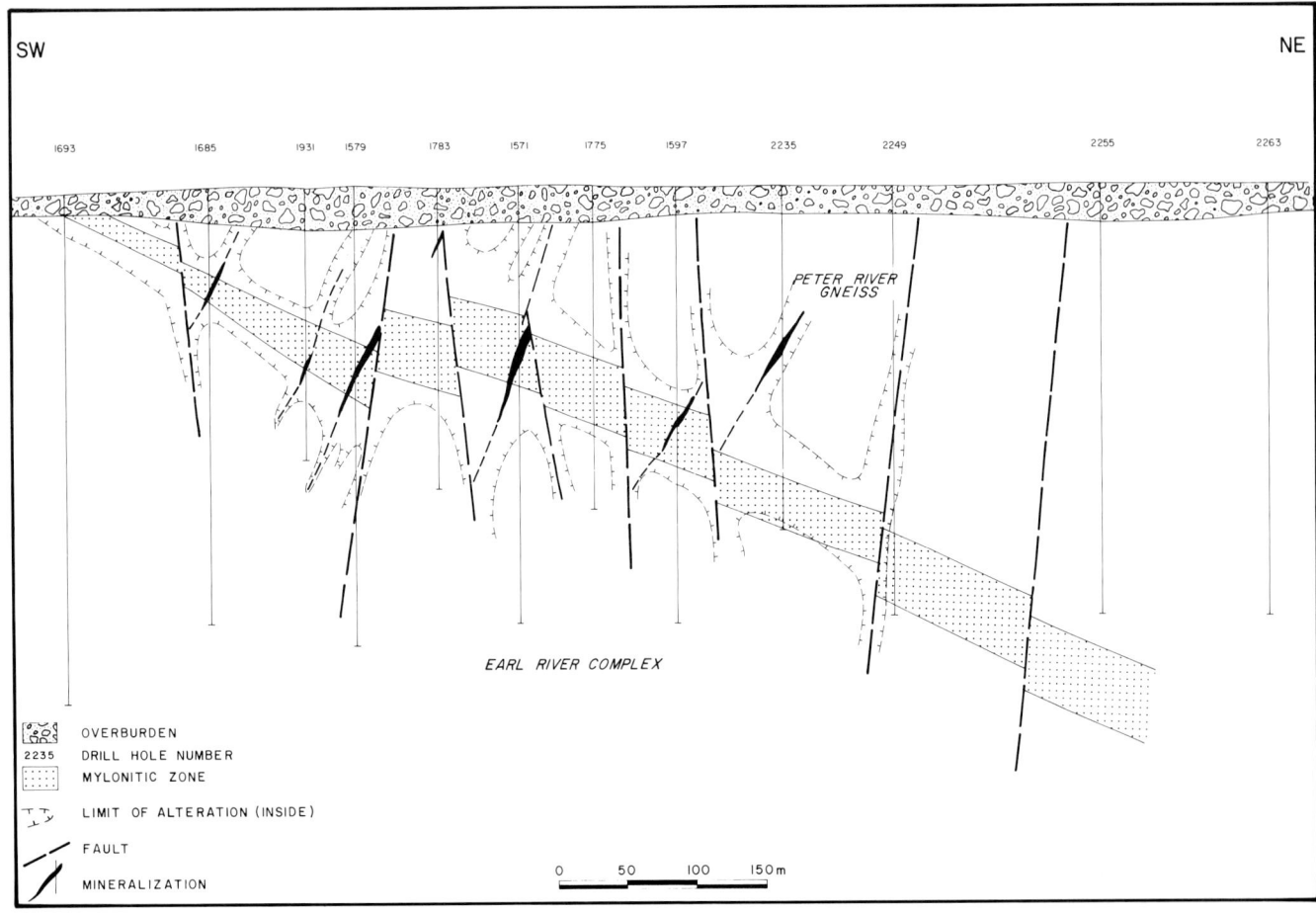

Figure 2. Schematic northeast-southwest geological cross section through the Dominique-Peter deposit along the eastern flank of the Dominique Dome (for section location see Figure 11).

Granitoids

Granitoids are present in both the Peter River gneiss and Earl River complex rocks, but are more abundant within the Earl River complex. They consist of coarse-grained leucocratic rocks which are seen on all scales, from centimetre-thick layers to lenses several metres to tens of metres thick. Their margins are, for the most part, parallel or subparallel to the foliation of the host rock. Granitoids are typically pink to grey on fresh surface, medium- to coarse-grained, massive to weakly foliated, and primarily composed of potash feldspar, plagioclase and quartz. Locally, small amounts of biotite are present. It is common to observe garnet and cordierite porphyroblasts and sillimanite in small to trace amounts in granitoids which exist within Peter River gneiss, and pyroxene in those present in the Earl River complex (Pagel and Svab, 1985). The probable origin of granitoid rocks seems to be associated with in-situ partial anatexis of predominantly feldspathic but also some aluminous gneiss (Pagel and Svab, 1985).

Mylonites

A zone of cataclastic deformation is developed along the transitional boundary between the Peter River gneiss and Earl River complex (Fig. 2). Although mylonitization seems to have predominantly affected the Peter River gneiss, it has also penetrated the Earl River complex in the northeast portion of the deposit. The mylonite zone represents a zone of ductile deformation almost parallelling regional foliation, and is best developed along the transitional boundary between the two lithologic units. This same mylonite zone has been traced by diamond drilling to the unconformity west of the 'D' deposit.

Using the classification and nomenclature proposed by Higgins (1971), the cataclastic zone is comprised of protomylonites, mylonites, and ultramylonites. The total zone is 40 to 60 m thick.

Protomylonites, the initial stage of cataclasis (Figs. 3 and 4) are more commonly observed in the granitoids of the transitional boundary. These protomylonitic granitoids are composed of megascopically visible fragments of potassium feldspar that are separated by gliding surfaces of fine granular material consisting of quartz and plagioclase. Porphyroclasts make up more than 50% of the rock.

Mylonites (Figs. 5 and 6) represent the greater proportion of the cataclastic zone and are characterized by streaky and ribbon-like textures; the planar structure developed is appropriately termed a fluxion structure. Porphyroclasts, ranging from 10% to 40%, are mainly cordierite or sillimanite aggregates. Garnets rarely survive the deformation. The feldspars are totally granulated. Quartz is partially to totally recrystallized and forms pressure shadows and tails around the porphyroclasts. Biotite is usually destroyed during the cataclastic process and combines with the granulated feldspar to form the microcrystalline groundmass.

Ultramylonites (Figs. 7 and 8) represent the highest degree of mylonitization. These rocks are very fine-grained, well laminated, and generally contain less than 5% porphyroclasts. Most porphyroclasts have been reduced to darker coloured streaks within the fine microcrystalline groundmass. Quartz is totally recrystallized.

Microfolding, primarily intrafolial folding, occurs within mylonites and ultramylonites (Figs. 9 and 10). Folding of this nature is the result of frictional drag between differentially moving fluxion planes or fluxion layers (Higgins, 1971) and is contemporaneous with the cataclastic process.

Athabasca Group Sediments

Exposed within the northwest portion of the Dominique-Peter area is a fault-bounded overturned block of red and green paleoweathered gneissic basement which overlies, and shows a normal stratigraphic contact with, basal Athabasca Group sediments (Fig. 11). From drill core interpretation, the sub-Athabasca unconformity is located approximately 500 m to the southwest of the deposit. Paleoweathering profiles were not observed directly within the deposit, however, the sub-Athabasca unconformity can be projected above the deposit. This implies that the Dominique-Peter deposit originally lay a short distance beneath the sub-Athabasca surface, suggesting that a genetic connection exists between the deposit and the unconformity.

Cluff Breccia

The basement succession and the Athabasca Group sediments near the unconformity have been affected by a younger tectonic event which is believed to be related to the formation of the Carswell structure (Currie, 1969; Harper, 1983; Pagel et al., 1985). Breccias, termed Cluff breccia (Currie, 1969), dated around 400 to 500 Ma (Bell, 1985), occur as irregular and anastomosing stringers, veins, and dykes, from a few millimetres to several metres thick. The breccia is a dark green or grey, and locally, black polymictic breccia carries fragments of the country rock into which it has intruded. The angular to sub-rounded fragments, up to several centimetres in size, are supported in a pulverized groundmass of aphanitic material. Distribution of the breccia is widespread and erratic, having been emplaced along pre-existing zones of weakness such as lithologic contacts, fractures, shear zones and faults.

MINERALIZATION

In the Dominique-Peter area, uranium mineralization is located within the zone of cataclastic deformation. Beyond the mylonite zone, mineralization is discontinuous and generally uneconomic. The deposit is structurally controlled and primarily consists of several northerly trending mineralized breccia zones and vein networks which have a known lateral extent of 800 m. The central portion of the deposit is dominated by east-northeast trending veins or vein stockworks which have been traced for a distance of 500 to 600 m (Fig. 11). The depth of the deposit is 120 to 300 m below the present topographic surface. An intense and pervasive pale green chloritic alteration and sericitic alteration is associated with the mineralization. The alteration halo, for the most part, follows the same trend as the mylonite zone (Fig. 2), but has an opposing dip to the west and northwest. Altera-

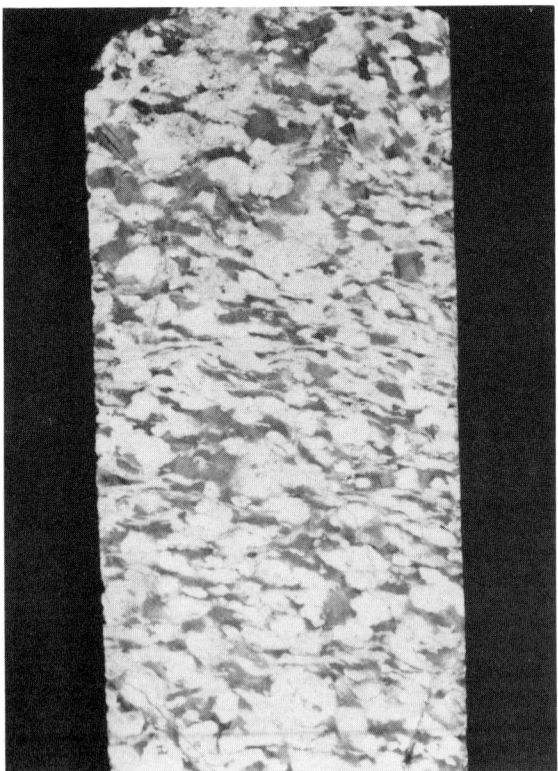

Figure 3. Protomylonite. Parent rock: granitoid from the transitional boundary. Drill Hole CLU 1897 − 116 m. Core width is 36.5 mm.

Figure 4. Protomylonite. Parent rock: granitoid from the transitional boundary. Drill Hole CLU 1807 − 155 m. Core width is 36.5 mm.

Figure 5. Mylonite. Parent rock: aluminous gneiss (Peter River gneiss). Drill Hole CLU 1799 − 174 m. Core width is 36.5 mm.

Figure 6. Mylonite. Parent rock: aluminous gneiss (Peter River gneiss). Drill Hole CLU 1787 − 253 m. Core width is 36.5 mm.

Figure 7. Ultramylonite. Parent rock: aluminous gneiss (Peter River gneiss). Drill Hole CLU 1547 – 118 m. Core width is 36.5 mm.

Figure 8. Ultramylonite. Parent rock: granitoid from the transitional boundary. Drill Hole CLU 1547 – 100 m. Core width is 36.5 mm.

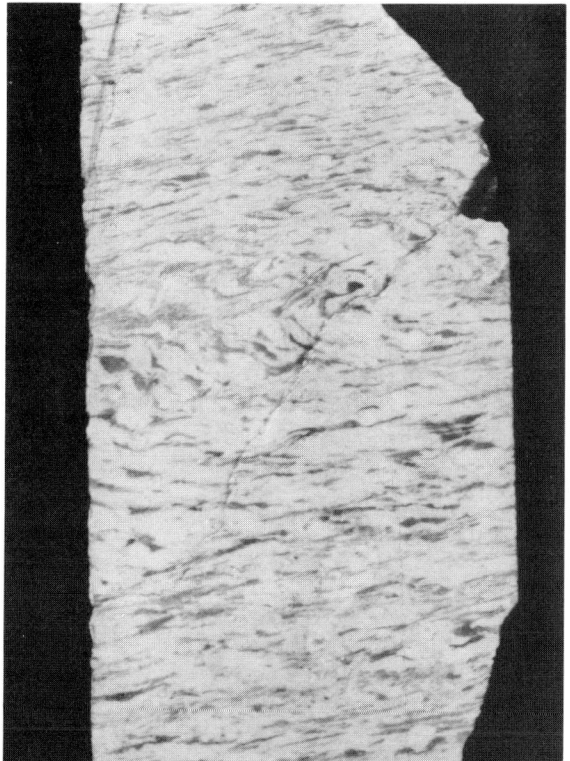

Figure 9. Ultramylonite displaying small scale intrafolial folds showing disruption of straight fluxion structure. Parent rock: granitoid from the transitional boundary. Drill Hole CLU 1897 – 100 m. Core width is 36.5 mm.

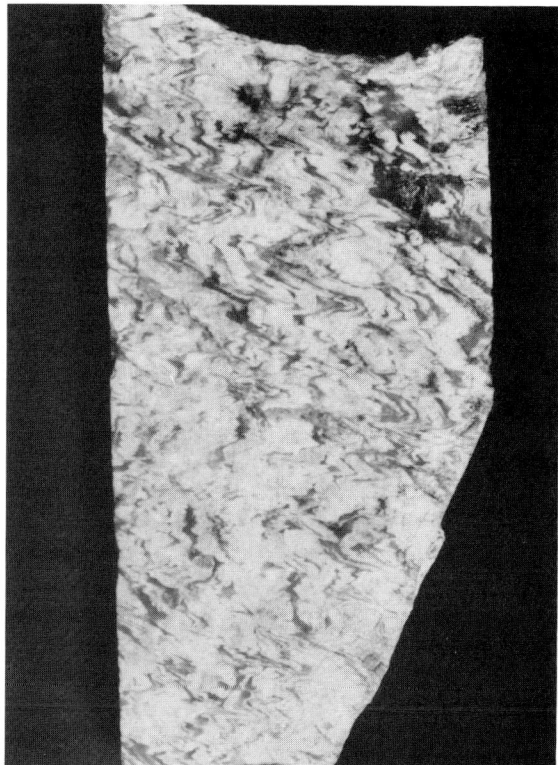

Figure 10. Intrafolial folding in ultramylonite. Parent rock: granitoid from the transitional boundary. Drill Hole CLU 1905 – 137 m. Core width is 36.5 mm.

tion is controlled by host rock fractures and faulting. Alteration in the central portion of the deposit extends in places to depths greater than 300 m, which is beyond the current depth of diamond drilling.

Types of Occurrences

The best intersections of mineralization are located in the pyritic and graphitic facies of the Peter River gneiss. Mineralized fractures in the Earl River complex show more discontinuity and contain lower and wider ranges of ore grade than those in the Peter River gneiss. Within this lithologic context, three distinct types of uraniferous occurrences have been differentiated: vein-type, breccia-type, and disseminations within Cluff breccia (Fig. 12).

The vein structures generally cross-cut the host rock fabric. These veins range from a few centimetres to tens of centimetres in thickness and are characterized by massive uranium mineralization. Commonly, a portion of the mineralization has also been forced for a few centimetres along fractures or joints paralleling the mylonitic laminae (Fig. 12a). This type of mineralization is principally observed in the east-northeast structures.

Breccia-type mineralization (north-south trending) is located within tectonic breccias (Fig. 12b). The brecciated zones which constitute channels for the hydrothermal solutions result from major north-south faulting which dextrally displaced the mylonite zone. The type of mineralization depends on the thickness of the brecciated zone. Within the smaller structures (less than 1 m thick) mineralization occurs as disseminations in an argillaceous matrix, or as a thin coating of uraninite on brecciated host rock fragments. Within the larger breccia zones most of the mineralization is concentrated at the upper and lower contact between the breccia and host rock; the breccia itself contains disseminations and thin uraniferous stringers, dipping from 15° to vertical.

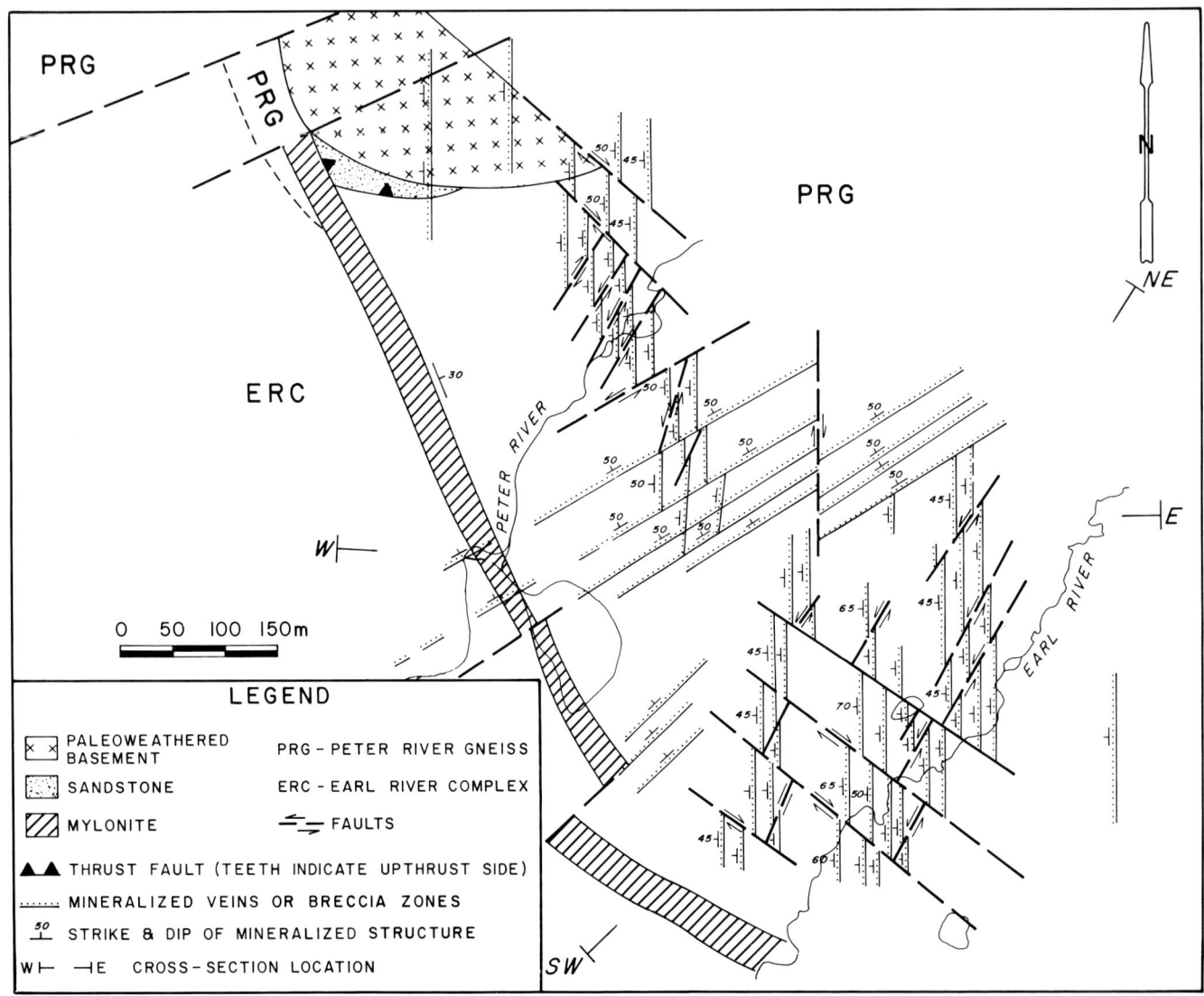

Figure 11. Schematic plan view of the Dominique-Peter deposit – 200 m level.

Mineralization contained within Cluff breccia is relatively minor (Fig. 12c). Their distribution is erratic and there are various grades of mineralization. Incorporated in the Cluff breccia are fragments from older veins and tectonic breccias. Only those Cluff breccias proximal to or cutting previous veins are mineralized.

Structures

Two main mineralized trends have been distinguished in the Dominique-Peter deposit: one oriented north-south, dipping to the west and another oriented east-northeast, dipping to the northwest (Fig. 11).

The Dominique-Peter deposit is divided into four zones on the basis of structural control: a Southern Zone and a Northern Zone with northerly trending mineralized veins and breccia zones, and a Central Zone and Southwestern Zone controlled primarily by east-northeast mineralized vein networks (Fig. 13, low angle mineralization).

The Southern Zone is dominated by several mineralized westerly dipping tectonic breccias, tensional structures resulting from large scale north-south faulting. From the west to the east, three subdivisions have been made. Two structures dipping 45° to 50° west contain sigmoidal or lenticular mineralization which is limited to the mylonite zone. In the central portion, a single structure dipping 60° to 70° west contains rich, vein-type mineralization also confined to the mylonite zone. In the east, three mineralized structures, approximately 10 m apart, dip 45° to 55° west. These brecciated structures are well developed in the mylonite zone but also extend above it into altered aluminous gneiss. The overall geometric shape of the mineralization is sigmoidal or lenticular.

Toward the north, the brecciated structures of the Southern Zone are terminated by east-northeast faults but remain open toward the south. The length of this zone is roughly 300 m, and mineralization is located between 110 m and 170 m below the topographic surface. Late northeast and northwest faults have displaced the earlier mineralized tectonic breccias. There is evidence, nevertheless, of an enrichment at the intersection between the later faults and the north-south mineralized structures.

The Northern Zone is dominated by a network of tectonic breccias and fractures which have a northerly orientation and dip 45° to 50° to the west. Mineralization is in lenticular pods well developed in the mylonite zone. Fractures containing massive mineralization have been intersected below the mylonite zone in the Earl River complex. Three main structures have been identified (Fig. 11) converging towards the south where they are intersected by east-northeast faults of the Central Zone. These structures have also been displaced by late northeast faults. Economic mineralization is located between 120 m and 200 m depths, over a lateral distance of 300 m, and remains open towards the north.

In the Central Zone, three east-northeast normal faults, dipping 60° to 70° to the northwest, offset and downfault the mylonite zone. Mineralization is found in fractures having roughly the same orientation but dip 50° to the northwest. The best intersections of mineralization are located in the mylonite zone along the footwall of these faults where fracturing is well developed. Most of the east-northeast mineralization is in open fractures. This system attains a length of 500 m to 600 m and appears to be open towards the northeast and the southwest. Most mineralization is located at depths between 110 m and 160 m, but deeper mineralization at 300 m has been intersected in the southwest portion of this zone.

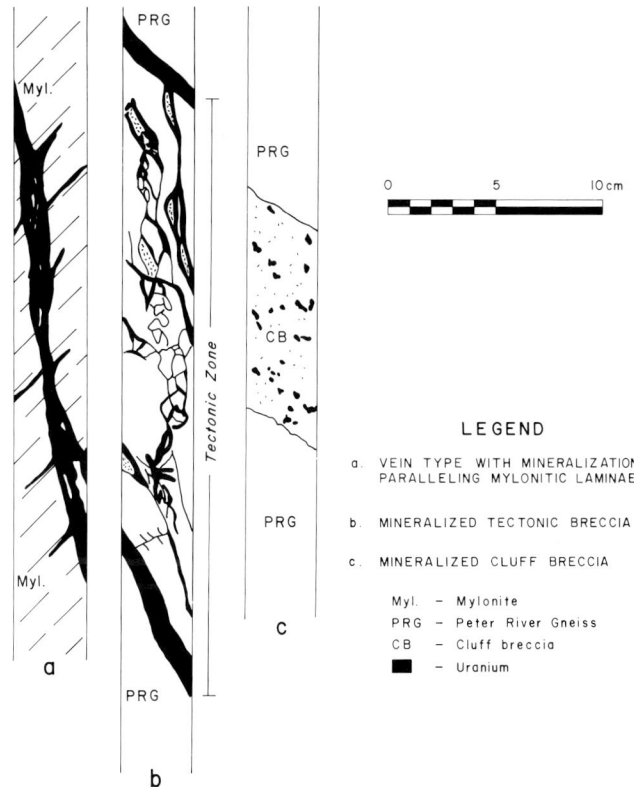

Figure 12. Types of mineralization observed in drill core in the Dominique-Peter deposit.

In the Southwestern Zone, four mineralized breccia zones oriented northeast-southwest and dipping 40° to 45° to the northwest have been recognized. Mineralization is associated with an altered and locally mylonitized zone within the Peter River gneiss. Although located in a similar geological framework to the other zones, mineralization occurs at depths of 35 m to 60 m. The known lateral extent of this zone is 100 m to 150 m and it remains open towards the southwest.

Uraniferous Assemblages

The uraniferous assemblages within the Dominique-Peter deposit are complex. From ore mineralogy (Ruhlmann, 1985) and chemical age determination studies (Pagel and Ruhlmann, 1985) conducted over the past years, three main types have been recognized: uraninite-sulphide-selenides, uraninite-sulphides, and pitchblende-carbonate. Table I summarizes the uraniferous and non-uraniferous assemblages observed in the Dominique-Peter orebody.

The uraninite-sulphide-selenide assemblage is associated with Mo, Bi, Au, and titanium oxides. This paragenesis is the most prominent, is observed mainly in the aluminous gneiss and is associated with a phyllitic gangue, which is commonly magnesium chlorite. It is U-Pb dated at 1050 Ma (Bell, 1985).

A second assemblage containing uraninite sulphides and minor gold occurs less commonly in the Dominique-Peter area and fills fractures within the feldspathic gneisses. It is associated with quartz veins and later magnesium tourmaline, similar to that of the OP deposit which is U-Pb dated at 900 Ma (Bell, 1985).

The pitchblende-carbonate association is the most recent paragenetic assemblage and has been observed over the entire Dominique-Peter deposit. This type of mineralization corresponds to a late episode U-Pb dated at 300 to 200 Ma (Bell, 1985).

An assemblage of coffinite and sulphides has been observed in all types of mineralization, and is particularly well developed in Cluff breccia. It is clearly a late stage deposition.

ALTERATION ASSOCIATED WITH MINERALIZATION

Alteration of the basement rocks in the vicinity of the deposit is of two distinct types: retrograde alteration subsequent to the Hudsonian metamorphism, and episodic hydrothermal alteration associated with the main mineralizing events. Time relationships between the different phases of alteration are uncertain. Recent work by Bell (1985) on samples from the altered zone indicate a wide range of K-Ar age dates: an isolated date of 1600 Ma reflecting possible deep weathering of the basement before the deposition of Athabasca Group sediments or an age indicative of the retrograde metamorphism, and dates from 1450 to 1000 Ma, corresponding to repeated incursions of hydrothermal fluids. Retrograde alteration is manifested by the pinitization of cordierite, chlorite replacing garnet and biotite, a weak sericitization of feldspar and sillimanite, and replacement of pyroxene by amphibole. Retrograde alteration is not the major alteration observed near the deposit.

The dominant and characteristic wall rock alteration is a pale green hydrothermal overprinting of original metamorphic minerals to fine-grained magnesium-rich chlorite, sericite and illite (Pagel and Svab, 1985). Pale green alteration is pervasive and affects a large volume of rock, developing a halo of bleached rock which wholly encloses the mylonite zone (Figs. 3 and 13). The relative proportion of chlorite and sericite reflects the original mineralogy and the degree of alteration (Table II). With increasing intensity, quartz becomes corroded and is the only original metamorphic mineral preserved. Pyrite seems to be associated with the pale green alteration and is commonly found as fine-grained globules or in fractures within the altered wall rock. Acicular dravite is locally abundant and is often in fractures in association with quartz veins. Tourmaline is considered to be a later mineral phase and may not be related to the mineralizing episode (Ruhlmann, 1985). Kaolinite, when present, appears in narrow zones surrounding faults. Pagel and Svab (1985) have observed that within the altered zone, a zonation of clays exist. Magnesium and aluminous chlorite is found adjacent to the mineralized veins and away from the mineralization, sericite becomes the dominant alteration mineral. Overall, wall rock alteration associated with the mineralization is a magnesium enrichment accompanied by anomalous concentrations of boron, fluorine, and arsenic.

TABLE I

URANIFEROUS ASSEMBLAGES IN THE DOMINIQUE-PETER OREBODY
(MODIFIED FROM RUHLMANN, 1985)

Features	Uraninite-Sulphide-Selenide	Uraninite-Sulphide	Pitchblende-Carbonate
Uraniferous Minerals	Uraninite Brannerite	Uraninite	Pitchblende
Non Uraniferous Minerals	Paraguanajuatite-Gold Guanajuatite-Altaite Clausthalite-Calaverite Freboldite-Gersdorffite Niccolite-Skutterudite Molybdenum Sulphide Galena-Chalcopyrite Pyrite-Marcasite	Chalcopyrite Galena Pyrite-Marcasite Gold	Pyrite-Marcasite Galena Chalcopyrite
Characteristic Associations	U-Bi-Mo-Se Au-Ti	U-Pb-Cu	U-Ca
Host Rocks	Aluminous Gneiss (Peter River Gneiss)	Feldspathic Gneiss (Earl River Complex)	All Basement Rock Types
Gangue Minerals	Magnesium Chlorite	Quartz, Dravite	Calcite
Dates	1050 Ma	900 Ma	300 - 200 Ma

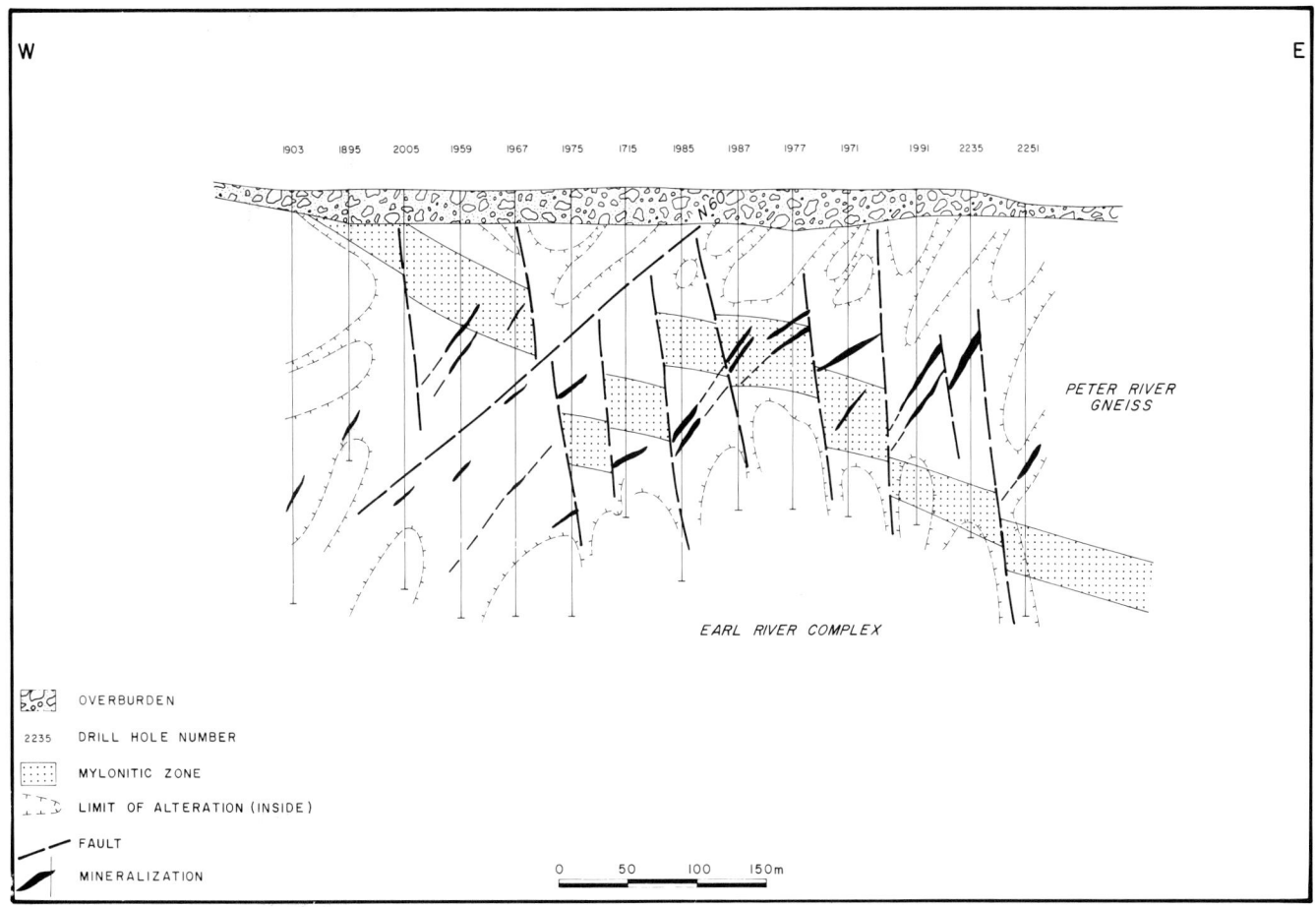

Figure 13. Schematic east-west geological cross section through the Dominique-Peter deposit along the eastern flank of the Dominique dome (for location see Figure 11).

ORE RESERVES

The Dominique-Peter orebody covers an area 800 m long and 600 m wide, and mineralization is mainly situated at depths between 100 m and 300 m. Within the zone of exploration drilling, comprised of 209 vertical drill holes at a grid spacing of 40 m, a development program of 123 inclined drill holes was undertaken in 1982 in order to improve the structural knowledge of the mineralization and to evaluate the deposit. The inclined holes were specifically located in order to provide intersection information at a distance of 15 m to 20 m along cross sections spaced 40 m apart. For the north-south mineralized structures, the structural interpretation of the Southern Zone was confirmed by two intermediate profiles spaced at 20 m between existing cross sections. The east-northeast mineralized structures have been confirmed only by a few inclined holes.

The frontal projection method, or longitudinal section of drill hole intercepts, was used for the evaluation of the Dominique-Peter deposit. This method requires a thorough knowledge of the structural setting of the mineralization. Each intersection has been studied on the basis of type of mineralization, alteration, and intersection angle in order to assess, with certainty, the structure to which each intersected mineralization belongs. Each hole has been sys-

TABLE II

TYPICAL ALTERATION PRODUCTS OF BASEMENT HOST ROCKS IN THE DOMINIQUE-PETER AREA (MODIFIED FROM SVAB, 1982)

Mineral	Characteristics of Alteration
Quartz	May show serrated grain edges and some corrosion.
Potassium Feldspar	Sericite. Usually only slightly sericitized until all mafics are chloritized.
Plagioclase	Commonly first to show alteration. Sericite ± white mica ± epidote ± chlorite (as observed in aluminous and feldspathic gneisses) or epidote ± sericite ± micaceous minerals (as observed in mafic gneisses).
Biotite	Usually the last to show alteration. Chlorite ± muscovite ± opaques along fracture planes.
Sillimanite	Sericite.
Cordierite	Pinite (sericite-chlorite).
Pyroxene	Chlorite ± opaques (in stronger alteration the pyroxene cleavage is obliterated — may alter to amphibole (uralitization).
Hornblende	Biotite ± chlorite ± opaques (magnetite).

tematically surveyed for deviation, making it possible to calculate the three dimensional true coordinates for each mineralized intersection. The mid-points of each intersection have been projected onto a vertical plane whose direction is parallel to the structure. The deposit can thus be represented by a two-dimensional model. The composite assays for each intersection were manually calculated for the thickness of the structure. The areas of influence for each intersection were determined by taking into account geological elements such as faults, limits of the mylonite, and limits of the alteration.

An evaluation, assuming 2.5 m minimum mining width and a cut-off grade of 0.2% U, gave the following results:

Tonnes of Ore: 1,761,000
Tonnes of U: 11,587
Average grade of U: 0.66%

To date, the perimeters of the deposit are still to be defined.

CONCLUSION

The Dominique-Peter deposit is located entirely within metamorphosed Aphebian rocks. The geological framework and controls for the mineralization in Dominique-Peter are numerous.

1) *Structural control.* Uranium mineralization is predominantly localized in rocks which have undergone ductile deformation. The mylonite zone plays an important role in the evolution of mineralization since it is mechanically suitable for the development of open fractures or fracture networks, and therefore is more susceptible to incursions of hydrothermal or diagenetic fluids. Mineralization is in the form of veins, veinlets, and disseminations of uraninite developed in a system of open fractures or breccia zones within or near faults. These faults have probably been reactivated after the deposition of the Athabasca Group and are classified according to three main trends: (1) a northerly structural trend where mineralization is located within fracture networks and breccia zones having a north-south orientation and dips of 45° to 70° to the west, dominant in the Northern and Southern Zones of the deposit, (2) the east-northeasterly structural trend, where mineralization is within fractures and fracture networks dipping from 45° to 75° to the northwest, prominent in the Central Zone of the deposit, and (3) late northeast and northwest to west-northwest fault movements which have horizontally displaced the mineralized veins and breccia zones and are considered to be of minor importance with respect to ore control. Displacements are in the order of tens of metres.

2) *Lithological control.* The richest and most homogeneous intersections of mineralization are found within the graphitic and pyritic Peter River gneisses. In the Earl River complex, the mineralized veins are small and the grades of mineralization are extremely variable. The contact between these two rock units appears to be an important control.

3) *Alteration.* Mineralization is closely associated with host rock alteration, a pervasive pale green chloritization, sericitization and illitization which are presumed to be hydrothermal in origin and related to the main mineralizing event dated at 1050 Ma. The alteration envelope, for the most part, follows the trend of the mylonite zone but is directly controlled by faulting and fracturing which predates alteration and mineralization.

4) *Unconformity.* The inferred sub-Athabasca unconformity is presently situated approximately 500 m to the southsouthwest. Although the deposit lacks certain criteria normally associated with unconformity deposits, such as the Athabasca cover rocks and paleoweathering profiles, a spatial and genetic relationship is believed to exist with the unconformity. The Dominique-Peter deposit as recognized to date is close to, or at least within tens of metres of, the unconformity as confirmed by the sandstone and paleoweathered basement northeast of Dominique and the fault-bounded sandstone slabs of the OP deposit to the east. Although the deposit is basement-hosted and hydrothermal in origin, Dominique-Peter may represent a large "root" extension of an unconformity-type deposit.

ACKNOWLEDGEMENTS

The authors acknowledge and express appreciation to the contributions of the geologists and technicians who participated in this project between 1980 and 1982. This study is the result of their continued interest in the project, diligent core logging, and numerous discussions. Special thanks are extended to the two anonymous reviewers for reading and criticizing the manuscript.

REFERENCES

Bell, K., 1985, Geochronology of the Carswell Area, Northern Saskatchewan: *in* Lainé, R., Alonso, D., and Svab, M., eds., The Carswell Structure Uranium Deposits, Saskatchewan: Geological Association of Canada Special Paper 29.

de Carle, A.L., 1980, Geology of the Key Lake Deposits, Saskatchewan: Canadian Institute of Mining District Annual Meeting, September, 1980, Flin Flon, Manitoba.

Currie, K.L., 1969, Geological Notes on the Carswell Circular Structure, Saskatchewan (74K): Geological Survey of Canada Paper 67-32, 69 p.

Harper, C.T., 1983, The Geology and Uranium Deposits of the Central Part of the Carswell Structure, Northern Saskatchewan, Canada: Unpublished Ph.D. Thesis, Colorado School of Mines, Golden, Colorado, 337 p.

Herring, B.G., 1976, The Metamorphism and Alteration of the Basement Rocks in the Carswell Circular Structure, Saskatchewan: Unpublished M.Sc. Thesis, University of British Columbia, Vancouver, 134 p.

Higgins, M.W., 1971, Cataclastic Rocks: United States Geological Survey Professional Paper 687, p. 97.

Lewry, J.F., and Sibbald, T.I.I., 1977, Variations in Lithology and Tectonometamorphic Relationships in the Precambrian Basement of Northern Saskatchewan: Canadian Journal of Earth Sciences, v. 14, p. 1453-1467.

_____, 1980, Thermotectonic Evolution of the Churchill Province in Northern Saskatchewan: Tectonophysics, v. 68, p. 45-82.

Pagel, M., and Ruhlmann, F., 1985, Chemistry of Uranium Minerals in Deposits and Showings of the Carswell Structure: *in* Lainé, R., Alonso, D., and Svab, M., eds., The Carswell Structure Uranium Deposits, Saskatchewan: Geological Association of Canada Special Paper 29.

Pagel, M., and Svab, M., 1985, Petrographic and Geochemical Variations within the Carswell Structure Metamorphic Core and their Implications with Respect to Uranium Deposition: *in* Lainé, R., Alonso, D., and Svab, M., eds., The Carswell Structure Uranium Deposits, Saskatchewan: Geological Association of Canada Special Paper 29.

Pagel, M., Wheatley, K., and Ey, F., 1985, The Origin of the Carswell Circular Structure: *in* Lainé, R., Alonso, D., and Svab, M., eds., The Carswell Structure Uranium Deposits, Saskatchewan: Geological Association of Canada Special Paper 29.

Powell, B., Koning, E., and Lainé, R., 1985, Geophysical Mapping of Gneiss Domes in the Carswell Structure and their Relationship to Uranium Mineralization: *in* Lainé, R., Alonso, D., and Svab, M., eds., The Carswell Structure Uranium Deposits, Saskatchewan: Geological Association of Canada Special Paper 29.

Ray, G., 1977, The Geology of the Highrock Lake – Key Lake Vicinity: Saskatchewan Geological Survey Report 197.

Ruhlmann, F., 1985, Mineralogy and Metallogeny of the Uraniferous Occurrences in the Carswell Structure: *in* Lainé, R., Alonso, D., and Svab, M., eds., The Carswell Structure Uranium Deposits, Saskatchewan: Geological Association of Canada Special Paper 29.

Svab, M., 1982, Petrographic Examination of the Mylonite Zone in Peter River: Amok Internal Report.

CHEMISTRY OF URANIUM MINERALS IN DEPOSITS AND SHOWINGS OF THE CARSWELL STRUCTURE (SASKATCHEWAN-CANADA)

M. Pagel
Centre de Recherches sur la Gèologie de l'Uranium, BP 23 – 54501 Vandoeuvre-lès-Nancy Cedex, France
and
Centre de Recherches Pétrographiques et Géochimiques BP 20 – 54501 Vandoeuvre-lès-Nancy Cedex, France

F. Ruhlmann
COGEMA – BU-DRM – 78140 Vélizy, France

ABSTRACT

A detailed electron microprobe study has been performed on uranium minerals from deposits and showings located in the Carswell structure (Saskatchewan-Canada). Samples of uranium oxides, coffinite, and brannerite were selected on the basis of their habit and their paragenetic associations, regardless of alteration processes. Among five groups of uranium oxides characterized by their mineral habit and paragenesis, consideration of elements such as Th, Ce, Y, Ca, Si, and Pb allows for the identification of at least six main generations of uranium oxides: (1) a Th-Y-Ce bearing uraninite in association with monazite is related to anatectic processes; (2) uranium oxides in cubes, radiating spherulites with automorphic edges, and in veinlets from D and Dominique-Peter deposits all have very similar chemical compositions and chemical ages around 1200 Ma in accordance with previous U-Pb isotopic ages; (3) a similar uranium facies from OP and Dominique-Peter deposits in association with dravite yields a more recent age; (4) the pitchblende from the Donna showing sandstones is especially enriched in calcium and has a PbO content up to 3.9%; (5) secondary collomorph pitchblende from the D deposit yields a chemical age close to 250 Ma, which corresponds to a remobilization age already evidenced by isotopic geochemistry; and (6) pitchblende from the Numac showing is rich in Ce, Y, and Ca, and is definitely a distinct generation. Coffinites have variable chemical compositions but two main groups are distinguished, depending on their yttrium and cerium content.

This data gives valuable information about mobilization and remobilization of uranium at different epochs. The data can provide criteria for selecting samples for more precise isotopic work, and show that quantitative electron microprobe analyses are useful in providing information for exploration purposes.

RÉSUMÉ

Une étude chimique détaillée des minéraux uranifères, en excluant les minéraux à U(VI) seul, dans les gisements et indices de la structure Carswell (Saskatchewan-Canada) a été réalisée à la microsonde électronique. Les échantillons d'oxydes d'uranium (uraninite et/ou pechblende), de coffinite, et de brannérite ont été sélectionnés d'après leurs morphologies, et leurs associations paragénétiques. Dans la mesure du possible, les effets de l'altération ont été écartés. A partir de cinq groupes d'oxydes d'uranium caractérisés par leur morphologie et leur paragénèse, six générations principales d'oxydes d'uranium ont été identifiées à partir des variations en Th, Ce, Y, Ca, Si, et Pb. (1) Pour la première fois, des concentrations d'uraninite, antérieures aux concentrations uranifères économiques, ont été mises en évidence dans l'indice Sophie. Ce sont des uraninites riches en Th, Y, Ce, et Pb, en association avec la monazite et sont reliées aux pegmatoïdes d'anatexie. (2) Les oxydes d'uranium, presents sous formes de cubes, de roues dentées, de sphérolites collomorphes ou à terminaisons automorphes, en filons et constituant les minéralisations principales, sont pauvres en Ca et ont des teneurs en Th, Y, et Ce en-dessous du seuil de détection de la microsonde électronique. Les âges chimiques, de l'ordre de 1200 Ma, sont en accord avec les données isotopiques. (3) Par rapport à la génération précédente, les oxydes d'uranium associés aux filons à dravite de l'indice OP ont des teneurs supérieures en Ca et inférieures en Pb. Le dépôt de ces oxydes est nettement postérieur au phénomène minéralisateur principal, mais se réalise toujours à partir de saumures. (4) Les pechblendes localisées dans les grès de l'indice Donna se caractérisent par leur richesse en calcium et leurs teneurs en PbO de 3.9% et sont en accord avec les données isotopiques attribuant un âge de 380 Ma. (5) La pechblende collomorphe liée à des minifronts d'oxydo-réduction en liaison avec des fractures du gisement D donne un âge chimique de 250 Ma qui correspond à l'âge de remobilisation déterminé sur les minéralisations principales en géochimie isotopique. (6) La pechblende de l'indice Numac se distingue de toutes les autres générations par sa richesse en Ce, Y, et Ca mais l'état d'altération des échantillons ne permet pas d'attribuer à cette génération sa place chronologique dans le schéma métallogénique des gisements et indices de la structure Carswell.

La formation de la brannérite est interprétée comme le résultat d'un phénomène hydrothermal en accord avec les résultats des inclusions fluides. Les coffinites ont des compositions chimiques très variables mais deux groupes principaux apparaissent, en considérant les teneurs en yttrium et cérium.

Ces données donnent des informations intéressantes sur le dépôt et la remobilisation de l'uranium à différentes époques, fournissent des critères appréciables pour la réalisation d'un travail isotopique plus précis, et montrent que l'utilisation quantitative de la microsonde électronique permet d'obtenir des réponses rapides et peu coûteuses en exploration.

INTRODUCTION

From U-Pb isotope geochemistry studies (Koeppel, 1968; Cumming and Rimsaite, 1979; Wendt et al, 1978; Gancarz, 1979; Bell, 1981; Hoeve et al., 1981; Worden et al., 1981), it appears that several stages of uranium mobilization and deposition occurred in areas where unconformity-type uranium deposits have been discovered in Saskatchewan. These ages have strong implications on genetic models. It is of great importance, therefore, to be able to relate these ages to mineralogy. Microscopic observations of uranium minerals, excluding U(VI) minerals, from showings and deposits of the Carswell structure (Ruhlmann, 1985) or from other deposits in Saskatchewan, have shown that there are distinct habits and paragenetic associations of uranium oxides.

The aim of this work is to try to solve the following problems in the metallogenic study of unconformity uranium deposits: (1) In a single deposit, is it possible to distinguish between several generations of uranium oxides and their alteration products?, (2) For all the deposits and showings, do these data indicate if uranium oxides were precipitated from the same solution or is it necessary to imply several metallogenic processes?, and (3) Can the origin of uraniferous fluids and physicochemical conditions of precipitation be investigated by using the study of the trace element contents of minerals?

The apparent age of mineralizing events can be defined by the Pb, U, and Th content of uranium oxides. This apparent age may be the true age if two conditions are met. Firstly, no common lead could have co-precipitated during formation of the uranium oxide. Secondly, radioactive daughters of ^{238}U and radiogenic lead must have remained in the lattice of the oxide. Specifically, U and Pb have been retained in the structure since the time of their formation. For Precambrian uranium oxides, the first condition is not very important in most cases because there is often only a small amount of common lead in carefully isolated uranium oxides, but the second one cannot be entirely ascertained. This necessitates a rigorous selection of samples for study. They must be devoid of secondary alteration which appears as zones of lower reflectivity.

The chemical ages have been calculated using the formula from Cameron-Schimann (1978):

$$t \text{ (years)} = \frac{Pb \times 10^{10}}{1.612\ U + 4.95\ Th}$$

It should be stressed that a chemical study of uranium minerals should be done before any geochronological U-Pb study. This prevents any mixing of different generations of uranium oxides. In the study of the Carswell area, the different generations of uranium oxides and their relative timing (except one showing) were determined before any geochronological studies (Ruhlmann, 1985).

ANALYTICAL CONDITIONS

Chemical analyses were performed on an automated electron microprobe CAMEBAX (Service Commun d'Analyses Interuniversitaire de Nancy) equipped with three wavelength-dispersive spectrometers. Analytical conditions were as follows: acceleration voltage = 15 KV; counting time = 6 seconds, the absorption current on thorite = 6 nA; beam diameter = less than a few microns. Corrections were made by a ZAF Cor 2 program. Standards used for the elements Ce, P, Zr, Y, U, Th, and Pb were as follows:

Ce: REE glass 3 ....(Ce = 3.41 wt %);
P: apatite ....(P_2O_5 = 41.83 wt %);
Zr: zircon ....(Zr = 49.76 wt%);
Y: YVO_4 ....(Y = 43.61 wt %);
U: synthetic uranium oxide UO_2
Th: synthetic thorium oxide ThO_2
Pb: galena ....(Pb = 86.6 wt%).

Mineral standards used for Ca, Fe, Mn, Ti, and Si were apatite, hematite, rhodonite, rutile, and albite respectively.

The sum of analyzed oxides in Table I ranges widely from 90.1 to 99.7%. This could be due to some or all of the following problems. (1) Some elements were not analyzed, and though using the energy dispersive spectrometer, attempts failed to find if other elements were present. These checks were not systematically applied. (2) Hydration of uranium oxides is also a possibility, since during heating, release of large quantities of water were observed on gas chromatography analyses of pitchblende samples; this was not quantified because the amount of water and hydroxyl ions cannot be evaluated by electron microprobe. (3) U and Fe are expressed as UO_2 and FeO, however, parts of U and Fe also correspond to UO_3 and Fe_2O_3 respectively.

This is a routine program, and no particular constraints or conditions were imposed. Due to the short counting time, statistics on minor elements are poor and data must be taken as only semi-quantitative. Since all the uranium oxides are very poor in, or contain no, thorium, no correction was made for any interferences between the uranium (Mα1) and thorium (Mβ1) lines.

URANIUM OXIDES

The uranium oxides have been studieddom the D, Dominique-Peter, and OP deposits, and the Donna, Numac, and Sophie showings (Tona et al., 1985). Results are presented in Table I. The classification was based on paragenesis and habit. Five groups were considered.

Group I: Uraninite Associated with Monazite

The association of uraninite with monazite was found in the Sophie area in a garnet-rich pegmatoid. It was particularly well preserved in the garnets. Uraninites and monazites were in contact, suggesting that they had crystallized at the same time.

Compared to all other uranium oxides, these uraninites have significant amounts of ThO_2 (1.9 to 0.9 wt %), Ce_2O_3 (0.3 wt %), and Y_2O_3 (1.1 to 1.5 wt %). They also contain the maximum lead content (19.6 wt % PbO) which corresponds to a chemical age of 1810 Ma. In the same garnets, monazite crystals are often zoned (Fig. 1). The core is transparent and birefringent, whereas the borders are opaque. This occurrence shows that, at equilibrium with monazite, the ThO_2 content of uraninite lies between 1% and 2%. The result is very similar to data obtained on leucogranites containing monazite and uraninite. On the contrary, in association with thorite, the ThO_2 content of uraninite is higher in granites (Pagel, 1981, 1982). In monazite, Y_2O_3 and ThO_2 show little variations, with values of around 3% and 8%, respectively. From the centre to the edges, Ce_2O_3 and P_2O_5 decrease abruptly by about half, whereas there is an increase of SiO_2, CaO, UO_2, and PbO. Significant amounts of FeO (total Fe) and Al_2O_3 are present only in the borders.

As discussed by Pagel and Svab (1985), the formation of pegmatoids has been attributed to generation by anatectic

TABLE I

CHEMICAL COMPOSITIONS OF URANIUM OXIDES (WEIGHT %) AS DETERMINED BY ELECTRON MICROPROBE IN DEPOSITS AND SHOWINGS OF THE CARSWELL STRUCTURE. N= NUMBER OF INDIVIDUAL ANALYSES. THE RANGES OF VALUES ARE INDICATED.

	N	UO_2	PbO	CaO	FeO	MnO	TiO_2	SiO_2	P_2O_5	ThO_2	Ce_2O_3	Total	Notes
Group I : *Uraninite associated with monazite*													
Sophie	5	73.9 / 70.0	19.6 / 13.4	1.0 / 0.7	0.4 / 0.2	0.1	0.1	1.7 / 0.0	—	1.9 / 0.9	0.3	93.8 - 90.1	Y_2O_3 = 1.5 - 1.1
Group II : *Euhedral uraninite in deposits*													
D (in chlorite)	2	86.2 / 83.8	8.2 / 7.1	2.1 / 1.9	0.6 / 0.5	0.2	0.2 / 0.0	0.6	0.1	—	—	97.2 - 95.3	
Dominique-Peter	3	80.1 / 79.7	15.2 / 14.3	0.6 / 0.5	—	0.2 / 0.0	0.5 / 0.0	—	—	—	—	96.0 - 95.8	
Dominique-Peter	6	82.9 / 79.8	17.3 / 11.3	1.0 / 0.6	0.1 / 0.0	0.1 / 0.0	1.4 / 0.0	1.0 / 0.0	0.1	—	—	98.9 - 94.7	Y detected
D (in Selenides)	4	82.4 / 74.9	14.1 / 5.4	2.3 / 1.0	0.3 / 0.0	0.1 / 0.0	0.8 / 0.2	0.3 / 0.1	—	—	—	97.5 - 93.2	
Group III : *Pitchblende-uraninite in deposits and showings*													
D	8	85.9 / 79.5	14.7 / 7.0	2.1 / 1.0	0.9 / 0.1	0.3 / 0.0	0.9 / 0.0	0.6 / 0.0	0.2 / 0.0	—	—	97.5 - 93.2	
Dominique-Peter	9	87.5 / 73.9	17.2 / 2.9	3.0 / 0.9	0.8 / 0.0	0.7 / 0.0	1.1 / 0.0	3.3 / 0.3	0.7 / 0.0	—	—	99.7 - 90.4	Large variations
OP	6	84.0 / 80.9	9.8 / 8.7	1.8 / 1.2	0.8 / 0.2	0.3 / 0.0	0.5 / 0.0	0.6 / 0.2	—	—	—	96.0 - 92.4	
Donna	5	80.9 / 77.0	15.3 / 10.3	1.3 / 0.9	0.3 / 0.0	—	0.4 / 0.0	0.1 / 0.0	0.1 / 0.0	—	—	93.9 - 90.7	
Group IV : *Pitchblende in strongly hematized basement rocks*													
Numac	4	80.5 / 78.5	5.7 / 3.8	5.7 / 4.4	1.2 / 0.5	1.0 / 0.4	0.9 / 0.8	1.0 / 0.5	—	2.0 / 0.2	—	94.3 - 91.7	Y present 0.3 - 0.5% Y_2O_3
Group V : *Collomorph pitchblende in deposits and showings*													
D	6	88.9 / 83.7	5.7 / 2.3	3.4 / 2.7	0.9 / 0.5	1.1 / 0.2	0.5 / 0.3	1.1 / 0.5	0.3 / 0.0	—	—	97.2 - 94.4	
Dominique-Peter (in carbonate veins)	7	86.7 / 81.2	0.7 / 0.0	6.2 / 5.0	0.2 / 0.0	0.8 / 0.0	0.2 / 0.0	4.8 / 2.5	0.2 / 0.0	—	—	95.4 - 90.2	Al_2O_3 : 0.2 - 0.0
(in cavities)	3	75.6	0.7	2.5	4.2	0.3	0.5	10.9	0.2	—	—	91.3 - 90.1	Al_2O_3 : 1.3 - 0.9
(cutting quartz-hematite)	1	73.8	2.9	3.3	0.2	0.1	—	8.9	0.1	—	—	90.3	Al_2O_3 : 1.0
Donna	4	83.3 / 78.0	3.9 / 2.8	8.9 / 7.7	—	—	—	1.1 / 0.1	0.3 / 0.1	—	—	95.6 - 90.8	

processes and U-Th enrichment by partial melting during Hudsonian metamorphism, a process which is similar to the generation of enriched uranium pegmatoids in areas such as at Mont Laurier, Quebec, Canada (Cuney, 1982). The presence of Th and Ce in uraninite is generally related to high temperature processes. They are absent or in lesser amounts in hydrothermally affected areas. The identification of primary uranium concentrations, predating the Athabasca sandstone deposition, is of primary importance for the genesis of unconformity-related uranium deposits. This is the first time it has been reported in the Carswell structure, but such early concentrations are known elsewhere in the Athabasca Basin (Charlebois Lake, Duddridge Lake, etc., Cameron-Schimann, 1978).

Group II: Euhedral Uraninite in the D and Dominique-Peter Deposits

In Group II, only isolated euhedral uraninites are considered. In the D deposit, the analyzed uraninites are located in chlorite or in selenides. For the Dominique-Peter deposit, they exhibit locally disturbed habits which seem to be due to deformation (Fig. 2).

Euhedral uraninites found in the D and Dominique-Peter deposits have a different chemical composition than the Group I uraninites because they contain no significant quantities of thorium, cerium, or yttrium. No U-Pb chemical age greater than 1200 Ma has been found in the D deposit (Fig. 3). This is in agreement with previously published ages of 1150 Ma (Bellon *et al.*, 1976) and 1050 Ma (Gancarz, 1979).

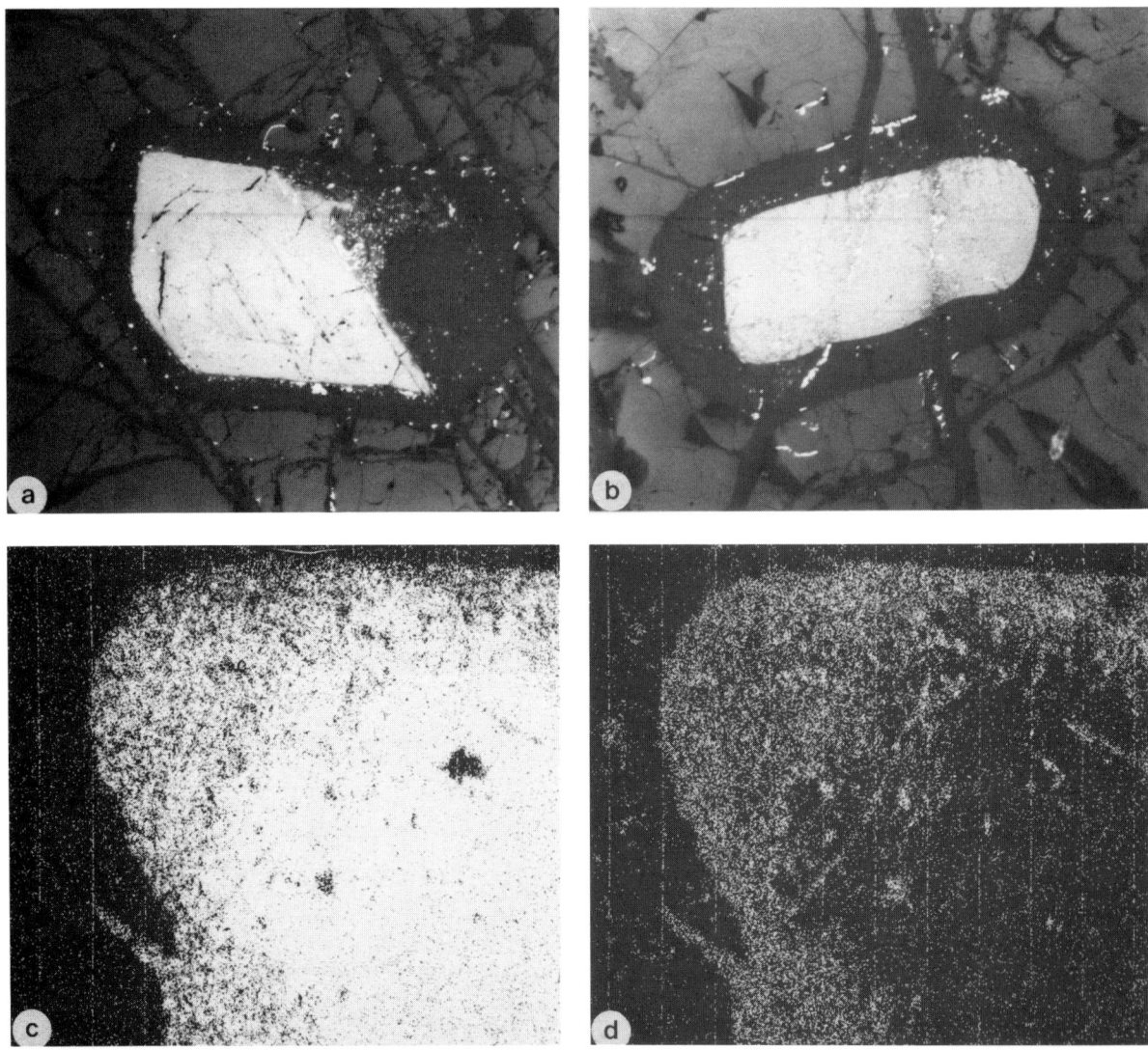

Figure 1. a) Sophie showing. Typical euhedral uraninite enclosed in garnet. The upper right part has probably been removed during sample preparation. Reflected light, × 235. b) Sophie showing. Other typical uraninite with more rounded edges. Reflected light, × 235. c) Sophie showing. X-ray plot of cerium. Part of a monazite crystal (× 200). Cerium is mainly depleted along the border. d) Sophie showing. X-ray plot of uranium in the same monazite. Uranium is mainly enriched along the border.

The age is close to the depositional age of other uranium deposits from Saskatchewan, such as Key Lake at 1270 Ma (Wendt et al., 1978; Hohndorf et al., 1981), Rabbit Lake at 1280 Ma (Cumming and Rimsaite, 1979), and Midwest Lake at 1330 Ma (Worden et al., 1981). In the Dominique-Peter area, higher lead contents of up to 17.3 wt % PbO have been found.

Group III: Pitchblende-Uraninite in all Deposits and Showings

In the Carswell structure deposits and showings, the usual morphological distinction between pitchblende and uraninite is not possible. Uraninite is a term reserved for euhedral crystals, whereas pitchblende is a term applied to spheroidal or collomorph textures. In these deposits, uranium is present as "cogwheels", "spheroids with automorphic edges", or "fibrous aggregates with automorphic ends". These particular textures are characteristic of many, if not all, unconformity-type uranium deposits from Saskatchewan and Australia (Pagel, 1983; Ewers and Ferguson, 1980).

Group III will not be discussed in detail because chemical compositions are identical to those of the preceeding group. FeO, MnO, TiO_2, and P_2O_5 are always very low (Table I). A PbO content not higher than 9.8 wt % at the OP deposit suggests a younger mineralizing event. The oxidation state of this facies is known from the D deposit (Cathelineau et al., 1982). U(VI)/U total varies from 0.30 to 0.42 with an a_o = 5.470 Å from X-ray data. This shows that the oxidation states are identical to those of Rabbit Lake, Canada and Jabiluka, Australia (Cathelineau et al., 1982).

Group IV: Pitchblende in Strongly Hematized and Albitized Basement Rocks

Mineralization is disseminated and affected by many alteration processes. Coffinite formation is widespread in these rocks, however, it is observed only in surface samples. Pitchblende from the Numac showing has a totally different chemical composition compared to Groups II and III. It is significantly richer in calcium, cerium, and yttrium. The PbO content ranges from 3.8 to 5.7 weight per cent and thus gives a chemical age of 390 ± 70 Ma. At this stage, two interpretations are possible. The presence of up to 2% Ce_2O_3 and up to 0.5% Y_2O_3 is comparable to Group I. However, thorium is absent, calcium is too high, and lead is low, suggesting an intense leaching of lead. In the second interpretation, the lead content suggests mineralization at about 400 Ma. The age of mineralization at Numac cannot be determined, but the data suggests that the U mineralization was due to a different event than the D and Dominique-Peter main mineralizing event.

Group V: Collomorph Pitchblende in Deposits and Showings

Group V contains typical pitchblende in half spherules or showing a "garland" texture. It occurs in different habits: as veinlets associated with mini-redox fronts related to fractures in the D deposit, in carbonate veins, in cavities, in cross-cutting quartz-hematite veins in the Dominique-Peter deposit, and in association with hematite in Donna boulders.

From all the data obtained from the various deposits or showings, there are wide variations in lead, calcium, iron, silica, aluminum, and phosphorus content. From the calcium content there is a clear distinction from Groups II and III. The calcium content is always greater in the collomorph pitchblende. The lead content is always low and suggests mineralizing periods younger than those of Groups II and III (Fig. 4). Extensive lead, calcium, and silica variations of the different pitchblende associations are attributed to variations in the chemistry of the solutions and probably different periods of deposition or remobilization (Fig. 5).

Three periods of mineralization have been recognized. In the Donna sandstones, where pitchblende is closely associated with hematite, a maximum chemical age of 330 Ma has been obtained. Bell (1985) gives an isotopic age of 380 ± 3 Ma on the same sample. In the D deposit, collomorph pitchblendes have relatively constant PbO contents, except for the analysis in which contamination by galena may have occurred; the Pb values indicate an age which agrees with the remobilization age of 250 Ma obtained from U-Pb isotope results (Bellon et al., 1976; Gancarz, 1979). Recent remobilization is inferred from field data, and some pitchblende from Dominique-Peter does not contain any lead that could be detected using electron microprobe.

COFFINITE STUDY

Data have been obtained on coffinite in sandstones from the D deposit, on coffinite associated with clays located in quartz – UO_{2+x} – tourmaline veins from OP, and on coffinite from pitchblende – coffinite veinlets cutting quartz – hematite rocks from Dominique-Peter (Table II).

The variations of chemical compositions are very large and it seems that there is a continuity in variation of SiO_2 from pitchblende to coffinite (Fig. 4). With microprobe analyses, it is not possible to determine if analyses correspond to pure pitchblende, coffinite, or a mixture of both.

The lead content is extremely low and variable, but the behaviour of radiogenic lead in coffinite is not sufficiently known to elaborate an interpretation.

Phosphorus is always present. There is a substitution between P_2O_5 and SiO_2, which indicates that the fluid phases have variable compositions during the coffinitization processes. CaO content does not reach values higher than 2.9%, which contrasts with the later pitchblende of Group V which is enriched in calcium.

Some coffinite grains are rich in Y_2O_3 or Ce_2O_3, especially in the OP deposit. As REE are generally not very mobile during alteration processes, the presence of REE in coffinite is probably due to a relict of REE from the primary uranium oxides. The presence of significant amounts of REE in uranium oxides has been described in the Alligator River deposits of Northern Australia by McLennan and Taylor (1980).

URANIUM-TITANIUM OXIDE STUDY

Some uranium-titanium oxide assemblages in the D deposit (Ruhlmann, 1985) have the same microscopic characteristics as brannerite: shape, reflectance, and internal reflection. Due to the very small quantities of material avail-

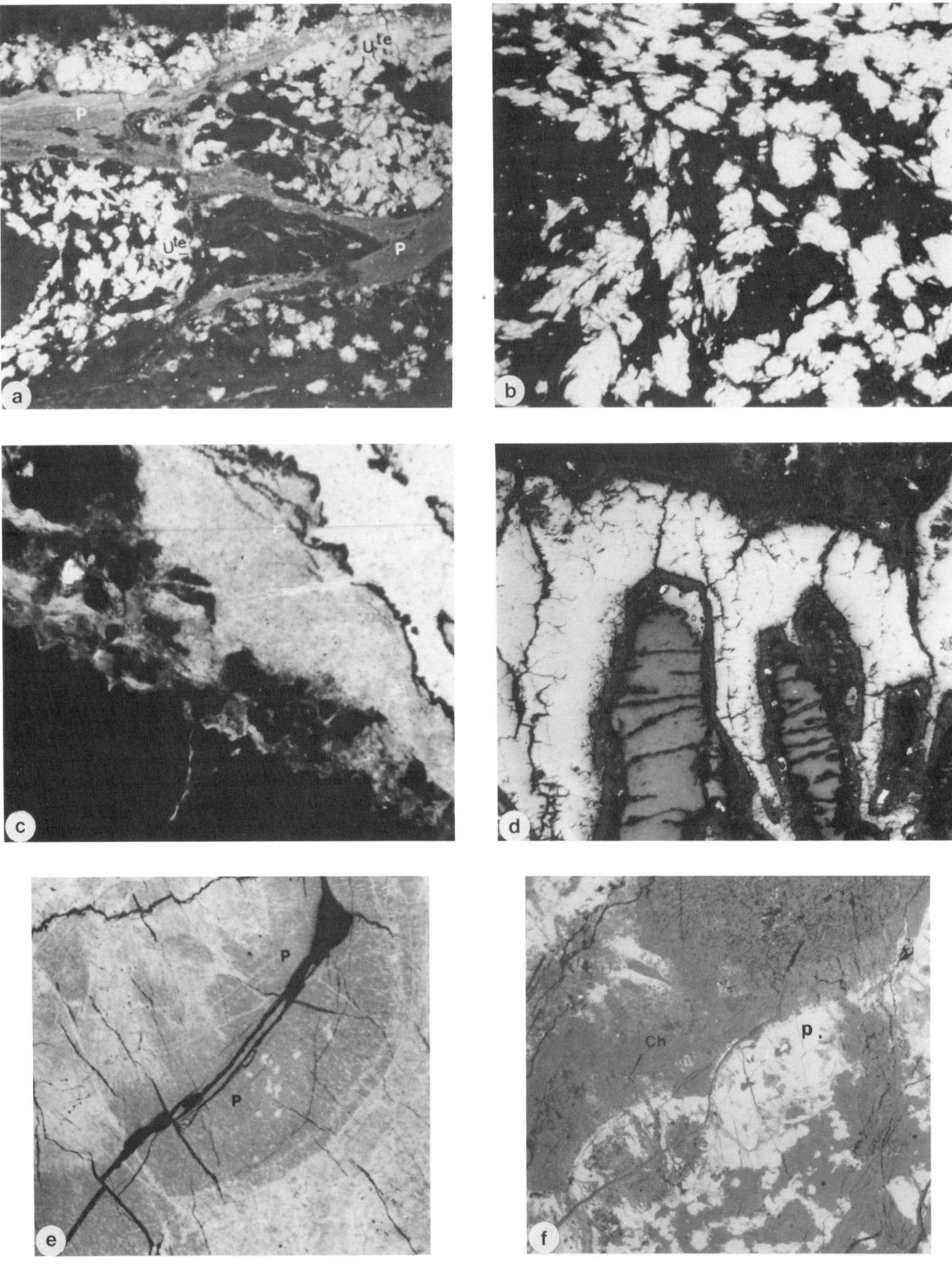

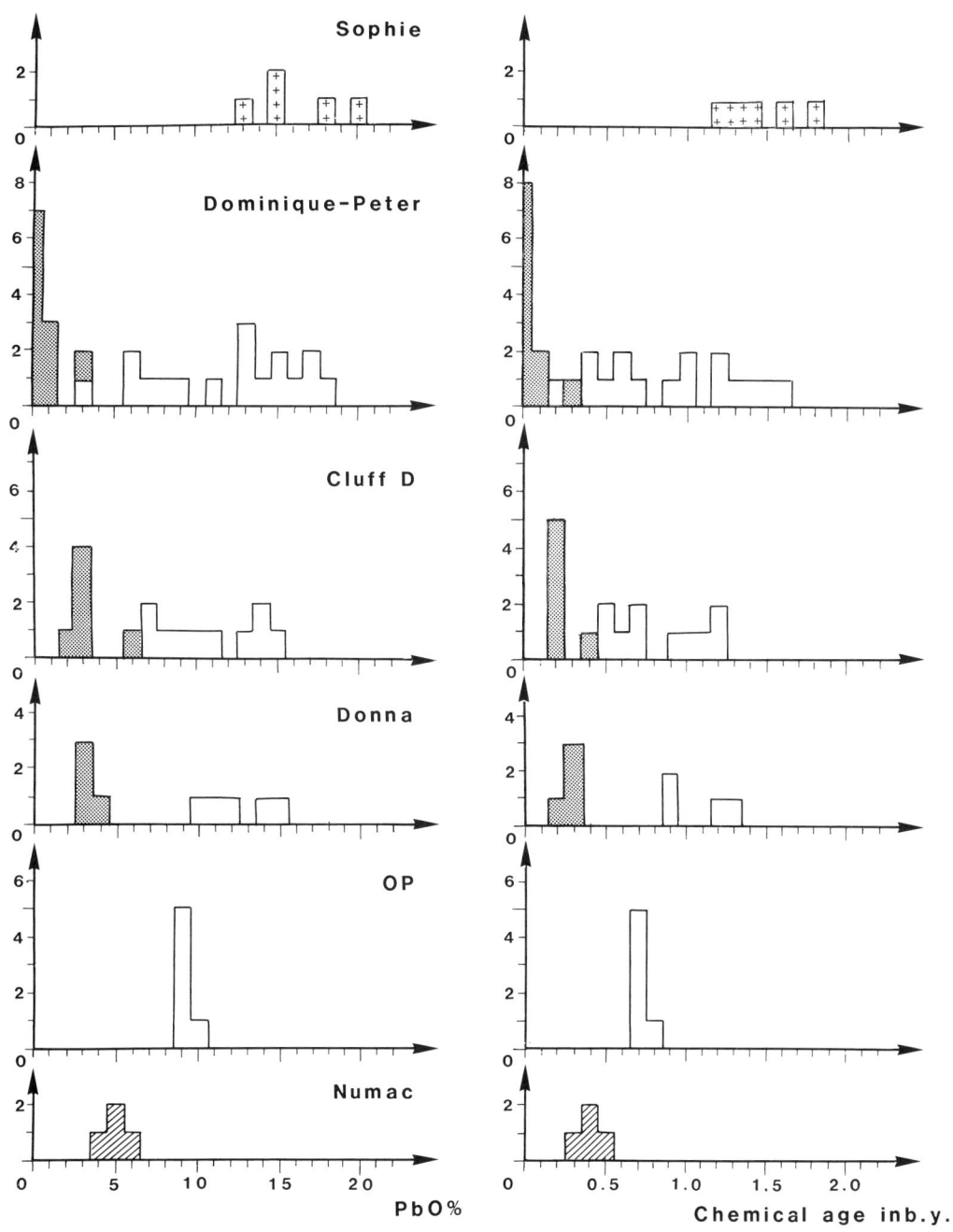

Figure 3. Histograms showing the evolution of the PbO content and inferred chemical ages of uranium oxides from different deposits and showings in the Carswell structure (crosses: Group I, white: Group II and III, hachured: Group V (see text)).

Figure 2. a) Presence of dispersed uranium oxides (U) which have a particular habit suggesting some deformation, and secondary pitchblende in veinlets (P). Reflected light, × 40. b) A detailed view of dispersed uranium oxides. Reflected light, × 85. c) Cluff Lake D deposit: Automorphic edges (lower part) of uranium oxide in chlorite (grey). Carbonate is present in the upper part. Transmitted light, × 100. d) Typical garland of uranium oxide on quartz crystals from the OP mine. Quartz has been partly replaced by sphalerite. Reflected light, × 65. e) Pitchblende (grey) and hematite (white) from the Donna showing. Reflected light, × 65. f) Pitchblende (P) surrounded by chlorite (Ch) from the Numac showing. Reflected light, × 25.

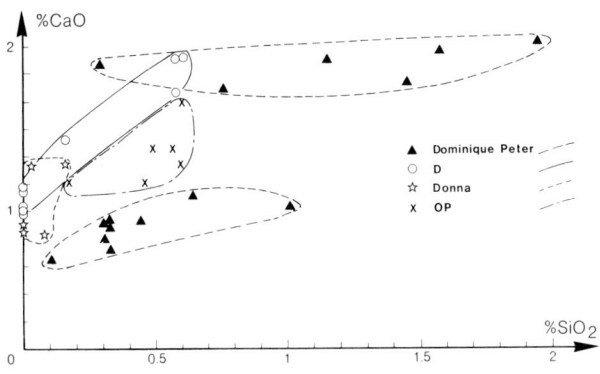

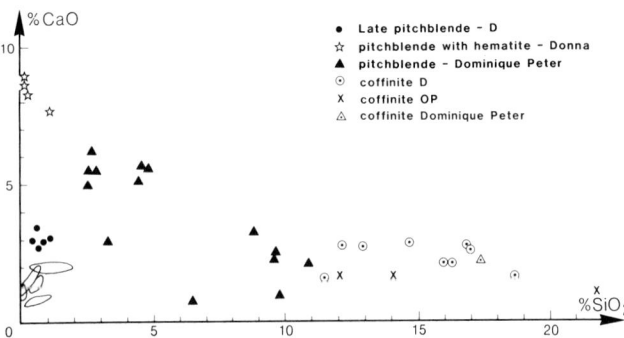

Figure 4. CaO-SiO₂ plot for uranium oxides and coffinites from the Dominique-Peter, D, and OP deposits, and from the Donna showing. Upper diagram: Groups II and III, lower diagram: Group V and coffinite.

able and to the alteration, no X-ray study was undertaken. Nevertheless, chemical compositions obtained by electron microprobe are compatible with brannerite (Table III).

Some indications concerning the genesis of this mineral can be obtained from chemical data. The absence of detectable thorium and the presence of significant quantities of vanadium are particularly noteworthy.

In the D deposit, titanium oxides are abundant. This is interpreted as a result of alteration of detrital Fe-Ti oxides. The formation of brannerite could result from circulation of uraniferous solutions, and their reaction with titanium oxides that were formed by the destruction of Fe-Ti oxides as proposed by Ramdohr (1957) in the Pronto mine. Such a reaction is supported by the occurrence of vanadium, which is also present in Fe-Ti oxides, and the absence of thorium in brannerite.

In conclusion, textural observations, automorphic shape, and chemical compositions indicate a hydrothermal origin for this brannerite.

DISCUSSION

Lead Content of Uranium Oxides

Chemical data, ages, and PbO% data obtained from various uranium oxides are presented in Figure 3. Taking the five different groups into consideration, several comments can be made.

Uraninite from the Sophie showing has the highest PbO content in the Carswell structure. This data can be compared with those of Charlebois Lake and Duddridge Lake, Saskatchewan (Fig. 6). These uraninites were present before the formation of the economic uranium deposits and are related

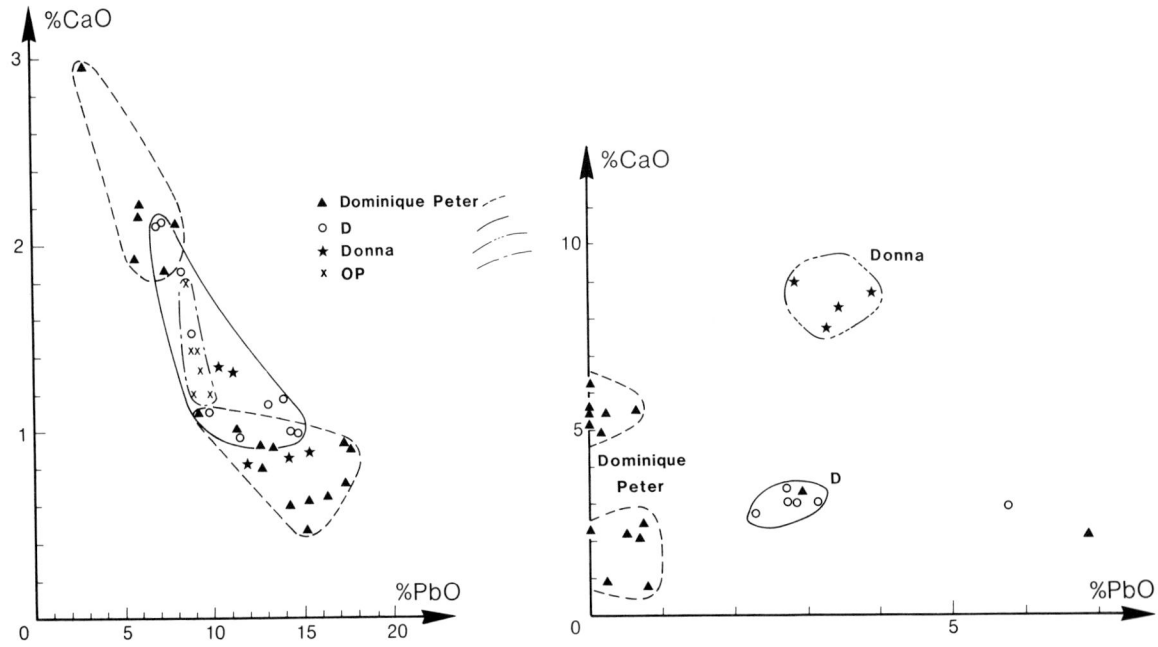

Figure 5. CaO-PbO plot for uranium oxides from the Dominique-Peter, D, and OP deposits, and from the Donna showing. Left diagram: Groups II and III, right diagram: Group V.

to Hudsonian metamorphism. They do not seem to be similar to the Beaverlodge-type of pitchblende dated at 1750 Ma (Koeppel, 1968) as is evidenced by their cerium, yttrium, and thorium content.

For the D, Dominique-Peter, and Donna areas, the PbO values range from 13% to 18% and correspond to the main mineralizing event recognized throughout northern Saskatchewan. Interpretation of the lead data suggests that mineralization in Dominique-Peter is older, at least in part, than in the D deposit. Caution must be used since higher PbO contents could also correspond to the presence of common lead or interaction with galena. If the highest PbO content obtained is considered in the Dominique-Peter deposit, a chemical age of about 1600 Ma is determined. This is surprisingly the same as some $^{39}Ar/^{40}Ar$ ages obtained on K-micas from alteration zones (Bell, 1985).

In the OP deposit, large quantities of uranium oxides are associated with tourmaline and have a significantly lower PbO content, which could mean that they are younger. Subsequent isotopic analyses by Bell (1985) on the same material give an age of 890 Ma. Such an age is also known in other deposits from Saskatchewan, but mineralogical descriptions indicate that this corresponds to a totally different process. At Key Lake, an age of 918 Ma is attributed to the formation of coffinite (Wendt *et al.*, 1978) whereas at Rabbit Lake, Cumming and Rimsaite (1979) interpret an age of 850 Ma as an episodic removal of radiogenic lead after the replacement of pitchblende by sulphides and arsenides.

Brannerite and its Significance

The occurrence of brannerite or related uranium-titanium oxides in the D deposit is interesting because this mineral is also present, always in minor amounts, in several unconformity-type uranium deposits outside of the Carswell structure. Brannerite has been described at the Rabbit Lake deposit (Ruzicka and Littlejohn, 1982) in Saskatchewan, at the Jabiluka deposit (Binns *et al.*, 1980) and the Ranger deposit (Eupene *et al.*, 1975) in Australia.

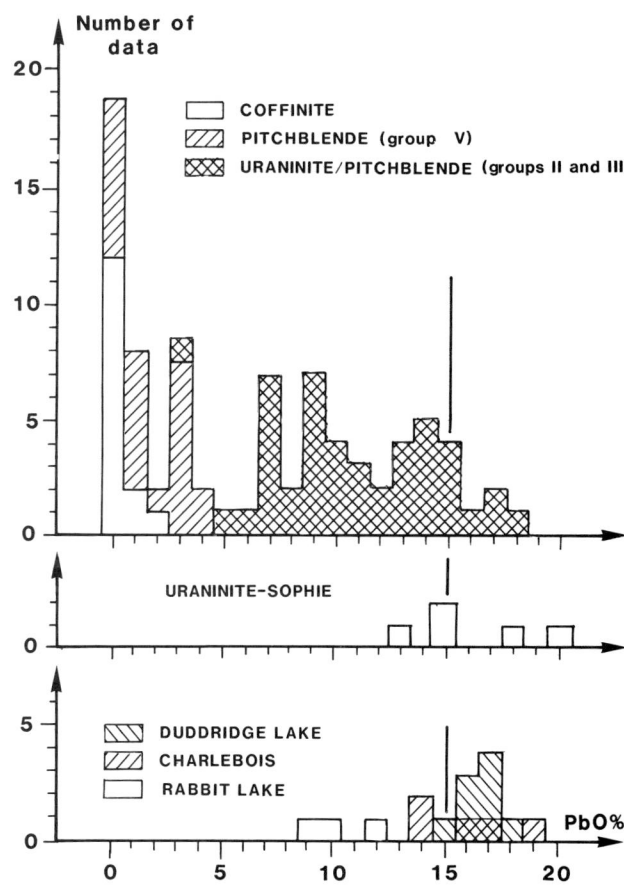

Figure 6. Distribution diagrams on the PbO content of uranium oxides and coffinite from northern Saskatchewan. Upper diagram: data obtained during this work on deposits and showings of the Carswell structure (except the Sophie showing). Middle diagram: data for uraninite from the Sophie showing. Lower diagram: data for pitchblende from Rabbit Lake (Rimsaite, 1977) and for uraninite from Charlebois and Duddridge Lake (Cameron-Schimann, 1978).

TABLE II

CHEMICAL ANALYSIS OBTAINED BY ELECTROMICROPROBE OF COFFINITES IN THE D (CLU 369 AND MP 74-10), OP (OP 3B) AND DOMINIQUE-PETER (DP) DEPOSITS

	Sandstones CLU 369				Sandstones MP-74-10					Vein OP 3b			DP
	1	2	3	4	1	2	3	4	5	1	2	3	1
UO_2	68.28	70.43	71.41	72.78	63.61	67.97	66.93	68.92	68.28	53.97	66.17	70.52	65.17
SiO_2	18.67	16.25	15.99	11.43	16.84	15.98	14.67	12.96	12.14	21.78	14.03	12.06	17.34
P_2O_5	0.36	0.51	0.36	0.22	2.18	1.80	2.04	1.31	1.38	0.98	1.40	0.81	0.69
Al_2O_3	2.22	1.07	1.17	1.16	2.83	1.36	1.10	1.07	0.78	6.29	2.09	1.09	1.29
CaO	1.65	2.11	2.11	1.57	2.72	2.69	2.86	2.70	2.76	1.03	1.70	1.69	2.20
FeO	0.20	0.33	0.28	0.31	0.29	–	0.04	0.26	0.04	1.00	–	–	0.57
MnO	0.20	0.04	–	0.15	–	–	–	–	–	0.12	–	–	–
TiO_2	0.12	–	0.13	–	–	0.59	0.67	0.54	0.86	–	–	–	0.04
PbO	–	0.38	–	1.75	0.09	0.10	0.72	–	0.49	0.13	–	0.33	0.53
Y_2O_3	–	–	–	–	0.62	0.69	0.63	0.51	0.53	1.81	2.81	1.91	0.53
Ce_2O_3	–	–	–	–	–	–	–	–	–	0.81	0.42	0.60	0.50
TOTAL	91.70	91.45	89.37	91.12	89.18	91.18	89.66	88.27	87.26	87.94	88.62	89.01	88.86

TABLE III

CHEMICAL COMPOSITIONS OF URANIUM-TITANIUM OXIDES FROM THE D DEPOSIT (THE RANGES OF SIX MICROPROBE ANALYSES ARE GIVEN)

OXIDE	WEIGHT %
UO_2	48.2 - 50.0
TiO_2	33.9 - 36.6 (24.9)
SiO_2	3.0 - 6.7
Al_2O_3	0.5 - 0.9
CaO	1.2 - 1.7
V_2O_3	0.7 - 1.4
PbO	0.2 - 1.2
FeO	0.2 - 4.1

TABLE IV

EPISODES OF URANIUM OXIDES DEPOSITION IN THE CARSWELL AREA

Chemical age in Ma	Mineralization	Isotopic data available
1800	Uraninite (Y, Th, Ce) – Monazite	
1200 to 1000	Uraninite	1050 (Bell, 1985)
	Pitchblende/Uraninite	1150 (Bellon et al., 1976)
	(Ca-poor)	1050 (Gancarz, 1979)
800	Pitchblende/Uraninite	890 (Bell, 1985)
400 or older	Pitchblende (Ce,Y)	
330	Pitchblende (Ca-rich)	380 (Bell, 1985)
250	Pitchblende (Ca-rich)	250 (Bellon, et al., 1976) (remobilization)
200 to present	Alteration of Pitchblende (Si-enrichment)	

As seen from their habit and location, these brannerites are not detrital in origin. At the present time, no data on the formation of brannerite in near-surface conditions are available. From a completion of natural occurrences and experimental data, brannerite is usually characteristic of medium to high temperatures (Pagel, 1981). The only indication of lower temperature is the description of brannerite in the Arlit deposit, Niger, located near a tectonic structure (Bonnamy et al., 1982), and where conditions of deposition or recrystallization are not known.

Habits of Uranium Oxides

Compared with microscopic observations on intragranitic veins, tabular and roll-type deposits in sandstones, disseminations in alaskites, pegmatites, and synmetamorphic occurrences, the uranium oxides of Groups II and III have a particular and characteristic habit.

Chemical data on cubes, "cogwheels" and spheroids with automorphic ends show that, in the same deposit, they have roughly the same composition. This excludes the theory that several generations of uranium oxides formed at different times. Factors controlling the texture and habit of uranium oxides are not totally clear. Temperature has been invoked as a determining factor for development of euhedral uranium oxides. The development of uraninite is observed in high temperature environments (granitoids and metamorphic rocks). On the contrary, in low to medium temperature environments such as intragranitic veins, and tabular and roll-type U deposits, collomorph pitchblende generally forms. The highest temperature found by fluid inclusion study for the precipitation of pitchblende is 350° C (Leroy, 1978). In the Penaran deposit, Vendee, France, where temperatures close to 360° C have been inferred by fluid inclusion studies, Cathelineau (1981) observed the local development of idiomorphic crystals of uranium oxide. Euhedral uraninite is present at 300° to 350° C in the Mistamisk deposit, Quebec, Canada (Kish and Cuney, 1981). In unconformity-related deposits, a compilation of data obtained by various techniques has shown that no such high temperatures have been reached (Pagel, 1983). Therefore, it appears that there is one or several other important factors for generation of such a habit. In these deposits, two processes are significant: precipitation from brines, and deposition in non-porous rocks. The chemical composition of uranium oxides does not appear to be a determinant factor since the composition of these uranium oxides does not differ from other oxides (Cathelineau et al., 1982).

Euhedral uraninites have no particular crystallographic parameters if a comparison is made with other hydrothermal uranium oxides. They support the suggestion of Cathelineau et al. (1982) to use a_0 as a tool to get an idea of the precipitation or recrystallization age of uranium oxides. This is obviously correct for pitchblende veins with low Th content but not for uraninite in, for example, pegmatoids and granites. Basham et al. (1982) have found a value of $a_0 = 5.475 \pm 0.003$ Å for uraninite from the Urgeirica granite, Portugal, whereas for adjacent pitchblende in veins, the value is 5.430 ± 0.006 Å. Chemical composition does not seem to be a determinant factor since Th and Ce are higher in pitchblende (Basham et al., 1982).

Some Consequences Regarding Genesis of Unconformity-Related Deposits

The discovery of a mineralizing event in the Carswell structure that occurred during Hudsonian metamorphism and anatexis appears to be very significant. This event is already known in the Cree Lake Mobile Zone which is the host of many deposits such as Rabbit Lake and Key Lake. Data from uraninites given by Cameron-Schimann (1978) show that they are also rich in REE and thorium. The generation of such U-Th-enriched pegmatoids is attributed to a small degree of partial melting (Parslow and Thomas, 1982; Cuney, 1981, 1982). Pagel and Svab (1985) also show that the basement is relatively rich in U for a granulite facies terrain.

Some alteration is probably related to retrograde metamorphism just after the Hudsonian orogeny, as evidenced by, for example, some quartz dissolution processes. However, until the present, no data has confirmed ages in the range of those of Beaverlodge at 1750 Ma (Koeppel, 1968). Data from Numac area is ambiguous, as it has low PbO contents but significant quantities of cerium and yttrium. A detailed comparison with the Gunnar deposit would be very significant.

The main mineralizing event, represented by the Dominique-Peter, D, and OP deposits, is probably a succession of mineralizing periods which may or may not occur in a deposit. Available isotopic data show at least four main periods: (1) 1330 ± 30 Ma (Gancarz, 1979) on disseminated low concentrations of uranium, (2) 1150 Ma on D (Bellon et al., 1976), (3) 1050 Ma on D (Gancarz, 1979), and Dominique-Peter (Bell, 1985), and (4) 800-900 Ma (Bell, 1985) on OP.

No mineralogical data have been given by Gancarz (1979), and the significance of the first event cannot be evaluated. If the chemical composition of uranium oxides, U-Pb isotopic data on uranium oxides (Bell, 1985), and K-Ar isotopic data on clays (Clauer et al., 1985) are considered, the major economic mineralization was deposited between 1250 Ma and 1050 Ma. At about 900 Ma, precipitation of uranium was more dispersed, generally in veins with no known massive concentrations.

The calcium content of uranium oxides increases from the older Dominique-Peter mineralization to the younger OP mineralization. This evolution is probably the result of the evolution of the mineralizing solution through time. Fluid inclusions are difficult to study in these deposits, since there was little quartz present prior to uranium oxide deposition. But in the few cases where it was possible, brines were observed, indicating that they were present over a long period of time. This is in agreement with the deduction that fluids observed in fluid inclusions at the Rabbit Lake deposit originate predominantly from sandstone formation waters (Pagel and Jaffrezic, 1977; Pagel et al., 1980; Pagel, 1983). Circulation can be attributed to tectonic processes and related in time to the emplacement of diabase dykes. Several late episodes of pitchblende deposition are evidenced by their CaO and SiO_2 contents and PbO content. This process is thought to be due to the leaching of pre-existing ore, at least for the D deposit, where the lower intersection of the isochron is about 250 Ma (Bellon et al., 1976).

Comparisons with other unconformity-related uranium deposits from Saskatchewan are difficult because of the lack of mineralogical data available on the geochronological work, except in the study by Cumming and Rimsaite (1979). However, all of these periods have been recorded, for example, in the study by Cumming and Rimsaite (1979), and a special metallogenic model need not be developed to explain the Carswell uranium deposits.

CONCLUSIONS

The principal conclusions for this study are:

1) Electron microprobe data, performed on five groups of uranium oxides characterized by their mineral habit and paragenesis, have shown that six different types of uranium oxides can be distinguished in the showings and deposits of the Carswell structure.

2) For the first time, concentrations of uranium oxides prior to economic mineralization have been unambiguously evidenced and characterized by the Sophie occurrence. Associations of the oxides with monazite and significant quantities of Th, Y, and Ce are in agreement with an origin during anatectic processes.

3) From textural observations and lead content, it follows that the main generation of uranium oxides are Ca-poor whereas more recent oxides are Ca-rich.

4) Calculation of chemical ages from the lead content of uranium oxides yields a succession of mineralizing events shown in Table IV.

5) The formation of brannerite indicates a thermal event in the D deposit which is in agreement with fluid inclusion studies.

6) Such a mineral evolution is also present in other unconformity-related uranium deposits in Saskatchewan, and no particular chemical mineralizing event could be related to the formation of the Carswell structure during Ordovician times. The structure induces only mechanical disturbances of previous uranium concentrations. Small chemical mobilizations are locally observed in breccias.

7) This study is a good example of the usefulness, due to the low cost and ease by which they can be done, of chemical analyses of uranium oxides in exploration. However, extreme caution must be used in the interpretation of the chemical ages because effects after the mineralizing event such as exposure to atmospheric weathering and recent mineralization can lead to erroneous results.

ACKNOWLEDGEMENTS

We would like to thank Bernard Poty and Michel Cuney for critically reading the manuscript and Margaret Svab for improving the english text.

REFERENCES

Basham, I.R., Bowles, J.F.W., Atkin, D., and Bland, D.J., 1982, Mineralogy of Uranium-Distribution in Samples of Unaltered and Sericitized Granite from Urgeirica, Portugal, in Relation to Mineralization Processes, A Preliminary Study: in Vein-Type and Similar Uranium-Deposits in Rocks Younger than Proterozoic: International Atomic Energy Agency, Vienna, p. 299-309.

Bell, K., 1981, A Review of the Geochronology of the Precambrian of Saskatchewan. Some Clues to Uranium Mineralization: Mineralogical Magazine, v. 44, p. 371-378.

———, 1985, Geochronology of the Carswell Area, Northern Saskatchewan: in Lainé, R., Alonso, D., and Svab, M., eds., The Carswell Structure Uranium Deposits, Saskatchewan: Geological Association of Canada, Special Paper 29.

Bellon, M., Devillers, C., Hagemann, R., and Touray, J.C., 1976, Dater les Minéralisations: Mémoire Hors Série de la Société Géologique de France, no. 7, p. 265-268.

Binns, R.A., McAndrew, J., and Sun, S-S., 1980, Origin of Uranium Mineralization at Jabiluka: in Ferguson, J., and Goleby, A., eds., Uranium in the Pine Creek Geosyncline: International Atomic Energy Agency, Vienna, p. 543-562.

Bonnamy, S., Oberlin, A., and Trichet, J., 1982, Two Examples of Uranium Associated with Organic Matter: Organic Geochemistry, v. 4, p. 53-61.

Cameron-Schimann, M., 1978, Electron Microprobe Study of Uranium Minerals and its Application to Some Canadian Deposits: Unpublished Ph.D. Thesis, University of Alberta, Edmonton, 343 p.

Cathelineau, M., 1981, Les Gisements Uranifères de la Presqu'île Guérandaise (Sud Bretagne). Approche Structurale et Métallogénique: Mineralium Deposita, no. 16, p. 227-240.

Cathelineau, M., Cuney, M., Leroy, J., Lhote, F., Nguyen Trung, G., Pagel, M., and Poty, B., 1982, Caractères Minéralogiques des Pechblendes de la Province Hercynienne d'Europe. Comparaison avec les Oxydes d'Uranium du Protérozoïque de Différents Gisements d'Amérique du Nord, d'Afrique et d'Australie: Reprint from "Vein-Type and Similar Uranium Deposits in Rocks Younger than Proterozoic", International Atomic Energy Agency, p. 159-177.

Clauer, N., Ey, F., and Gauthier-Lafaye, F., 1985, K-Ar Dating of Different Rock Types from the Cluff Lake Uranium Ore Deposits (Saskatchewan - Canada): in Lainé, R., Alonso, D., and Svab, M., eds., The Carswell Structure Uranium Deposits, Saskatchewan: Geological Association of Canada Special Paper 29.

Cumming, G.L., and Rimsaite, J., 1979, Isotope Studies of Lead-Depleted Pitchblende, Secondary Radioactive Minerals and Sulphides from the Rabbit Lake Uranium Deposit, Saskatchewan: Canadian Journal of Earth Sciences, v. 16, p. 1702-1715.

Cuney, M., 1981, Comportement de l'Uranium et du Thorium au Cours du Métamorphisme: Rôle de l'Anatexie dans la Génèse des Magmas Riches en Radioéléments: Thèse Doctorat Es-Sciences, Université de Nancy, 511 p.

―――, 1982, Processus de Concentration de l'Uranium et du Thorium au Cours de la Fusion Partielle et de la Cristallisation des Magmas Granitiques: in Compte-Rendu du Symposium sur les Méthodes de Prospection de l'Uranium: Agence de l'OCDE pour l'Energie Nucléaire, p. 277-291.

Eupene, G.S., Fee, P.H., and Colville, R.G., 1975, The Ranger One Uranium Deposits: in Economic Geology of Australia and Papua, New-Guinea: Australian Institute of Mining and Metallurgy Monograph Series no. 5, 1 - Metals, p. 308-317.

Ewers, G.R., and Ferguson, J., 1980, Mineralogy of the Jabiluka, Ranger, Koongarra, and Nabarlek Uranium Deposits: in Ferguson, J., and Goleby, A., eds., Uranium in the Pine Creek Geosyncline: International Atomic Energy Agency, Vienna, p. 363-374.

Gancarz, A.J., 1979, Chronology of the Cluff Lake Uranium Deposit, Canada (Abstract): International Uranium Symposium on the Pine Creek Geosyncline, N.T., Australia, Extended Abstracts, p. 91-94.

Hoeve, J., Cumming, G.L., Baadsgaard, H., and Morton, R.D., 1981, Geochronology of Uranium Metallogenesis in Saskatchewan: Canadian Institute of Mining, Uranium Symposium, September 8-13, 1981, Saskatoon, Technical Program, p. 11-12.

Hohndorf, A., Lenz, H., Wendt, I., von Pechmann, E., and Voultsidis, V., 1981, Radiometric Age Determination on Samples of the Key Lake Uranium Deposit: Canadian Institute of Mining Uranium Symposium, September 8-13, 1981, Saskatoon, Technical Program p. 24.

Kish, L., and Cuney, M., 1981, Uraninite − Albite Veins from the Mistamisk Valley of the Labrador Trough, Quebec: Mineralogical Magazine, v. 44, p. 471-483.

Koeppel, V., 1968, Age and History of the Uranium Mineralization of the Beaverlodge Area, Saskatchewan: Geological Survey of Canada Paper 67-31, 111 p.

Leroy, J., 1978, The Margnac and Fanay Uranium Deposits of the La Crouzille District (Western Massif Central, France), Geology, Mineralogy, and Fluid Inclusions Studies: Economic Geology, v. 73, p. 1611-1634.

McLennan, S.M., and Taylor S.R., 1980, Rare Earth Elements in Sedimentary Rocks, Granites, and Uranium Deposits of the Pine Creek Geosyncline: in Ferguson, J., and Goleby, A., eds., Uranium in the Pine Creek Geosyncline: International Atomic Energy Agency, Vienna, p. 175-190.

Pagel, M., 1981, Facteurs de Distribution et de Concentration de l'Uranium et du Thorium dans Quelques Granites de la Chaîne Hercynienne d'Europe. Thèse d'Etat, Université de Nancy, 566 p.

―――, 1982, Successions Paragénétiques et Teneurs en Uranium des Minéraux Accessoires dans les Roches Granitiques: Guides pour la Recherche de Granites Favorables à la Présence de Gisements d'Uranium: in Compte-Rendu du Symposium sur les Méthodes de Prospection de l'Uranium, Paris, 1-4 Juin, 1982: Agence de l'OCDE pour l'Energie Nucléaire, p. 445-456.

―――, 1983, Les Gisements d'Uranium Liés Spatialement aux Discordances: Géologie et Géochimie de l'Uranium, Mémoire 1, Nancy, 380 p.

Pagel, M., and Jaffrezic, H., 1977, Analyses Chimiques des Saumures des Inclusions du Quartz et de la Dolomite du Gisement d'Uranium de Rabbit Lake (Canada): Aspect Méthodologique et Importance Génétique: Comptes Rendus de l'Académie des Sciences, Paris, t. 284, Série D, p. 113.

Pagel, M., and Svab, M., 1985, Petrographic and Geochemical Variations within the Carswell Structure Metamorphic Core and their Implications with Respect to Uranium Deposition: in Lainé, R., Alonso, D., and Svab, M., eds., The Carswell Structure Uranium Deposits, Saskatchewan: Geological Association of Canada Special Paper 29.

Pagel, M., Poty, B., and Sheppard, S.M.F., 1980, Contribution to some Saskatchewan Uranium Deposits Mainly from Fluid Inclusions and Isotopic Data: in Ferguson, J., and Goleby, A., eds., Uranium in the Pine Creek Geosyncline: International Atomic Energy Agency, Vienna, p. 639-654.

Parslow, G.R., and Thomas, D.J., 1982, Uranium Occurrences in the Cree Lake Zone, Saskatchewan, Canada: Mineralogical Magazine, v. 46, p. 163-171.

Ramdohr, P., 1957, Die Pronto-Reaktion: Neues Jahrbuch für Mineralogie Monatshefte, p. 217-222.

Rimsaite, J., 1977, Mineral Assemblages at the Rabbit Lake Uranium Deposit, Saskatchewan: A Preliminary Report: Report of Activities, Part B: Geological Survey of Canada Paper 77-1B, p. 235-246.

Ruhlmann, F., 1985, Mineralogy and Metallogeny of Uraniferous Occurrences in the Carswell Structure: in Lainé, R., Alonso, D., and Svab, M., eds., The Carswell Structure Uranium Deposits, Saskatchewan: Geological Association of Canada Special Paper 29.

Ruzicka, V., and Littlejohn, A.L., 1982, Notes on Mineralogy of Various Types of Uranium Deposits and Genetic Implications: in Current Research, Part A: Geological Survey of Canada Paper 82-1A, p. 341-349.

Tona, F., Alonso, D., Svab, M., and Wheatley, K., 1985, The Carswell Structure Uranium Deposits (Saskatchewan): in Lainé, R., Alonso, D., and Svab, M., eds., The Carswell Structure Uranium Deposits, Saskatchewan: Geological Association of Canada Special Paper 29.

Wendt, I., Hohndorf, A., Lenz, H., and Voultsidis, V., 1978, Radiometric Age Determination on Samples of the Key Lake Uranium Deposits: in Short Papers on the 4th International Conference on Geochronology, Cosmochronology and Isotope Geology: United States Geological Survey Open File Report 71-701, p. 448-449.

Worden, J.M., Cumming, G.L., and Baadsgaard, H., 1981, Geochronological Setting and Mineralization Ages of the Midwest Uranium Deposit, Northern Saskatchewan (Abstract): Canadian Institute of Mining Uranium Symposium, September 8-13, 1981, Saskatoon, Technical Program, p. 10.

The Carswell Structure Uranium Deposits, Saskatchewan,
edited by R. Lainé, D. Alonso and M. Svab,
Geological Association of Canada Special Paper 29, 1985

A CHEMICAL STUDY OF THE CARBONACEOUS MATERIAL FROM THE CARSWELL STRUCTURE

P. Landais
Centre de Recherches sur la Géologie de l'Uranium, 3, rue du Bois de la Champelle, 54500 Vandoeuvre-lès-Nancy, France

J.M. Dereppe
Laboratoire de Chimie Physique et de Cristallographie, Bâtiment Lavoisier - 1, place Louis Pasteur, 1348 Louvain-La-Neuve, Belgium

ABSTRACT

Carbonaceous material is frequently associated with Precambrian uranium deposits such as Oklo (Gabon) and Witwatersrand (South Africa). Nevertheless, their study is very difficult and their origin is not well known. Solid state ^{13}C nuclear magnetic resonance (NMR) analyses have been carried out on the carbonaceous material of the Carswell structure in order to define their chemical structure. The carbonaceous material from the D and Claude areas are similar and can be considered to be bitumen. They show a weak maturity in spite of the high temperatures assumed for the solutions carried along the unconformity. Other classical geochemical investigations such as Rock-Eval pyrolysis and element analysis have been carried out.

RÉSUMÉ

Les matières carbonées sont très fréquemment associées aux gisements Précambriens tels Oklo (Gabon) ou le Witwatersrand (Afrique du Sud). La maturité élevée, les altérations radiolytiques ou tardives de ces matières carbonées sont des obstacles à une bonne connaissance géochimique.

La RMN du ^{13}C solide semble être une technique performante pour définir leur structure et tout particulièrement dans le cas des gisements de Cluff Lake.

Les analyses infrarouge (Fig. 2) et RMN (Fig. 5) ont montré l'identité chimique des matières carbonées des zones D et Claude. Celles-ci peuvent être considérées comme étant des reliquats d'hydrocarbures migrés ou bitumes. En outre, il ne semble pas que l'altération radiolytique ait sérieusement modifié les caractéristiques des bitumes (Figs. 4 et 5).

La comparaison des résultats RMN obtenus sur la structure Carswell avec des analyses concernant des charbons tertiaires de type III (Fig. 6) montre la faible maturité des bitumes de Cluff et suggère une aliphaticité initiale importante. De plus elle permet de caractériser un phénomène d'oxydation s'attaquant essentiellement aux carbones aliphatiques.

Le problème de l'origine des bitumes de la structure Carswell a été abordé. Parmi les hypothèses envisagées, deux semblent plus plausibles: 1) la genèse d'hydrocarbures à partir des dolomies organogènes de la formation Douglas, ou 2) la production d'hydrocarbures légers lors de l'altération des graphitoïdes du socle gneissique.

La maturité des dolomies de la formation Douglas a pu être approchée par la RMN^{13}C (Fig. 6). Leur aromaticité indique, en la comparant avec d'autres données RMN, un enfouissement qui aurait pu dépasser 3000m. Au cours de son évolution, la matière organique de Douglas a donc probablement généré des hydrocarbures qui auraient migré jusqu'à la discordance et seraient piégés dans les zones d'altération.

Néanmoins, l'absence de matière organique dans les grès Athabasca et la nature des eaux issues des grès ne semblent pas favoriser cette hypothèse qui ne doit pourtant pas être négligée.

Les graphitoïdes du socle gneissique n'ont pu être analysés par RMN mais nous avons pu établir leur composition élémentaire qui révèle que ces graphitoïdes contiennent de l'hydrogène et de l'oxygène.

La disparition des graphitoïdes dans les zones d'altération et la liaison spatiale des zones à bitumes avec un socle graphiteux incite à penser que les graphitoïdes du socle pourraient avoir joué un rôle de source dans la formation des bitumes.

D'autre part, on notera que les bitumes se localisent exclusivement dans les zones d'altération, ce qui indiquerait que leur formation est intimement liée au phénomène d'altération.

Enfin, le caractère initial aliphatique des matières carbonées de Cluff leur a probablement conféré une grande réactivité ainsi qu'une large capacité réductrice face aux solutions chargées en uranium. On peut donc penser que leur rôle dans le phénomène minéralisateur n'a pas été négligeable.

INTRODUCTION

Carbon is frequently associated with uranium in Precambrian deposits. Thucholite, a carbonaceous material containing uranium and thorium was first described in pegmatitic material from Parry Sound, Ontario (Ellsworth, 1928). In Precambrian conglomerate deposits, for example Witwatersrand, South Africa, and Elliot Lake, Canada, such uraniferous carbon occurs in many different ways: as mudstone veneer, wrapped around pebbles, entrapping sand grains and as detrital accumulations (Minter, 1978). It looks like vitrinite but the origin of the material is not well known. Many authors favour the idea that the carbon may have two origins: algal mats and polymerized migrated products, or biochevically decayed primitive micro-organisms (Button and Adams, 1981). One should also consider that Precambrian organic matter undergoing increasing burial can generate petroleum or petroleum-like material.

Investigations on the Oklo deposit, Gabon, clearly indicate the source rock – reservoir relationships between the pelitic rocks of the Francevillian series and the bitumens (Cassou et al., 1975; Alpern, 1978). Kerogens have reached a maturation stage quite similar to that of anthracite (Cassou et al., 1975) and pyrolysis parameters show a high degree of maturity (Vandenbroucke et al., 1978). Elementary composition of the Oklo kerogens and bitumens indicates an oxidation process (Rouzaud, 1979), shown by a shifting of the samples from the evolution path I organic matter of algal or bacterial origin, to the evolution path III of land derived organic matter. Rouzaud (1979) and Bonnamy (1981) observed, using high resolution electron microscopy, the complexing of uranium by the Oklo organic matter.

In these two types of Precambrian deposits, organic geochemistry, petrographic, and microscopic studies have defined some organic parameters, however, structural organic geochemistry has seldom been used in this field. It is challenging research to investigate the origin and evolution of carbonaceous material in the uranium deposits of the Carswell structure by carrying out new experimental techniques.

This paper summarizes the major results obtained on these deposits. They have been examined by a routine geochemical approach and by a structural study. It should be noted that this technique has seldom been used on this type of Precambrian material as described herein and that the results and the hypothesis proposed must be discussed further and supported with geological information.

NATURE AND OCCURRENCE OF PRECAMBRIAN ORGANIC MATTER

Bacterium-like rods and algal-like spheroids in rocks with ages ranging from 2100 to 3300 Ma have been documented in South Africa, Siberia, and Canada. These organisms are called procaryote because their genetic material is disarranged and they are asexual (Hunt, 1979). They are similar to the modern blue-green algae and are adapted to a reducing atmosphere where the partial pressure of oxygen was very low (less than 0.1 atm).

Photosynthesis produced oxygen and caused the beginning of the evolutionary history on earth (Schopf et al.,

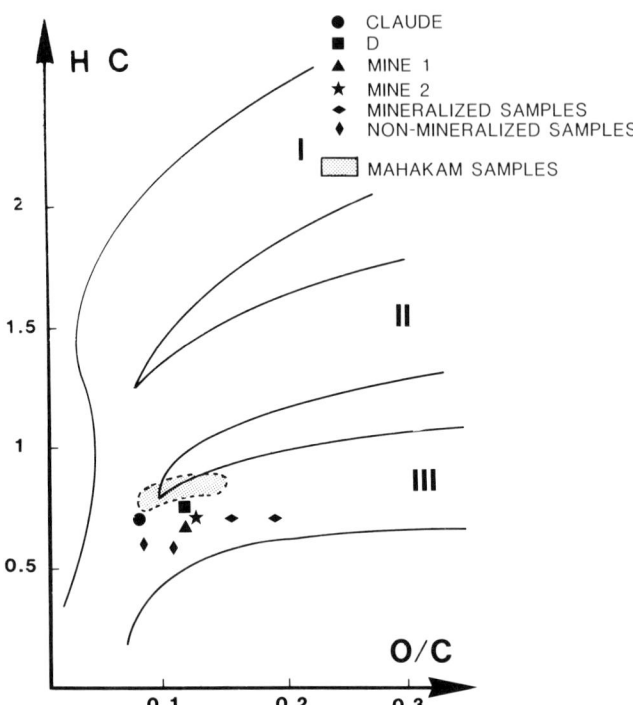

Figure 1. Carswell structure samples plotted in a Van Krevelen diagram – Comparison with the samples of the Mahakam Delta (Dereppe et al., 1983). Evolution path I: algal kerogen. Evolution path II: marine kerogen. Evolution path III: continental plant kerogen.

1973). It is marked by the appearance of eucaryotes around 1000 to 1500 Ma.

The organic content of sedimentary Precambrian rocks is generally low but essentially it is composed of algal type organic matter. It belongs to the type I evolution path in a Van Krevelen diagram (Fig. 1) which refers to algal material with a high initial H/C atomic ratio and a low initial O/C ratio (Tissot and Welte, 1978). These kerogens generate and release large amounts of gaseous and liquid hydrocarbons as evidenced by their lower H/C value during burial.

Nevertheless, most of the Precambrain sedimentary rocks have reached a high degree of metamorphism and are barren in terms of hydrocarbon content, and presumably many Precambrian petroleum accumulations have been destroyed.

CARBONACEOUS MATERIAL IN THE CARSWELL STRUCTURE

Carbonaceous Material Distribution

Two different types of carbon can be found in the Precambrian of the Carswell structure: graphitoids in the altered gneissic basement, and carbonaceous matter associated with the alteration zones in the sandstones of the D area or in the basement in Claude area. No carbon has been recognized in the unaltered Athabasca sandstones.

Several modes of occurrence of carbonaceous material from the altered areas have been noted: millimetric spheroids trapped in the chloritic matrix, centimetric to decimetric accumulations, argilo-carbonaceous associations, coating of fault planes, and impregnations in the sandstone remains. These occurrences suggest either dissemination type carbon or migration type carbon. This material can be associated with mineralized or non-mineralized areas. There is no continuity in the carbonaceous material distribution. Carbon seems to be only concentrated in the altered areas but it may also be absent.

One may assume that the distribution of the carbon in the Carswell structure does not suggest a syngenetic origin. Geological observations support the idea that the carbon accumulations may be of bitumen type in spite of the absence of a real reservoir. A hydrocarbon generation process in a bitumen-source rock relationship similar to the one found in the Oklo deposit is not probable, due to the lack of source rocks and the mode of occurrence in the deposits.

Microscopic and Geochemical Approach

It is quite difficult to apply geochemical techniques to the study of Precambrian organic matter. Most of the carbonaceous materials are chloroform insoluble and have been thermally altered. The most efficient tool uses the analysis of the kerogen solid phase.

Rouzaud (1979) and Bonnamy (1981) studied Precambrian carbonaceous material by high resolution transmission microscopy and energy dispersive Xray analysis. They chose samples having high and low uranium content from the D and Claude areas. Rouzaud observed that the D samples were associated with a pelitic facies and considered them to be sedimentary organic matter, whereas the Claude samples, found in faults in the basement, were classified as bitumen type organic matter. He favoured a source-reservoir relationship between the "D source rock" and the "Claude bitumens."

New elementary analysis (Table I) has shown that the representative points of the Cluff samples when plotted in a Van Krevelen diagram (Fig. 1), lie on the oxygen rich path, or evolution path III. This high oxygen content implies an oxidation process. Furthermore, samples with a high uranium content are the most oxidized. Rouzaud (1979) and Bonnamy (1981) explain the anomalies in the oxygen content by fixation of uranium, observing the trapping of UO_2 crystals inside the pores of the carbonaceous matter and the occurrence of organo-uranium complexes. They also note that after heat treatment, up to 1000°C, these carbonaceous particles are filled with uraninite microcrystals which are produced by the destruction of organo-uranium complexes. They conclude that during the first stages of diagenesis, free radicals of the organic matter fix uranyl cations UO_2^{2+}, and that under the action of the geothermal gradient, uranium is released as UO_2 crystals while organic matter loses hydrogen and fixes oxygen (Bonnamy et al., 1982).

This paper deals with the study of five samples, four of which were selected from the different altered facies near the sandstone-basement unconformity. The Claude sample is a massive bitumen type carbonaceous material sampled in a barren area of the altered basement, D was collected in a low uranium content, altered zone of the Athabasca sandstones, and Mine 1 and Mine 2 were collected in a mineralized zone of the Claude open pit. All these samples are shiny black, brittle, and have massive carbonaceous material containing more than 70% carbon. This sampling was completed with an organic-rich dolomite (DGL 3) from the Douglas Formation.

Most of the common geochemical techniques are restricted to chloroform-soluble organic matter. Extractions carried out on the Cluff samples give total organic carbon contents below 0.5% so it seems very difficult to perform any classical geochemical technique using chloroform extract such as constitution, gas chromatography or mass spectrometry. The results would not provide any helpful answer regarding the origin and evolution of the carbonaceous material.

Rock-Eval pyrolysis has been used on three samples: D, Claude, and Mine 1. They show similar spectra and yield the same quantities of hydrocarbons during pyrolysis. Their temperatures of maximal hydrocarbon generation (T_{max}) are comparable and range from 435° to 442°C (Table I). These temperatures correspond to the beginning of the oil generation zone for type I kerogen, or to a "theoretical" burial which cannot exceed 4000 m. These results are confirmed by carbon-ratio values (Gransch and Eisma, 1970) obtained by Connan (written communication) on other barrren material from the D and Claude zones which range from 0.8 to 0.85.

TABLE I

ANALYTICAL RESULTS AND ROCK-EVAL PYROLYSIS PARAMETERS OF THE CARSWELL STRUCTURE SAMPLES

	% C	% H	% O	% Ash	H/C at	O/C at	T_{max} °C	IH	IO
Claude	78.9	4.3	6.6	3.3	0.66	0.063	435	116	15.7
D	60.4	3.6	10.0	–	0.71	0.120	435	108	18.5
Mine 1	77.8	4.0	12.4	2	0.62	0.12	442	96	16.2
Mine 2	78.5	4.2	12.9	1.8	0.64	0.123	–	–	–

H/C at:	Hydrogen/Carbon – Atomic ratio
O/C at:	Oxygen/Carbon – Atomic ratio
T_{max} (C):	Temperature of maximum hydrocarbon generation
IH:	mg hydrocarbons/g organic carbon yield during pyrolysis
IO:	mg CO_2/g organic carbon yield during pyrolysis
(–):	Not determined

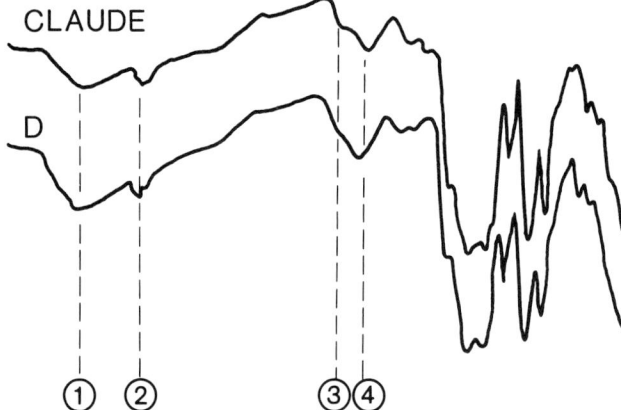

Figure 2. Infrared spectra of Claude and D samples. IR BANDS ASSIGNMENT: 1) OH groups; 2) CH_2 and CH_3 aliphatic groups; 3) carbonyl and carboxyl groups; 4) aromatic carbons and carbonyl groups.

C, H, and O analytical results (Table I, Fig. 1) show striking anomalies. The shifting of the samples on the evolution path III indicates a strong oxidation process although this Precambrian carbonaceous material ought to be poor in oxygen. This oxidation is confirmed by infrared spectra (Fig. 2) which shows strong absorption attributed to C = O groups, ketones, acids, and esters, a wide band mainly related to aromatic structures and a very weak band for CH_2 and CH_3 groups indicating the relatively low aliphatic content of the Carswell structure samples. Furthermore, one may establish that the D and Claude spectra are quite similar. The oxidation and relatively weak apparent maturity seems to contradict the geological history of the Athabasca Basin.

Nevertheless, after these routine analyses, there remains some controversy in connection with the origin, structure, and maturity of the carbonaceous material. Solid state ^{13}C NMR analyses were then performed on these samples in order to define a gross chemical composition for this material.

^{13}C NUCLEAR MAGNETIC RESONANCE STUDY

High resolution nuclear magnetic resonance (NMR) has often been utilized to study soluble fossil materials, primarily petroleum fractions, liquids derived from coals, and humic and fulvic acids. The parameters extracted from ^1H and ^{13}C NMR spectra combined with other information, such as elemental analysis or infrared (IR) data, lead to basic information such as aromaticity factor, aromatic ring size, aliphatic chain lengths, branching parameters, and average molecular weights.

Solid material is characterized by a very large peak broadening due to dipole-dipole interactions, chemical shift anisotropies, and long T_1 relaxation time. Most of the earlier studies on solids dealt with measurements of relaxation time by pulsed NMR or line shape by continuous wave techniques, giving only partial information which excluded detailed analysis such as that made on liquid or soluble material. However, several sophisticated techniques based on high power multiple pulse sequences have been developed during the last decade, which to some extent, overcome the problems associated with the observation of NMR on solid samples.

The observation of ^{13}C by cross polarization (CP) with high power, proton decoupling (Pines et al., 1973) and magic angle spinning (MAS) is much easier and rewarding. By combining these techniques, true high resolution spectra are obtained: the ^{13}C-H dipole-dipole broadening is removed by high power proton decoupling, the $^{13}C-^{13}C$ interaction is negligible due to isotopic dilution of ^{13}C, and the chemical shift anisotropy is removed by fast spinning at the magic angle. Moreover, the sensitivity problems associated with long T_1 ^{13}C relaxation times are solved by cross polarization. Due to the large range of ^{13}C chemical shift ($\pm$ 230 ppm), a useful resolution is obtainable even with complex samples such as fossil material.

EXPERIMENTAL TECHNIQUE

The basic principles of the cross polarization, magic angle spinning experiments will not be discussed here. The reader is referred to the articles by Pines et al. (1973), Schaefer et al. (1977), and Stejskal and Schaefer (1975). Carbon 13 spectra were obtained at room temperature using a Bruker CXP 100 spectrometer, operating at 22.63 MHZ for the ^{13}C, equipped with an external ^{19}F field stabilizer and a magic angle spinning device using an Andrew type rotor. Matched spin lock, single contact, cross polarization was performed according to the procedure summarized in Figure 3.

The best available rotor made of deuterated, polymethyl metacrylate (PMMA) gives a residual CP-MAS spectrum which is probably due to incomplete deuteration (Melchior et al., 1980). This artifact was taken into account and each spectrum presented is the difference between spectra obtained in the same conditions for the rotor, plus the sample, and for the rotor alone. This precludes any useful measure of the C = O band which interferes with the same band for the PMMA.

QUANTITATIVE ASPECTS OF THE ^{13}C CP-MAS SPECTRA

The ^{13}C CP-MAS spectra of some typical samples are shown in Figure 4. In contrast to spectra of pure substances of low to moderate molecular weight that show well defined sharp peaks, spectra of not pure and complex materials such

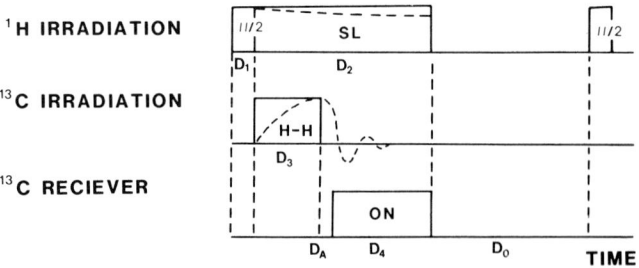

Figure 3. Timing sequence for ^1H and ^{13}C irradiations and ^{13}C observation in a CP experiment.

as lignite, coal, and bitumens are poorly resolved and show only "absorption bands" rather than peaks. It is now well established that in the absence of strong conformational effects, the isotropic chemical shift observed in solids is very close to the one observed in liquid samples (Dereppe et al., 1978). The interpretation of the ^{13}C CP-MAS spectra has been based mainly on chemical shift data reported in the literature. Aliphatic carbons resonate in the 0 to 55 ppm region, oxygen and nitrogen substituted carbons between 50 and 110 ppm, aromatic carbons from 110 to 160 ppm, carboxyl carbons between 160 and 190 ppm, and aldehydic and ketonic carbons resonate even downfield at 200 to 230 ppm.

It has generally been assumed that CP-MAS yields lineshapes with reasonably true relative intensities (Resing et al., 1978) when considering closely related samples. The observed intensity for the ^{13}C nucleï depends on the magnetization sequence. This in turn depends on the degree of equilibrium of the protons in the rotating frame (T_1). The cross polarization dynamics are further complicated by MAS (Stejskal et al., 1977). Therefore, not all carbons polarize uniformly fast and there is no guarantee that the same cross polarization parameter setting ($D_3 = 1$ mS and $D = 3$ S) will serve equally well for the various types of carbons and for all the samples. These conditions were accepted in order to obtain an acceptable S/N ratio in a reasonable period of time.

Despite the uncertainty related to the quantitative aspects of the ^{13}C CP-MAS spectra, the entire range of chemical shifts was divided as is usual in classical high resolution spectroscopy on heavy petroleum products, into several regions characteristic of various carbon types. The limits of these regions, given in Table II, are those proposed by Bartle et al. (1979) in their study of super-critical-gas extract of coal. The integrated intensities of the various bands are normalized to 100. They are listed in Table III.

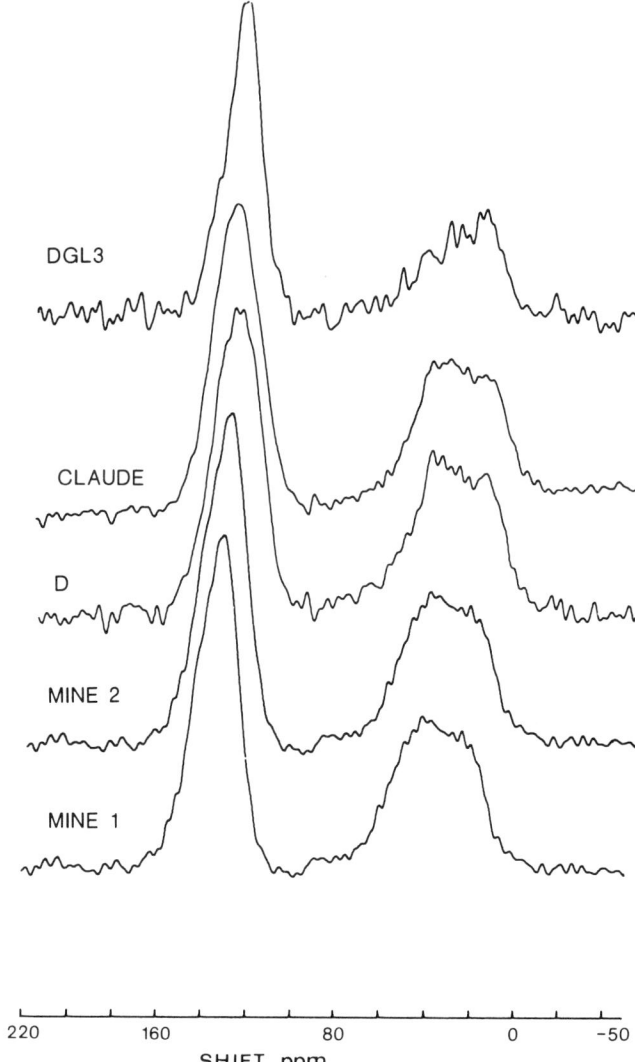

Figure 4. ^{13}C NMR spectra of the Carswell structure carbonaceous matter.

TABLE II

ASSIGNMENT OF ^{13}C NMR BANDS

Band (a)	^{13}C shift range (ppm, TMS)	Major Assignments
A_1	164 - 148	Oxygen substituted aromatic C in phenols and aromatic ethers
A_2	148 - 137	Alkyl substituted aromatic C
A_3	137 - 130	Bridgehead aromatic C
A_4	130 - 107	Unsubstituted aromatic C
S_1	107 - 53	Oxygen substituted aliphatic C in ethers, alcohols, and methoxy groups
S_2	53 - 37	Methylene bridge and bridge-head C of naphtenes
S_3	37 - 23	Methylene group C in saturated chains
S_4	23 - 0	Methyl group C

(a) The oxygen doubly bonded C in COOH, CHO and C = O groups, which appears in the 230-164 ppm range, is not observed here for reasons given in the experimental section.

TABLE III

INTEGRATED INTENSITIES OF THE VARIOUS NMR BANDS AND NMR PARAMETERS OF THE CARSWELL STRUCTURE SAMPLES

	CLAUDE	D	MINE 1	MINE 2	DGL 3
A_1	3.5	4.5	2.6	1.9	1.3
A_2	15.9	15.3	12.6	12.5	9.9
A_3	15.7	16.0	15.8	16.1	17.7
A_4	24.7	26.6	29.6	30.0	46.8
S_1	7.4	7.9	5.0	5.7	0.4
S_2	14.1	12.0	12.7	12.2	2.2
S_3	11.6	11.0	12.7	13.3	7.8
S_4	6.7	7.0	9.2	9.2	13.4
F_A	59.8	62.4	60.6	60.5	75.7
C_A	11.8	11.7	11.8	12.0	10.8
σ	0.44	0.42	0.34	0.34	0.19
β	0.26	0.29	0.36	0.36	1.34

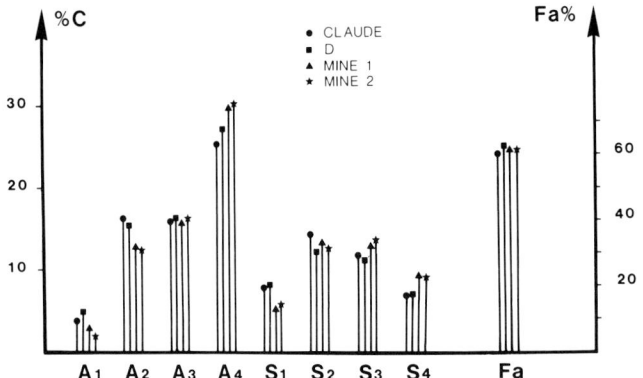

Figure 5. Histogram of the integrated intensities of the various NMR bands from the Carswell Structure carbonaceous matter.

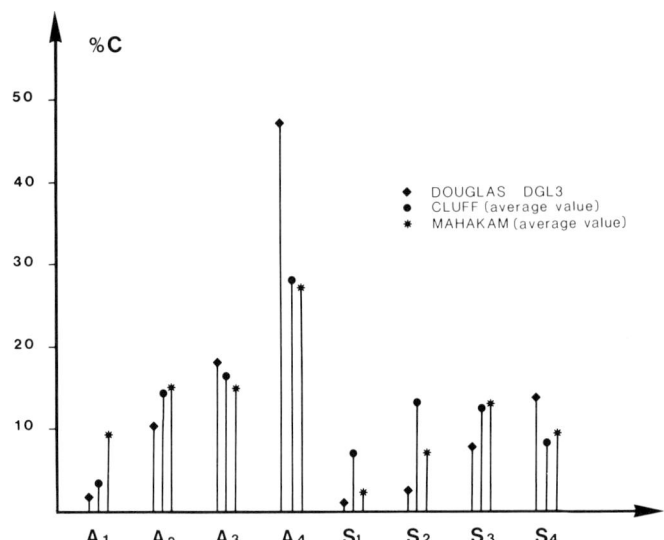

Figure 6. Histogram of the integrated intensities of the various NMR bands — Comparison between the average value of Carswell structure samples, Douglas kerogen, and the average value of Mahakam coals (Dereppe et al., 1983).

DISCUSSION

The intensity of the most important bands and the aromaticity factor F_A which is given by:

$$F_A = A_1 + A_2 + A_3 + A_4/100$$

are depicted in Figure 5.

Fossil materials such as coal, asphaltene, and bitumen are thought to consist of aromatic rings substituted by or linked with aliphatic chains and naphthenic rings combined in a three dimensional network, and the concept of molecules cannot be applied. One possible way to chemically characterize such compounds is to use an ultimate or minimum unit which can reproduce the three dimensional network by polymerization or association. The basic units may be pictured as discs or aromatic nucleï substituted by aliphatic chains and naphthenic rings. The estimation of the mean aromatic disc dimension by NMR is based on the fact that the ratio of the peripheral to the total number of aromatic carbons is clearly a function of the degree of condensation of the atomic rings. A good relation between the absolute number of aromatic carbons per disc (C_A) for hexagonal structure is given by:

$$C_A = 7 \frac{\text{(total aromatic)}^2}{\text{(peripheral)}} - 1$$

which yields, using the A_1 to A_4 notation:

$$C_A = 7 \frac{(A_1 + A_2 + A_3 + A_4)^2}{(A_1 + A_2 + A_4)} - 1$$

The degree of substitution of the aromatic rings either by oxygen or by saturated chain is given by:

$$\sigma = \frac{A_1 + A_2}{A_1 + A_2 + A_4}$$

The branching index of saturated chains which represent the methyl/methylene ratio is given by:

$$\beta = \frac{S_4}{S_2 + S_3}$$

The calculated values are given in Table III.

COMPARISON BETWEEN THE DIFFERENT CARBONACEOUS MATERIALS IN THE CARSWELL STRUCTURE

The purpose of the following NMR discussion is not to give quantitative results but to compare different samples in order to define the structure of the carbonaceous material. The values of Table III and Figures 5 and 6 are given for guidance and must be used only to compare different organic matters analyzed in the same technical conditions. It would be speculative to consider the results of the integration of the different band areas as absolute values. The classical NMR parameters such as C_A, F_A, σ, and β are used in a comparative way and the band integration is obtained by a deconvolution method which permits a better analysis of the spectra.

We have depicted the integrated intensities of the various NMR bands on a histogram (Fig. 5). One may notice that all four samples exhibit nearly the same aromaticity factor (Table III) and that the D and Claude histograms (Fig. 5) and spectral responses (Fig. 4) are rather similar. Minor but measurable differences appear between carbonaceous material sampled in mineralized (Mine 1 and Mine 2) and non-mineralized areas (Claude, D): A_4, S_3, and S_4 band intensities increase in Mine 1 and Mine 2 while A_2 and S_2 band intensities decrease for these two samples. This leads to a higher degree of substitution of aromatic rings and a lower branching index for Claude and D (Table III).

These results are consistent and could be related to radiolysis which can destroy aliphatic chains, produce CH_4 or CO_2, and diminish the substitution ratio of aromatic rings (Landais and Connan, 1980). Nevertheless, radiolysis does not seem to provoke, in this case, striking transformations in the carbonaceous network. The number of aromatic carbons (C_A) and the number of total carbons ($C = C_A/F_A$) per structural unit are similar for these carbonaceous materials.

These observations constitute evidence for the presumed similar gross chemical structure of D and Claude organic matter. Consequently, it is probable that the carbon associated with the altered areas near the unconformity must be considered as bitumen since Claude has been geologically recognized as a migrated product. The reader should note that this chemical evidence is contrary to the differences between D and Claude organic matter established by Rouzaud (1979).

COMPARISON BETWEEN THE CARSWELL STRUCTURE CARBONACEOUS MATERIAL AND A COAL SERIES FROM THE MAHAKAM DELTA

A considerable amount of work has been done these last years to provide some information about NMR studies of natural organic matter, but NMR results essentially deal with soluble fossil material. There are new measurements on solid material (Maciel et al., 1981; Pugvire et al., 1982) but to our knowledge, no data qualifying Precambrian organics is available as yet. Nevertheless, it is interesting to compare the Cluff carbonaceous material with coals presenting similar elemental compositions. A tertiary coal series from the Mahakam delta, Indonesia, studied by Boudou (1981) and Dereppe et al. (1983), was chosen for comparison. Representative points reflecting analysis of these coals were plotted on a Van Krevelen diagram. The points lie on the evolution path III (Fig. 1), indicating their land-derived origin. Although it can be speculative to compare NMR parameters of such different carbonaceous materials, one may admit that this is one possible way to characterize the Carswell structure samples. These results must be interpreted with caution even if all the analyses have been performed in the same technical conditions and on the same NMR spectrometer.

Integrated intensities of NMR bands and NMR parameters of the coals are listed on Table IV. The aromaticity factors of the reference coals range from 58.4% to 71.1%. The mean aromaticity factor of Carswell structure materials, at 60%, lies within this range. Average NMR data for Mahakam coals and Carswell structure samples are depicted on a histogram (Fig. 6). One can notice that the divisions of the integrated band values are quite analogous except for the A_1, S_1, and S_2 bands. The A_1 band represents the oxygen substituted aromatic carbon while the S_1 band represents the oxygen substituted aliphatic carbon.

The A_1/S_1 ratio yields markedly different results: 4.2 for the coals and 0.48 for the Carswell structure samples. This clearly implies that the chemical distribution of oxygen in the carbonaceous structure from the Carswell area is different from that in the coal structure. This difference may be attributed to an oxidation process affecting the aliphatic carbons. Detailed inspection of Tables III and IV reveals that if we consider coal No. 28581 which has nearly the same aromaticity factor as Claude, alkyl substituted and bridgehead aromatic carbons are less important in the coal. Furthermore, the methyl/methylene ratio is lower in the Carswell structure samples. This suggests the presence of longer aliphatic chains in the Precambrian material.

A rough comparison of F_A, C_A, and average values indicate that the aromaticity of the Carswell structure samples is weak and corresponds to the aromaticity of a tertiary land-derived coal buried at less than 3000 m. This is corroborated by similar Rock-Eval T_{max} values for both types of samples: 430° to 450° C for coals, 435° to 442° C for the Carswell structure samples.

Careful spectral examination of A_1/S_1, σ, β, S_2, and S_4 suggests that the Carswell structure carbonaceous material is not syngenetic organic matter and that an oxidation process has led to the alteration of the aliphatic chains.

From the data given here, specifically concerning the methyl/methylene ratio (β), it is possible that the Carswell structure bitumens initially were made up most often by aliphatic chains. Effectively, β is less important in the Carswell structure carbonaceous matter ($0.26 < \beta < 0.36$) than in the Mahakam coals ($0.40 < \beta < 0.52$). During thermal alteration, aromaticity increased and reached the present stage. The apparent maturity deduced from the comparison with the Type III coal is weak if one considers the geological history of the Athabasca sandstones, but is reasonable according to a low initial aromaticity of the carbonaceous products. Type I organic matter, algal derived organic matter, or bitumens are initially less aromatic than Type III organic matter and need a higher thermal alteration to reach the same aromaticity factor as Type III organic matter. The range of T_{max} values for Carswell structure carbonaceous matter and investigated Mahakam coals is comparable, i.e., 430° to 445°C. This suggests a higher maturity for the Carswell structure carbonaceous matter.

An F_A value of 60% could be consistent with the temperatures indicated by fluid inclusions and geological reconstitution in the Carswell structure (Pagel, 1975) if one admits the high initial aliphaticity of the Carswell structure carbonaceous matter.

The bitumens also have been diagenetically oxidized. The amount of oxygen cannot be attributed to mineral oxygen (low ash content, see Table I) or water oxygen (the samples are dried for 24 hours before analysis). This oxidation produces the shifting of the samples from evolution path I to evolution path III in the Van Krevelen diagram (Fig. 1) and contributes to the length reduction of the aliphatic chains.

GENETIC PROPOSALS FOR THE CARSWELL STRUCTURE CARBONACEOUS MATERIAL

It has previously been established that all of the Carswell structure material studied here is likely to be the oxidized remains of migrated aliphatic products. There is still a problem concerning the origin of these bitumens.

The succession of main steps in the evolution of organic matter (diagenesis, catagenesis, and metagenesis) leads to the generation of liquid and gaseous products. Oil or bitumens are generated during catagenesis, the temperature of maximum oil generation ranging from 70°C to 120°C (Tissot and Welte, 1978). However, the amounts of hydrocarbons and their composition depend on parameters such as the nature of the organic matter and the temperature versus time relationship. A correlation between gas or oil and a source rock must be established to reconstitute the geochemical process.

TABLE IV

ATOMIC RATIOS, INTEGRATED INTENSITIES OF THE VARIOUS NMR BANDS AND NMR PARAMETERS OF COALS FROM THE MAHAKAM DELTA (DEREPPE ET AL., 1983)

Samples	28875	28907	28581	29188	29194	29201
A_1	11	9.1	9.6	8	9.6	6.1
A_2	13	16.3	11.8	15.6	14.9	15.9
A_3	12.3	16.8	11.8	16.0	13.8	16.6
A_4	22.1	27.2	25.6	26.6	28.0	32.5
S_1	3.2	0.1	4	1.1	3.4	1
S_2	9.6	5.7	8.3	7.7	6.4	4.9
S_3	15.6	9.8	15.3	13.2	12.9	11.4
S_4	7.7	9.6	10.3	9.4	8.7	10.9
F_A	58.4	69.4	58.8	66.2	66.3	71.1
C_A	10.2	11.2	10	11.2	10.2	10.9
σ	0.31	0.62	0.44	0.45	0.45	0.67
β	0.52	0.48	0.46	0.47	0.47	0.40
H/C at	0.89	0.77	0.87	0.86	0.91	0.75
O/C at	0.13	0.12	0.12	0.07	0.08	0.05

In uranium deposits like Oklo or Lodève, France, a good source rock-petroleum relationship has been demonstrated and evidence for oil and gas formation has been shown (Vandenbrouke et al., 1978; Landais and Connan, 1980). The real problem in the Carswell structure is to find a source rock for the bitumens located near the unconformity.

Several hypotheses must be taken into account. Organic matter from the Douglas Formation dolomites may have generated bitumens, or it may have been generated from pelitic rocks situated near the unconformity. There may have been a secondary migration of heavy oils from the Athabasca tar sands in western Canada, or possibly a generation of hydrocarbons during the alteration of graphite in the basement.

The Douglas Formation Dolomites

NMR measurements and elemental analyses have been performed on the kerogen extracted from a sample of the Douglas Formation (DGL3). Contrary to the results found by Bonnamy (1981), the Douglas kerogen is not highly oxygenated; its O/C atomic ratio is only 0.075 and it can be considered as a Type I kerogen (Fig. 1). Solid state NMR results are shown in Figure 6 and Table III, and the DGL3 NMR spectrum is depicted in Figure 4. One should note that in Figure 4, the aliphatic carbon chemical shift region at 0 to 100 ppm is less important in the DGL3 spectrum than in the other Carswell structure sample spectra. This leads to a higher aromaticity factor of 75.7% which indicates a considerable burial for the Douglas Formation of more than 3000 m, which is consistent with fluid inclusion results (Pagel, 1975) and the illite crystallinity (Hoeve et al., 1981). Thus, one should consider that during their thermal alteration, the Douglas kerogens may have generated liquid and gaseous hydrocarbons as in any other Type I organic matter undergoing catagenesis. As far as geological patterns are concerned, it seems possible that an oil migration from the Douglas Formation dolomites through 1000 m of Athabasca sandstones to reach the unconformity could have occurred at the beginning of the catagenesis of the Douglas kerogen. No occurrence of bitumens has been noted in the sandstone pile, but the alteration process affecting the Athabasca sandstones is also restricted to the vicinity of the unconformity.

If the brines which are responsible for the alteration of the sandstones and the basement are diagenetic brines originating in the sandstones, one can assume that they have transported the Douglas hydrocarbons through the sandstones to the unconformity. Furthermore, the alteration process affecting both sandstones and basement which involved the meeting of the diagenetic brines and waters originating in the basement, could create a disequilibrium which is able to provoke the precipitation of the hydrocarbons. This phenomenon could explain the close association between altered areas and carbonaceous material.

It should be noted that the Athabasca diagenetic brines are oxidizing brines which are not favourable to the transport of hydrocarbons. Nevertheless, the hypothesis of a bitumen-source rock relationship between the Douglas Formation and the carbonaceous material, such as occurs in Claude and D, cannot be discounted. Solid state NMR ^{13}C analysis has shown that the Douglas kerogens have reached the catagenesis zone and that they have probably both lost and generated hydrocarbons.

A Precambrian Source Rock

It does not seem reasonable to consider a generation of hydrocarbon from Precambrian pelitic rocks located near the unconformity. Neither black shales nor organic carbonates have been found in this area. We have previously shown that the "coals" of the D zone described by Rouzaud (1979) are in fact bitumens associated with the argillaceous alteration. If these pelitic rocks responsible for the generation of hydrocarbon had been destroyed, one should observe that their break down products such as liquid hydrocarbons would have also been destroyed.

The Athabasca Sandstones

A secondary migration from the Athabasca tar sands of western Canada is not geologically unreasonable. The major Cretaceous-Paleozoic unconformities provide favourable migration paths which are available for oils. However, the Athabasca oils are younger than the mineralization in the Athabasca deposits (Bellon et al., 1976; Gancarz, 1979; Bell, 1985) and they never present geochemical characteristics which can be compared with the Carswell structure bitumens. Athabasca oils are asphaltic, viscous, highly degraded, and sulphur rich. These materials do not correspond to the geochemistry of the Carswell bitumens. Furthermore, it would be a real coincidence if these heavy oils had chosen only the altered areas as reservoirs.

Gneissic Basement Graphitoids

The last hypothesis for the genesis of the Carswell structure bitumens deals with the occurrence of carbonaceous material in the gneissic basement. Petrography has defined this material as graphite. Elemental analysis performed on moderate and strongly altered facies of the basement has shown that the gross composition of the carbonaceous material is not purely graphite. It is either severely, but not completely, metamorphosed organic matter or graphitoids, or a mixture of graphite and another type of carbonaceous material. Analytical results reveal oxygen and hydrogen content ranging from 3% to 8% and 1% to 1.5%, respectively, for low ash content (especially pyrite) carbon concentrates. This clearly indicates that the carbonaceous material of the basement surrounding altered areas has been affected by hydrothermal solutions. Two hypotheses can be considered: bitumen impregnations have mixed with the graphite of the basement, or the original carbonaceous material of the basement has been partly transformed during the alteration process.

One should note that bitumens often occur in the altered areas located near the most graphitic zones of the basement. Wherever the alteration phenomenon occurs, the basement can always be considered as a source for solutions or mineral phases (Pagel, 1975) and one can assume that the carbonaceous material in the basement has also been a source for the formation of the bitumen during the alteration. There is no chemical evidence for such a process, but the close mixing of bitumens and altered areas from where the graphitoids have disappeared, fully supports the hypothesis of a common phenomenon for the transformation of the rocks and the precipitation of hydrocarbons. This process must be taken into account as long as a ^{13}C NMR analysis has not been successfully performed on basement carbon. It has been impossible, to the present time, to detect NMR spectra on the material that has been strongly dehydrogenated to yield good polarization of the ^{13}C atoms.

Investigations into the carbon isotopic composition of the Carswell structure carbonaceous material have indicated strong negative ^{13}C values: -48.5 to -48.8 ± 0.1 per mil relative to PDB (analysed by J. Rosenbaum and L. Turpin, Laboratoire de Géochimie Isotopique, CRPG, Nancy). Even for Precambrian kerogens, these values are too negative to correspond to values from sedimentary organic matter. This high enrichment in light isotopic ^{12}C suggests that the carbonaceous matter of the Carswell structure deposits is derived from rich C-H linkage hydrocarbons which contain relatively light carbon. One should note that these values are not fully consistent with the results compiled by Button and Adams (1981) concerning carbon of the Precambrian quartz pebble conglomerates but correlate well with the results of Grauch et al. (in press) on the organic material and entrapped methane of the Carswell structure. These results suggest that the source of the carbonaceous matter of the D and Claude deposits may be the graphitoids of the gneissic basement and that these bitumens could be formed by the condensation of light and complex hydrocarbons enriched in ^{12}C isotopes. The disappearance of graphitoids in the alteration zones noted by Harper (1983) and confirmed in this work must also be taken into account, and favours this genetic process.

CONCLUSION

The carbonaceous material associated with altered zones of the basement and with the Athabasca sandstones is bitumen, migrated products generated during diagenesis. Chemical structures define a low aromaticity factor implying a high initial aliphaticity, if one considers the temperatures found by the fluid inclusion study (Pagel, 1975). The absence of real source rocks in the Athabasca sandstones suggests that their formation is likely to be linked to the alteration process.

Two carbon sources can be contemplated: the organic dolomites of the Douglas Formation and the carbon in the gneissic basement.

The close relationships between uranium and bitumen demonstrated by Rouzaud (1979) could result from the oxidation affecting the carbonaceous material. The hypothesis of a syngenetic pre-enrichment or complexing of uranium by sedimentary organic matter (Rouzaud, 1979; Bonnamy, 1981) does not seem to be realistic. The uranium-bitumen interaction must be diagenetic, and it is possible to consider that the formation of the bitumen is contemporaneous with the mineralization. Finally, aliphatic structures are very efficient during the oxidation-reduction process and especially in the reduction of uranium-carrying solutions.

ACKNOWLEDGEMENTS

The authors want to thank Jeffrey Rosenbaum and Laurent Turpin for performing the isotopic investigations and J. Connan from SNEA(P) for the carbon ratio and chloroform extract data. The authors also gratefully acknowledge Amok Ltd. for providing the bitumen samples, the Centre de Recherches sur la Géologie de l'Uranium (Nancy) for their support, and M. Pagel for his scientific assistance.

REFERENCES

Alpern, B., 1978, Etude Pétrographique de la Matière Organique d'Oklo: in "Les Réacteurs de Fission Naturels", TC 119/11: International Atomic Energy Agency, Vienna, p. 333-351.

Bartle, D.D., Martin, T.G., and Williams, D.F., 1979, Structural Analysis of Supercritical Gas Extracts of Coals: Fuel, v. 58, p. 143.

Bell, K., 1985, Geochronology of the Carswell Area, Northern Saskatchewan: in Lainé, R., Alonso, D., and Svab, M., eds., The Carswell Structure Uranium Deposits, Saskatchewan: Geological Association of Canada Special Paper 29.

Bellon, H., Devillers, C., Hagemann, R., et Touray, J.C., 1976, Dater les Minéralisations: Mémoire Hors Série de la Société Géologique de France no. 7, p. 265-268.

Bonnamy, S., 1981, Relations Matières Organiques – Métaux (U-V-Ni): Etude en Microscopie et Diffraction Electronique: Thèse 3è Cycle, Université d'Orléans, 82 p.

Bonnamy, S., Oberlin, A., and Trichet, J., 1982, Two Examples of Uranium Associated with Organic Matter: Organic Geochemistry, v. 4, p. 53-61.

Boudou, J.P., 1981, Diagénèse Organique de Sédiments Deltaïques (Delta de la Mahakam, Indonésie): Thèse de Doctorat, Université d'Orléans, 196 p.

Button, A., and Adams, S.S., 1981, Geology and Recognition Criteria for Uranium Deposits of the Quartz-Pebble Conglomerate Type: United States Department of Energy Report (GJBX-3(81), 389 p.

Cassou, A.M., Connan, J., Correia, M., et Orgeval, J.J., 1975, Etudes Chimiques et Observations Microscopiques de la Matière Organique de Quelques Minéralisations Uranifères: in "The Oklo Phenomenon", SM 204-19: International Atomic Energy Agency, Vienna, p. 195-206.

Dereppe, J.M., Moreau, C., and Castex, H., 1978, Analysis of Asphaltenes by Carbon and Proton Nuclear Magnetic Resonance Spectroscopy: Fuel, v. 57, p. 435.

Dereppe, J.M., Boudou, J.P., Moreau C., and Durand, B., 1983, Structural Evolution of a Sedimentologically Homogeneous Coal Series as a Function of Carbon Content by Solid State ^{13}C NMR: Fuel, v. 62, p. 375.

Ellsworth, H.V., 1928, Thucholite – A Remarkable Primary Carbon Mineral From the Vicinity of Parry Sound, Ontario: American Mineralogist, v. 13, p. 419-441.

Gancarz, A.J., 1979, Chronology of the Cluff Lake Uranium Deposit, Canada (Abstract): International Uranium Symposium on the Pine Creek Geosyncline, N.T., Australia, Extended Abstracts, p. 91-94.

Gransch, J.A., and Eisma, E., 1970, Characterization of the Insoluble Organic Matter of Sediments by Pyrolysis: in Hobson, G.D. and Speers, G.C., eds., Advances in Organic Geochemistry, 1966: New York, Pergamon, p. 407-426.

Grauch, R.I., Leventhal, J.S., and Harper, C.T., in press, Organic Material from the Cluff Lake (Canada) Uranium District – Its Nature and Paragenetic Implications of the Included Mineral Assemblages: 27th International Geological Congress, Moscow (1984), Volume IX, Part 1, p. 351-352.

Harper, C.T., 1983, The Geology and Uranium Deposits of the Central Part of the Carswell Structure Northern Saskatchewan, Canada: Unpublished Ph.D. Thesis, Colorado School of Mines, Golden, Colorado, 337 p.

Hoeve, J., Cumming, G.L. Baadsgaard, H., and Morton, R.D., 1981, Geochronology of Uranium Metallogenesis in Saskatchewan: Canadian Institute of Mining Uranium Symposium, September 8-13, 1981, Saskatoon, Technical Program, p. 11-12.

Hunt, J.M., 1979, Petroleum Geochemistry and Geology: New York, W.H. Freeman and Co., 617 p.

Landais, P., and Connan, J., 1980, Relations Uranium – Matière Organique dans Deux Bassins Permiens Français: Lodève (Hérault) et Cerilly-Bourbon l'Archambault (Allier): Bulletin du Centre de Recherches SNEA(P), v. 4, no. 2, p. 709.

Maciel, G.E., and Dennis, L.W., 1981, Comparison of Oil Shales and Kerogen Concentrates by ^{13}C Nuclear Magnetic Resonance: Organic Geochemistry, v. 3, no. 4, p. 105.

Maciel, G.E., Sullivan, M.J., Szeverenyi, N.M., and Miknis, F.P., 1981., C-13 NMR on Solid Samples and its Application to Coal Science: in Cooper, B.R., and Petrakis, L., eds., American Institute of Physics Conference Proceedings, Chemistry and Physics of Coal Utilization, 1980, no. 70, p. 66-81.

Melchior, M.T., Rose, K.D., and Miknis, F.P., 1980, Artefacts in Magic Angle Spinning NMR: Comments on Potential of Proton Enhanced ^{13}C NMR for the Classification of Kerogens: Fuel, v. 59, p. 594.

Minter, W.E.L., 1978, A Sedimentological Synthesis of Placer Gold, Uranium and Pyrite Concentrations in Witwatersrand Sediments: Canadian Society of Petroleum Geologists Memoir 5, p. 801-829.

Pagel, M., 1975, Cadre Géologique des Gisements d'Uranium dans la Structure Carswell (Saskatchewan, Canada): Etude des Phases Fluides: 3è Cycle Docteur de Spécialité, Université de Nancy, France, 157 p.

Pines, A., Gibby, M.G., and Waugh, J.S., 1973, Proton Enhanced NMR of Dilute Spins in Solids: Journal of Chemistry and Physics, v. 59, p. 569.

Pugmire, R.J., Zilm, K.W., Woolfenden, W.R., Grant, D.M., Dyrkacz, G.R., Bloomquist, C.A.A., and Horwitz, E.P., 1982, Characterization of Dissolved Organic Materials in Surface Waters within the Blast Zone of Mount St. Helens, Washington: Organic Geochemistry, v. 4, no. 2, p. 85.

Resing, H.A., Garroway, A.N., and Hazlett, R.N., 1978, Determination of Aromatic Hydrocarbon Fraction in Oil Shale by 13C NMR with Magic Angle Spinning: Fuel, v. 37, p. 450.

Rouzaud, J.M., 1979, Etude Structurale de Matières Carbonées Associées à des Minéralisations Uranifères: Thèse de Spécialité, Université d'Orléans, 90 p.

Schaefer. J., Stejskal, E.O., and Buchdahl, R., 1977, Magic Angle ^{13}C NMR Analysis of Motion in Solid Glossy Polymers: Macromolecules, v. 10, p. 384.

Schopf, J.W., Haugh, B.N., Molmar, R.E., and Satterthwaitt, D.F., 1973, On the Development of Metaphytes and Metazoans: Journal of Paleontology, v. 47, no. 1, p. 1.

Stejskal, E.O., and Schaefer, J., 1975, Removal of Artefacts from Cross Polarization NMR Experiments: Journal of Magnetic Resonance, v. 18, p. 560.

Stejskal, E.O., Schaefer, J., and Waugh, J.S., 1977, Magic Angle Spinning and Polarization Transfer in Proton Enhanced NMR: Journal of Magnetic Resonance, v. 28, p. 105.

Tissot, B.P. and Welte, D.H., 1978, Petroleum Formation and Occurrence: New York, Springer-Verlag, 538 p.

Vandenbroucke, M., Rouzaud, J.N., and Oberlin, A., 1978, Etude Géochimique de la Matière Organique Insoluble (Kérogène) du Minérais Uranifère d'Oklo et des Schistes Apparentés du Francevillien: in "Les Réacteurs de Fission Naturels", TC-119/10: International Atomic Energy Agency, Vienna, p. 307-332.

EXPLORATION IN THE CARSWELL STRUCTURE

J.S. Wilson

Amok Ltd., 817-825 45th St. W., P.O. Box 9204, Saskatoon, Saskatchewan S7K 3X5

ABSTRACT

Geochemical prospecting was first applied to the Carswell structure to investigate airborne radiometric anomalies and mineralized boulders. A regional survey of organic sediments outlined uraniferous zones, some of which defined known areas of mineralized bedrock whereas others were new prospects. The anomalies were the expression of the composition of the underlying glacial drift. The need to extend the level of investigation from the surface into the drift led to a program of overburden sampling to develop a three-dimensional image of the geochemical pattern. Overburden sampling by mobile drill succeeded in finding buried mineralization by detecting a primary mechanical dispersion in the basal till and tracing it to a bedrock source. The dispersion of mineralized bedrock extends only a short distance from the source. It is strongest in the basal overburden immediately above the bedrock and becomes more diffuse with increasing distance from the source.

In general, areas of basal till can be expected to give the simplest type of glacial dispersion, whereas areas with more complex stratigraphy and history will present a complex dispersion pattern. The landform/sediment association of the Cluff Lake area suggests a late phase glacial re-advance that developed the Cluff Lake Moraine. Mineralized boulder trains in this area were investigated unsuccessfully by drift geochemistry. They are unusual in that, although they define a trend which roughly parallels the last glacial movement, they are not associated with a mineralized dispersion within the drift. The relationship between these boulder trains and the Cluff Lake Moraine is important and suggests that they represent supraglacial deposits. Basal material was elevated by thrusting at the ice margin and developed into a cover of ablation deposits.

RÉSUMÉ

L'échantillonage des sédiments organiques dans la région de Carswell avait pour but de faire ressortir les zones potentiellement favorables à des minéralisations uranifères. Le choix des sédiments organiques était basé sur leur propriété de fixation des éléments traces mobiles dans les eaux superficielles et de leur abondance et répartition homogène sur l'ensemble de la région. La campagne stratégique confirma les zones minéralisées et souligna plusieurs autres zones anomales.

Les anomalies géochimiques des sédiments organiques reflettent la composition du recouvrement glaciaire sousjacent. Le besoin d'étendre la profondeur d'investigation depuis la zone superficielle au glaciaire sousjacent a nécessité un échantillonage détaillé tri-dimensionnel des anomalies. L'échantillonnage par une sondeuse percutante à gros diamètre, fut mis au point pour l'étude systématique du recouvrement et de la roche saine. A Claude ouest la technique permit de mettre en évidence une minéralisation cachée et de démontrer l'éfficacité de la prospection géochimique du till: une anomalie diffuse developpée dans le till de base sur une distance de seulement 150 m en aval de la source avec la plus forte anomalie se retrouvant à la base du recouvrement immédiatement au dessus de l'affleurement; l'anomalie devenant plus diffuse à mesure que l'on s'éloigne de la source et que l'on monte dans le recouvrement.

En général, les zones de till basal devraient donner des halos de dispersion les plus simples, alors que les zones stratigraphiquement plus complexe donneront un halo de dispersion glaciaire plus compliqué. La morphologie dominante du relief dans la région de Carswell est celle d'une alternance entre les plaines à moraine en "flutes" reposant sur un recouvrement glaciaire et les plaines de sédiments fluvio-glaciaires. L'association morphologie-sediments dans la région de Cluff suggère une phase tardive avec nouvelle avance de la glace formant ainsi une moraine frontale locale: la moraine de Cluff Lake. La formation de la moraine et la stratigraphie locale du glaciaire jouent un rôle important dans la constitution des trains de galets: tel celui de Donna. La distribution superficielle des galets mineralisés, selon une direction parallèle à la direction d'écoulement de la glace, est singulière du fait de l'absence d'annomalies géochimiques associées. L'environnement de dépôt et la tectonique glaciaire de la bordure du glacier, créèrent des dépôts supraglaciaires, composés essentiellement de débris de fond couvrant la surface de la glace. Le train de galets en bordure du glacier est caractéristique de dépôts supraglaciaires, et est alors sans liaison avec le recouvrement sur lequel il finit par se poser. La compréhension de l'environnement du dépôt glaciaire est la clé de l'interprétation des trains de galets et des anomalies géochimiques superficielles dans les régions à terrains glaciaires.

INTRODUCTION

This paper presents a history of the role and evolution of geochemistry applied to uranium exploration in the Carswell circular structure, northern Saskatchewan. Exploration in the area was initiated by an airborne radiometric survey flown in 1967 and was intensified when pitchblende cobbles were found near Cluff Lake the following year. Initially, geochemical prospecting was designed to find and assess surface anomalies. With increased levels of investigation came more specific prospecting techniques that eventually led to the discovery of a uranium showing beneath a cover of glacial drift.

Drift prospecting relies on the development of a mechanical dispersion of mineralized bedrock. An understanding of the glacial history and processes of glacial dispersion are the key to the interpretation of boulder trains and geochemical anomalies in the Carswell area.

PHYSIOGRAPHY

The Carswell area lies within the Athabasca Plain (Acton et al., 1960) – a physiographic region underlain by the Athabasca Group sandstones. Relief is gentle with a regional slope towards the northwest. The most prominent topographic feature is the ring of dolomite outcrops, forming cliffs up to 70 m high, which define the outer limits of the Carswell structure (Currie, 1969). Outcrops occupy about 1% of the surface, and commonly there is a relatively continuous cover of glacial deposits averaging 5 m in thickness, locally reaching thicknesses of up to 100 m. The landscape is dominated by a drumlinized till plain of low relief, interspersed with glaciofluvial deposits. The consistent orientation of the streamlined forms indicates a final ice flow direction from the northeast. The surface drainage network is strongly controlled by the linear landforms and bedrock structure. The gentle topography results in poor drainage and the development of wetlands. The area is forested mainly by coniferous with subordinate deciduous trees. Podzolic soils are developed throughout the area. The degree of their development depends upon the composition of the overburden and local environment. The area lies within the zone of discontinuous permafrost (Brown, 1967). Ice encountered at shallow depths in mineral and organic terrain in the early summer has all but disappeared by late summer.

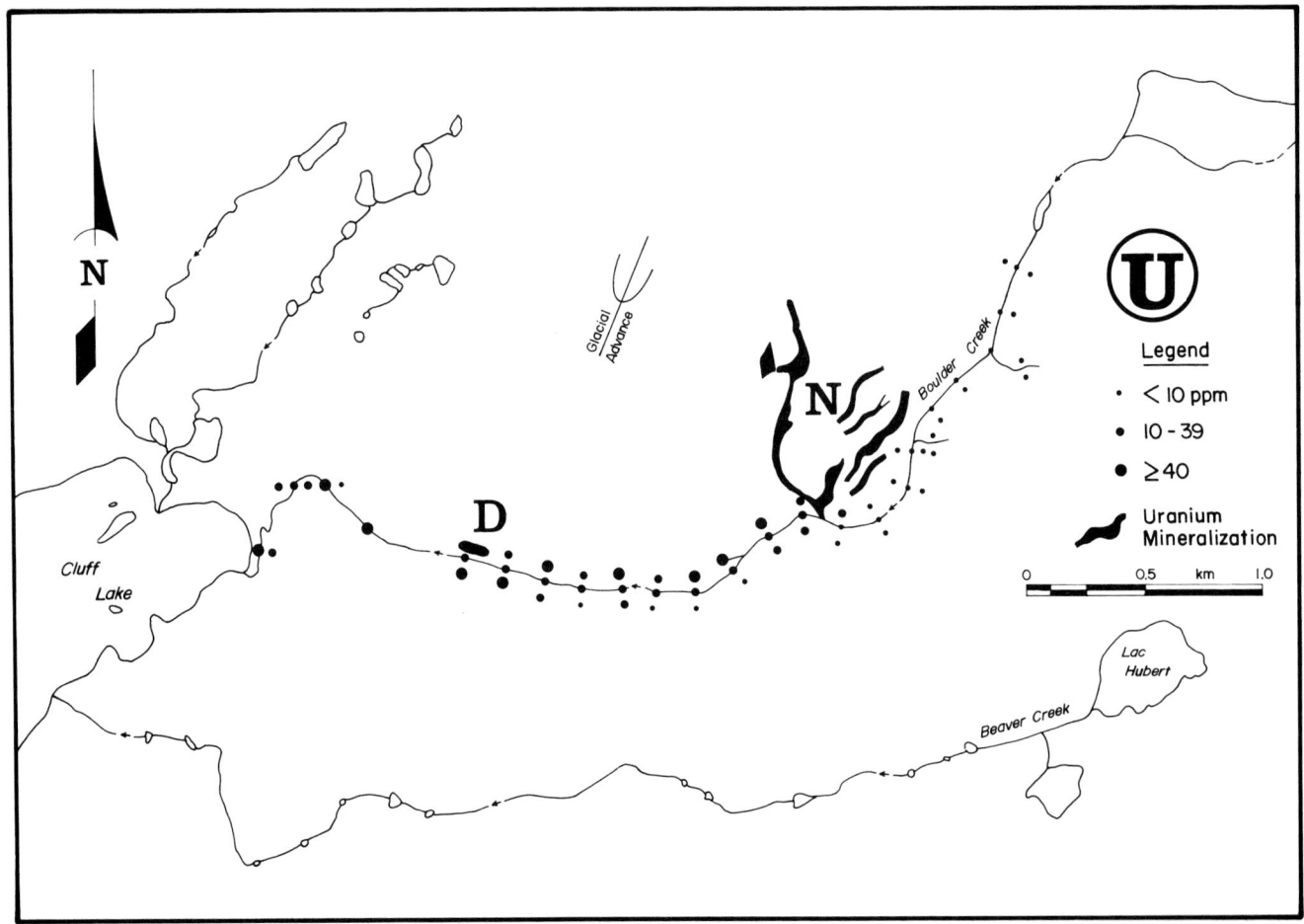

Figure 1. U anomalies in stream sediments along Boulder Creek, Cluff Lake area.

SOIL AND HUMUS GEOCHEMISTRY

The earliest geochemical surveys in the Cluff Lake area were conducted to investigate airborne radiometric anomalies and pitchblende boulders. Throughout the paper, the word "boulder" will be used in place of the more appropriate term "clast". In this context, a boulder defines a mineralized rock fragment and does not relate to size. The first surveys consisted of more or less random sampling of stream sediments along Boulder Creek, the site of the initial boulder discoveries. By the time organized geochemical programs were introduced, the main mineralizations of the area (the D and N deposits) had been found. Nevertheless, the surveys revealed that uranium anomalies had developed downstream from the N area (Fig. 1) and extended at least to the D zone. The stream sediment survey was expanded to investigate radiometric anomalies and mineralized boulders elsewhere in the Carswell structure. Unfortunately, there was no consistency in sample collection and both mineral and organic sediments were included, making the interpretation of results difficult. As a result, the survey was discontinued. Despite these difficulties, uranium anomaly areas were identified for further prospecting.

It was considered essential to continue with a regional geochemical survey of the Carswell structure to eliminate unfavourable areas. For the survey to be effective, it was necessary to select a suitable sampling material. The sampling medium had to be easily recognized and collected in the field and occur uniformly throughout the area. It was also important that its chemical composition be representative of the elemental abundances of the surrounding mineral terrain. For these reasons, organic sediments were considered the most suitable.

Before undertaking a massive sampling program of organic sediments, an orientation survey was conducted to evaluate organic sampling as an exploration tool in the area (Granier et al, 1977). Samples were collected at various depths along a profile through a shallow swamp and analyzed for uranium and lead (Fig. 2). The resulting pattern indicated a wide uranium dispersion extending into the swamp from the margin, with maximum values at depth. This pattern was attributed to the migration of uranium from the adjacent mineral terrain into the swamp, where it was captured by the organic matter. The lead distribution was much less extensive and was restricted to the margin of the swamp. This partitioning

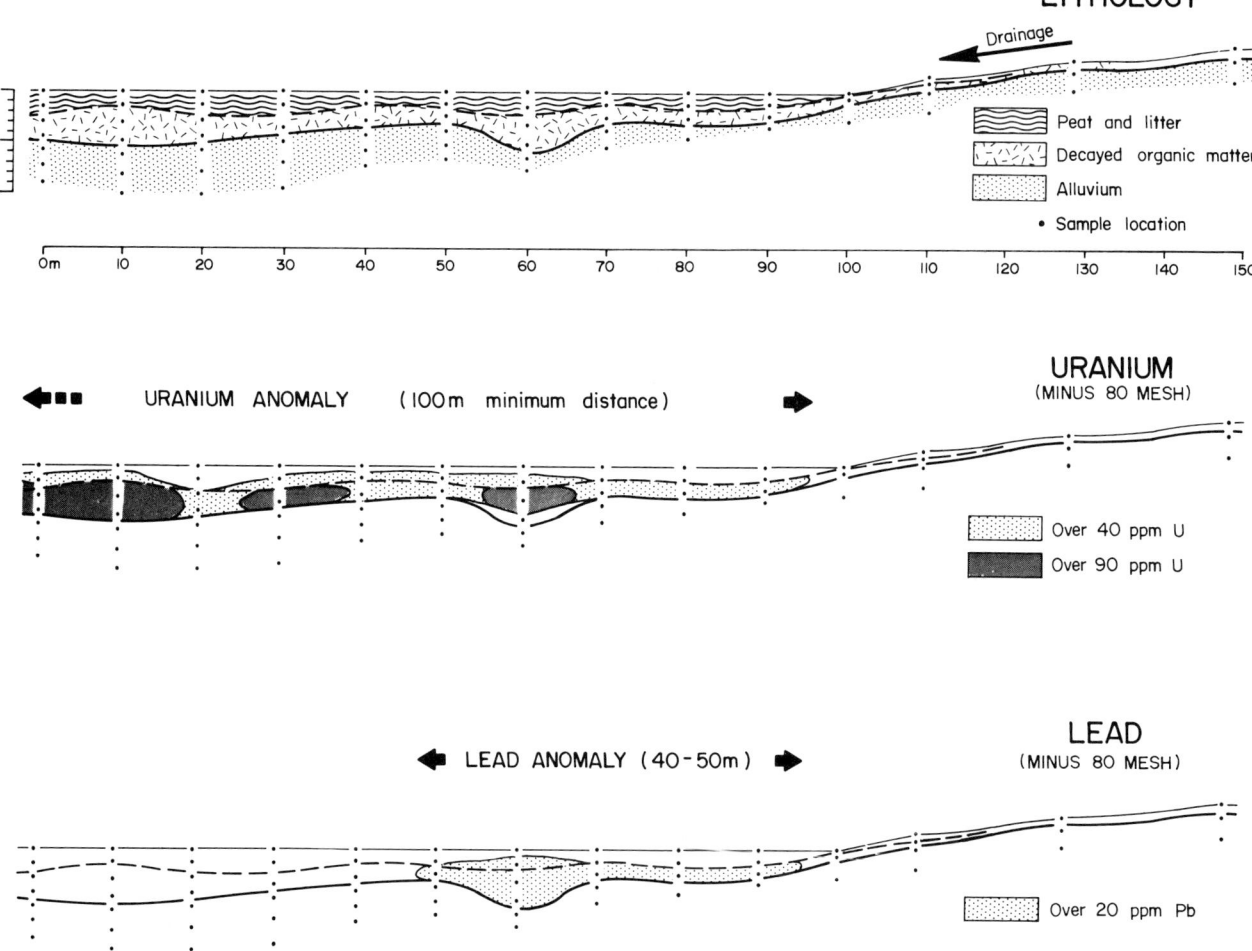

Figure 2. Comparison of U and Pb anomalies developed in organic sediments from a nearby source, Cluff Lake area (modified from Granier et al., 1977).

of uranium and lead anomalies was due to contrasting geochemical mobilities, where both originated from a common nearby source. From this study, it was concluded that sampling along swamp margins would be adequate to detect locally mineralized overburden or bedrock and that samples from the organic-inorganic interface would be of greatest value.

Several methods of sample preparation were considered and ashing was found to be the most satisfactory (Monteil, 1974). The samples were air dried and then sent to a commercial laboratory for analysis. All samples were disaggregated and sieved to minus 80 mesh (180 μm), then ashed at 500°C for 3 hours. Following a hot nitric acid extraction, uranium was analyzed fluorimetrically and lead by atomic absorption.

Regional Geochemical Survey

The search for unconformity and basement type mineralization restricted the regional geochemical survey to the basement core of the Carswell structure (see map in pocket). The survey covered an area of about 360 km² from which nearly 3000 samples of organic sediment were collected.

The regional uranium distribution outlined several anomalous zones (Fig. 3). In the Cluff Lake area, the mineralization of D and N is enclosed in a wide zone which includes areas to the north where bedrock outcrops with minor mineralization are found. The zone extends a short distance southward to the area underlain by Athabasca sandstones. The Claude orebody, on the other hand, produces only a very localized surficial anomaly. A distinct pattern of broad uraniferous zones continues northwest to the edge of the basement core. A high concentration of anomalies appears along the northwestern edge of the basement core, where values reach up to 1000 ppm uranium. These values are higher than those of the Cluff Lake area, but no mineralization has yet been found.

A striking feature of the regional pattern is the lack of uranium anomalies on the northeast side of the Carswell structure compared to the southwest (Fig. 3). This is probably related to the structure and lithology of the basement core. A major northwest to southeast trending zone of magnetic anomalies divides the core into two general lithological units (Tona et al., 1985). The northeast part consists predominantly of mixed feldspathic and mafic gneisses, often enriched in thorium. The southwest part consists mainly of aluminous gneisses, in some places uraniferous.

The regional geochemical survey was designed to outline the pattern of uranium distribution and to identify areas of mineral potential. It revealed several areas of interest, some of which were known mineralized zones while others were new prospects. The existence of anomalies in the northwest which are equal to or greater in extent and magnitude than those of the Cluff Lake area holds promise for significant potential mineralization elsewhere in the Carswell structure.

Since the landscape is not homogeneous with respect to element concentrations, it was difficult to assess the significance of the surficial anomalies without knowledge of the nature and composition of the underlying drift and bedrock.

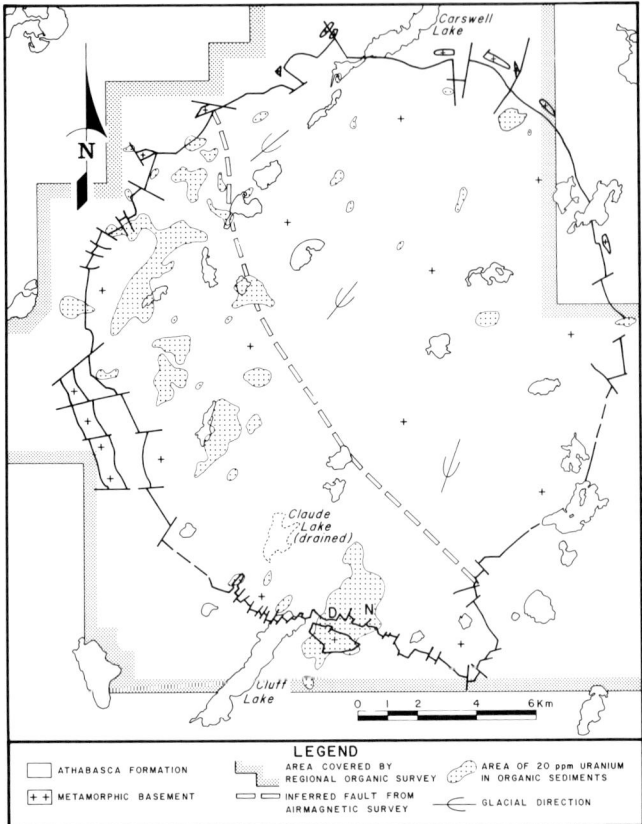

Figure 3. Regional U distribution in organic sediments of the Carswell structure.

The apparent northeast to southwest trend of most of the anomalous areas suggests that these areas were probably related to glacially transported overburden. At the regional level, therefore, the geochemical anomalies may be considered to be an expression of the composition of the underlying drift and to reflect a glacial dispersion. In order to explain the anomalies, it was necessary to extend the investigation to the underlying drift.

DRIFT GEOCHEMISTRY

During glaciation, material is eroded from the glacier bed and transported in the direction of ice movement. The resulting distribution of debris, formed by the continuous diffusion of material, is much like a plume of smoke from a chimney and defines a dispersion train. Where mineralized bedrock is eroded, the distribution of particles produces a mineralized dispersion consisting of ore fragments mixed with finely ground debris. In general, the large clasts appear close to the source and the fine particles dominate farther away (Dreimanis and Vagners, 1971). The dispersion envelope originates at the bedrock source, spreading out laterally and climbing upwards in the drift, with increasing distance from the source in a down-glacier direction. Eventually, mineralized debris will appear at the surface where it can be detected by surface prospecting. In order to relate the

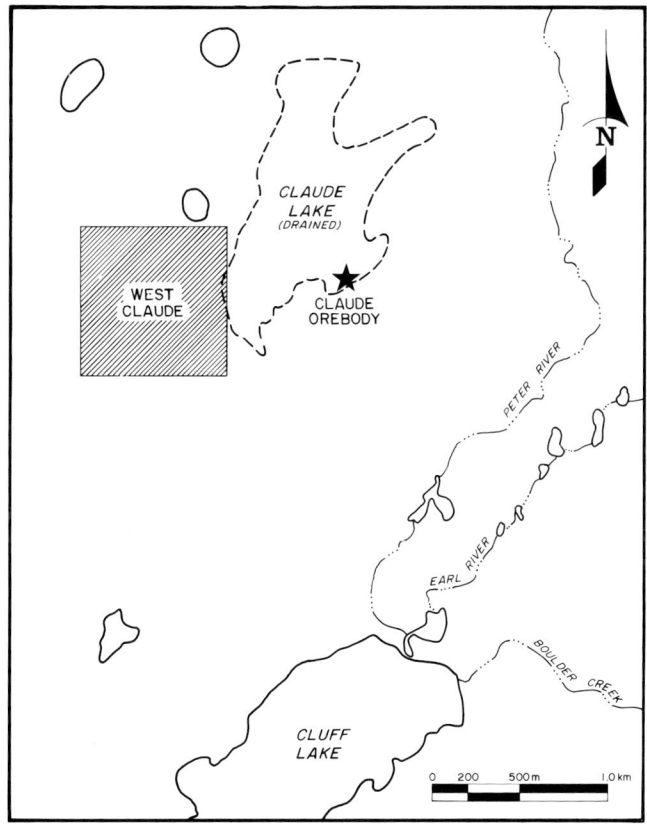

Figure 4. Location of Claude orebody and West Claude overburden drilling survey, Cluff Lake area.

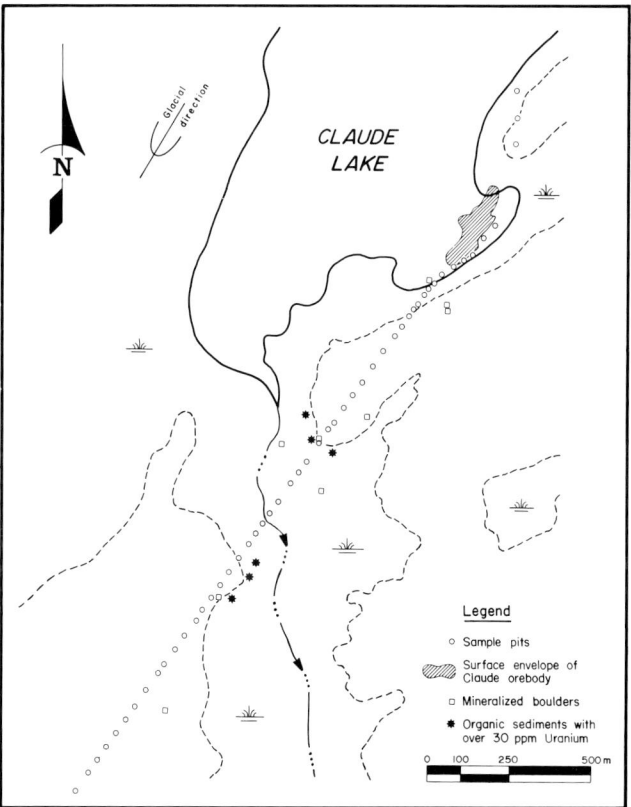

Figure 5. Location of the overburden sampling profile passing over the Claude orebody and oriented in approximately the ice direction.

mineralized particles to a bedrock source, an understanding of the development of glacial dispersions is required. A dispersal train from a known orebody, the Claude deposit, was examined and provided the basis for interpretations of boulder trains and surficial geochemical anomalies in the region.

Claude – a Model of a Mineralized Dispersion

The Claude orebody is located below the water level along the eastern shore of Claude Lake (Fig. 4), where it occurs in a zone of tectonized and altered aluminous gneisses (Tona *et al.*, 1985). The ore is widely disseminated with an average grade of 0.3%, but with sections up to several per cent uranium. The first indication of mineralization was an airborne radiometric anomaly along the south shore of Claude Lake, extending to the southwest. Subsequent ground prospecting led to the discovery of a boulder train consisting of 10 mineralized boulders and in-place mineralization. The landscape is dominated by streamlined landforms, well-developed to the north and west, which record the glacial direction of N30°E.

A study of the surficial deposits and development of the mineralized dispersion was undertaken by Granier *et al.* (1981a). It consisted of a series of 50 pits aligned in the glacial direction and extending for about 2 km down-ice of the Claude ore zone (Fig. 5). In each pit, the overburden was logged and sampled at 25 cm intervals. The minus 80 mesh (180 μm) fraction was analyzed for 13 elements: Ag, As, Au, B, Cu, Mo, Ni, Pb, Se, Th, U, V, and Zn. Bedrock was also sampled and analyzed where possible.

The overburden in the Claude Lake area is composed of till overlain locally by thin stratified drift. It generally comprises a massive, indurated, stony material with abundant, locally derived, subangular clasts. There is a gradual vertical textural variation from a matrix-supported till to a till with high clast density and coarser matrix near the base. The contact with the underlying bedrock is in many places indistinct, the till grading into a zone of fractured and crushed bedrock.

The sampling profile intersected a 75 m long section of the Claude ore zone. Discontinuous bedrock mineralization was encountered in only one of the four sites within this ore zone. Bedrock geochemistry identified the zone as an area of anomalous As, B, Mo, V, and U, where maximum values were associated with the mineralized structures. The remaining elements were also anomalous but their patterns were inconsistent and abundances much lower. The geochemistry of the overlying till indicated a strong dispersion from the mineralized zones, southwest for a distance of about 250 m (Fig. 6). The dispersion train was relatively continuous for about 150 m beyond the limit of the ore zone. About 350 m farther along the profile, anomalies reappeared at the edge of a local bedrock depression. A significant reappearance of strong polymetallic anomalies at a distance of

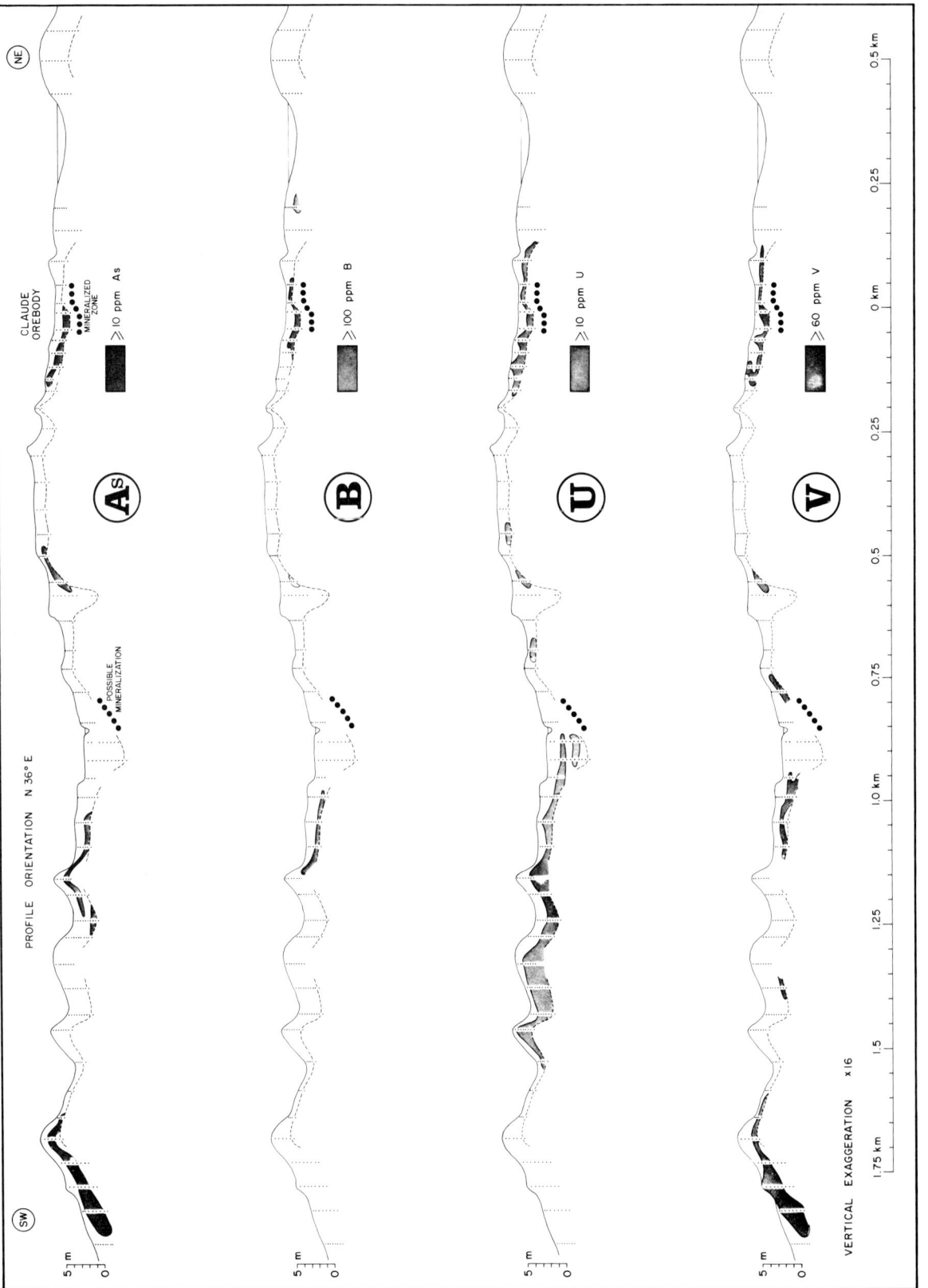

Figure 6. As, B, U, and V dispersion in overburden profile, showing anomalies generated from the Claude orebody.

about 900 m from Claude could be a sign of possible local mineralization (Fig.6). This has yet to be verified, although recent diamond drilling indicates the presence of uranium mineralization at depth (215 m) in the area.

The till is derived from erosion of the underlying bedrock and transported a short distance in the direction of ice movement. The length of the mineralized dispersion is influenced, in part, by the extent of till cover and bedrock relief, but it appears that element abundances are reduced to about twice background after about 250 m. The elements that give the best patterns are those that are enriched in the ore zone: As, B, U, and V. An important feature of the dispersion is that it gives a relatively uniform pattern even though it is derived from a very diffuse mineralization. The effect of the till development is to blend the mineralized debris from numerous narrow sources to give a homogeneous dispersion. Considering the overall form and dimensions of the dispersal train, it is recognized that the detection of a mineralized dispersion requires detailed geochemical sampling and knowledge of the surficial deposits of the area.

The significance of geochemical anomalies in organic terrain is confirmed by the development of uranium anomalies in organic sediments bordering mineralized overburden (Fig. 5). Uranium, and presumably other mobile elements, are leached from the overburden into the swamps where they are trapped by organic materials. Geochemical anomalies in organic terrain can therefore be considered as an indication of anomalous overburden and indirectly of local mineralization.

The study of the Claude dispersion is an assessment of the ability of drift geochemistry to define a mineralized train and is important in designing future exploration programs to investigate other boulder trains and surficial geochemically anomalous areas.

West Claude – Application of the Model

At the west side of Claude Lake, opposite the Claude orebody (Fig. 4), about 50 mineralized boulders, some with up to 54% uranium, were found. Most were clustered in a small area and no trend was apparent in their distribution. Prospecting of the area was limited to organic sediment surveys owing to extensive organic terrain. Uranium anomalies were revealed but no clear pattern emerged. Shallow diamond and percussion drilling in the area encountered minor mineralization but none of the type or grade indicated by the boulders. Since mineralized bedrock, however minor, was encountered in the area, it was not unreasonable to assume that the boulders could be part of a locally derived boulder train. The challenge for continued exploration was to link the boulders to a mineralized dispersion and follow it to a bedrock source. To accomplish this, a program of detailed overburden sampling was carried out (Granier et al., 1981b). The area of investigation was selected on the basis of the expected glacial direction, inferred from streamlined landforms, position of the mineralized boulders, and previous geological work.

The sampling was performed by a track-mounted (Fig. 7) overburden drill capable of collecting disturbed but relatively uncontaminated samples at any desired depth in overburden and bedrock. The technique was effective to depths

Figure 7. Overburden drill featuring an Atlas-Copco drill mounted on the back of a Foremost tractor.

of up to 30 m, however, the optimum depth was between 10 m and 15 m. The drill was capable of penetrating all types of overburden but sample quality varied with the nature of the drift. In general, sample recovery was poorer in loose, saturated sands and gravels than in compact clay-rich material. Bouldery tills did not present serious problems because the drill simply cored through large boulders. A description of the technique is given by Monteil (1976). Samples were collected at 60 cm intervals in the overburden and at 1.5 m into the bedrock. The collection of bedrock samples was important for both the lithological identification and geochemistry of the underlying bedrock. In order to maximize the area covered, yet allow for a tight enough spacing to detect even a fine dispersion, a 40×40 m grid was used.

The overburden sampling outlined a discontinuous northeast to southwest uraniferous dispersion, beginning west of Claude Lake and extending southwest to include the mineralized boulders. In general, the dispersion was most continuous and best developed within the lower levels of the overburden, immediately above bedrock (Fig.8). With increasing height above bedrock, the dispersion became more diffuse, showing the effect of greater dilution and secondary remobilization by surface leaching (Fig. 9). The maximum uranium anomalies were located in the northeast as a small zone west of the lake and more significantly as a narrow northwest to southeast trend in the central part of the grid. The latter zone coincides with a steep bedrock depression and is associated with major uranium, arsenic, and boron anomalies in the bedrock (Fig. 10). This elemental association can be taken as an indication of Claude style mineralization in the area. The bedrock depression may be the expression of a mineralized structure, a mechanically weak zone which had been deeply eroded during glaciation.

The general dispersion pattern and the association of both bedrock and overburden anomalies with the northwest to southeast trending bedrock structure suggested a local source for the mineralized overburden. A second phase of overburden drilling was carried out to inspect certain areas along this trend as well as to investigate the anomalous zone to the northeast which seemed to form the apex of the dispersion train (Fig. 11). The approach was to complete five narrow sections over the favourable areas, four cutting across the bedrock feature and one located over the apparent northeast apex of the dispersion train. The sections were oriented northeast to southwest and consisted of 3 lines, 10 m apart, with spacing at 10 m along each line.

Geochemical anomalies in bedrock indicated that the central bedrock feature contained several very narrow anomalies with up to 900 ppm uranium, suggestive of thin mineralized fractures. The potential zone to the northeast did not produce anomalies and was rejected as a possible source. The area of the bedrock depression was the most likely source of the mineralized dispersion, since it was a zone of geochemical anomalies in both bedrock and overburden.

The follow-up to these anomalies was an investigation of the geology and structure of the underlying bedrock by deep percussion drilling. Three drill fences were placed along the previous northeast to southwest overburden lines (see Fig. 11). Since a diffuse and tectonized mineralization similar to Claude was expected, the drill holes were inclined and ar-

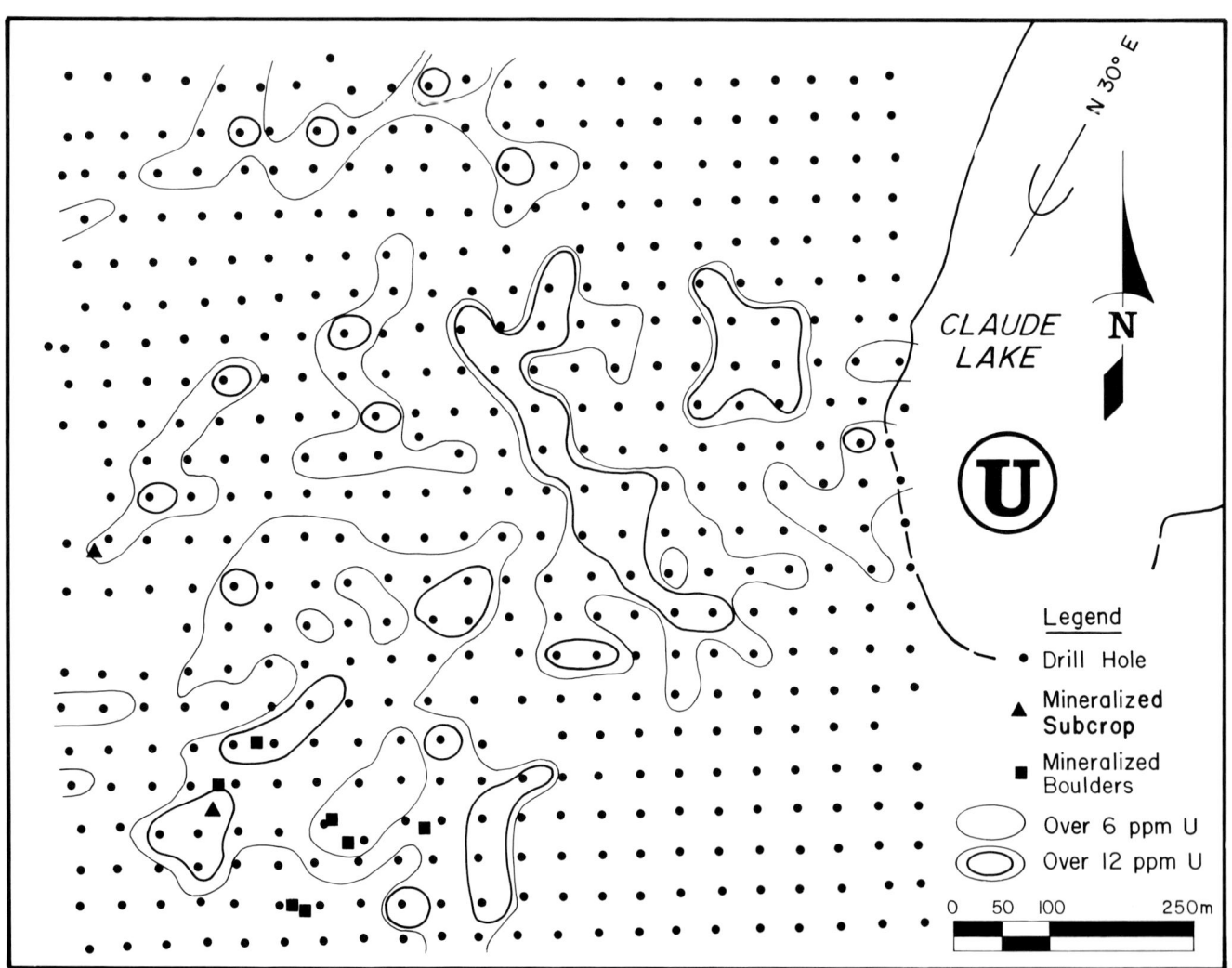

Figure 8. U distribution in overburden at 60 cm above the bedrock, West Claude.

ranged in pairs to create an "X" pattern. This was to ensure that all structures were intersected and to provide maximum coverage.

Mineralization up to 1.5% uranium (radiometric grade) was intersected in the southwest end of the centre drill fence. It appeared at the contact between aluminous and feldspathic gneisses and defined a narrow zone of 2 m width, dipping to the southwest. The mineralization subcropped along the southern edge of the bedrock depression. A later diamond drill program indicated that the uranium was found in altered feldspathic gneisses associated with Cluff and tectonic breccias. The uranium was generally finely disseminated but in places was visible as massive black clots or stringers. Unfortunately it is not of sufficient size or grade to be considered of economic interest. Although the very rich mineralization of some of the boulders was not found in the exploratory drilling, it is conceivable that very small but rich pods of mineralization occur within the mineralized zone.

The discovery of mineralization at West Claude, buried beneath glacial deposits, demonstrated the effectiveness of drift geochemistry and was an important contribution to exploration in the area. The application of the technique to other areas produced, in most cases, only disappointing results. It became apparent that the simple model of glacial dispersion developed from the Claude and West Claude dispersal trains did not apply to all situations and that boulder trains must be interpreted in terms of their glacial history.

GLACIAL GEOLOGY

The importance of glacial deposits in mineral exploration has long been realized in the Carswell area and through the course of exploration, several studies to locate the source of mineralized boulders were undertaken (Langford, 1972; Doak, 1972; Alley, 1979; Soyer, 1980). These advanced the understanding of the glacial history of the area but were generally incomplete and of limited scope. A comprehensive program was therefore initiated to examine the regional glacial setting of the boulder trains and to interpret the glacial history. The study is continuing and only preliminary observations are presented here.

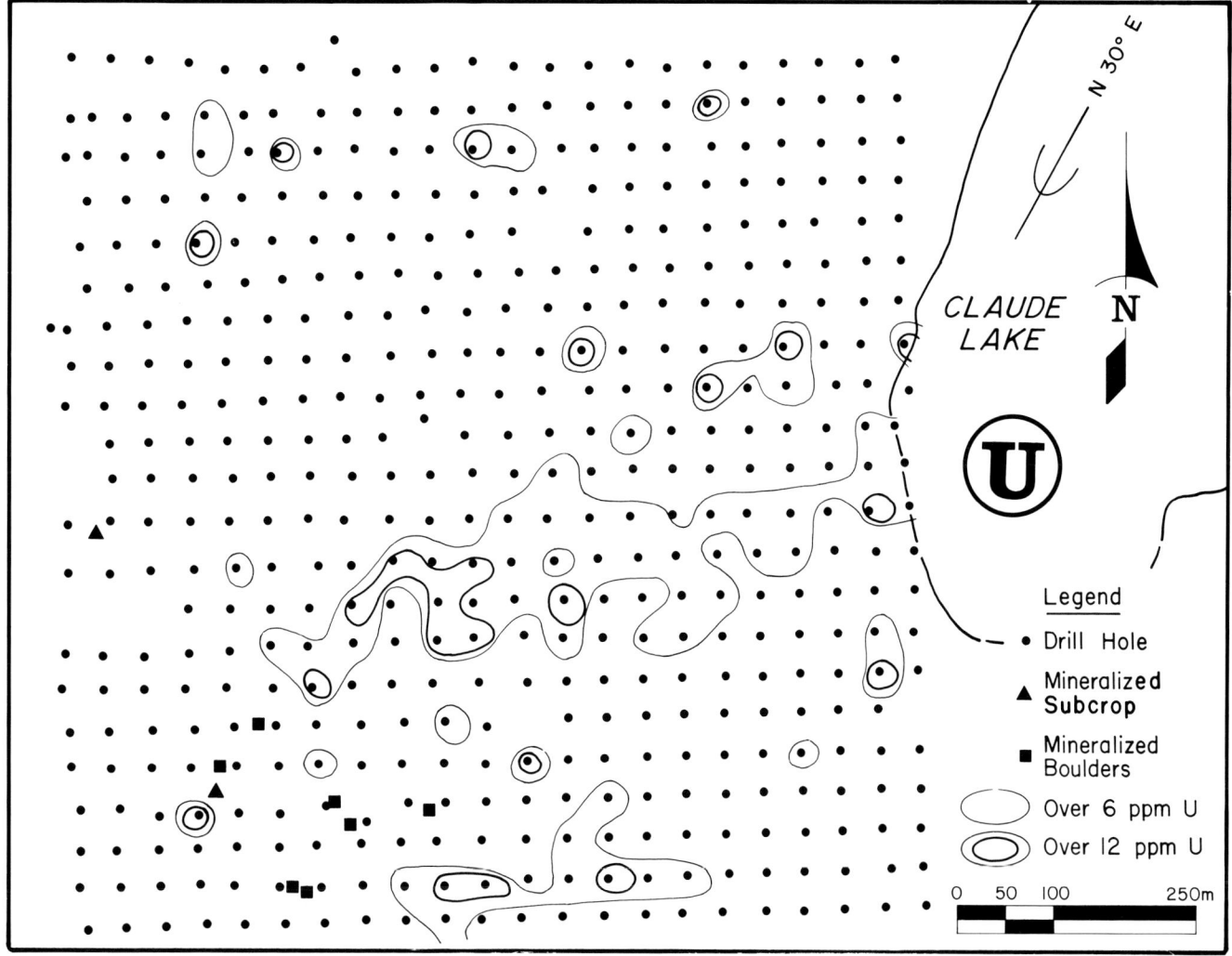

Figure 9. U distribution in overburden at 180 cm above the bedrock, West Claude.

The Carswell area is characterized by a generally low relief, streamlined till plain underlain by basal till and interspersed with stratified drift. The regional stratigraphy is relatively simple and consistent. It comprises a single, homogeneous sandy till. The till texture and composition is influenced by local bedrock and shows a regional variation in response to the changing bedrock of the Carswell structure. In general, the till is light pink-purple to brown-grey and ranges from very indurated to non-compacted and loose. It is generally matrix-supported with subrounded clasts, many of which are striated and dispersed in a very sandy matrix. In most exposures the till is massive and homogeneous, locally displaying a slight layering at higher levels, defined by thin sub-horizontal sand stringers and lenses. Till thickness averages 5 m with a maximum of about 15 m. The till rarely rests on a smooth and striated bedrock but generally passes into a zone of bedrock rubble consisting of angular rock fragments. In the Cluff Lake area, a second till is observed. It is distinguished from the sandy till by its yellow-brown colour and stony texture. It generally occurs as a thin and discontinuous cover throughout most of the Cluff Lake area.

Glaciofluvial deposits are distributed throughout the region and generally occupy bedrock-controlled drainage systems. Extensive deposits fill major bedrock valleys to depths of up to 100 m. They consist mainly of stratified sands and fine gravels but can include finely laminated silts, clays, and boulder gravels. The deposits take the form of irregular mounds and ridges associated with buried stagnant ice, eskers, or flat sandy plains formed of outwash sands. Glaciolacustrine deposits are generally absent, although some fine-grained sediments can be found in lowland areas.

These deposits appear to represent a single episode of glaciation. The structure and texture of the sandy till, and the streamlined forms which it develops suggest a subglacial origin with possibly a melt-out phase in the late stage of deposition (e.g., Boulton, 1970; Shaw, 1982). The direction of the streamlined landforms and bedrock striations indicates that the ice movement during the deposition of the till was from the northeast. The uppermost stony till is generally associated with hummocky and irregular terrain and probably relates to supraglacial deposition (e.g., Boulton, 1967, 1968).

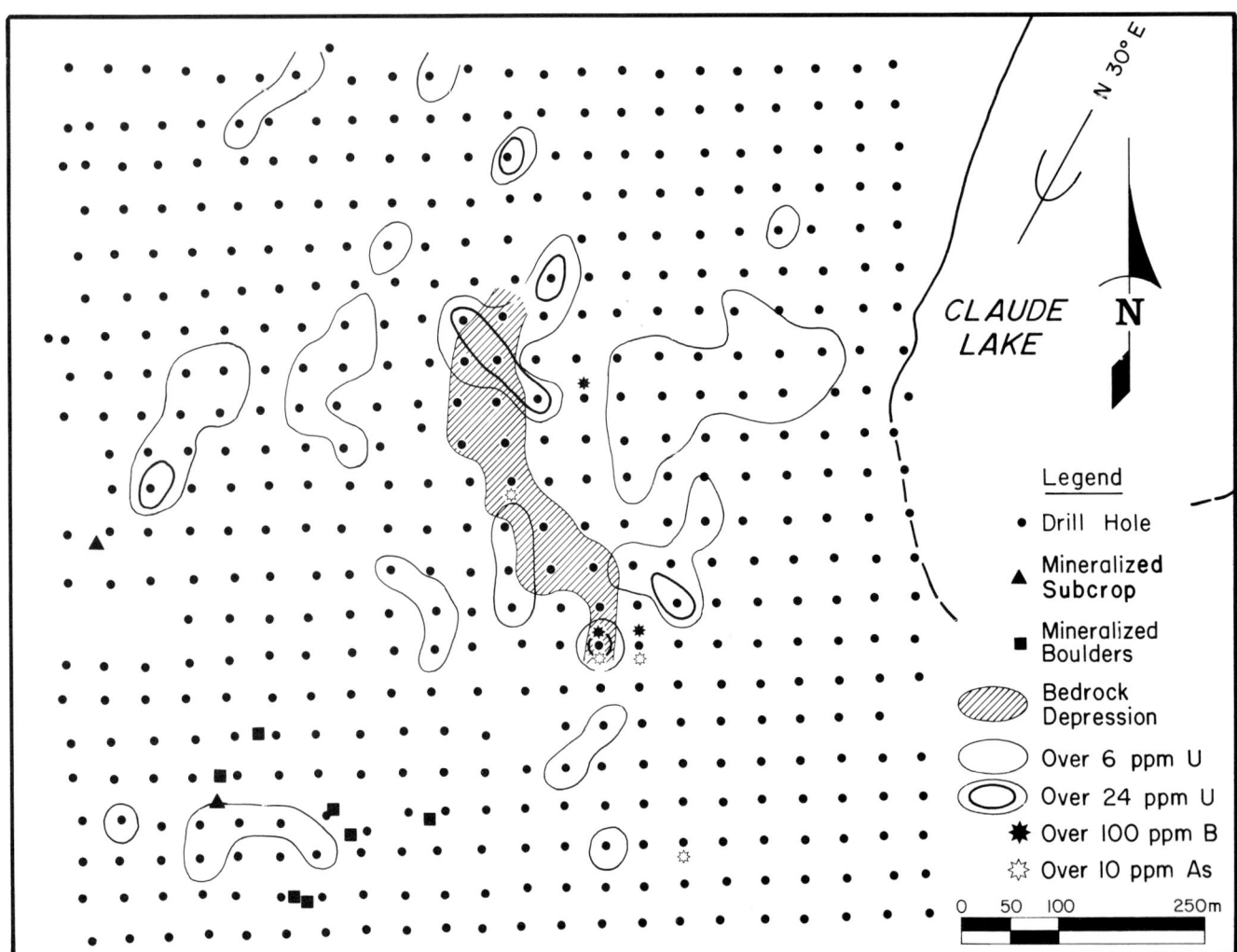

Figure 10. As, B, and U anomalies in bedrock and their relationship to the bedrock depression, West Claude.

THE CLUFF LAKE MORAINE

A prominent landform of the Cluff Lake area is a narrow relatively continuous ridge, stretching for about 10 km across the south end of Cluff Lake. It begins rather abruptly at the west side of the lake and continues eastward across the lake where it terminates into a zone of hummocky terrain (Fig. 12). The ridge system is unusual in its continuity, especially the eastern portion where it forms an almost unbroken single-crested arcuate ridge. Topographically, the ridge is not a prominent feature, generally rising only 3 m to 5 m above the surrounding terrain, although in places it may be up to 15 m high. The feature may not have been important in the context of mineral exploration had it not been associated with mineralized boulder trains. In this respect the origin of the ridge and an interpretation of the local glacial history were essential to the investigation of the boulder trains.

A complete section of the ridge is exposed at one location, revealing a complex overburden stratigraphy (Fig. 13A). The ridge represents a local thickening of the overburden and is unrelated to bedrock. The underlying sandstone presents a very smooth and level surface which is overlain directly by a layer of massive, pink coloured sandy till. This layer thickens from about 50 cm in the south to over 1 m in the north. The till is in turn overlain by stratified deposits which volumetrically are the most important component of the ridge. These sorted deposits grade from poorly sorted matrix-supported boulder gravels in the centre to fine gravels and horizontally laminated sands in the south. The contact between the stratified deposits and the underlying sandy till is sharp and planar. On the north flank of the ridge, the coarse gravels are overlain by a thin wedge of sandy till that extends from the base, where it merges with the main body of sandy till, up to the crest. The sandy till in turn is overlain by a layer of brown stony till that also ends at the ridge crest where it appears to merge with the gravels. It continues down the north slope and is present locally as a discontinuous covering. The sandy till continues southward beyond the limit of the ridge, while the stony till appears only north of the ridge.

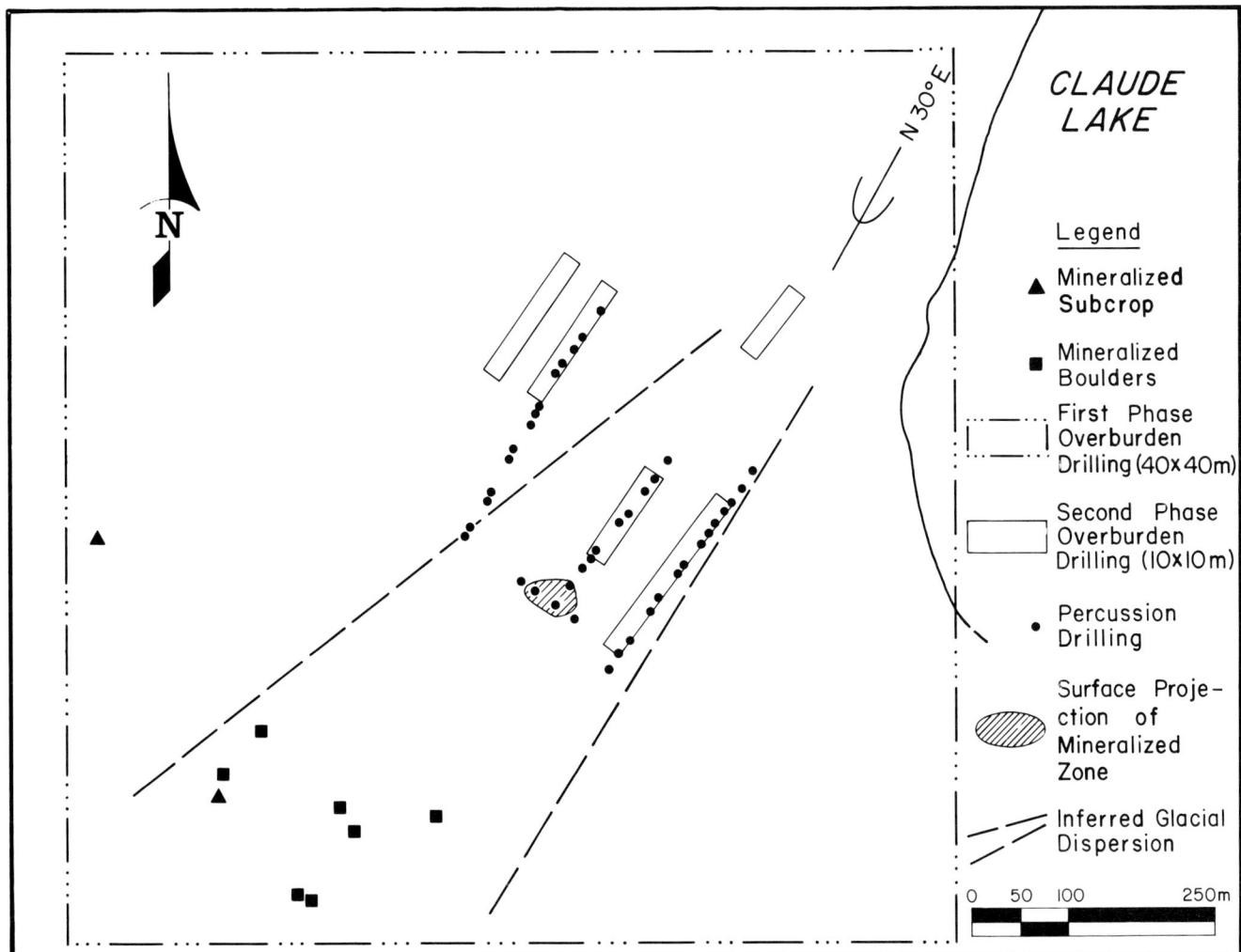

Figure 11. Sequence of follow-up drilling leading to the discovery of uranium mineralization, West Claude.

A second section, which did not reach bedrock, exposes a similar sequence with lesser amounts of stratified drift (Fig. 13B). A massive to slightly sorted sandy till is the dominant component, with minor sands and some gravels. The till appears to be arranged in slabs, one on top of the other and dipping towards the north.

On the basis of its internal structure and composition, the ridge is interpreted to be an end moraine, referred to here as the Cluff Lake Moraine, and is comparable with frontal deposits of some modern glaciers (e.g., Boulton 1967, 1968).

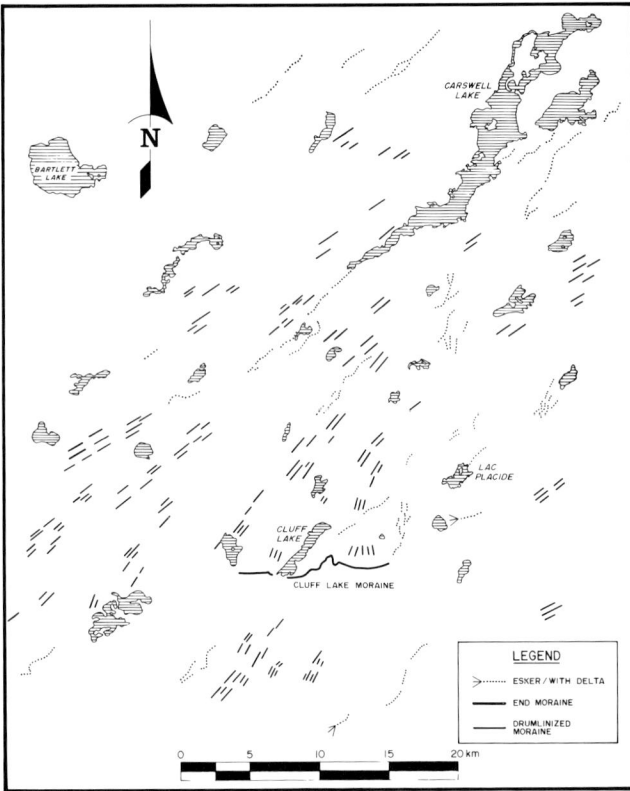

Figure 12. Glacial geomorphology of the Carswell area.

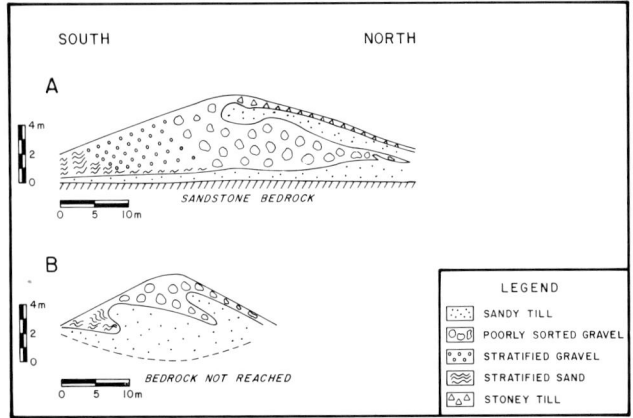

Figure 13. The internal structure and composition of the Cluff Lake Moraine.

The structure and disposition of the sandy till indicate thrusting or folding of subglacial sediments whereas the stratified drifts indicate active glaciofluvial sedimentation at the ice margin. The thin cover of stony till on and behind the moraine can be related to supraglacial debris derived by thrusting of basal material to the ice surface. Once exposed at the surface, the debris is released by melting and spreads out over the ice front where it can be reworked into outwash deposits. Eventually, upon complete deglaciation, the supraglacial debris is deposited as an irregular cover over the sandy till. A ridge containing stratified drift and till marks the position of the former ice front.

The landform association supports the end moraine interpretation, suggesting that the moraine marks the terminal position of a local re-advance into the Cluff Lake area (see Fig. 11). The regional direction of streamlined landforms indicates ice flow from the northeast with a gentle shift to a more southerly flow in the Cluff Lake area. This local flow pattern is directed towards and terminated by the Cluff Lake Moraine. The moraine could indicate only a standstill during the recession. However, the development of sole casts at the base of a till overlying stratified deposits in the area, suggests that a re-advance had occurred. During this phase, glaciofluvial deposits, probably outwash sands, were overridden by active ice. The sediments were undisturbed except in their upper portions where they were cut by low angle thrusts inclined in the direction of ice flow. These structures can be produced by active ice overriding the sediments and are similar to those described by Shaw (1982), who interprets them as subglacial deformation features.

THE DONNA TRAIN – A NEW ATTEMPT AT BOULDER TRACING

About 3 km east of Cluff Lake, a group of about 200 mineralized boulders, some containing up to 25% uranium, were discovered during routine ground prospecting. These boulders, referred to as the Donna boulders, define an approximate north to south trend that extends from at least Beaver Creek southward to the Cluff Lake Moraine (Fig. 14). The discovery of the boulder train was unexpected since an airborne radiometric survey of the area gave no radiometric anomalies. Subsequent geochemical prospecting produced no geochemical anomalies in the underlying drift despite extensive profile sampling in about 60 shallow pits. The boulder train was unusual in that the boulders were found at or very near the surface (none at depth) and furthermore were not associated with a geochemical dispersion in the drift. These features cannot be reconciled in terms of a classical glacial dispersion, and it appears that the Donna boulder train has a more complex glacial history.

The boulder distribution parallels the local streamlined forms of the Cluff re-advance and appears to terminate at the Cluff Lake Moraine. Most boulders are strewn about the surface behind the moraine, although some are found within the stratified drift of the end moraine complex. These features suggest that the boulder train and the Cluff Lake Moraine are intimately associated. Considering the glaciodynamics and depositional environment of an ice margin, it appears that the boulders are a part of a supraglacial

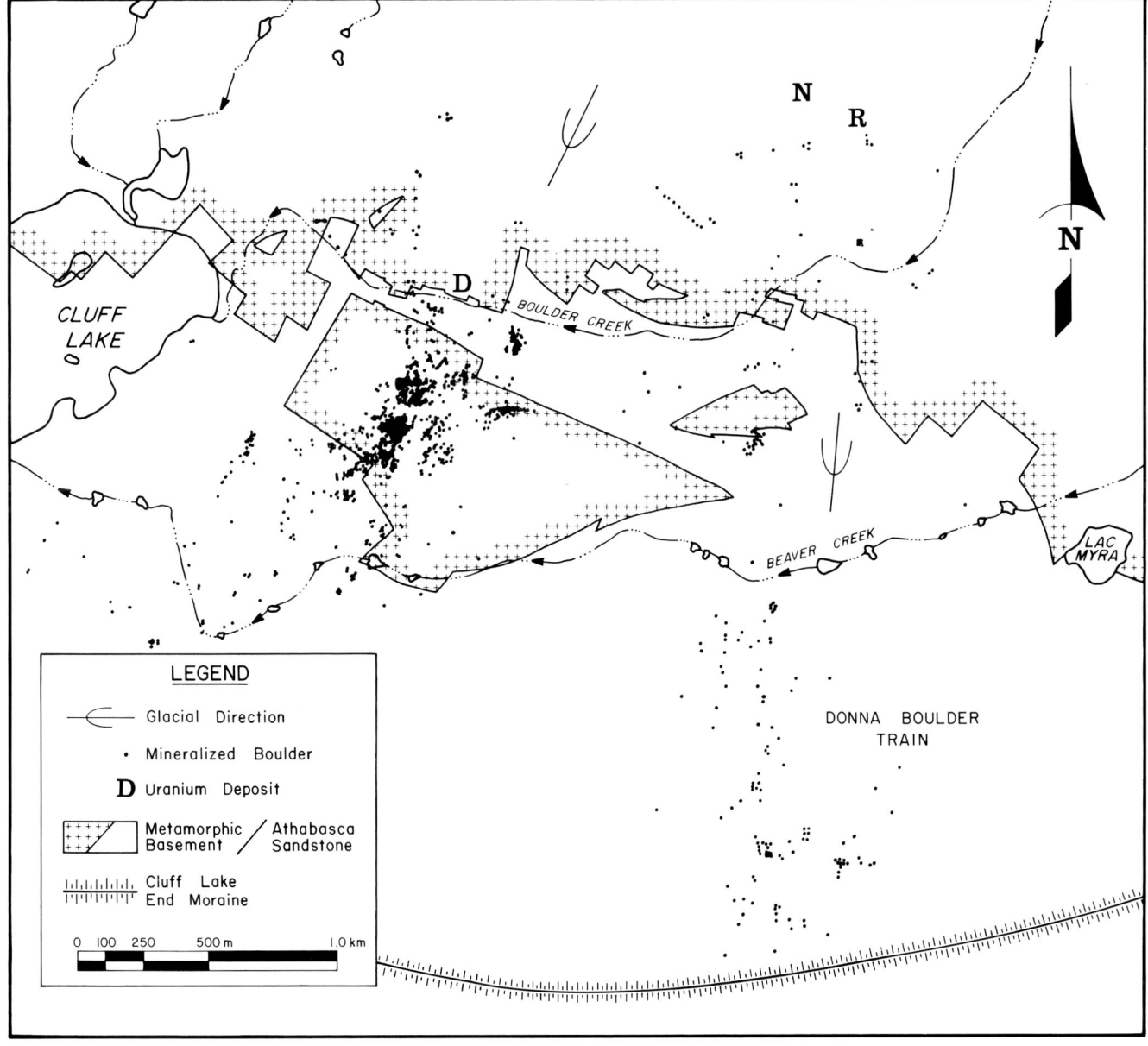

Figure 14. Location of the Donna boulder train, also showing the boulder fan from the D deposit, Cluff Lake area.

deposit. Compression at an ice margin, caused by deceleration, can thrust basal debris into the ice where it becomes exposed at the surface. It is released from the ice and distributed over the ice surface by either gravity flow or meltwater transport. The extent to which the final boulder distribution reflects secondary supraglacial rather than primary glacial dispersion is uncertain and depends upon ice slope and surface topography. It is likely that part of the boulder distribution represents mass movement down the ice slope and towards the ice front. The thin layer and generally stony character of the surface ablation deposits in most locations may suggest that the large boulders, including the mineralized boulders, remained as lag deposits on the decaying ice surface. The absence of a geochemical dispersion associated with the boulders is explained by the supraglacial deposition. There is consequently no relation between the boulder train and the sandy till upon which they were deposited.

The source of the Donna boulders has yet to be located but the recognition of the glacial environment has shed new light on the problem. Since most of the boulder train consisted of sandstone, the source must lie south of the main sandstone-basement contact (Fig. 14). The recognition of a final north to south direction suggests that the contact area south of the N mineralization was the most likely source. Subsequent bedrock drilling, however, failed to locate significant mineralization at the surface, although there were encouraging indications at depth.

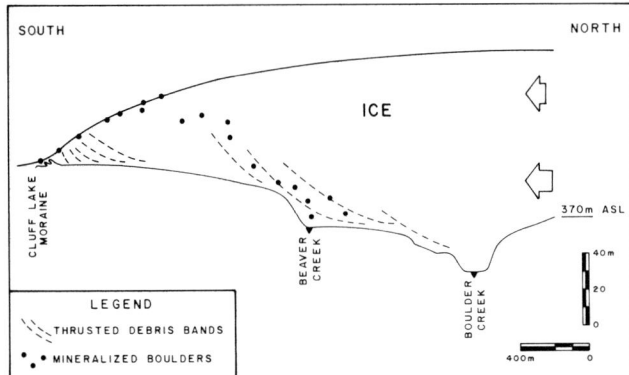

Figure 15. Idealized cross section through the glacier at the time of the development of the Cluff Lake Moraine showing a possible mechanism for the formation of the Donna boulder train, Cluff Lake area.

Glacial geology of the area suggests that the source of the boulders is farther south, in the area of Beaver Creek. In this region, a steep ridge of up to 30 m relief defines the south side of the Beaver Creek valley and trends almost east to west across the glacial direction. The steep slope may have enhanced glacial thrusting (eg: Moran, 1971) in the area and carried local debris to the ice surface (Fig. 15). The boulders were subsequently distributed over the glacier surface and finally deposited as supraglacial debris upon final melting of the ice. The search for the source of the Donna boulders is continuing, and it is hoped that the recent interpretations will prove to be correct.

CONCLUSION

Geochemistry has been an important part of uranium exploration of the Carswell structure. From the regional organic sediment surveys to the overburden sampling, the objective was to detect and trace mineralized dispersions. The organic surveys were designed to assess the mineral potential of a wide area in order to eliminate unfavourable regions and concentrate efforts on the more encouraging areas. Since it was expected that these anomalous zones were, in part, expressions of the composition of the underlying glacial drift, the next phase was drift geochemistry. The discovery of mineralization at West Claude, following a model developed from the Claude dispersal train, demonstrated the effectiveness of drift geochemistry. The success, however, was based on relatively simple glacial deposits consisting of basal till. Application of the technique to other areas such as the Donna area, was unsuccessful, indicating that a simple model of glacial dispersion was not applicable to all situations. This prompted an investigation of the local glacial history to explain mineralized boulder trains. It revealed that some boulder trains are not the result of classic subglacial erosion and transport but may represent supraglacial debris, elevated from the glacier bed by thrusting.

ACKNOWLEDGEMENTS

The author is grateful to Amok Ltd. for their support of the studies on the glacial geology of the Cluff Lake area, which are part of an MSc thesis, and for the opportunity to publish this paper. Thanks are due to W.O. Kupsch, thesis advisor, and J. Shaw for critically reviewing the manuscript and for fruitful discussions on the subject of glacial geology. Special thanks to C.C. Smart for valuable suggestions and improvements to the text. Finally, thanks to G. Stoesz for drafting the figures, and L.L. McNally for typing the manuscript.

REFERENCES

Acton, D.F., Clayton, J.S., Ellis, J.H., Christiansen, E.A., and Kupsch, W.O., 1960, Physiographic Divisions of Saskatchewan: Saskatchewan Research Council Geology Division Map 1.

Alley, D., 1979, Drift Prospecting and Quaternary Geology in Zone Indicielle: Amok Internal Report.

Boulton, G.S., 1967, The Development of a Complex Supraglacial Moraine at the Margin of Sorbreen Ny Friesland, Vestspitsbergen: Journal of Glaciology, v. 6, no. 47, p. 717-735.

_____, 1968, Flow Till and Related Deposits on Some Vestspitsbergen Glaciers: Journal of Glaciology, v. 7, no. 51, p. 291-412.

_____, 1970, The Deposition of Subglacial and Melt-Out Till at the Margin of Certain Svalbard Glaciers: Journal of Glaciology, v. 9, no. 56, p. 231-245.

Brown, R.J.F., 1967, Permafrost in Canada: Geological Survey of Canada Map 1246A.

Currie, K.L., 1969, Geological Notes on the Carswell Circular Structure, Saskatchewan (74K): Geological Survey of Canada Paper 67-32, 60 p.

Doak, R.S., 1972, Surficial Geology and Boulder Tracing in the Carswell Area, Saskatchewan: Amok Internal Report.

Dreimanis, A., and Vagners, U.J., 1971, Bimodal Distribution of Rock and Mineral Fragments in Basal Tills: in Goldthwait, R.P., ed., Till: A Symposium: Ohio State Press, p. 237-250.

Granier, C., Monteil, G., and Trichet, J., 1977, Relative Behaviour of Uranium and Lead in Some Acidic and Chelating Environments: in Second Symposium on the Origin and Distribution of Elements AIGC - UNESCO Paris, p. 148.

Granier, C., Monteil, G., and Trichet, J., 1981a, Eléments pour une Meilleure Conception de la Géochimie Tridimensionnelle Appliquée à la Recherche d'Anomalies Polymétalliques en Recouvrement Glaciaire: VRSS-DGRST 78, no 70741-70742.

Granier, C., Monteil, G., Wilson, S., 1981b, West Claude Overburden Drilling – Progress Report: Amok Internal Report.

Langford, F., 1972, Cluff Lake and Carswell Area, Report on the Glacial Geology: Amok Internal Report.

Monteil, G., 1974, Rapport de Fin de Mission, Géochimie: Amok Internal Report.

_____, 1976, Uranium Prospecting: The Use of a Mobile Percussion Drill for Geochemical Sampling and Geological Mapping in a Glacial Overburden Environment: Saskatchewan Geological Survey Special Publication no. 3, p. 390-396.

Moran, S.R., 1971, Glaciotectonic Structures in Drift: in Goldthwait, R.P., ed., Till: A Symposium: Ohio State Press, p. 127-148.

Shaw, J., 1982, Melt-Out Till in the Edmonton Area, Alberta, Canada: Canadian Journal of Earth Sciences, v. 19, p. 1548-1569.

Soyer, B., 1980, Cluff Lake - Etude de Géologie du Quaternaire et Boulder Tracing, Secteur Claude Ouest et Donna: Amok Internal Report.

Tona, F., Alonso, D., and Svab, M., 1985, Geology and Mineralization in the Carswell Structure – A General Approach: in Lainé, R., Alonso, D., and Svab, M., eds., The Carswell Structure Uranium Deposits, Saskatchewan: Geological Association of Canada Special Paper 29.

CASE HISTORIES OF THE RADON TUBE SAMPLER IN THE CARSWELL STRUCTURE

B. Powell
Amok Ltd., 817-825, 45th St. W., P.O. Box 9204, Saskatoon, Saskatchewan S7K 3X5

ABSTRACT

Contamination by radon daughters and thoron (^{220}Rn) is a common problem associated with radon emanometry. An original sampling technique is described which minimizes these problems by collecting radon daughter products on plastic films inside aluminum tubes planted vertically in the ground. Examples of results from ten locations within the Carswell structure illustrate the various types of anomalies and radon sources encountered in the area. The advantages of large areal coverage, high sampling density, insensitivity to thoron, and sampling at more than one depth are illustrated. Radon sources are placed in three categories: 1) bedrock, 2) overburden, and 3) spring water. Bedrock sources tend to form X or Y shaped anomaly patterns where fracture systems intersect and occur adjacent to contacts between granitic gneisses and conductive metasediments. Overburden sources are of two types: mechanical dispersions of mineralization in till which usually give rise to ribbon-like anomalies stacked along the direction of glacial transport, and chemical precipitates of radium with manganese and iron hydroxides which give rise to small isolated anomalies. Erratic radon anomalies have been observed associated with spring water in one area of the Carswell structure.

RÉSUMÉ

Faible productivité et mauvaise qualité des données due à la contamination par les descendants du radon ou du thoron (^{220}Rn) sont des inconvénients inhérents à l'émanométrie radon. Une technique originale est décrite dans cet article; la méthode des tubes à radon diminue l'importance de ces inconvénients en piègeant les descendants du radon sur un film inséré dans un tube en aluminium planté verticalement dans le sol. La lecture des descendants du radon piégés sur le film polyester se fait dans une chambre scintillométrique RD200 légèrement modifiée. Par ailleurs l'emploi des tubes à radon augmente considérablement le rendement jusqu'à 150 prélèvements par jour.

Dix examples d'étude en provenance de la structure de Carswell illustrent bien les différents avantages de la technique d'échantillonage de même que les différents types d'anomalies qui ont été rencontrées. L'insensibilité de la méthode au thoron permet une interprètation rapide des résultats. La maniabilité et l'élimination de la contamination de la cellule scintillométrique par les descendants solides du radon, permettent de couvrir rapidement de vastes zones avec une forte densité de points de prélèvement. Il est par ailleurs possible de faire des prélèvements à différentes profondeurs. Tous ces facteurs facilitent l'interprétation des résultats et des anomalies par l'utilisateur.

Les sources de radon sont classées dans trois catégories: 1) celles en provenance de la roche mère, 2) celles en provenance du recouvrement, et 3) celles en provenance des eaux de source.

INTRODUCTION

Prior to 1978, limited radon emanometry data had been collected from seven areas within the Carswell structure using ratemeters initially and then digital radon detectors (Morse, 1976) after 1975. One experimental Track Etch survey was carried out over the N orebody in 1974 by the Saskatchewan Geological Survey (Beck and Gingrich, 1976). These early surveys had detected uranium mineralization which would have been otherwise undetectable. It became apparent, however, that the complex geology of the Carswell structure required a radon method which could be implemented more economically in order to maximize the sampling density and coverage. This gave rise to the radon tube sampler (Powell, 1982), depicted in Figure 1, which allows a production rate of about 150 samples per detector per day compared with 50 samples or less per day for the pumped gas methods. In 1978 it was employed for the first time by Amok Ltd. within the Carswell structure and has been used for all subsequent radon surveys in that area, primarily because of its inherent speed and economy. All but one of the case histories to be described below involve this device. The locations of the surveys are given in Figure 2.

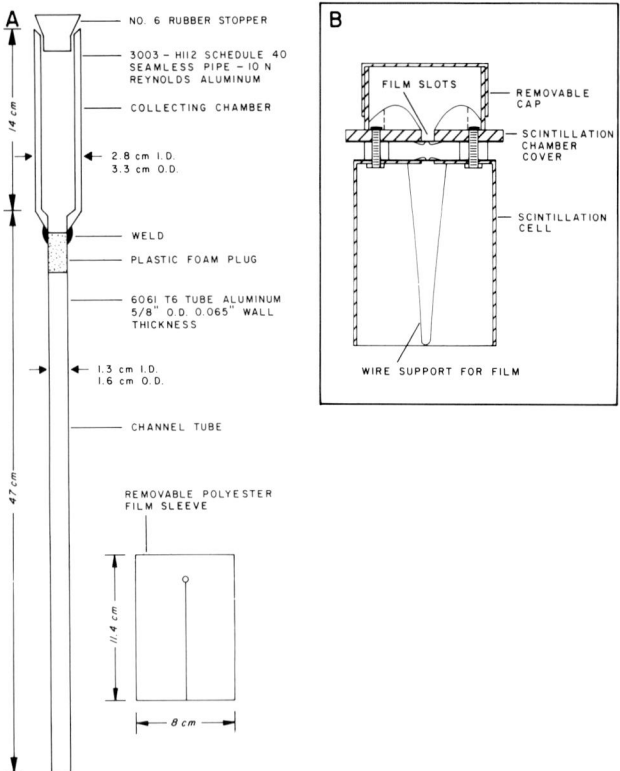

Figure 1. A) Sampler for collecting radon daughter products on polyester film. B) Modified scintillation cell for counting alpha emissions from the polyester film.

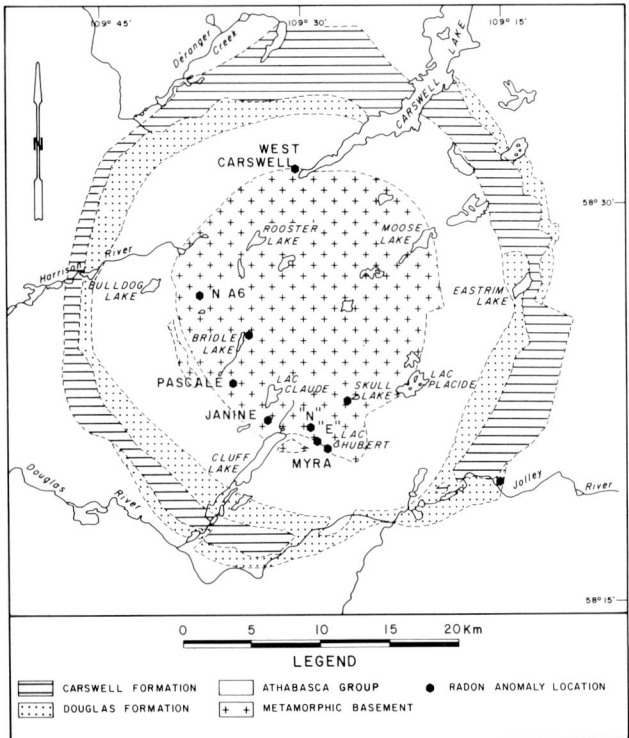

Figure 2. Locations of radon anomalies in the Carswell structure.

Uranium mineralization in the metamorphic core of the Carswell structure occurs as veins or fracture fillings in the contact zone of granitic domes, comprising the Earl River complex, with an overlying sequence of conductive metasediments called collectively the Peter River gneiss. After the emplacement of these domes, fracturing along major fault zones enabled the circulation of hydrothermal fluids and mineral deposition primarily around 1050 Ma (Tona et al., 1985). The fault directions which are known to be associated with mineralization are north-south, N45° E and east-west. Radon anomalies resulting from in situ mineralization will tend to follow one or more of these directions and will often form distinctive X or Y shaped patterns. This phenomenon results from the deposition of uranium at the intersections of two or more fracture systems and provides a means of distinguishing in situ sources from those caused by radioactive tills and mineral springs.

THE RADON TUBE SAMPLER

For a conventional radon determination with a digital radon detector, the soil gas sample is pumped directly into a zinc sulphide scintillation cell within the instrument. The radon abundance in the sample is indicated by the rate of alpha emission which is measured by counting scintillations from the cell in a fixed period of time with a photomultiplier tube and electronic counter. The cell is subsequently flushed out with atmospheric air to remove the radon gas. However, the daughter products produced by the decay of the radon while it is in the cell are difficult to remove because, being solids, they become attached to the surfaces of the scintillation cell. This form of contamination is substantial and is cumulative when more than one sample is taken. Frequent changes of the scintillation cell are necessary to minimize this effect. If the radon could be introduced as a solid into the reading instrument, this contamination problem would not occur. In the radon tube sampler, the solid daughter products of radon are collected on a small plastic film which is then introduced into the cell instead of the soil gas. The rate of alpha emission from the film can then be used as a measure of the abundance of the radon gas in the sampler. The cell noise level typically increases by about 0.2% of the sample count immediately after a two-minute reading with this technique because either some daughters are dislodged by alpha recoil or because some radon escapes from the film. This compares with 25 to 50% for a similar pumped gas reading. Figure 1B depicts the modified scintillation cell which adapts the digital radon detector to read the polyester films. A slot at the upper end accepts the polyester sleeve for alpha counting and a wire support holds the film centrally within the cell. Alpha particles emitted from daughter products on the film strike the phosphor coating of the cell, producing scintillations which are counted in the normal fashion by the radon detector.

The mapping of radon abundance in soil by collecting the daughter products is the principle of operation of the radon tube sampler. Tubes are left in the ground overnight, during which time radon diffuses into them and reaches an equilibrium state with its daughter products which are collected on polyester films. The tubes are subsequently read by counting

CASE HISTORIES OF THE RADON TUBE SAMPLER 191

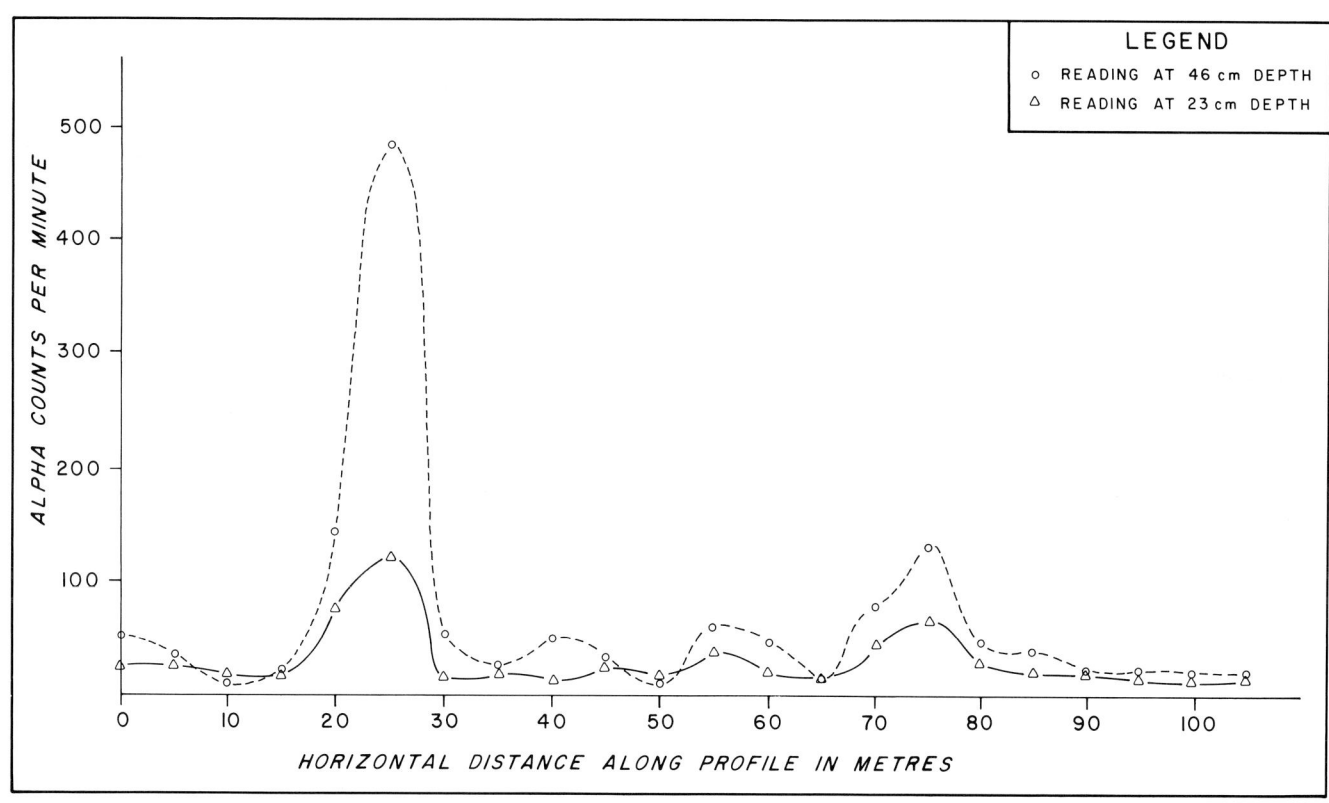

Figure 3. Measurement of radon abundance at two depths along a profile in the Pascale area.

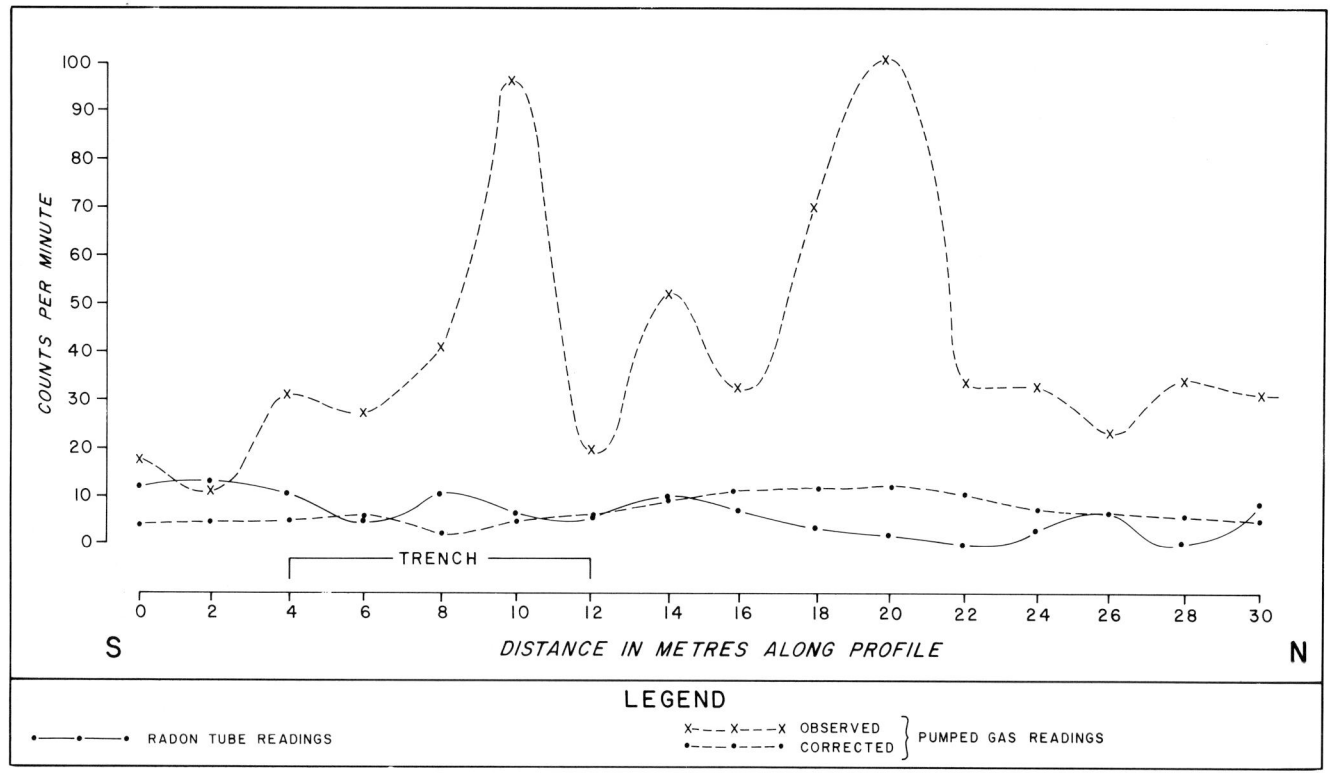

Figure 4. Radon profiles over a thoriferous conglomerate showing west of Carswell Lake.

alpha emissions from the films for a fixed period of time (usually two minutes). Similar surveys have been carried out by other workers using different procedures (e.g., Card and Bell, 1982). The collection of radon daughters is also the basis for the methodology of the Alpha Card system introduced in 1979 by Alpha Nuclear Ltd. The abundances of the daughter products are related to the abundance of radon when they are at equilibrium by:

(1) $X_1 N_1 = X_2 N_2 = X_3 N_3 \ldots$ etc.

where X_i and N_i are respectively the decay constant and the abundance of the ith member of the decay series. Only ^{218}polonium and ^{214}polonium are collected in quantities which are sufficient to produce significant alpha emissions. While alpha decays can occur from the elements ^{218}radon, ^{218}astatine, and ^{214}bismuth, they are extremely rare. ^{210}Lead and those alpha-emitting radioelements below it in the decay series would require a period of time comparable to the 22 year half-life of ^{210}lead to accumulate in sufficient quantities.

The unusual design of the radon tube evolved from a series of simple tests which began in 1977. One early test employed 10 cm lengths of electrical wire suspended in vertical holes in the ground to collect the daughter products. Progressively larger collecting surfaces were employed in subsequent tests to maximize collection. Polyester film was ultimately selected for its impermeability, stability, and durability. It lines the interior wall of the collecting chamber in the form of a removable sleeve, thus minimizing the loss of daughters to the chamber wall. Experiments reported by McCorkell and Card (1978) indicate that the daughter products precipitate rapidly and uniformly on the inner surfaces of a small collection chamber like that of the radon tube.

It is desirable to sample the soil gas from as great a depth as is practicable because the atmosphere is a nearly perfect sink for radon. The long narrow channel tube of the sampler allows the soil gas sample to be collected by diffusion from 46 cm below the ground surface without the necessity of digging open holes as is done with many other radon survey systems. Instead, a vertical hole is punched to the required depth with a steel rod which is then removed and the channel tube is dropped into place. The advantage of this approach is demonstrated in Figure 3 which shows two sets of simultaneous measurements made in the Pascale area with pairs of radon tubes in juxtaposition at depths of 23 and 46 cm along the same profile. The anomaly at 25 m along profile which was caused by a uraniferous pegmatite vein at a depth of 1.5 m, has a signal to noise ratio of about 16 at the 46 cm depth but only about 5 at the 23 cm depth. The length of the channel tube is, therefore, a compromise between obtaining the optimum signal and the practical problems of maintaining uniform survey conditions. A longer tube is less likely to reach its full depth capability consistently because of obstructions in the overburden, and is more likely to encounter the water table in wet areas.

The long channel tube also prevents any significant amount of thoron, a member of the thorium decay series with a half life of only 54.5 seconds, from reaching the collecting chamber. The approximate attenuation of the thoron signal was estimated by monitoring the alpha activity within the collecting chamber of a radon tube after it had been con-

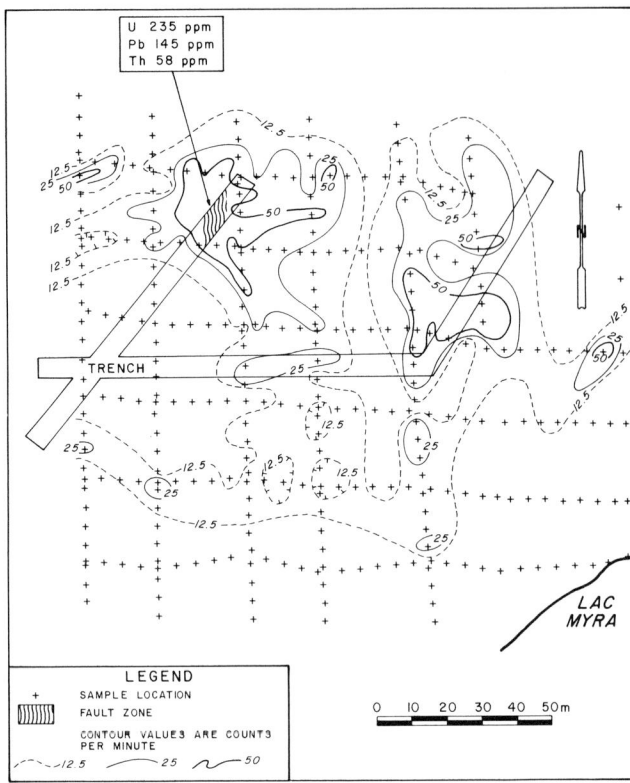

Figure 5. The Myra showing and radon anomaly near the N orebody.

nected to a ^{222}radon source. The time between connecting the source to the tube and first detection of anomalous alpha activity at the collection chamber was then taken to be the minimum transit time. This was found to be about 25 minutes. Thoron (^{220}Rn) is an isotope of radon and, therefore, would be expected to make the trip in essentially the same time. This amounts to 27.5 half lives, resulting in an attenuation of about 10^{-8}. The radon tube is therefore inherently unresponsive to thoron. Other methods require either a special field procedure or a correction to remove the thoron signal (e.g., Fleischer and Mogro-Campero, 1979). It is desirable to remove the thoron before the reading is taken because it will contribute to the overall noise in the data even if a correction is applied.

Figure 4 shows the results of two radon profiles made with the same digital radon detector along a line across a thoriferous conglomerate zone in the west Carswell area of Figure 1. The first profile was obtained in the conventional fashion by pumping the soil gas through a 50 cm extraction probe. The observed data show two strong peaks of 101 and 96 counts per minute (cpm) over the conglomerate. However, there are no significant anomalies after correction for thoron by the procedure described by Morse (1976). The second profile, obtained with radon tube samplers, shows no anomalies correlatable with the pumped gas peaks.

Other examples of radon anomalies observed within the Carswell structure are given below. These have been selected to illustrate the different kinds of results which have been obtained and are divided into three groups: those originating from mineralization in place within the bedrock, those originating from radium either mechanically dispersed or precipitated within the overburden, and those originating from radon emanating from spring water.

BEDROCK ANOMALIES

Figure 5 depicts the radon anomaly obtained in 1978 over the Myra showing which is situated about 1.5 km southeast of the N orebody and 3 km east of Cluff Lake (see Fig. 2). This was the first showing detected solely by a systematic survey employing the radon tube sampler. No radioactivity was detectable on the surface. The initial survey, comprising 2,700 readings on a 20 m by 50 m grid, covered an area of roughly three square kilometres along the southern boundary of the metamorphic core. Once the anomalous area was identified, detailed readings were taken at 5 m intervals on north-south and east-west lines spaced 25 m apart. The area is comprised of rubbly, subcropping sandstone to the southwest which is in contact with Earl River gneisses to the northeast. The source of the radon in the west lobe of the anomaly was found to be a zone of radioactive fault gouge trending N170° E, about 2 m below surface. A sample of this material was analysed and found to contain 235 ppm of uranium. The east lobe was also trenched but it was not possible to reach the source. The predominant north-south or N170° E trend of these anomalies parallels a similar trend of mineralization in the N, R, F, and Dominique-Peter mineralized zones, which is known to be controlled by major N-S faulting (Tona *et al.*, 1985). This direction veers slightly to about N170° E in the Myra area. The trend of the radon anomalies may therefore reflect an origin similar to that of the nearby N orebody. Drilling in the Myra area has not yet encountered any economic mineralization.

Figure 6 shows the results of the first field profile carried out with the radon tube sampler. It was completed across the north end of the N orebody in June of 1978 to confirm that the equipment was operating as expected. The results are presented here because it is the only profile over an established ore zone in the Carswell structure and affords some comparison with a Track Etch survey in the same area (Beck and Gingrich, 1976). Both radon surveys were carried out after the deposit had been substantially delineated by drilling. Also shown in Figure 6 are the drill hole accumulations along the radon tube profile. These match the radon tube profile very closely but there is a shift of about 20 m between the two. The north-south extension fractures which control the mineralization are known to be dipping at about 45° to the

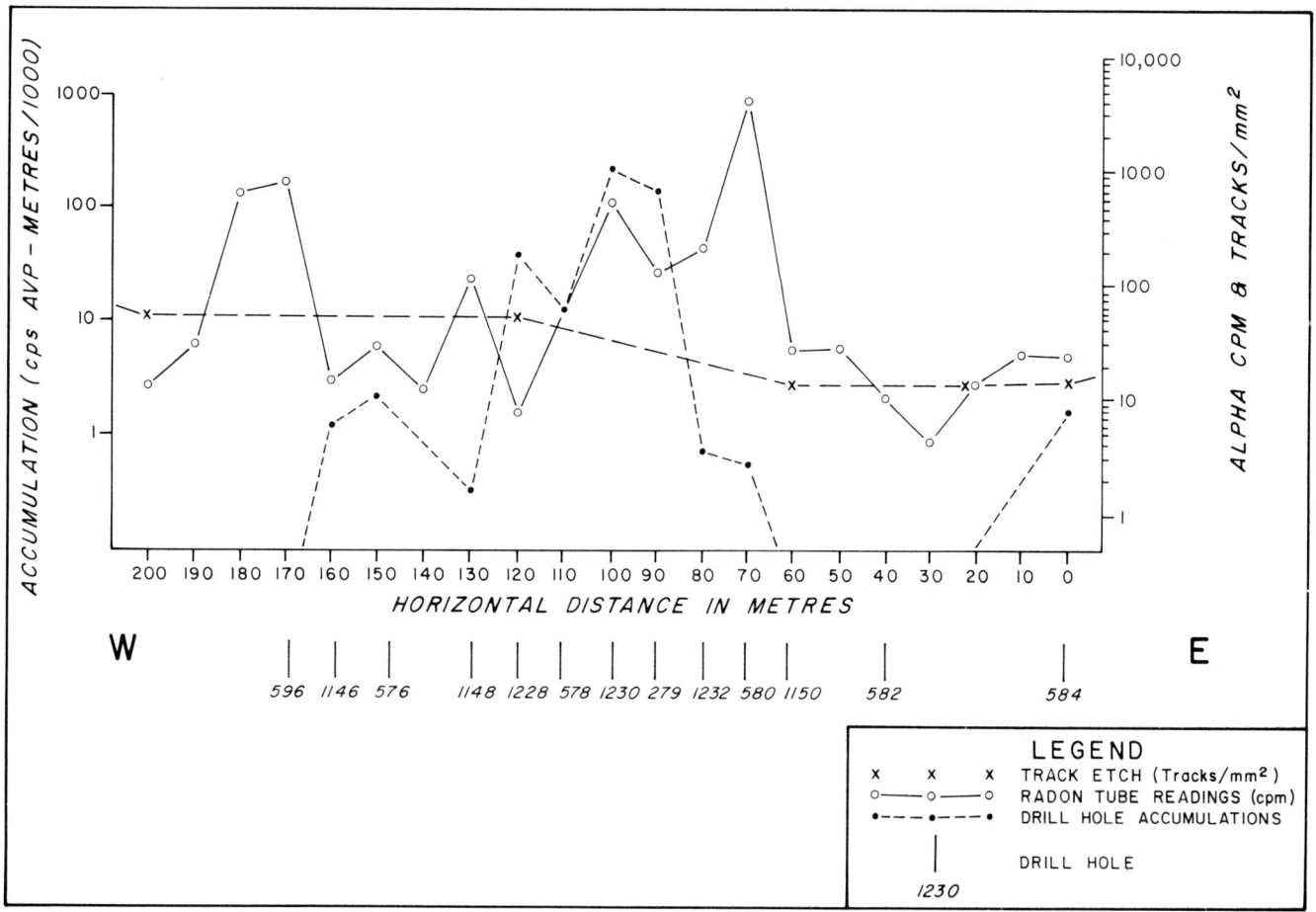

Figure 6. Radon tube profile across the N orebody.

west. The shift, therefore, may be a migration path effect, but more likely reflects the position of the surficial mineralization because the deposit comes to within a metre of surface. The Track Etch readings do not reflect the pattern of mineralization as faithfully as the radon tube profile but it is equally clear that this is mainly due to the larger sampling interval of the former. Note also that the Track Etch readings are up to 30 m out of the profile.

Figure 7 shows a portion of a systematic radon survey of 4,600 readings in the north A6 area, which is situated on the western border of the metamorphic core of the Carswell structure. The first showing was identified during systematic prospecting in 1979 when a reading of 5,000 cps (counts per second) was obtained with an SPP2 scintillometer on a rubbly subcrop near the centre of Figure 7. Extensive till cover prevented mapping extensions of the mineralization with a scintillometer. A systematic radon survey on a 10 m by 50 m grid was therefore carried out as a prelude to trenching. This delineated a swarm of linear radon anomalies covering an area of approximately 2 km^2 along the sandstone gneiss contact but almost entirely over the gneisses. Most of these anomalies trend N50° E, parallel to an extensive fracture system. Other anomalies are oriented N180° E or N140° E. All are commonly observed directions of faulting within the Carswell structure. The radon anomalies in Figure 7 are shown extending from the eastern flank of a MaxMin conductor, which is caused by a band of subvertical, graphitic Peter River gneiss. This relationship was observed throughout much of the north A6 area.

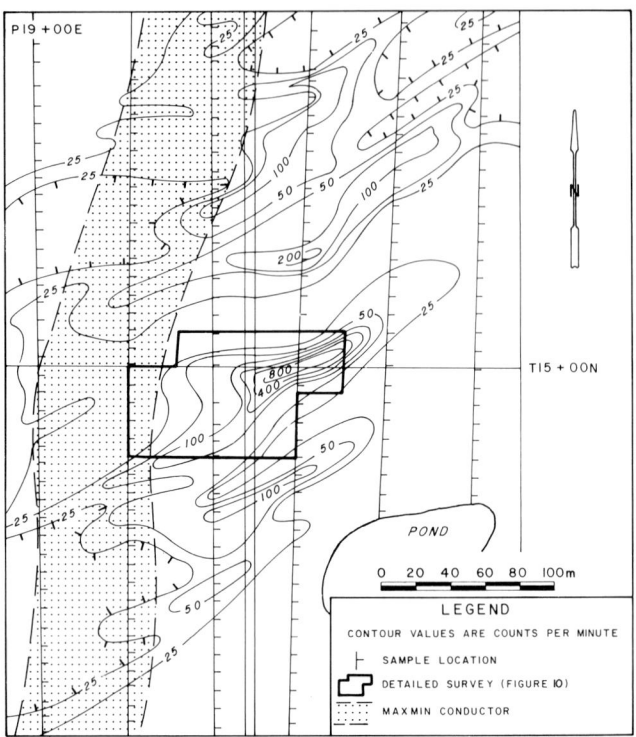

Figure 7. North A6 area systematic radon survey and MaxMin conductor.

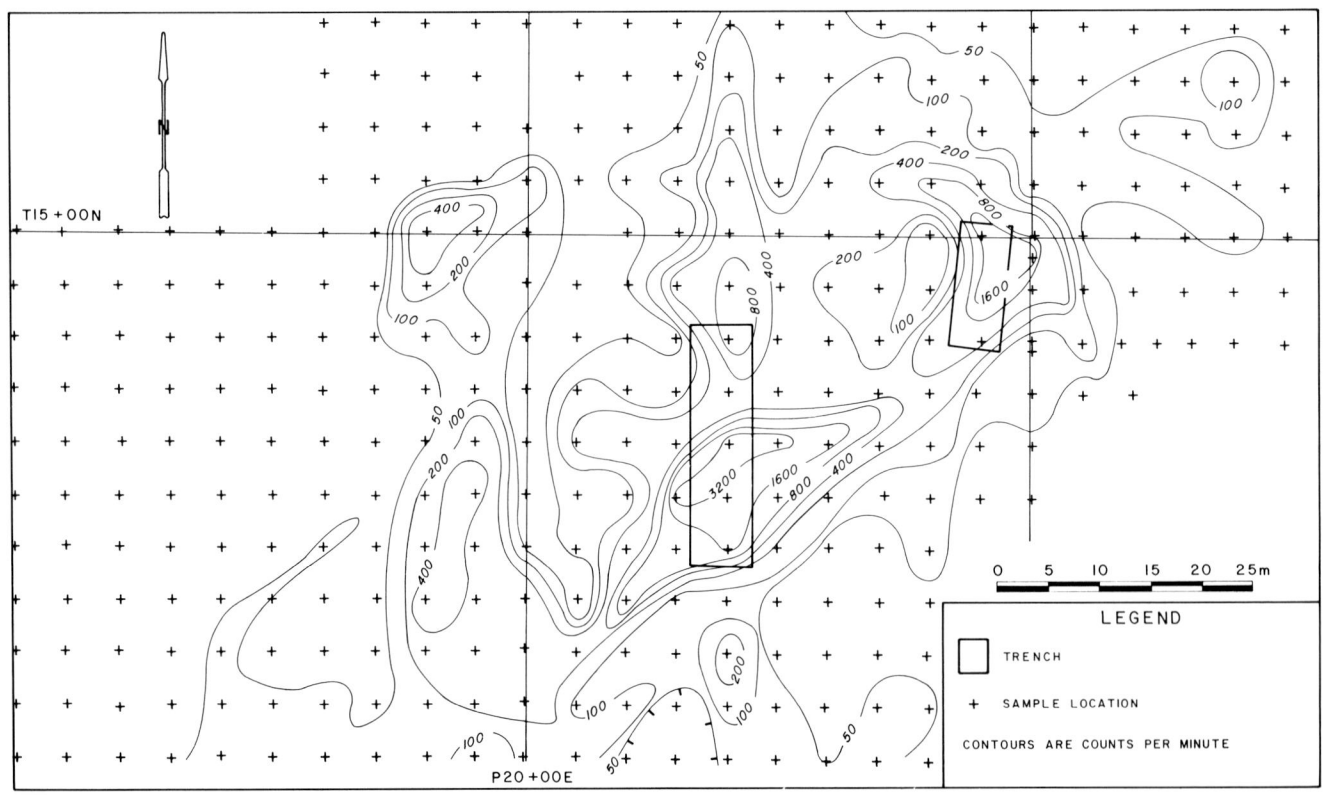

Figure 8. North A6 detailed radon survey of an anomaly in Figure 7.

Figure 8 shows the results of the 5 m by 5 m detailed survey outlined in the centre of Figure 7. This confirmed the anomalies of the systematic survey and revealed secondary northwest and north-south trends forming X or Y shaped anomaly patterns typical of bedrock mineralization found in place in the Carswell area. The two trenches revealed weak mineralization at depths of 1 to 2 m in a northeast trending breccia dyke with intersecting shear zones. Sporadic off-scale scintillometric readings were observed.

Figure 9 illustrates another radon anomaly found adjacent to conductive Peter River gneiss. On this occasion, the conductor was defined by a tripole resistivity survey which was caused by the chemical precipitation of radium with iron and manganese from spring water. It was obtained in the conventional fashion with a soil gas extraction probe and hand pump a year prior to the introduction of the radon tube sampler. Figure 11b depicts the southeastern wall of the trench which was cut to investigate the radon anomaly. A radioactive layer of manganese-iron hydroxide was revealed plunging from this wall through the floor of the trench. It occurs immediately above a layer of water saturated, thixotropic silt which directs the spring water to the site of the anomaly. Spectrometric analysis of a sample of the radioactive layer yielded a value of 540 ppm equivalent uranium, but followed by a small radon survey of 720 readings along the margin of the conductive zone. This is an exceptional case since the majority of radon tube discoveries in the Carswell area were the result of systematic surveys of several thousand readings carried out simultaneously with the ground geophysics. It illustrates the value of preselecting the area of a radon survey.

The four trenches which were subsequently cut across the anomaly encountered pegmatites and granitoid gneiss with bands of Peter River gneiss about one and a half to 2 m beneath a glaciofluvial outwash. Radioactivity of up to 4,000 cps was noted with an SPP2 scintillometer along fractures in the granitoid gneiss. The uranium values of grab samples, shown in Figure 9, correlate well with the radon results. The highest radon reading of 350 cpm was obtained at the location of the highest uranium value of 1200 ppm in trench C.

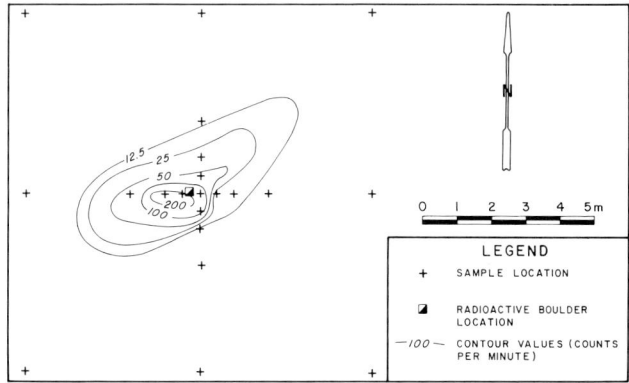

Figure 10. "E" zone detailed radon survey over a mineralized boulder occurrence.

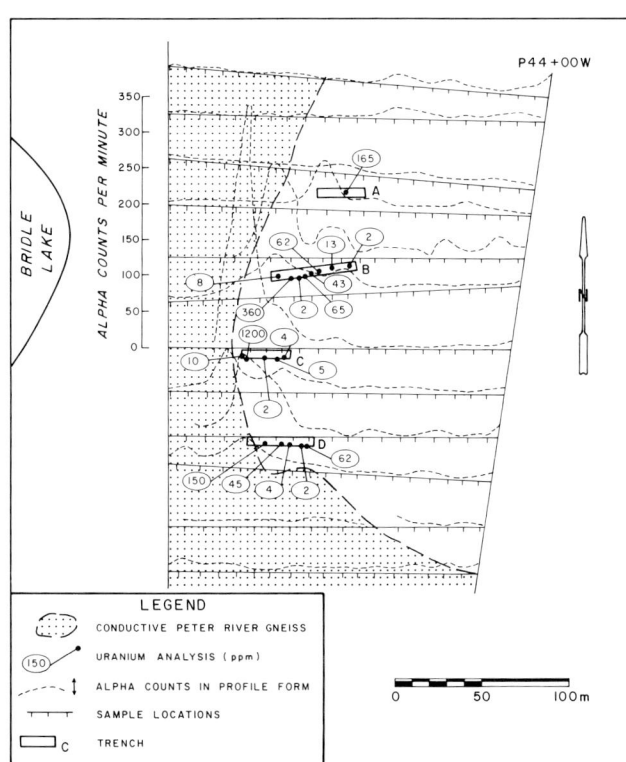

Figure 9. Bridle Lake radon survey.

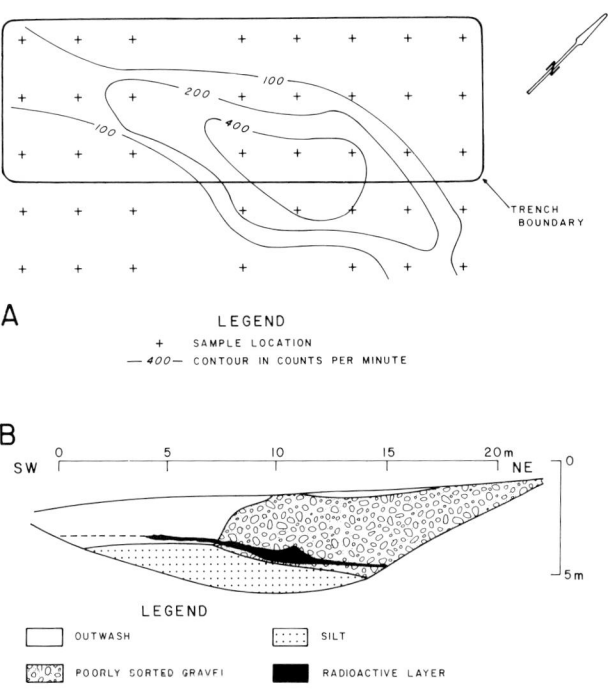

Figure 11. A) Radon anomaly and plan view of a trench at the Jolley River area. B) Southeast wall of the trench at the Jolley River area.

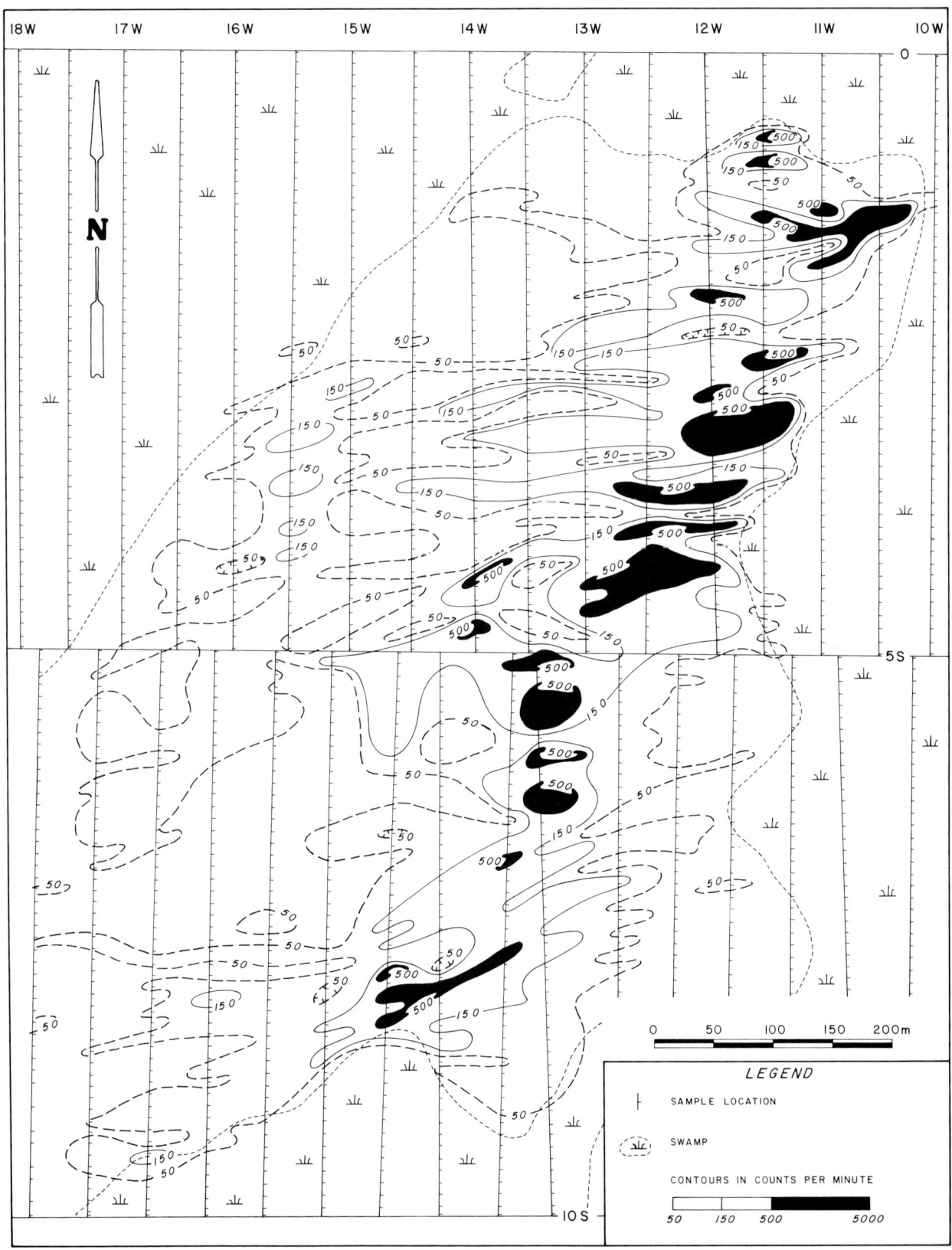

Figure 12. Janine area systematic radon survey of a mechanical dispersion of mineralization in till.

The 10 m westward displacement of the radon peaks relative to the highest uranium values in trenches A, B, and D might reflect either glacially transported rubble or the downhill flow of radon-rich ground water towards Bridle Lake. Water emanating from fractures in trench D was found to contain a high level of radon (4,000 cpm - RD200) and probably contributed to the observed radon tube anomaly. Subsequent drilling encountered weak scattered mineralization similar to what was found in the trenches.

ANOMALIES ORIGINATING WITHIN THE OVERBURDEN

Figure 10 depicts a radon anomaly over a mineralized boulder occurrence. This example is unusual because the boulder was found essentially by itself in a glaciofluvial outwash. Mineralized boulders are commonly found within an extensive radioactive till horizon which is more amenable to detection by a radon survey. The boulder was discovered by a single anomalous tube which emphasizes the value of a high sampling density. Additional tubes were planted at 5, 2, 1, and 0.5 metre intervals around this position to determine the lateral extent of this anomaly. At the time, no anomalous radioactivity could be detected on surface. Trenching subsequently revealed a pitchblende boulder at a depth of approximately 50 cm and displaced slightly from the peak of the anomaly as shown in Figure 10. The displacement may reflect the distribution of the halo of slightly radioactive material found around the boulder or perhaps a channeling of the radon gas through a more permeable portion of the till.

Figure 11a shows a small radon anomaly near the Jolley River on the southeastern edge of the Carswell structure geochemical analysis found no significant uranium. It was therefore concluded that radium was present but in complete disequilibrium with the parent uranium, a phenomenon commonly observed in mineral springs (Cadigan and Felmlee, 1977).

The spring water was ultimately traced to a radioactive fault about 400 m to the southeast. However, very little is known about the frequency, extent, or distribution of radon anomalies in the area because only 356 soil gas samples were actually taken. It is probable that a systematic radon tube survey of a few thousand readings would have facilitated exploration of the area.

Figure 12 shows an extensive pattern of radon anomalies caused by a mechanical dispersion of radium with uranium in the glacial till at the Janine area. The axial alignment of the anomaly with the regional direction of glacial transport of N30° E and ribbon-like or stacked character of the peaks are clues that the source material has been glacially transported. Trenching revealed a radioactive till at a depth of 2 m on the bedrock surface at the north end and essentially on the surface to the south of the area shown. The mineralization was initially believed to be derived from the Claude orebody which lies about 1.3 km to the northeast. However, chemical analysis of the till revealed a profile of elemental abundances which was inconsistent with that hypothesis. Subsequent drilling at the north end of the anomaly encountered mineralization at depth.

SPRING WATER ANOMALIES

Figure 13 shows a group of transient radon anomalies. As depicted, they have been smoothed by means of a 5 point averaging filter in order to be able to present the contours at the scale shown on the figure. The outline of each anomaly is approximately two standard deviations above the mean for swamps and about one for elevated ground in the general area. The peak values of the smoothed anomalies are also given. It can be seen that the anomalies are generally oriented in a north-south or N170° E direction, like those of the Myra showing. It is assumed that they have a similar fault related origin. The highest field reading of 1,592 cpm (322 cpm filtered) occurs in the swamp in the middle of the area shown. This is curious because wet areas are generally noted for extremely low radon levels. However, extensive seepage of spring water from adjacent higher ground had been noted. Trenching failed to encounter any anomalous radioactivity either in the overburden or the bedrock. But water samples revealed dissolved radon levels of up to 60 times background in the areas of the radon tube anomalies.

Three attempts to reproduce portions of the radon tube anomalies failed completely. The anomalous readings cannot be attributed to operator error or instrument malfunction because the original results were recorded by three instument operators on two radon detectors over a period of three days. Although it seems clear that the original anomalies were caused by radon in spring waters, it is not known why the tubes responded on the first occasion but not on subsequent attempts. Kristiansson and Malmquist (1982) have suggested a model in which rising gas bubbles in ground water along permeable structures might provide a mechanism for radon transport from large depths. Such a phenomenon may also be time dependent. If detection by the radon tube sampler is dependent on a chance capture below the water table of one or more of these bubbles, a large number of tubes would have to be planted to ensure detection of an anomaly. The original anomalous readings at Skull Lake fit this model well because they were generally erratic. The attempts to repeat the anomalies may have failed because only 20 tubes or less were used each time.

Resistivity and magnetometer surveys of the area delineated a body of Earl River complex gneisses in contact with Peter River gneiss beneath the radon anomalies as shown in Figure 13. Graphitic conductors were also detected along the contact. Drilling of the contact encountered no anomalous radioactivity, but values of dissolved radon in water samples from the three holes in Figure 13 were up to 8 times the average value from a nearby hole which was outside the area of radon anomalies. The highest values generally occurred in the vicinity of breccias and alteration zones.

CONCLUSION

The radon tube sampler is suitable for conducting large systematic surveys at high sampling densities which are particularly well suited to the narrow and discontinous anomalies commonly encountered in the Carswell structure and metamorphic environments in general. It provides the interpreter with a larger and better resolved overview of

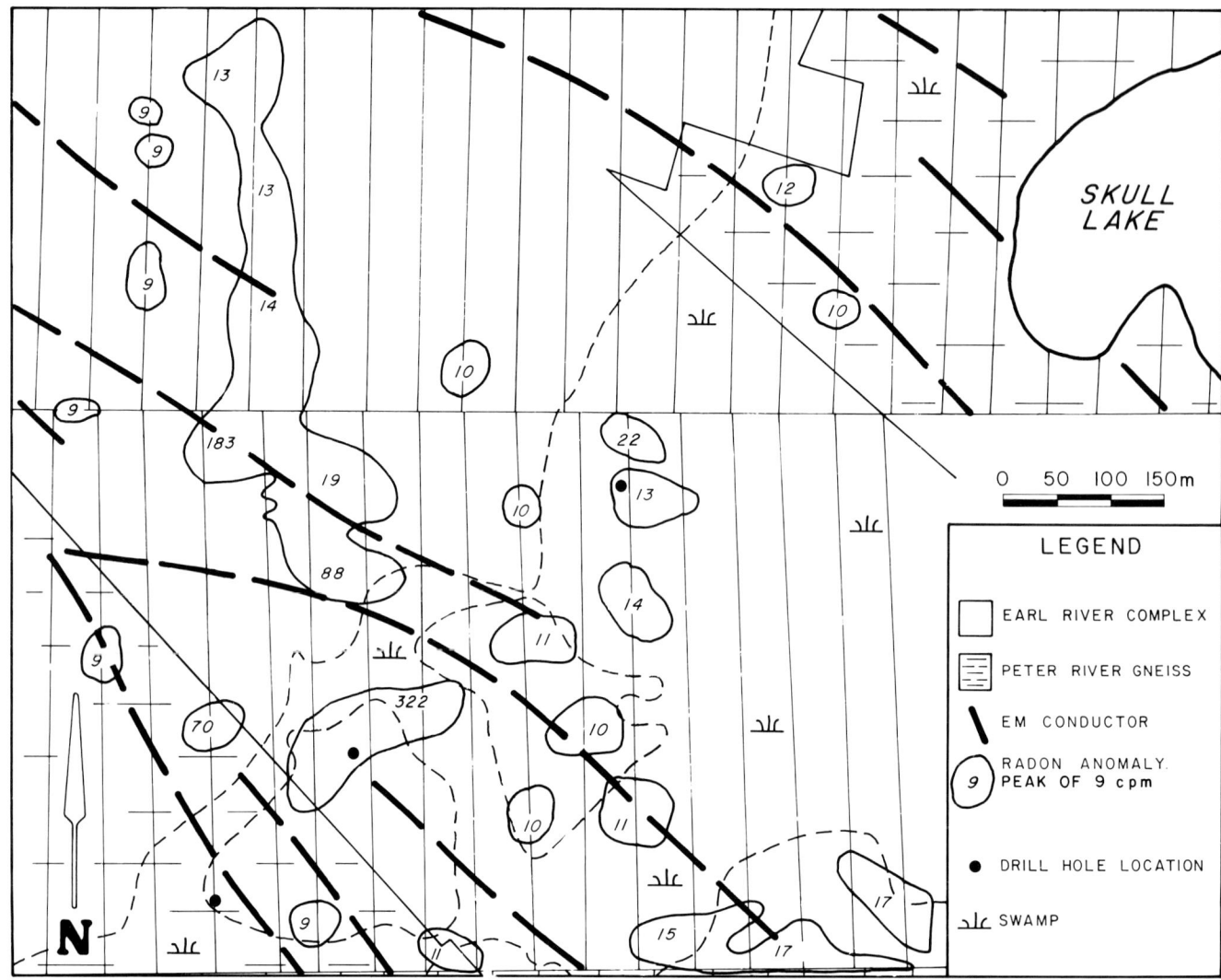

Figure 13. Radon anomaly caused by spring water near Skull Lake.

radon abundance which allows a more precise correlation with geological and geophysical information.

It is important to assess radon anomalies thoroughly before any drilling takes place because few such anomalies lead directly to a discovery. A thorough understanding of the cause, and a sound geological model are essential prerequisites to launching a drilling program. Geophysical information is particularly helpful in this regard.

Sampling of anomalous zones at more than one depth, as was done in the Pascale area example, may help to provide some preliminary assessment of the radon source. An anomaly which decreases or does not increase significantly with depth is probably caused by an overburden source which is close enough to the surface to allow hand trenching. An anomaly which increases rapidly with depth is more likely to be caused by a deeper source. Trenching with heavy equipment is usually an effective means of identifying the radon source. Most radon sources have been conclusively identified within 3 m of surface by trenching. Those which could not be explained were inaccessible to trenching by virtue of being wholly within the bedrock or beneath swamps.

ACKNOWLEDGEMENTS

The author would like to acknowledge the enthusiastic support and able assistance of the many geologists who made use of the radon tube sampler over the years, notably Messrs. D. Yaychuk, G. Roy, D. Chipley, D. Alonso, B. Laporte, A. Lau, K. Tapaninen, A. Durand, E. Koning, F. Tona, and many others too numerous to mention. Thanks are also extended to the students and technicians who collected the many thousands of samples in spite of the tedium, the heat, the rain, and the bugs.

REFERENCES

Beck, L.S., and Gingrich, J.E., 1976, Track Etch Orientation Survey in the Cluff Lake Area, Northern Saskatchewan: Canadian Institute of Mining Bulletin, no. 769, v. 4, p. 104-109.

Card, J.W., and Bell, K., 1982, Collection of Radon Decay Products – A Uranium Exploration Technique: Journal of Geochemical Exploration, v. 17, p. 63-76.

Cadigan, R.A., and Felmlee, J.K., 1977, Radioactive Springs Geochemical Data Related to Uranium Exploration: Journal of Geochemical Exploration, v. 8, p. 381-395.

Fleischer, R.L. and Mogro-Campero, A., 1979, Integrated Radon Mapping in the Earth – Assessment of the ^{220}Rn Signal and its Exclusion: Geophysics, v. 44, no. 9, p. 1541-1548.

Kristiansson, K., and Malmquist, L., 1982, Evidence for Nondiffusion Transport of ^{222}Rn in the Ground and a New Physical Model for the Transport: Geophysics, v. 47, no. 10, p. 1444-1452.

McCorkell, R.H., and Card, J.W., 1978, The Decay Products of ^{222}Rn in Etched Track Radon Detection: Journal of Geochemical Exploration, v. 10, p. 277-293.

Morse, R.H., 1976, Radon Counters in Uranium Exploration: Exploration for Uranium Deposits: International Atomic Energy Agency, Vienna, p. 229-239.

Powell, B.W., 1982, Sampler and Cell for Radon Detectors and Method of Using Same: Canadian Patent 112-8216.

Tona, F., Alonso, D., and Svab, M., 1985, Geology and Mineralization in the Carswell Structure – A General Approach: *in* Lainé, R., Alonso, D., and Svab, M., eds., The Carswell Structure Uranium Deposits, Saskatchewan: Geological Association of Canada Special Paper 29.

GEOPHYSICAL MAPPING OF GNEISS DOMES IN THE CARSWELL STRUCTURE AND THEIR RELATIONSHIP TO URANIUM MINERALIZATION

B. Powell, E. Koning, and R. Lainé
Amok Ltd., 817-825 45th St. W., P.O. Box 9204, Saskatoon, Saskatchewan S7K 3X5

ABSTRACT

There are two major rock groups in the metamorphic core of the Carswell structure. A series of pelitic metasediments called the Peter River gneiss overlies a mixed series of granitic gneisses called the Earl River complex. The latter are known to intrude the Peter River gneiss as ovoid or ridge-shaped domes which constitute an important control on uranium mineralization in the Cluff Lake area. Geophysical survey results are discussed in terms of the geology and rock properties. A diapiric model is proposed to account for these domes and for the regional similarities of the uranium deposits of the Athabasca Basin.

RÉSUMÉ

A Carswell, les gneiss de Peter River reposent sur les gneiss du complexe Earl River. Les mesures effectuées sur carottes en 1981 dans le zone minéralisée de Dominique-Peter révélèrent un contraste de densité important entre les gneiss de Peter River (2,670 kg/m³) et les gneiss du complexe Earl River (2,580 kg/m³). C'est à la suite de cette découverte qu'une compagne de gravimétrie au sol, couvrant environ les 25 km² de la bordure sud du coeur de la structure de Carswell, et comprenant tous les gisements d'uranium de Cluff Lake, fut entreprise. Quatres anomalies négatives couvrant quatres masses de gneiss du complexe Earl River sur les zones de Claude Lake, Dominique, OP-NRF, et Hubert furrent soulignées. Une comparaison entre la géologie et la carte de résistivité apparente permit de conclure que les quatres masses constituent un contrôle stratigraphique majeur des gisements. D'après leur forme, leur densité, leur réponse magnétique, et les foliations des roches susjacentes il semble que ces quatres corps soient des diapires granito-gneissiques résultant d'une séparation anatexique à partir de granito-gneiss ancien plus lourd, appèlé le complexe basique de Earl River, qui affleure sur la moitié est du coeur métamorphique de la structure.

Il est demontré qu'il existe une relation spatiale entre les gisements d'uranium connus et un système d'anomalies gravimétriques négatives linéaires, orientées NE et ESE, dans le bassin Athabasca. Ces anomalies négatives sont reliées à des zones de granitisation.

INTRODUCTION

The Carswell circular structure depicted in Figure 1 has been described by Tona et al. (1985), Harper (1981), and Currie (1969). The map enclosure (in pocket) is a more detailed compilation and interpretation of the available geological and geophysical data. On the surface, the structure is target-shaped. There is an outer ring, 39 km in diameter, comprised of the Carswell and Douglas Formations resting upon the William River Subgroup which in turn forms a zone about 5 km wide between the outer ring and the core. A metamorphic basement complex comprises the cylindrical inner core, 18 km in diameter, which extends vertically some 600 m to 1200 m higher than the surrounding basement below the Athabasca Basin. The core is, for the most part, in faulted contact with the surrounding sediments which are highly deformed for a distance of about 10 km beyond the core. There is a zone, about 2 km wide, immediately surrounding the inner core where bodies or inliers of basement gneisses have been encountered or detected within the Athabasca Group at depths ranging from zero to 700 m below the surface. However, beyond this zone, there is no geological or geophysical evidence of significant bodies of basement gneisses within the Athabasca Group. Bodies of sandstone within the metamorphic core are small (their thickness is usually less than 10 m) and rare compared to basement inliers and are associated with multiple fault contacts near the core boundary.

From 2 to 15 km beyond the metamorphic core, there is evidence of subsidence of the basement. The basement depth contours shown in Figure 1 are compiled primarily from depths to the magnetic basement interpreted by Guillon

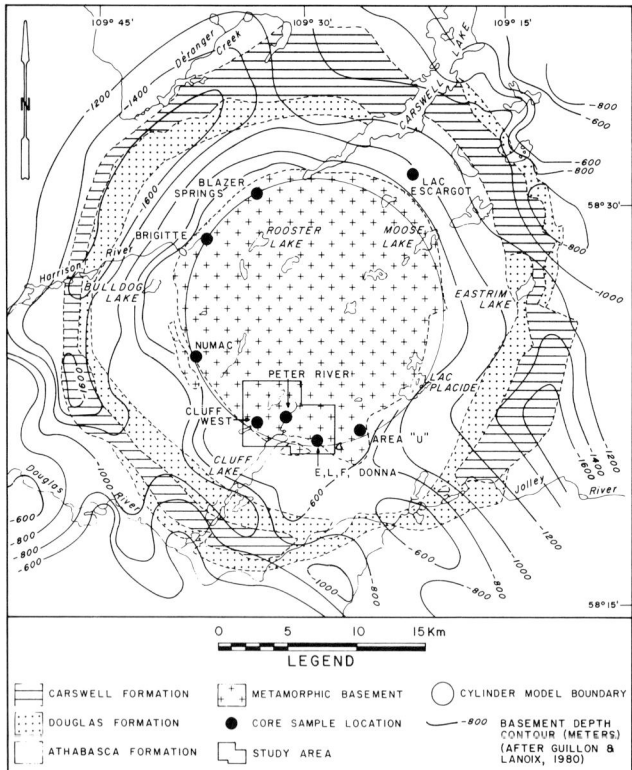

Figure 1. The configuration of the Carswell structure. A central core of uplifted metamorphic basement is surrounded by a zone of subsidence, evidenced by the depth to basement contours and the preservation of the Carswell and Douglas Formations.

and Lanoix (1980) but include some depths reported by Spector (1978). Depths tend to reach a maximum beneath the Carswell and Douglas Formations, an observation which is consistent with the hypothesis presented by Currie (1969) that these were preserved as a result of subsidence. An area of shallow basement where there is a paucity of Carswell and Douglas Formations at the south edge of the structure may reflect a local uplift after the formation of the Carswell structure.

Geophysical exploration in the Carswell structure has historically pursued three objectives: initial "grass roots" exploration, delineation of the shape of the Carswell structure as a whole in three dimensions, and the definition of environments of uranium deposition within the core area. This paper is primarily concerned with the last objective. Among the ground geophysical methods, MaxMin, VLF, resistivity, magnetometry, and gravity predominate. Those methods commonly used elsewhere in the Athabasca Basin to sound through the sandstones, such as magnetotellurics, time and frequency-domain electromagnetic surveys with large loops, and seismic reflection/refraction, have been only occasionally employed because the unconformity surrounding the core is generally too deep to find deposits of economic value.

THE GEOLOGY AND GEOPHYSICAL PROPERTIES OF THE CORE ROCKS

There are two major rock units within the metamorphic core: the Peter River gneiss and the Earl River complex. The uranium deposits in the core of the Carswell structure occur

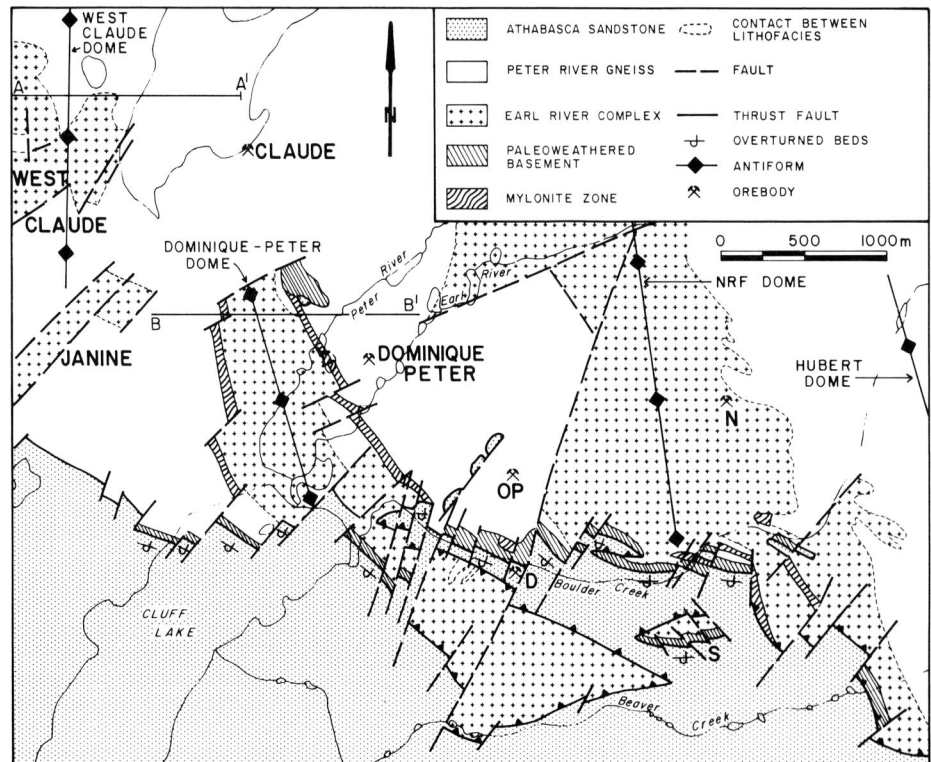

Figure 2. Geology of the study area. See Figure 1 for location. The locations of the Cluff Lake ore deposits are shown with their respective granitic gneiss domes (antiforms) as they appear on the surface.

in a transitional zone at or within 200 m of the contact between domes of Earl River complex and the overlying Peter River gneiss. There are four such domes in the study area shown in Figure 2: the West Claude dome; the Dominique-Peter dome; the NRF dome, and the Hubert dome. The Peter River gneiss, which is the younger of the two according to Bell (1985) and Tona et al. (1985), is composed of aluminous gneisses characterized by a garnet-cordierite-sillimanite assemblage with minor amounts of interfoliated granitoids and layers of garnet-rich (up to 40%) aluminous gneiss. The Peter River gneiss is known to be at least 472 m thick (DDH 1623) in the area north of the Dominique-Peter area and east of Claude Lake. Granitoids are abundant at the interface between the Peter River gneiss and the Earl River complex and probably mark an anatectic front (Pagel and Svab, 1985). The Earl River complex, which has been drilled to depths of 545 m (DDH 425), is comprised of feldspathic gneisses, mafic gneisses, granitoids and amphibolites.

A felsic Earl River complex, which is comprised predominantly of granitoids and feldspathic gneisses, is encountered mostly in the west half of the core. The mafic gneisses of the Earl River complex are the least remobilized facies (Pagel and Svab, 1985). In the Peter River area, Bell et al. (1985) have noted a general increase in the mafic content of the Earl River complex with depth and have described the mafic gneiss as komatiitic in composition. A mafic Earl River complex, which is comprised predominantly of mafic gneiss and mafic feldspathic gneiss, is encountered on surface in the east half of the core and below the felsic Earl River complex on the west side. Bell (1985) has dated the felsic and mafic Earl River complex rocks at circa 1870 Ma and 2005 Ma. respectively, and the Peter River gneiss at circa 1760 Ma. Harper (1981) reports a somewhat older age of circa 2300 Ma for samples of "granodiorite" (taken from the Marie-Andrée area in the northern perimeter of the core) which are apparently the equivalent of Earl River complex. The mafic Earl River complex is therefore interpreted to be stratigraphically older and deeper than the felsic Earl River complex which is in turn older than the Peter River gneiss. It is concluded that the east half of the core, where the mafic Earl River complex is abundant, has been uplifted and eroded relative to the west half, where there are greater abundances of Peter River gneiss and granitoids.

Table I gives some average values of magnetic susceptibility, bulk density, and resistivity for the two major rock units of the core. The values of magnetic susceptibility and bulk density were determined from measurements on core samples while the resistivities and chargeabilities were estimated from borehole and surface resistivity measurements. For the density measurements, the samples were presoaked in water and towel dried to return them to their natural state prior to weighing. All values should be considered as representative but not necessarily definitive because there are considerable variations of all properties within each unit.

The average magnetic susceptibility of the Peter River gneiss at 20×10^{-6} C.G.S. is sharply lower than the Earl River complex. The higher average (and standard deviation) of the Earl River complex is attributable to occasional zones of amphibolite, biotite feldspathic gneiss, and mafic gneiss where susceptibilities can be as high as 10^{-2} C.G.S. The average densities of both units are higher in the east half of the core. For this reason, they are quoted separately for the west and east halves. In both halves the average density of the Peter River gneiss is greater than the Earl River complex. The Peter River gneiss have resistivities of less than 300 ohm-metres, averaging about 35 ohm-metres in zones in contact with domes of Earl River complex and about 100 ohm-metres elsewhere. The Earl River complex averages about 2,500 ohm-metres and is therefore easily distinguished from the Peter River gneiss by resistivity and EM techniques. The chargeability of the Peter River gneiss is about six times that of the Earl River complex, but this property has not been exploited because the units are adequately mapped with resistivity surveys.

GEOPHYSICAL SURVEYS OVER THE CLUFF LAKE DEPOSITS

Gravity

In 1981, density measurements were made on core samples from eleven diamond drill holes in the Peter River discovery zone as part of a larger program of density measurements. Table II summarizes the densities of the three major rock groups by area in the Carswell structure: Athabasca Group sandstones, Earl River complex, and Peter River gneiss. A significant density contrast of 80 kg/m³ was found to exist between the Peter River gneiss and the Earl River complex. A gravity survey was therefore carried out in the spring of 1982 with the intention of mapping the extent of the Earl River complex at depth.

TABLE I

AVERAGE PROPERTIES OF THE ROCKS OF THE METAMORPHIC CORE

Unit	Magnetic Susceptibility (10^{-6} C.G.S. Units)	Bulk Density (kg/m³) West Half of Core	Bulk Density (kg/m³) East Half of Core	In Situ Resistivity (ohm-metres)	In Situ Chargeability (mV/v)
Peter River Gneiss	20 ± 14 (537)*	2,650 ± 160 (391)	2,740 ± 90 (49)	35 - 100	60
Earl River Complex	154 ± 975 (1502)	2,580 ± 50 (felsic) (498)	2,650 ± 130 (mafic) (37)	2500	10

*The number of samples for each determination is given in brackets. Standard deviations are shown as ±.

TABLE II
AVERAGE CORE SAMPLE DENSITIES* FROM THE CARSWELL STRUCTURE

Area	Athabasca Group	Earl River Complex	Peter River Gneiss
Numac	2,550 ± 30 (56)	2,580 ± 50 (150)	2,620 ± 50 (94)
Brigitte	2,530 ± 90 (26)	–	–
Cluff West	2,490 ± 50 (44)	2,590 ± 110 (34)	–
Dominique-Peter	–	2,580 ± 50 (276)	2,660 ± 120 (290)
Blazer Springs	2,570 ± 40 (12)	2,530 ± 50 (17)	2,650 ± 90 (7)
E,L,F, Donna	2,550 ± 70 (122)	2,590 ± 60 (21)	–
OP (DDH 2757)	–	2,630 ± 80 (157)	–
"U"	–	2,650 ± 130 (37)	2,740 ± 90 (49)
Lac Escargot	2,470 ± 50 (40)	–	–

*Units are kg/m^3.
Number of samples is quoted in brackets.
Standard deviations are shown as ±.

An area of approximately twenty-five square kilometres, along the south edge of the metamorphic core, was surveyed which encompassed all of the known ore deposits in the study area and comprised 2,380 readings on a nominal 100 m by 250 m grid. A La Coste & Romberg, Inc. Model G gravity meter was used to record the gravitational field at each station, from which Bouguer values were computed using the 1967 International Gravity Formula (Nagy, 1978) and a Bouguer density of 2,670 kg/m^3. Terrain corrections were applied using Hammer's (1939) charts.

The observed anomalous field was found to be dominated by the core which rises vertically for more than 1000 m through the Athabasca sandstones. A residual Bouguer map was therefore computed by subtracting out the field of a right vertical cylinder, computed by the method of Nagy (1965) representing a core which is 9.25 km in radius (the cylinder model boundary is shown in Fig.1). For the purpose of this computation, a uniform density of 2,670 kg/m^3 was assumed for the core and 2,550 kg/m^3 for the Athabasca sandstones. The position of the cylinder has been placed slightly off-centre to obtain the best fit to the southern edge of the core where this was most critical. The cylinder model does not constitute a complete representation of the entire Carswell structure. Nevertheless, it was expected to represent adequately the regional field created by the core of the structure in the survey area. In addition, the gravitational effect of Cluff Lake itself has been removed by computer modelling a bathymetric map according to the method of Nagy (1980), assuming a density contrast of -1,670 kg/m^3. The resulting map is presented in Figure 3.

Four gravity lows can be seen in Figure 3, corresponding to the domes of Earl River complex in Figure 2: a large low in the northwest corner of the map which extends under Claude Lake and is bounded by a N45°E fault through Claude Lake, a smaller low under the Dominique-Peter area which is joined by a narrow bridge to the Claude Lake body, a long narrow north-south trending low between the OP and N deposits, and a modest low at the east edge of Figure 3 coincident with the Hubert dome. Also shown are the outlines of the five ore zones in the area. One observes that the three major basement-hosted deposits (Claude, Dominique-Peter, and N) reside on areas of high horizontal gradient corresponding to the margins of the known domes of Earl River complex. This is made clearer in Figure 4 which depicts two east-west geological cross sections through the West Claude and Dominique-Peter domes. These show the domal shape of the two bodies of Earl River complex which correlate well with their respective gravity lows. While the distribution of Earl River complex, as it is known from drilling in the area, is consistent with the pattern of gravity lows, some minor lows can be attributed to faults and alteration zones such as the Cluff Lake fault which trends N45°E through the lake and through the Dominique-Peter dome. These can be seen by applying a high-pass, digital filter to the data (not shown). Six major directions of faulting have been identified: north-south, N45°, N60°, N80-90°, N120°, and N140°E. All of these directions appear to be related to regional faulting discernible on Landsat photos. Gregory (1983a) has obtained similar directions from Landsat photos in the eastern part of the Athabasca Basin. The N60°, east-west and north-south directions are known to be the most important for controlling uranium mineralization in the Cluff Lake area.

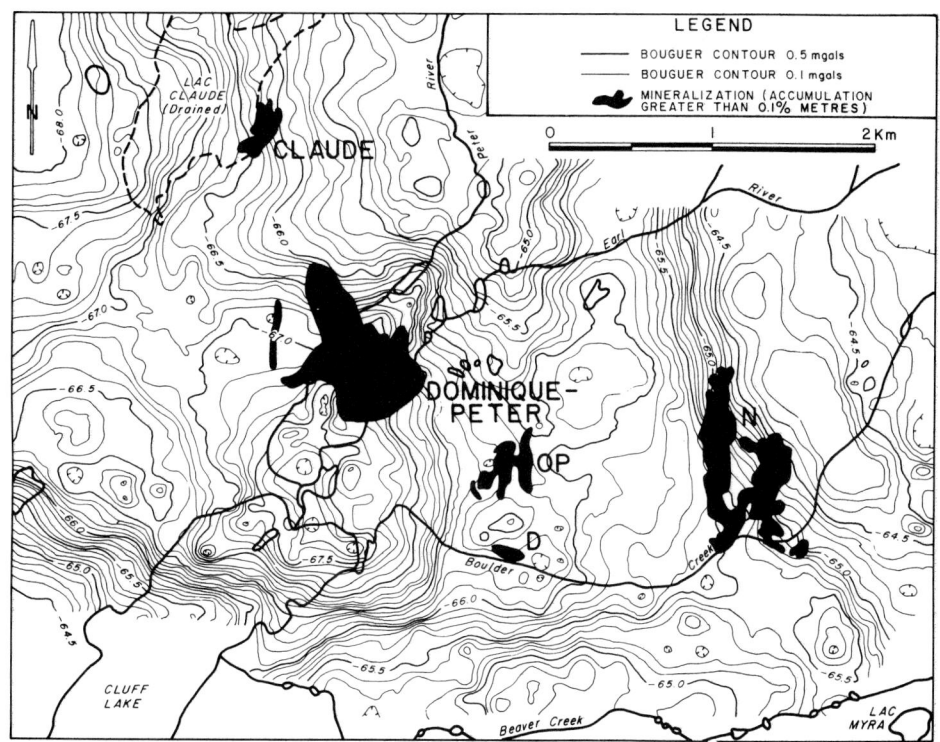

Figure 3. Residual Bouguer contour map of the study area. The mineralized zones in black fall on the margins of domes of Earl River complex defined here as negative residual Bouguer anomalies.

Airborne Resistivity

A halfspace apparent resistivity map of the same area (Fig.5), computed by Dighem Ltd. (Fraser, 1978) from a helicopter-borne EM survey by Aerodat Ltd., reflects the distribution of the Earl River complex at the surface. Areas where the apparent resistivity is less than 300 ohm-metres generally reflect Peter River gneiss, while those which are greater than 300 ohm-metres are either Earl River complex

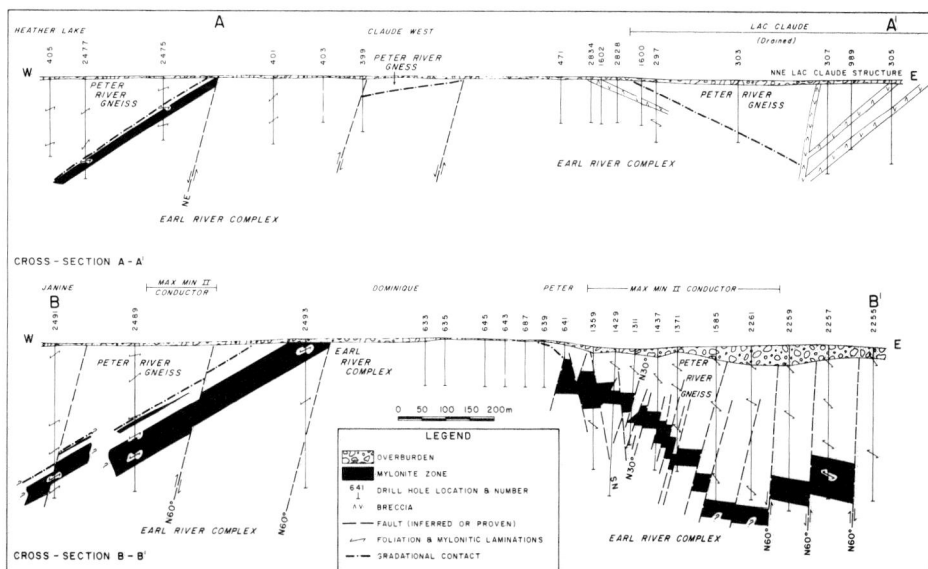

Figure 4. Geological cross sections of the Claude Lake (above) and Dominique-Peter domes (below). See Figure 2 for location. Both domes intersect the surface at their crests. Foliations and mylonitic laminations parallel the dome surfaces. MaxMin conductors flank the Dominique-Peter dome.

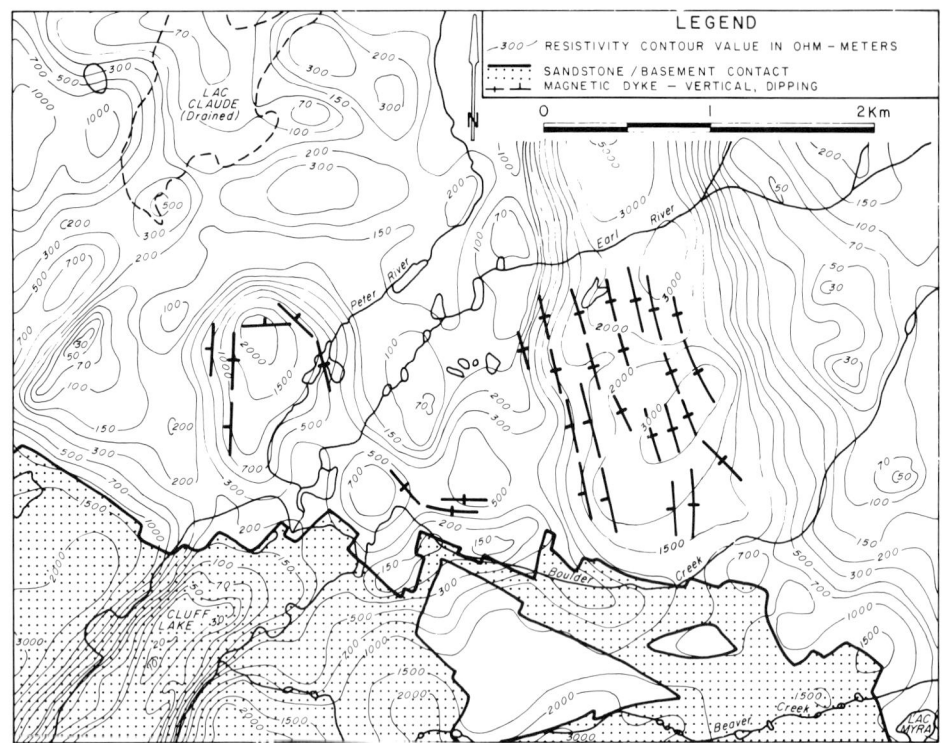

Figure 5. An apparent resistivity map of the study area. Areas where the apparent resistivity is less than 300 ohm-metres generally reflect Peter River gneiss, while those greater than 300 ohm-metres are either Earl River complex or sandstone.

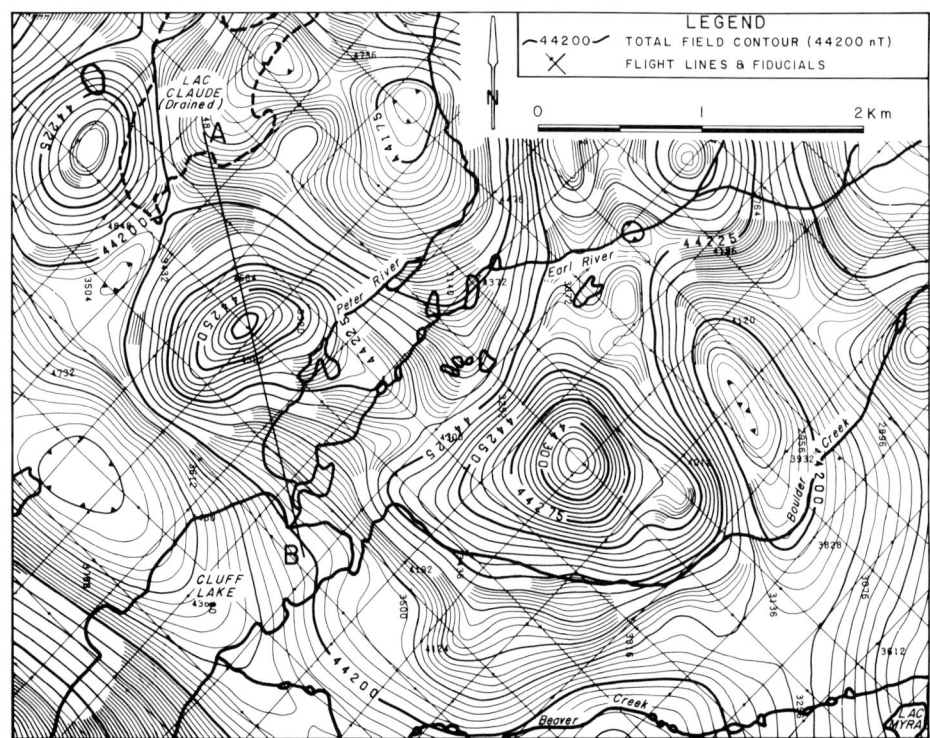

Figure 6. Aeromagnetic survey of the study area. The distribution of positive magnetic anomalies is less extensive but quite similar to the negative residual Bouguer anomalies in Figure 3.

or sandstone. If one compares Figures 3 and 5, it can be seen that substantial portions of the West Claude, Dominique-Peter, and Hubert domes are covered by Peter River gneiss, whereas the NRF dome is not. This is consistent with the geological observations: the Claude ore deposit occurs within Peter River gneiss but adjacent to the West Claude dome which extends under Claude Lake, the Dominique-Peter ore deposit is found at a mylonitized contact with Earl River complex beneath 140 m to 220 m of Peter River gneiss, the N and OP deposits begin essentially at the bedrock surface within and adjacent to the Earl River complex, respectively. The Hubert Lake dome does not reach the surface and no associated ore deposits have been found to this date. It would thus appear that mineralization is found on or near the surfaces of domes of Earl River complex. There also appears to be a halo of more conductive Peter River gneiss immediately surrounding the domes. This is further illustrated in Figure 4 (cross section B-B') which shows the locations of two MaxMin II conductors on the flanks of the Dominique-Peter dome. The conductivity is attributed to an increase in the graphite and sulphide content along fractures in the Peter River gneiss.

Magnetometry

Shallow, dyke-like, magnetic bodies which were outlined by ground magnetic surveys are shown in Figure 5 in the Dominique-Peter area and to the east between Earl River and Boulder Creek (those in the Claude Lake area and elsewhere have not been shown because of incomplete coverage). All bodies strike within and parallel to the edges of the domes of Earl River complex, as defined by the halfspace resistivity contours, and dip subvertically. In general, their positions and orientation are consistent with an anticlinal or domal structure.

Figure 6 illustrates the total magnetic field in the Cluff Lake area at 200 m above the ground level from a high sensitivity aeromagnetic survey of the Carswell structure conducted by Geoterrex Ltd. in 1979. The distribution of positive magnetic anomalies is less extensive but quite similar to the distribution of negative gravity anomalies. If the gravity data are an accurate reflection of the extent of the granitic domes, it is apparent that some portions of the Earl River complex have a significant magnetic susceptibility, while other portions do not. Guillon and Lanoix (1980) observed that the magnetic anomalies in the core area could be separated into two groups: small, shallow magnetic bodies and large, subvertical, intrusive-looking magnetic bodies which are deeply rooted in the basement beneath the core of the Carswell structure. To illustrate this, Figure 7 shows a two-dimensional, semi-infinite, two-body, thick-dyke model which has been fitted to a profile through the center of the magnetic anomaly over the Dominique-Peter dome. The computations were performed by the method of Gay (1967). Body A constitutes a deeply rooted magnetic body about 230 m below surface on the north half of the Dominique-Peter dome. Body B reflects the shallow magnetic dykes depicted in greater detail in Figure 5, situated at the peak of the resistivity high where Earl River complex is found at the

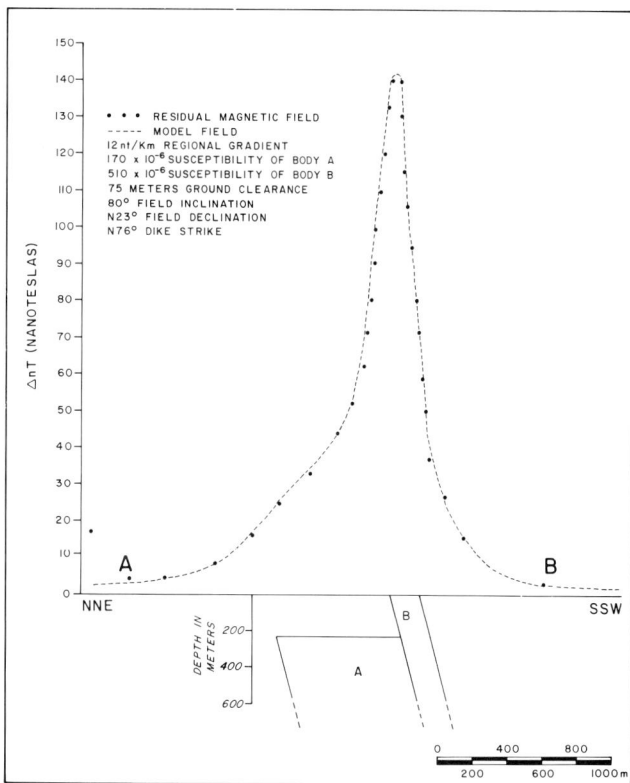

Figure 7. Two-dimensional magnetic model of the Dominique-Peter dome. Guillon and Lanoix (1980) observed that the magnetic bodies in the core could be separated into large, subvertical, intrusive-looking bodies (A) and smaller, shallow magnetic bodies (B).

surface. Both bodies are about 1 km north of the centre of the Dominique-Peter gravity low.

Synthesis

A three dimensional block model (Fig. 8) was fitted to a gridded version (200 m × 200 m centres) of the gravity map in Figure 3. The computations were performed by the method of Nagy (1980). The units represented are West Peter River gneiss (2,670 kg/m^3), East Peter River gneiss (2,740 kg/m^3), granitoid and felsic Earl River complex (2,570 kg/m^3), mafic Earl River complex (2,670 kg/m^3), Athabasca sandstone (2,570 kg/m^3), altered sandstone or regolith at the core boundary (2,350 kg/m^3), and overburden (2,000 kg/m^3). The densities used in the modelling were partly determined by the need to keep the number of blocks to a manageable size. The blocks were generally fitted to the gridded values within a tenth of a milligal. About 5% of the computed values deviated between 0.1 and 0.4 milligals. Drilling and resistivity information were used wherever possible to fix the surface positions of the blocks and minimize the ambiguity of the results.

Figure 9 shows an interpretive cross-section through the study area. It is based largely on the three-dimensional gravity block model, known geology from outcrops and drill holes (Fig.8), and the previously noted characteristics of the magnetic anomalies. The bodies of the mafic Earl River complex, depicted in the root zones of the domes, reflect the

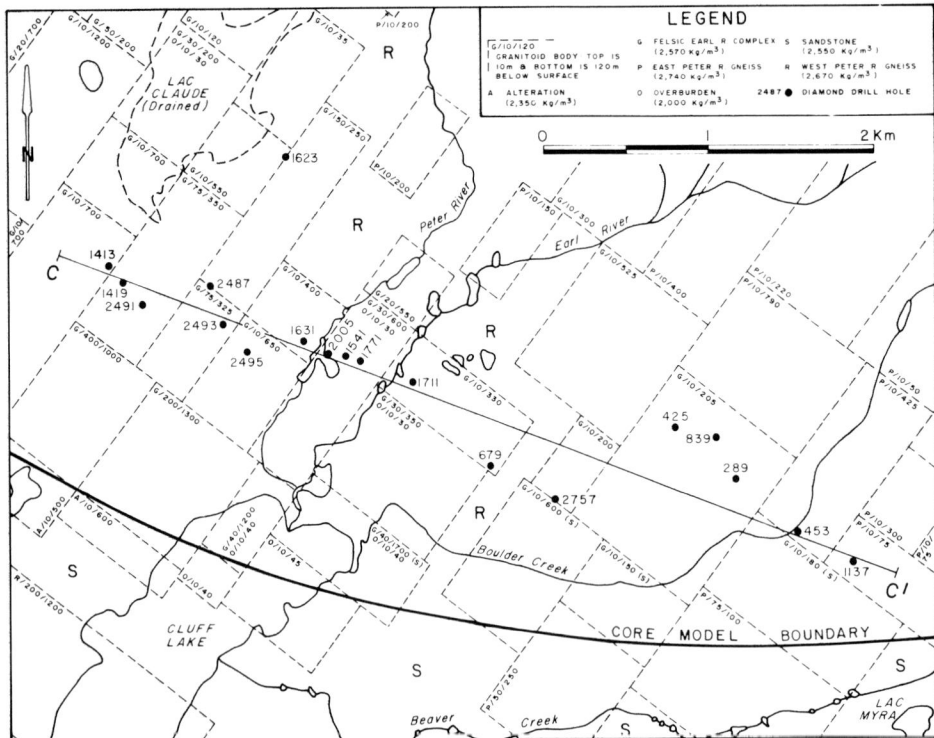

Figure 8. Three-dimensional block model of the residual Bouguer map in Figure 3.

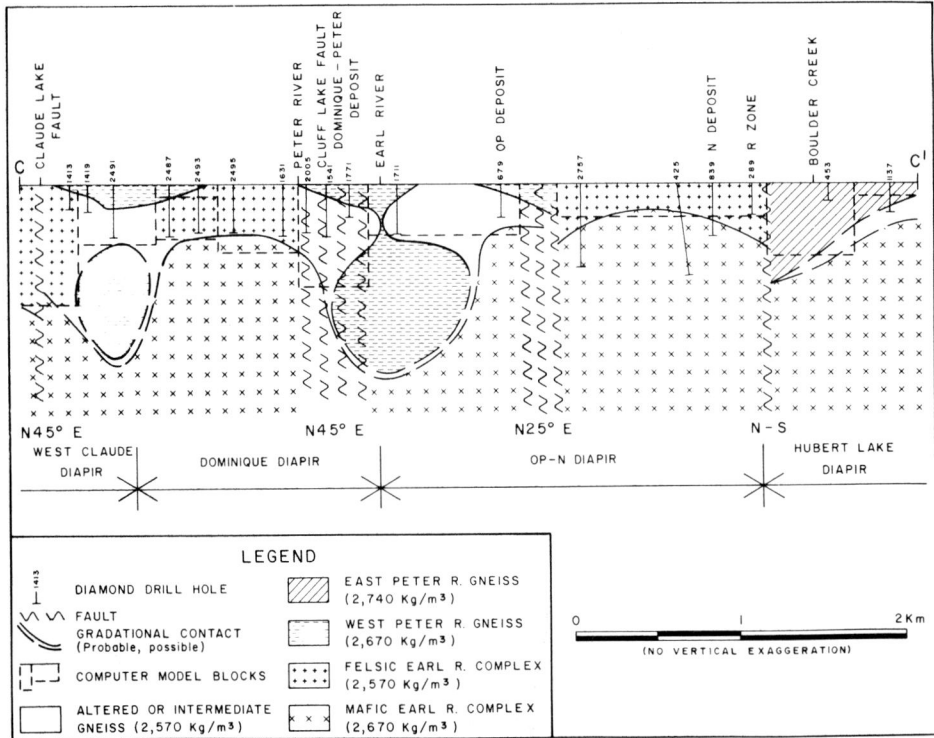

Figure 9. An interpretive cross section through the metamorphic core. A simple diapir model is employed to explain the distribution of Earl River complex established by the gravity model of Figure 8 and the aeromagnetic and resistivity surveys (Figs. 4 and 5).

observed increase in mafic content of the Earl River complex, and the subvertical, deeply rooted nature of the magnetic bodies. The mafic Earl River complex with depth, the finite depths of the bodies of felsic Earl River complex is expected to be detectable magnetically, but at 2,650 kg/m³, it is too dense to produce a significant gravity low in a host of Peter River gneiss. Diamond drill hole CLU 2757 (Fig. 8), which is situated between OP and N, and about 200 m south of the section shown in Figure 4, encountered the mafic Earl River complex gneisses at a depth of 207 m. These have an average density of 2,650 kg /m³ from 300 m to the bottom of the hole at 480 m, which is the same as the mafic Earl River complex gneisses observed to the east in the "U" area (Table II). The average density in the upper 300 m of the same hole is 2,610 kg/m³, a density which is intermediate between Earl River complex gneisses in the west half of the core and those to the east. Since there are large fluctuations in densities down the hole, the lighter zones may simply become less common at depth.

Some features of the domes of the Earl River complex suggest a diapiric process:

1) In plan view the domes are ovoid to ridge-like in shape and occur predominantly at intervals of 1.5 km throughout the core. They exhibit a similar orientation locally, but there are substantial variations in orientation from one part of the core to another. They are similar in that aspect to salt ridges and domes, notably those of NW Germany (Ramberg, 1981, p. 249, and Jaritz, 1973).

2) The felsic Earl River complex, which occupies the upper portions of the domes, has a density which is significantly lower than that of the surrounding Peter River gneiss. This is true in both the east and west halves of the core even though rock densities are higher in the east.

3) Mafic content increases with depth in the Earl River complex. This is an observed and well understood characteristic of granitic gneiss diapirs (Ramberg, 1981, p. 297-309). A buoyant layer of anatectic granitoid basement will pull a non-buoyant, non-rigid basic substratum upward with it as the diapir forms.

4) Magnetic dykes within the Earl River complex strike parallel to the dome surfaces, dip steeply, and generally have large depth extents much like the marker beds in the cores of model diapirs presented by Ramberg (1981).

5) The domes are mantled by a layer of Earl River complex granitoids and a conductive, graphitic-pyritic zone of Peter River gneiss which is detectable by MaxMin II (Fig. 4). This is consistent with a doming process from initially horizontal layers.

6) Foliations, internal and external to the domes, tend to parallel the dome faces (Fig. 4). This is a characteristic of diapirism as opposed to buckle folding (Schwerdtner and Lumbers, 1980, and Dixon and Summers, 1983).

7) Radiometric ages in the upper felsic Earl River complex are intermediate between the younger Peter River gneiss above and the older mafic Earl River complex below.

The rocks of the core were subjected to extreme tectonism related to the Carswell cryptoexplosion event. It is therefore surprising that dome structures are preserved. However, geological cross sections as exemplified in Figure 4 constitute strong evidence. This is not unreasonable if the displacement of the core at the time of the event was almost entirely in an upward direction and relative movements of blocks of core were small compared to the dimensions of the diapirs. All of the faults perceived inside the core by gravity appear to be regional faults. This suggests that radial and concentric faulting associated with the Carswell event within the metamorphic core are relatively minor by comparison.

REGIONAL OBSERVATIONS

Figure 10 is a gravimetric map of the Carswell structure (Department of Energy, Mines and Resources, Canada, 1963). Within the basement core the gravitational field is significantly lower on the west side where there is a significantly greater abundance of Earl River complex granitoids and Peter River gneiss. Both observations are consistent with substantial diapir development. The field on the east side, however, is uniformly high and corresponds to an area of predominantly mafic Earl River complex. Also shown in Figure 10 are the uranium occurrences of the Carswell structure with values of greater than 1000 ppm U. These occurrences cluster around two areas within the west half of the metamorphic core where the gravitational field is apparently lower because of the presence of felsic Earl River complex. On the east side, where mafic Earl River complex

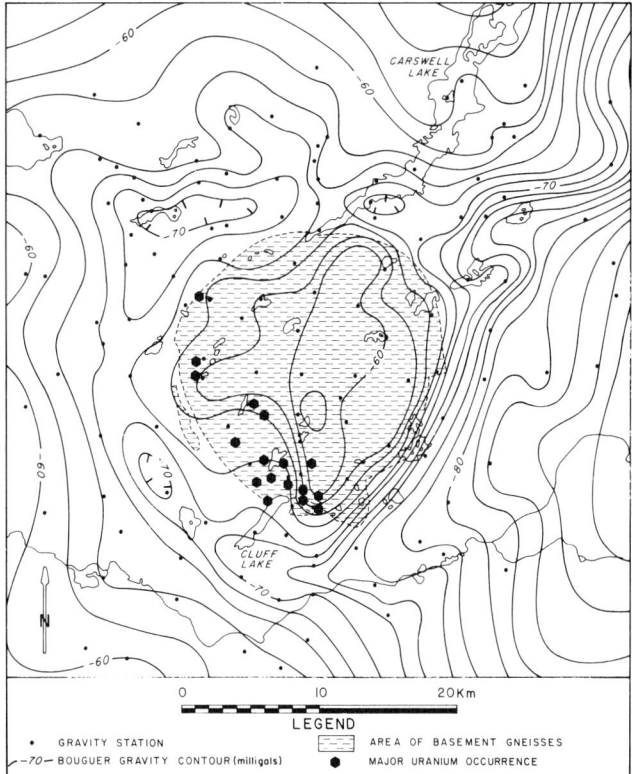

Figure 10. Gravity map of the Carswell structure (Energy, Mines and Resources, Canada 1963, and Spector, 1978). The lower gravitational field of the west side of the metamorphic core is attributed to gneiss domes. The low surrounding the core is due to subsidence. The large low along the right side of the figure reflects the Clearwater Domain.

predominates, there are no significant uranium occurrences.

The concentric gravity low just beyond the core of the Carswell structure is due to the previously mentioned subsidence of the sandstone-basement unconformity in that zone (Currie, 1969). The large gravity low along the east side of the figure is a northeast trending linear regional anomaly called the Athabasca axis by Darnley (1981) who notes that it can be traced for 2,000 km from Baker Lake, N.W.T. to the foothills of the Western Cordillera. The Carswell structure sits on the west flank of this gravity low, which coincides with the Clearwater Domain (Fig. 11) and along a left-lateral northeast fault displacement of it of about 3.5 km. It would seem reasonable to ask if there is any relationship between the gravity lows observed within the core of the Carswell structure and the gravity low associated with the Clearwater Domain.

Most of the regional gravity lows in the Athabasca region have the same N45° E orientation and are distributed across the basin at intervals of about 60 km (Fig. 11). Walcott (1968) interpreted some of these as granitic intrusions. Gregory (1983b) notes that the gravity low on the west shore of Wollaston Lake corresponds to a "synform of Aphebian supracrustals" whereas the corresponding high immediately to the west represents "granulitic rocks in the basement". This pattern is similar to what is observed in the metamorphic core of the Carswell structure. Hudsonian age Peter River gneiss and Earl River complex granitoids predominate over the west half of the core where the gravitational field is low relative to the east half where older granulite facies (Herring, 1976) Earl River complex predominates.

The uranium deposits of the Athabasca Basin (Fig. 11) exhibit a tendency to fall either directly over, or immediately adjacent to one or another of the outlined gravity lows. The Key Lake deposits are the only exception to this observation since they occur some 50 km from the nearest of these gravity lows. However, they do occur in an area of Hudsonian age supracrustals and granites (Tremblay, 1982; Ray, 1977). It can be seen in Figure 11 that the gravity low on the west shore of Wollaston Lake should extend to Key Lake to complete the observed pattern of a low approximately every 60 km.

Many of the characteristics of the Earl River complex in the Cluff Lake area have been observed in similar granitoid domes or "massifs" found in association with uranium deposits at the east rim of the Athabasca Basin: they tend to form ovoid or elongated ridges; foliations in both the pelitic and granitoid gneisses parallel the dome interface (Sibbald, 1983; Ray, 1977); the granitoids appear to be derived partially from migmatization and/or anatexis of overlying supracrustals and partly by remobilization of the older basement (Sibbald, 1983); there is a mass excess associated with the metapelites relative to the granitoids (Sobczak, 1983); and positive aeromagnetic anomalies occur over the domes while negative anomalies occur over adjacent pelitic gneisses (Sibbald, 1983; & Gregory, 1983b).

DISCUSSION

The regular pattern and 60 km intervals of the northeast trending gravity lows in Figure 11 are consistent with a model of regional diapiric emplacement of granite batholiths from an initial, horizontal layer of buoyant granite. Ramberg (1981, p. 102) described a specific model comprised of equal layers of unmetamorphosed sediments, metasediments, and buoyant granite on a basic substratum. He determined that the ratio of the diapir interval (dominant wavelength) to the layer thickness would be 4.5:1. For this model, a 60 km interval between batholiths would develop from a granitoid layer 13 km thick at a depth of 27 km. Walcott (1968) fitted model batholiths with vertical dimensions of 34 km to 43 km to the same gravity lows. These values are consistent with an initial 27 km depth to the granitoid layer in the Ramberg model if one allows for a thickening of the interdiapiric crust as a consequence of diapiric emplacement, as well as the uncertainties in both of the Ramberg and Walcott models.

Granitoid diapirs typically exhibit higher orders of diapirs superimposed upon them (Schwerdtner and Lumbers, 1980). The smaller gneiss diapirs observed in the Carswell structure and along the east rim of the Athabasca Basin may have developed in this fashion. The local preservation of Aphebian supracrustal gneisses might then be a direct result of their subsidence associated with the development of higher order diapirs. Clark and Burrill (1981) and Darnley (1981) have argued that granite batholiths, rich in the radioelements, were a source of radiogenic heat driving the low temperature hydrothermal activity of the Athabasca Basin. In such a model, fault zones which cross the higher order diapirs may have acted as conduits for uranium-bearing fluids rising to the sandstone-basement unconformity.

CONCLUSION

The uranium deposits in the Cluff Lake area in the metamorphic core of the Carswell structure are associated with granitic gneiss domes which comprise the Earl River complex. These domes intrude an overlying series of pelitic metasediments known collectively as the Peter River gneiss. The observable characteristics of these domes are consistent with a diapiric process involving a buoyant layer of anatectic granitoid basement. Negative residual gravity anomalies in the west half of the core are attributed to masses of light felsic Earl River complex in their crests. Mafic Earl River complex, which is encountered beneath the felsic type and at surface over the east half of the core, is characterized by higher rock densities and ages. It is concluded that the east half of the core may have experienced an uplift and erosion relative to the west half.

A regional pattern of linear northeast trending gravity lows every 60 km across the Athabasca Basin is consistent with diapiric emplacement of batholiths involving a deep, buoyant layer of granitoid basement. If the batholiths were a source of radiogenic heat driving the low temperature hydrothermal activity associated with the formation of the uranium deposits, this would explain the correlation of the deposits with the regional gravity lows. The domes observed in the core of the Carswell structure and similar domes observed at the east rim of the Athabasca Basin are interpreted as higher order diapirs superimposed on these regional diapiric batholiths.

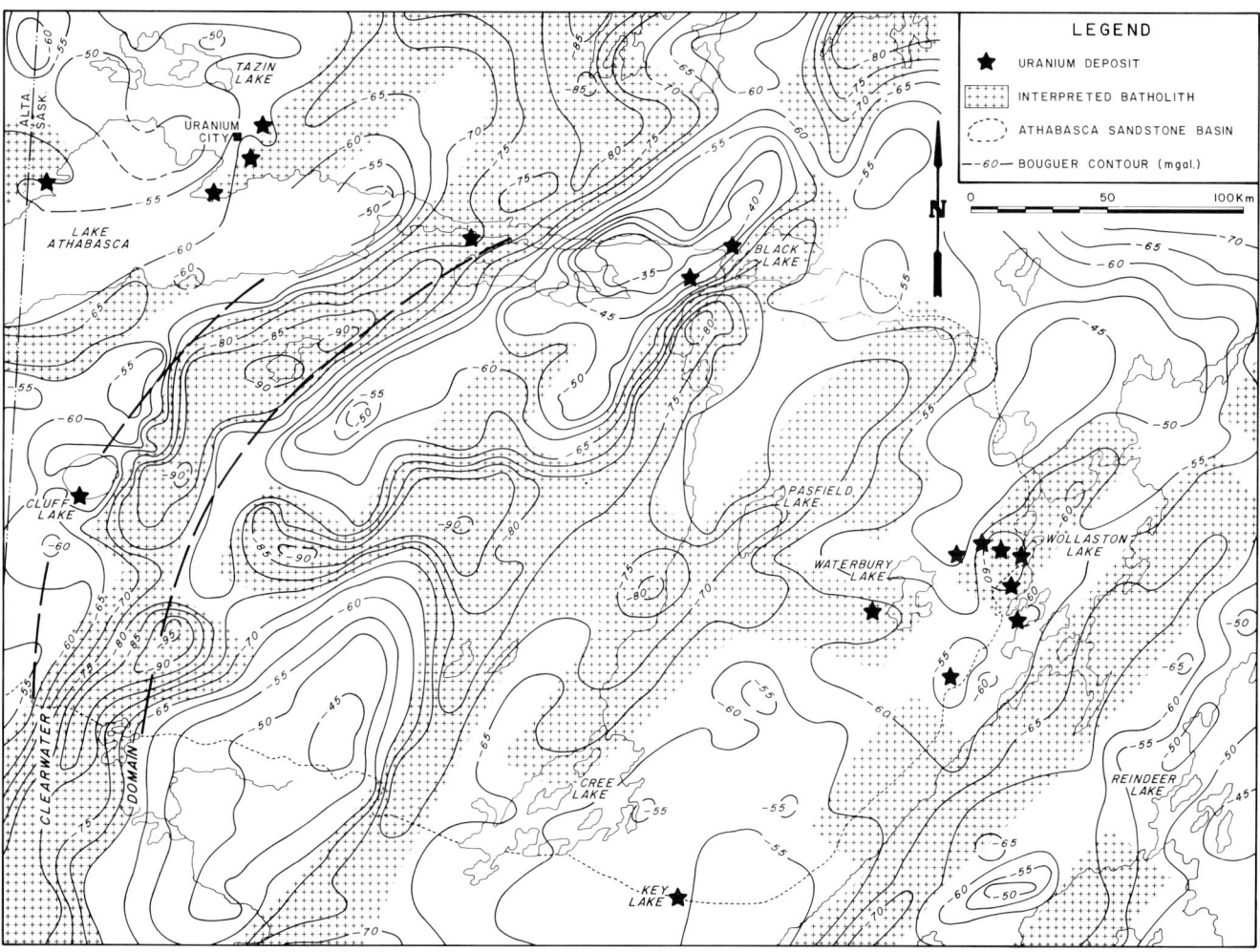

Figure 11. Regional gravity lows and their relationship to the uranium deposits of the Athabasca Basin (Energy, Mines and Resources, Canada, 1981, Gravity Map Series no. 56108 and 56102).

ACKNOWLEDGEMENTS

The authors would like to acknowledge the contribution of the many outstanding geologists, geophysicists, and technicians who contributed to the knowledge and understanding of the Carswell structure and its uranium deposits over the years. Notable among these are Messrs. W. Rose, M. Proulx, D. Yaychuk, B. Laporte, K. Metcalfe, A. Lau, T. Durand, F. Tona, K. Tapaninen, C. McAleenan, D. Alonso, C. Andrews, J.R. Blaise, J. McAuley, S. McNamara, J. Preciado, M. Svab, K. Wheatley, S. Wilson, M. Yip, M. Gabette, M. Giroux, and J.C. Cheveau.

REFERENCES

Bell, K., 1985, Geochronology of the Carswell Area, Northern Saskatchewan: in Lainé, R., Alonso, D., and Svab, M., eds., The Carswell Structure Uranium Deposits, Saskatchewan: Geological Association of Canada Special Paper 29.

Bell, K., Cacciotti, A.D., and Schnessl, J.H., 1985, Petrography and Geochemistry of the Earl River Complex, Carswell Structure, Saskatchewan: A Possible Proterozoic Komatiitic Succession: in Lainé, R., Alonso, D., and Svab, M., eds., The Carswell Structure Uranium Deposits, Saskatchewan: Geological Association of Canada Special Paper 29.

Clark, L.A., and Burrill, G.H.R., 1981, Unconformity-Related Uranium Deposits, Athabasca Area, Saskatchewan, and East Alligator Rivers Area, Northern Territory, Australia: Canadian Insitute of Mining and Metallurgy Bulletin, v. 74, no. 831, p. 63-72.

Currie, K.L., 1969, Geological Notes on the Carswell Circular Structure, Saskatchewan (74K): Geological Survey of Canada Special Paper 67-32, 60 p.

Darnley, A.G., 1981, The Relationship between Uranium Distribution and some Major Crustal Features in Canada: Mineralogical Magazine, v. 44, p. 425-436.

Dixon, J.M., and Summers, J.M., 1983, Patterns of Total and Incremental Strain in Subsiding Troughs: Experimental Centrifuged Models of Interdiapir Synclines: Canadian Journal of Earth Sciences, v. 20, p. 1843-1861.

Fraser, D.C., 1978, Resistivity Mapping with an Airborne Multicoil Electromagnetic System: Geophysics, v. 43, no. 1, p. 144-172.

Gay, S., 1967, Curves for Interpretation of Magnetic Anomalies: *in* Mining Geophysics, v. II: Society of Exploration Geophysicists, Tulsa, Oklahoma, p. 512-548.

Gregory, A.F., 1983a, An Application of Landsat Data to Exploration in the NEA/IAEA Athabasca Test Area: *in* Cameron, E.M., ed., Uranium Exploration in the Athabasca Basin, Saskatchewan, Canada: Geological Survey of Canada Paper 83-11, p. 111-115.

_____, 1983b, Interpretive Geological Mapping as an Aid in the NEA/IAEA Athabasca Test Area: *in* Cameron, E.M., ed., Uranium Exploration in the Athabasca Basin, Saskatchewan, Canada: Geological Survey of Canada Paper 83-11, p. 171-178.

Guillon, J.C., and Lanoix, M., 1980, High Sensitivity Airborne Magnetometer Survey in the Cluff Lake Area, Saskatchewan: Geoterrex Ltd., Ottawa, 44 p.

Hammer, S., 1939, Terrain Corrections for Gravimeter Stations: Geophysics, v. 4, no. 3, p. 184-194.

Harper, C.T., 1981, Uranium Metallogenic Studies: Cluff Lake Area: *in* Summary of Investigations 1981, Saskatchewan Geological Survey Miscellaneous Report 81-4, p. 57-61.

Herring, B.F., 1976, The Metamorphism and Alteration of the Basement Rocks in the Carswell Circular Structure, Saskatchewan: Unpublished M.Sc. Thesis, Unversity of British Columbia, Vancouver, 134 p.

Jaritz, W., 1973, Zur Entstehung der Salzstrukturen Nordwestdeutschland: Geologische Jahrbuch, v. 10, 77 p.

Nagy, D., 1965, The Gravitation Attraction of a Right Vertical Circular Cylinder: Department of Energy, Mines, and Resources, Earth Physics Branch, Unpublished Manuscript.

_____, 1978, Direct Gravity Formula for the Geodetic Reference System − 1967: Bulletin Géodésique, v. 52, p. 159-164.

_____, 1980, The Gravitational Effect of Three-Dimensional Bodies of Arbitrary Shape: Earth Physics Branch, Department of Energy, Mines, and Resources Internal Report 80-4.

Pagel, M., and Svab, M., 1985, Petrographic and Geochemical Variations within the Carswell Structure Metamorphic Core and their Implications with Respect to Uranium Mineralization; *in* Lainé, R., Alonso, D., and Svab, M., eds., The Carswell Structure Uranium Deposits, Saskatchewan: Geological Association of Canada Special Paper 29.

Ramberg, H., 1981, Gravity, Deformation, and the Earth's Crust: London, Academic Press, p. 102-110, 297-309.

Ray, G.E., 1977, The Geology of the Highrock Lake − Key Lake Vicinity, Saskatchewan: Saskatchewan Geological Survey, Department of Mineral Resources, Report 197.

Schwerdtner, W.M., and Lumbers, S., 1980, Major Diapiric Structures in the Superior and Grenville Provinces of the Canadian Shield: *in* Strangway, D.W., ed., The Continental Crust and its Mineral Depostis: Geological Association of Canada Special Paper 20, p. 149-180.

Sibbald, T.I.I., 1983, Geology of the Crystalline Basement, NEA/IAEA Athabasca Test Area: *in* Cameron, E.M., ed., Uranium Exploration in the Athabasca Basin, Saskatchewan, Canada: Geological Survey of Canada Paper 82-11, p. 1-14.

Sobczak, L.W., 1983, Gravity Surveys in the NEA/IAEA Athabasca Test Area: *in* Cameron, E.M., ed., Uranium Exploration in the Athabasca Basin, Saskatchewan, Canada: Geological Survey of Canada Paper 82-11, p. 151-166.

Spector, A., 1978, Report on Aeromagnetic Interpretation, Cluff Lake Area, Saskatchewan: Allen Spector and Associates Ltd., Toronto, 21 p.

Tona, F., Alonso, D., and Svab, M., 1985, Geology and Mineralization in the Carswell Structure − A General Approach: *in* Lainé, R., Alonso, D., and Svab, M., eds., The Carswell Structure Uranium Deposits, Saskatchewan: Geological Association of Canada Special Paper 29.

Tremblay, L.P., 1982, Geology of the Uranium Deposits Related to the Sub Athabasca Unconformity, Saskatchewan: Geological Survey of Canada Paper 81-20, 56 p.

Walcott, R.I., 1968, The Gravity Field of Northern Saskatchewan and Northeastern Alberta: Department of Energy, Mines, and Resources, Earth Physics Branch, Ottawa, Gravity Map, Series 16 to 20.

The Carswell Structure Uranium Deposits, Saskatchewan,
edited by R. Lainé, D. Alonso and M. Svab,
Geological Association of Canada Special Paper 29, 1985

THE ORIGIN OF THE CARSWELL CIRCULAR STRUCTURE

M. Pagel
Centre de Recherches sur la Geologie de l'Uranium. B.P. 23 - 54501 Vandoeuvre les Nancy Cedex, France
and
Centre de Recherches Petrographiques et Geochimiques. B.P. 20 - 54501, Vandoeuvre les Nancy Cedex, France

K. Wheatley
Amok Ltd., 817 - 825 45th Street W., P.O. Box 9204, Saskatoon, Saskatchewan S7K 3X5

F. Ey
Institut de Geologie, 1, rue Blessig STRASBOURG 67084 Cedex, France

ABSTRACT

The Carswell circular structure is characterized by the following features: (1) it is located along the synclinal axis of the Athabasca Basin at a site where the underlying basement rocks exhibit a complex structural pattern, (2) the area was tectonically active during Proterozoic sedimentation of the Athabasca Group, (3) it has a central basement core which has been uplifted about 2 km into the overlying sediments, (4) multiple "slabs" of basement and sandstone are present, oriented in normal, subvertical, or overturned positions in the contact area, (5) arcuate faults and, to a much lesser extent, radial faults are present in the sedimentary cover, (6) shock features, such as planar features in quartz, deformation lamellae, shatter cones, and split recemented quartz pebbles, are present in the basement rocks and along the unconformity contact, but are absent in the overlying Carswell Formation dolomites, (7) several types of breccia are present; some are older than and some are related to the development of the structure, and (8) melt-derived cryptocrystalline breccias were emplaced during Cambro-Ordovician time. The origin of the structure is considered in terms of two models, namely impact and endogenic.

RÉSUMÉ

La structure circulaire de Carswell est localisée dans une zone complexe d'un point de vue structural, et unique dans le bassin Athabasca, nord Saskatchewan. Elle est située à l'intersection de deux sutures protérozoïques, près du complexe plutonique de Clearwater, qui disparaît au sud de la structure decalé vers l'Est par une structure faillée. La structure de Carswell est bordée et recoupée par de nombreuses failles et linéaments.

La structure Carswell consiste en une partie centrale de socle métamorphisé entourée par des grès et des dolomies affectés par des failles circulaires et verticales. Des écailles de grès et de socle sont communes près de la discordance entre le socle et la couverture sédimentaire. Les zones de faiblesse antérieures, telles que les failles et les contacts, sont recoupées par les nouvelles failles, créés durant la formation de la structure circulaire.

Les figures de choc sont fréquentes. Les figures planaires dans les grains de quartz et les shatter cones, peu développés, ont été observés dans le socle près de la discordance. Des galets tronçonnés existent dans les conglomérats au-dessus de la discordance.

Les brèches sont nombreuses dans la structure de Carswell. Elles ont été divisées en 3 types principaux: (1) brèches cryptocristallines magmatiques en dykes ou filonnets, rares dans la partie nord de la structure, absentes dans la couverture sédimentaire, et pouvant être alignées suivant une direction N.E.; (2) brèches de fragmentation, les plus communes en filons dans le socle et dans les grès, près du contact; (3) pseudotachylites qui se présentent en filonnets, montrant des textures fluidales.

Trois groupes de données sont disponibles sur l'âge de la structure. Wanless et al., (1968) ont obtenu 467 ± 28 et 486 ± 55 Ma par la méthode K-Ar. Von Einsiedel (1981) a déterminé des âges allant de 416 à 513 millions d'années par la méthode $^{40}Ar/^{39}Ar$, tandis que Bell (1985) donne des âges allant de 365 à 515 millions d'années.

Une étude d'inclusions fluides (Pagel, 1975) a permis de reconnaitre 4 types d'inclusions qui ont pu être reliées à la formation de la structure. Ce sont uniquement des inclusions aqueuses, piégées à basses pressions et températures décroissantes avec le temps.

Deux modèles pour la formation de la structure circulaire

Carswell sont présentés et discutés: (1) le modèle d'impact météoritique et (2) le modèle endogène. Le modèle d'impact est basé sur la présence de figures planaires, de shatter cones, d'écailles, et sur la géochimie des brèches. Les brèches de Cluff et certaines inclusions fluides sont interpretées comme étant le résultat de l'impact.

Le modèle endogène considère que la localisation de la structure Carswell est à relier à une zone structurale unique du bassin Athabasca, et à une activité tectonique et hydrothermale associée. Dans cette hypothèse, la structure serait le résultat d'une cryptoexplosion couplée avec une tectonique verticale. Les figures de choc sont formées par l'emplacement rapide du noyau central de socle.

INTRODUCTION

The origin of the Carswell circular structure has been debated for two decades (Innes, 1964; Currie, 1969; Pagel, 1975; Wheatley, 1982; and Harper, 1983). The purpose of this paper is to consider two possible modes of origin for the Carswell structure. Accordingly, the structural, petrological, mineralogical, and geochemical features that have a bearing on the origin of the structure are described.

CARSWELL STRUCTURE – LOCATION IN THE ATHABASCA BASIN

The Carswell structure is located on the east-west synclinal axis of the Athabasca Basin. Gibb et al. (1983) have proposed a number of Proterozoic sutures in the Canadian Shield (Fig. 1), and it is noted that there is an intersection of two major sutures in the vicinity of the Carswell structure (Fig. 1, sutures 5a and 5b). The outcrop locations of dolomites in both the Athabasca and Thelon Basins overlie one of these sutures (Figs. 1, suture 5a). The Patterson high and the Jackfish sub-basin (Fig. 2) are topographic features that have been defined by Hobson and MacAulay (1969) and Ramaekers (1981). The Carswell structure lies on the boundary line separating these two features. The structure is also located immediately west of the Clearwater plutonic complex. The complex trends NNE-SSW and transects the basement below the Athabasca Basin. The Clearwater complex may be the Proterozoic suture 5a, as described by Gibb et al. (1983). This complex is defined on surface by a magnetic high (Fig. 3) and a gravity low, and is offset about 15 km to the west by a fault immediately south of the Carswell structure. This WNW-ESE fault has been intruded by a dyke dated at 949 Ma (Wanless et al., 1968). Other dykes in the basin, however, trend N-S to NW-SE and were emplaced from 1360 to

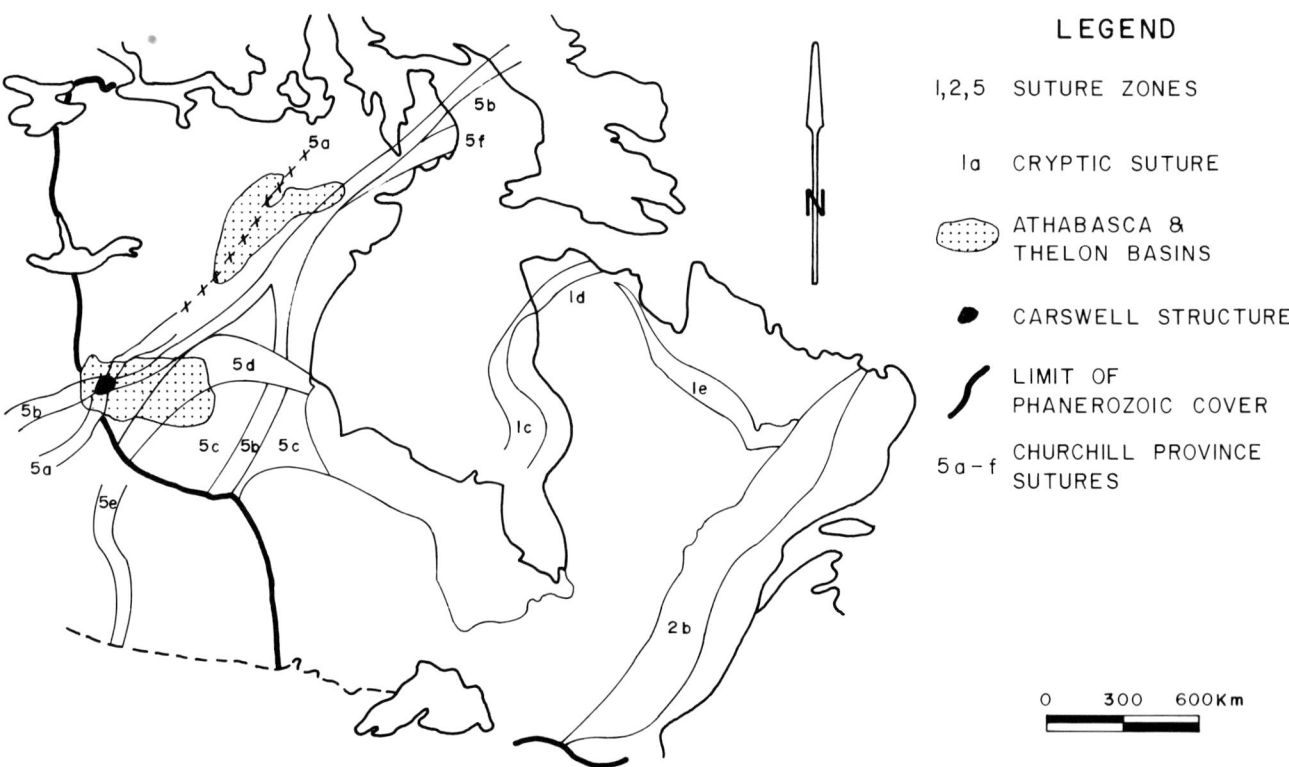

Figure 1. Locations of proposed Proterozoic sutures in Canada. 1(a) – 1(e) = circum-Superior suture. 2(b) = Grenville province suture. 5(a) – 5(e) = Churchill province sutures. (Modified from Gibb et al., 1983).

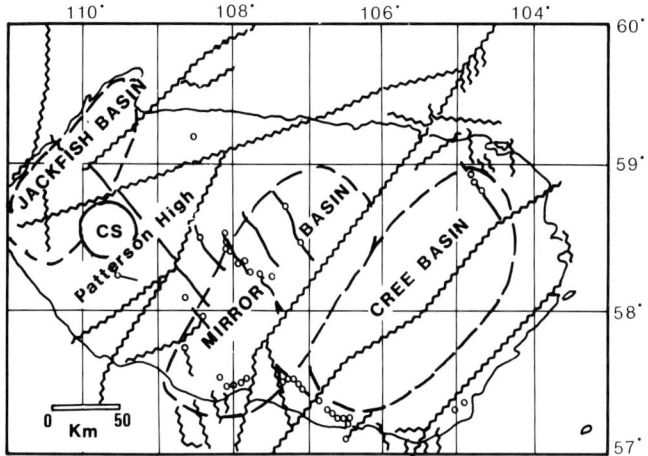

Figure 2. Location of the Carswell structure in the Athabasca Basin and relations with the major structural pattern (Ramaekers, 1981).

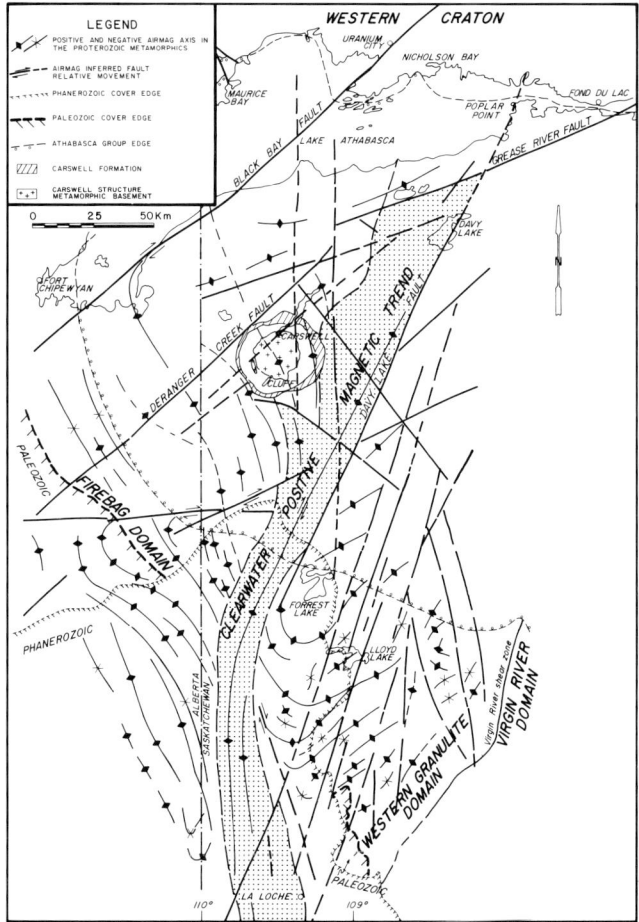

Figure 3. Structural location of the Carswell structure. (Tona et al., 1985).

1000 Ma (Ramaekers, 1981). The Carswell structure is also enclosed by all of the formations of the Athabasca Group in a lopsided "bull's-eye" outcrop pattern. The structure is bounded and transected by numerous faults and lineaments, which also partially define the limits of the surrounding Tuma Lake Formation.

STRUCTURAL OBSERVATIONS

The Carswell structure is roughly circular in outline and is approximately 39 km in diameter (Fig. 4). The structure consists of a circular central basement complex composed of Aphebian gneisses and granitoids, 18 km in diameter. This is surrounded by Athabasca sandstones which in turn are encompassed by the Carswell Formation dolomites which describe the outer limit of the structure. Immediately outside the structure to the east, the depth to basement under the sandstones is about 1.3 km.

At the unconformity between the basement complex and the overlying Athabasca Group, the structure is very complex. The contact may be horizontal, subvertical or overturned, and may be intensely faulted with several repetitions. Sandstone and basement "slabs" or wedges have resulted from these faults and are common around the circumference of the basement complex (Fig. 5). These slabs are quite variable in size and orientation. Basement slabs up to several hundred metres long may overlie sandstones in faulted contact, or sandstone slabs may be found well within the basement complex. The normal sandstone-basement contact may or may not be preserved within the slabs. The faults bounding the slabs typically do not occur along preexisting fault planes or lineaments, suggesting that the Carswell event created a new set of faults that overprinted the older structures. These older faults appear to have been re-activated after the Carswell event, perhaps as a late readjustment stage.

A number of arcuate faults occur within the sandstones and dolomite rings but have not been observed in the basement complex (Fig. 4). The Carswell Formation dolomites have been folded and strongly faulted (Hendry and Wheatley, 1985). The folds usually occur on a large scale and are open — their axial trace being irregularly cut by the arcuate faults. Nevertheless, some folds could have an arcuate axial hinge line. The emplacement of the basement complex caused most of the formation to be rotated into near vertical ridges that are up to several hundred metres wide and several kilometres long. A set of dip-slip arcuate faults were formed either during or immediately after the fold development. Radial faults are described in the southeast part of the dolomite ring (Alonso and Rouve, 1973), but are absent throughout the rest of the ring (Fig. 4).

SHOCK FEATURES

Shock features are those features ascribed to small-scale deformation of rocks usually found only at meteorite impacts or large explosion sites. The features discussed here are planar features, shatter cones, and split pebbles.

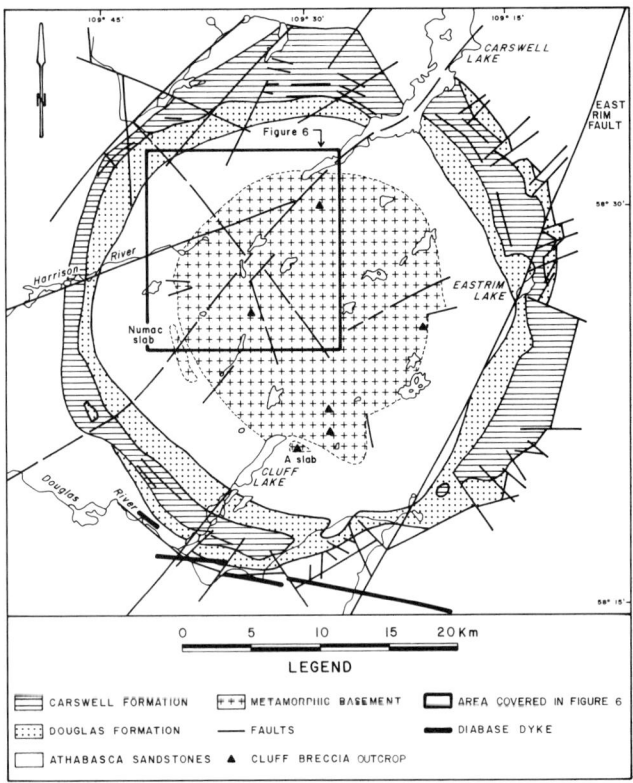

Figure 4. Geological pattern of the Carswell structure.

Planar Features

Decorated planar features have been observed in quartz from basement rocks, breccias, melt-derived cryptocrystalline rocks, and more rarely Athabasca sandstones (Currie, 1969; Harper, 1983). Pagel (1975) and Harper (1983) have shown that planar features are present at the centre of the basement core, as well as along the border of the core as described by Currie (1969). In the sedimentary cover, they are only rarely present in the basal part of the Athabasca sandstones near the unconformity, and in one location where sandstone fragments are enclosed in a polymictic fragmental breccia (Harper, 1983). In sandstones, planar features have been observed in the detrital quartz grains and they may or may not be continuous into the surrounding quartz overgrowths.

The planar features are defined by small fluid inclusions (Pagel, 1975). Lambert and Pagel (1977) have shown by scanning electron microscope that these inclusions are locally formed as an aggregate of very small quartz crystals (some tens of microns wide) with the same crystallographic orientation at the scale of the observed area. Up to five sets of planar features have been observed in a single quartz grain. These planar features do not cross the whole grain, but are preferentially developed at the grain boundary. Locally, these planar features may be obliterated by quartz recrystallization. Short (1968) and Pagel (1975) have shown that the fluid inclusion plane is usually coincident with the $(10\overline{1}3)$ crystallographic plane in quartz. The data of Currie (1969) suggest several crystallographic planes (Fig. 6).

Particular characteristics on the distribution of planar features have been noted: an absence of these features in

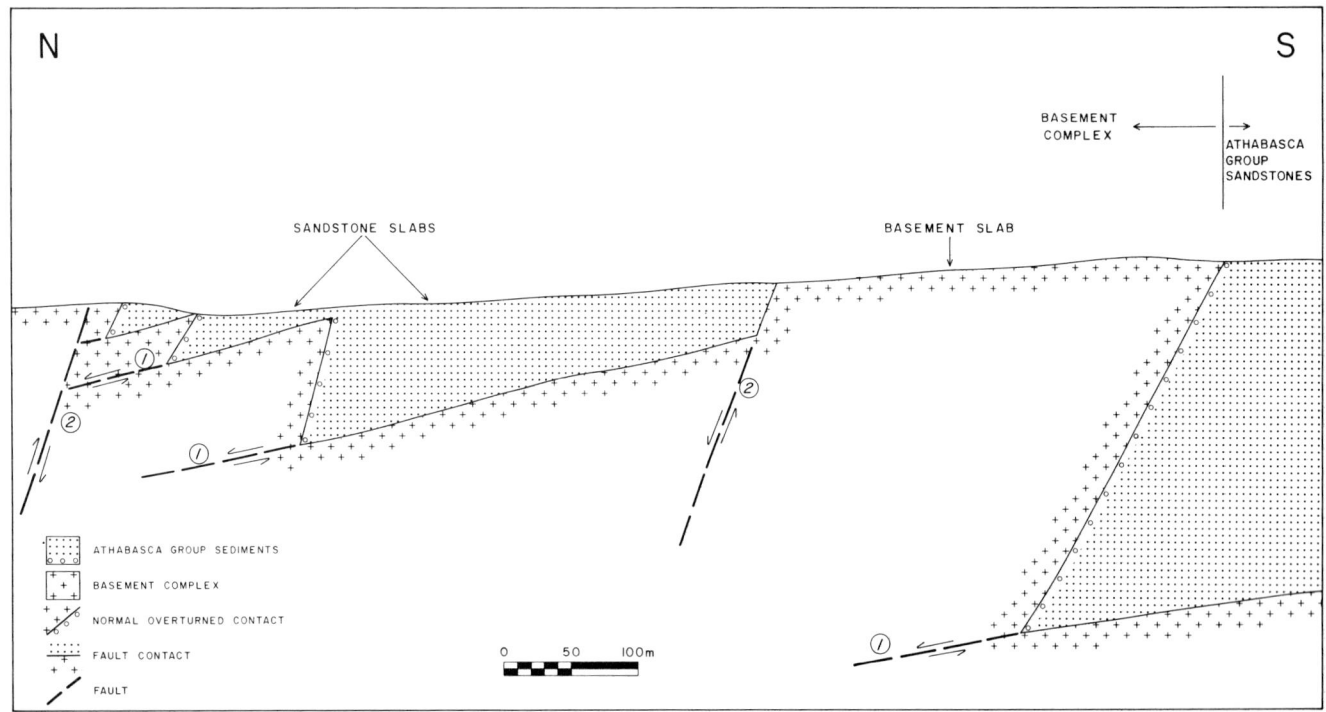

Figure 5. North-south schematic cross section of the southern border of the basement complex. Model is based on ten diamond drill holes. "1" represents initial subhorizontal faults, "2" represents a later set of subvertical faults.

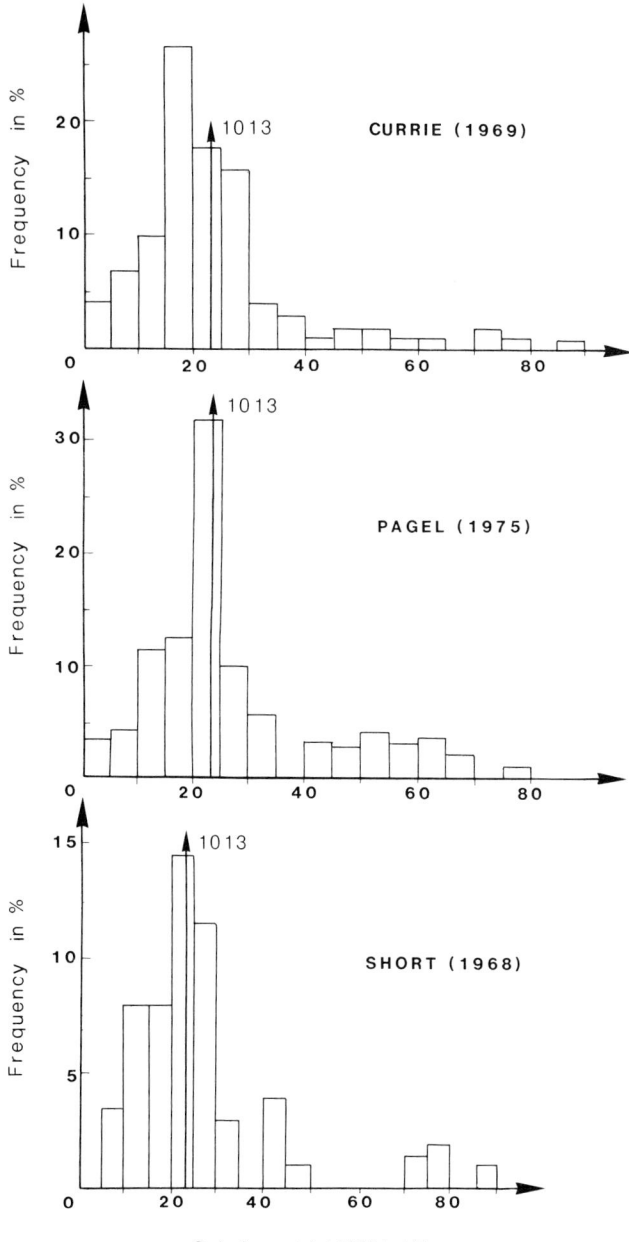

Figure 6. Histograms showing the angular relationship of planar features with respect to the C-axis of quartz.

quartz which is enclosed in garnet although the surrounding quartz may display them, and the parallel orientation of planar features in quartz from myrmekites. Finally, the presence of quartz with or without planar features in breccias suggests that their formation was prior to, or contemporaneous with, the formation of the polymictic fragmental breccias. It may be mentioned that planar features are relatively rare throughout the structure and are not common in quartz from a large mylonitic zone at one of the orebodies (Lainé, 1983).

Shatter Cones

Shatter cones have been known in the Carswell structure since their description by Innes (1964) but Currie (1969) has

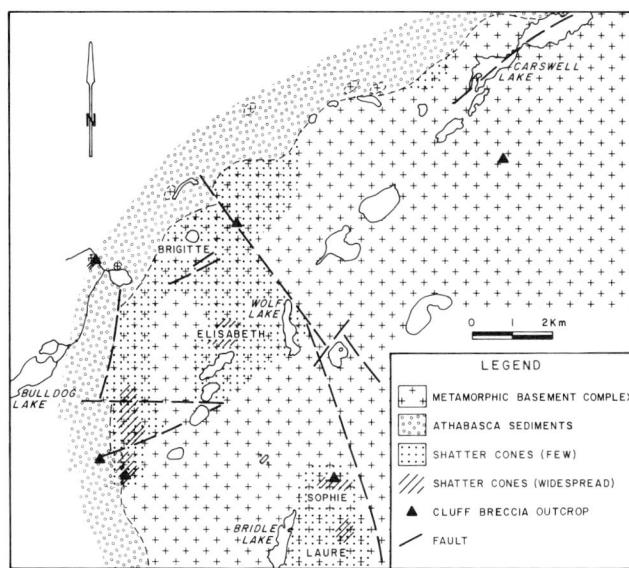

Figure 7. Areal distribution of shatter cones in the northwest part of the Carswell structure (modified from Bassaget and Camps, 1973).

questioned the exact nature of these fractures. With the excavation of trenches and mines, structures have been observed showing radiating striae and conical surfaces when they are broken along the cone axis (Harper, 1983).

The structures are usually one to several centimetres in length and typically have a random orientation. Lillié (1982) examined shatter cones in the Claude experimental pit; the orientation of the cones are locally consistent but vary from one side of the pit to the other and none point towards the centre of the basement complex. A reconnaissance mapping of the shatter cones was carried out by Bassaget and Camps (1973) (Fig. 7). It appeared that their distribution was governed by the unconformity contact, major fractures in the area, and rock type. In one area the shatter cones occur only in the aluminous gneisses and not in the migmatites or granitoids. Two shatter cones were noted within quartzite pebbles in the conglomerates immediately adjacent to the basement contact. Preferential development in the basement rocks adjacent to the unconformity contact has been noted by Currie (1969). Harper (1983) described horse-tail striations in fine-grained, massive Athabasca sandstones near the unconformity contact. However, when shatter cones from the Carswell area are compared with those observed in Sudbury, Ontario, for example, they appear to be very poorly developed.

Split and Recemented Quartz Pebbles

In the Cluff Lake area, split and recemented quartz pebbles are present in the basal conglomerates of the Athabasca Group where they are in contact with the basement complex. These pebbles contain a series of sub-parallel fractures displaced in the same sense. The displacement along the fractures is usually in the order of a few centimetres. A second oblique set of fractures is locally present but is weakly developed. The fractures have been cemented by microcrystalline quartz.

Other Possible Shock Features

Other possible shock features in the Carswell structure include recrystallization of metamorphic or granitic quartz into a very fine polycrystalline aggregate, the presence of kink bands in biotite, and fractured garnets. Kink bands in biotite occur in rocks which contain quartz with abundant decorated features; they often exhibit tiny opaque lines that are discordent with cleavage. Harper (1983) also describes planar features in apatite. The presence of silica glass was noted (Currie, 1969). Microscopically, this isotropic material has been described as being clear, having a lower index than quartz, and occurring in or at the ends of cleavage fractures.

BRECCIAS

The description of breccias in the Carswell structure was first made by Currie (1969) who determined they were related to the formation of the structure. The formation of the structure is here unofficially termed the Carswell event. Other data were obtained later by Pagel (1975), von Einsiedel (1981), and Harper (1983). Several classifications have been proposed, but always with a genetic connotation. In this paper, three main types of breccias are distinguished: (1) the melt-derived cryptocrystalline breccias, (2) the polymictic and monomictic fragmental breccias, and (3) pseudotachylites, collectively known as the Cluff breccias.

Melt-Derived Crystalline Breccias

Melt-derived breccias (red, brown, or green in colour) occur as rare irregular dykes, up to tens of metres wide, throughout the basement rocks of the structure; small stringers have also been observed in drill core. They were termed impactites by Pagel (1975), volcanic-like breccias by von Einsiedel (1981), and karnaite by Harper (1983). They are absent in the sedimentary cover. Their distribution is poorly known but it appears that they are extremely rare in the northern part of the structure. The major outcrops define a NE-SW trend from Lac Thibout in the east to Cluff Lake in the south (map in pocket, Fig. 4).

The breccias show flow textures, vesicles or amygdules, quench features, and spheroidal bodies with radial textures which indicate crystallization from a melt. They also contain various amounts of rock and mineral fragments which have undergone varying degrees of assimilation. Megascopic observation suggests that the flow of the fused melt was upward (Currie, 1969). Recent drilling and observations confirm this, even in the southern extent of the basement complex where the basement-sandstone contact has been extensively examined (Tona et al., 1985; Blaise and Koning, 1985). Sandstone fragments in Cluff breccias do not occur in basement rocks. The matrix is composed of orthoclase, quartz, and elongated skeletal crystals. From microprobe data, these latter crystals are composed of biotite or chlorite but their original mineralogy is not certain. Clay minerals present in these breccias include chlorite, illite, and kaolinite. Cesium and thorium ratios have been determined for this breccia type (Fig. 8). Chemical analyses are presented in Table I and Table II, and a comparison with basement rocks

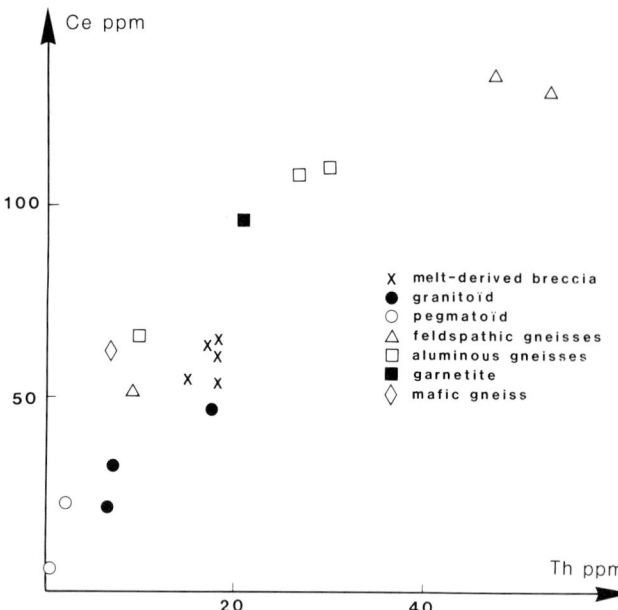

Figure 8. Plot of Th vs Ce for the melt-derived breccias (data from von Einsiedel, 1981) and representative lithology of the basement complex.

is provided in Table III. Table III shows that the Cluff breccias are potassium enriched and calcium and sodium poor relative to the basement rocks.

Polymictic and Monomictic Fragmental Breccias

Polymictic and monomictic fragmental breccias are volumetrically the most important breccias and include all breccias made up of fragments of local rocks and minerals, either polymict or monomict. They correspond to the classic Cluff breccia and tectonic-like breccia (von Einsiedel, 1981) or suevite (Harper, 1983). They occur as irregular veins from centimetres to several metres wide and are composed mainly of fragmented rocks and minerals from the nearby basement. They are generally grey, green, sometimes red or brown, and are found mainly in the basement core. In two locations they occur in the sandstones and conglomerates of the inner ring at the contact with the basement.

Sorting of the clasts has been observed; typically the smaller clasts occur near the wall of the dyke and the larger clasts at the center. The clasts are angular, and may be mono-or polycrystalline. Elongated clasts and minerals typically exhibit an orientation parallel to the dyke walls.

Metamorphic quartz, quartz with planar features, or recrystallized quartz is present in the same breccias. Pseudotachylite fragments are often found in these breccias. Alteration is widespread, and minerals include chlorite, illite, kaolinite, iron oxides, and carbonates.

Pseudotachylites

Pseudotachylites form irregular veins (0.5 to 5 centimetres wide) of fine-grained, dark green material. They exhibit flow textures, and also contain basement rock fragments. To date, the pseudotachylites have not been studied in detail.

TABLE I

CHEMICAL COMPOSITIONS OF MELT-DERIVED CRYPTOCRYSTALLINE ROCKS FROM THE CARSWELL STRUCTURE, VALUES IN %, SAMPLES COLLECTED FROM THE SOUTHWEST HALF OF THE BASEMENT CORE
(VON EINSIEDEL, 1981)

	1	2	3
SiO_2	69.12	64.11	63.36
Al_2O_3	13.30	16.63	13.82
Fe_2O_3†	3.94	3.56	6.92
MnO	0.01	0.03	0.06
MgO	1.34	1.33	2.94
CaO	0.08	0.12	0.12
Na_2O	0.99	0.58	0.52
K_2O	6.79	10.47	7.01
TiO_2	0.44	0.37	0.83
L.O.I.	3.37	2.77	3.42
Total	99.38	99.97	99.00

†Fe_2O_3 as total iron

TABLE II

SELECTED TRACE ELEMENT COMPOSITIONS OF MELT-DERIVED CRYPTOCRYSTALLINE ROCKS FROM THE CARSWELL STRUCTURE, VALUES IN PPM, SAMPLES COLLECTED FROM THE SOUTHWEST HALF OF THE BASEMENT CORE
(VON EINSIEDEL, 1981)

	1	2	3	4	5	6
Rb	210	324	143	179	178	179
Sr	–	42	50	58	35	43
Th	18	15	18	–	18	17
U	7.6	6.0	4.6	–	7.8	4.0
Pb	–	24	20	24	26	26
Ce	54	55	66	–	61	64
Zr	–	145	167	142	119	155

TIME OF THE FORMATION OF THE BRECCIAS

Radiometric age dating studies have been performed on the melt-derived crystalline breccias. The time of the formation of these breccias may possibly indicate the time of the formation of the Carswell structure.

Three sets of data have been produced (Fig. 9). Wanless *et al.* (1968) have published two ages using the potassium-argon method: 467 ± 28 Ma and 486 ± 55 Ma. It should be noted that these two dates have been obtained on two parts of an inhomogeneous sample (Currie, 1969).

Von Einsiedel (1981) gives $^{40}Ar/^{39}Ar$ ages by the single-step laser fusion method. On six samples taken in the south and west of the Carswell structure, individual ages fall between 416 and 513 Ma. Von Einsiedel (1981) states that these ages exhibit scatter outside of analytical uncertainties. Recent dating by the same method gives ages of 365 to 515 Ma (Bell, 1985). These large variations in age could be attributed to one or several of the following factors: (1) presence of minerals and rock fragments from the basement, (2) post-emplacement alteration that has widely affected the breccia, (3) argon leakage after crystallization of the magma-derived, cryptocrystalline rocks, and (4) intrusion of the breccias over an extended period of time.

Little could be done concerning post-emplacement alteration and argon leakage, but both von Einsiedel (1981) and Bell (1985) took precautions to avoid sampling the host rock fragments within the breccias. Bell (1985) suggests that the breccias intruding over an extended period of time was the main cause for the large variation in age dates.

FLUID INCLUSIONS RELATED TO THE FORMATION OF THE STRUCTURE

Fluid inclusions that could be associated with some stages of the formation and evolution of the structure are: (1) those which are arranged in sets of planes and which constitute the planar decorated features, (2) those located in quartz-cemented split pebbles, (3) those located in vuggy quartz from melt-derived cryptocrystalline rocks or in relict basement quartz which has been recrystallized, and (4) those from carbonate cement in breccias.

TABLE III

COMPARISON BETWEEN CHEMICAL AVERAGES FOR THREE ROCK TYPES IN THE SOUTHERN PART OF THE CARSWELL STRUCTURE BASEMENT CORE (PAGEL AND SVAB, 1985), AN AVERAGE BASEMENT AS DERIVED FROM THE FIRST THREE COLUMNS, AND AN AVERAGE CLUFF BRECCIA FROM THE BASEMENT CORE (LAINÉ, 1983)

	ALUMINOUS GNEISS	FELDSPATHIC GNEISS	GRANITOID/ PEGMATOID	AVERAGE BASEMENT	AVERAGE CLUFF BRECCIA n = 33
SiO_2	62.57	68.13	73.75	68.15	67.72
Al_2O_3	16.77	14.54	14.82	15.38	14.21
Fe_2O_3†	8.26	4.14	1.12	4.51	4.36
MnO	0.09	0.04	0.03	0.05	0.04
MgO	2.49	2.04	0.38	1.64	2.03
CaO	0.31	1.27	0.77	0.78	0.29
Na_2O	0.94	1.99	3.28	2.07	0.70
K_2O	3.96	5.35	4.45	4.59	7.13
TiO_2	0.47	0.33	0.03	0.28	0.42
P_2O_5	0.12	0.26	0.18	0.19	0.20

†Fe_2O_3 as total iron

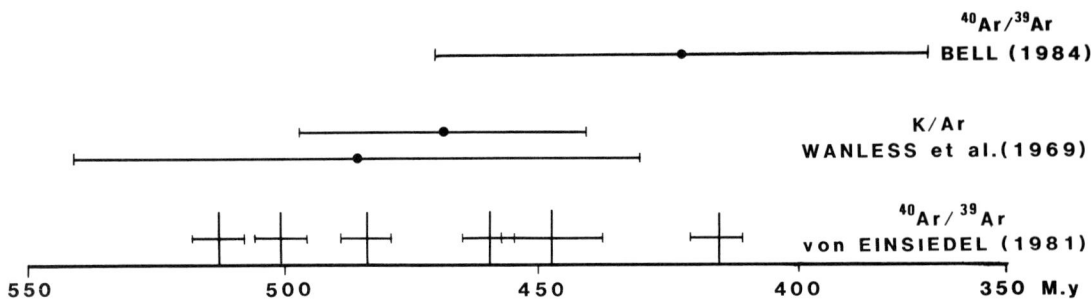

Figure 9. Isotopic ages on the melt-derived cryptocrystalline breccias.

Three types of fluid inclusions have been observed: liquid aqueous inclusions, low density gas inclusions, and two phase inclusions with a wide range of ratios between the gas bubble and the liquid aqueous phase.

The salinity of aqueous solutions, estimated by the freezing point temperature, is quite variable. The values range from about zero to 17 weight per cent equivalent NaCl, with even higher local salinities. Homogenization temperatures are widespread, but suggest the trapping of aqueous fluid at decreasing temperatures under low pressures (Pagel, 1975).

It should also be noted that fluid inclusions present before the structure formation, i.e., those related to metamorphism and retrograde metamorphism, have undergone several modifications and severe fracturing and are quite distinctive from the later inclusions described above.

INTERPRETATION AND CONCLUSIONS

There are two main models for the origin of the Carswell structure, namely, the impact model and the endogenic model.

The Impact Model

Shock metamorphism features such as planar features and shatter cones are generally considered as sufficient criteria to support the impact model, since they are interpreted to require very high pressures to form. Other features which may be compatible with this model are: (1) the circular structural pattern and formation of slabs, (2) fluid inclusion data, and (3) the chemistry of the melt-derived, cryptocrystalline rocks.

Structural Pattern. It has been noted that the arcuate and radial faults associated with the Carswell event cross-cut older faults and this, along with the development and rotation of slabs, indicates that the circular structure was formed by a short-lived event. This event created a new structural pattern and did not simply re-activate the older structures. Another feature noted is the rotation of the slabs with an absence of folding. This could not have occurred at depth and suggests that basement and cover were near the paleotopographic surface during their ejection. Ey (1984) has shown that the formation of slabs in the Cluff Lake area implies compression of the rocks in a horizontal sense. This could be due to sliding along the uplifted basement complex or may be explained by radial movement away from the centre of the plug related to the Carswell event.

Fluid Inclusions. Pagel and Poty (1975) interpreted planar decorated features from the Charlevoix structure as healed microfractures formed over a long period of time, beginning at high temperatures soon after formation of the structure and continuing, with time, to lower temperatures. The composition of the trapped fluids in the Carswell area is compatible with an impact model and may correspond to a mixing of formation water with meteoric water (Pagel, 1975). The abundance of gaseous inclusions suggests that fractures were healed in the presence of a gas at very low pressure.

Nature and Chemistry of Breccias. Rock fragments in breccias, the relatively homogeneous chemistry of the melt-derived cryptocrystalline breccias, and a geochemical similarity with an average basement rock with an enrichment in K and a depletion in Ca and Na may be explained by the impact model. In this case, these breccias were derived from melting and fragmentation of host rocks. Enrichment in K and depletion in Ca and Na is common at meteorite impact sites (Dence, 1971). Von Einsiedel (1981) has suggested a higher Ce content for these breccias, however, more systematic data presented here on the different lithologies in the Carswell structure indicate that this is not the case. Cesium and thorium data of melt-derived breccias do not plot outside the field of the country rocks (Fig. 8) and could be interpreted as originating from these country rocks. However, strontium isotope data derived from the breccias do not match with a simple derivation of basement rocks in the Carswell area. Bell (pers. commun., 1983) calculated the initial strontium isotope ratio to be between 0.7418 and 0.7698, assuming an approximate age of 500 Ma for the breccias, thus implying a large crustal component. However, at that time, aluminous gneisses of the basement complex had an average $^{87}Sr/^{86}Sr$ ratio of 0.89 (0.8138 to 1.0901) and feldspathic and mafic gneisses and granitoids of the basement complex had a ratio of 0.82 (0.7366 to 1.0233). The somewhat higher $^{87}Sr/^{86}Sr$ ratios in the basement rocks suggest that the Cluff breccias were not produced simply by reworking of average basement as we now know it.

Coesite and stishovite have not been detected in the Carswell structure, but few attempts have been made to locate such minerals. Their stability field is determined by temperature, conditions after the impact, and by the availability of fluid phases (Lambert, 1977). One or more of these conditions may have been unsuitable for the formation of these minerals. Stöffler (1971) noted that coesite and stishovite are present in structures where planar features in quartz are undecorated.

The impact model is proposed to have the following succession of events: a compression stage due to meteorite impact, formation of slabs by radial movement during the excavation stage, and a readjustment stage with vertical movement. During the second and third stages, folding and downfaulting of the Carswell and Douglas Formations account for their preservation and unique presence in the Carswell area. If vertical lithological profiles through the Athabasca Basin are considered, it is obvious that the Carswell and Douglas Formations could not be observed elsewhere, since they overlie the Tuma Lake Formation which is restricted to an area surrounding the Carswell structure. However, if their combined thickness is estimated at about 800 m, it is difficult to consider their deposition only in the Carswell area.

The Douglas and Carswell Formations are generally in fault contact. Wheatley (1982) presented evidence to indicate that, prior to faulting, the contact was gradational. If all contacts within the Athabasca Group are, or were, conformable in the Carswell area, then the total thickness of the sedimentary sequence would have been at least 2450 m: 1650 m (sandstones from Ramaekers, Fig. 3.5, 1981) plus 800 m (estimated Carswell and Douglas Formations). If the formation temperature of the fluid inclusions located between detrital quartz grains and their quartz overgrowths is considered with this maximum thickness, a geothermal gradient of about 80° C/km is implied. This is unrealistic, and extensive erosion before the impact has been proposed (Pagel, 1975). The absence of fall-back breccias in the Carswell structure may be ascribed to erosion following the impact. No readjustment of the fluid inclusions has yet been documented.

The Endogenic Model

The Carswell structure is the host of five known uranium orebodies. It is located at the intersection of two proposed Proterozoic sutures, and lies along the E-W synclinal axis of the Athabasca Basin. The erosional pattern of the Athabasca Group sediments also happens to form a lopsided "bull's-eye" with the Carswell structure at the center. Pacquet and McNamara (1985) have observed that the basal conglomerates overlying the basement complex in the Cluff Lake area indicate deposition on an uneven palaeotopography affected by synsedimentary tectonic activity. Moreover, a contemporaneous acid volcanism supplies part of the detrital material. Bell (1985) suggested that there was considerable hydrothermal activity in the area at the time of deposition of the Athabasca sediments. This, combined with the presence of the dyke along the southern border of the structure, the major faults and lineaments intersecting the area, and the proximity of the Clearwater plutonic complex, all emphasize a unique setting for the Carswell structure.

The range of age dates on the breccias argues against a single event model (Bell, 1985); they may have been intruded over an extended period of time. The upward flow direction noted by Currie (1969) and Amok geologists suggests that their source was from below, not above. These breccias are relatively homogeneous, however Peredery (1972) argues in favor of a meteorite impact for the Sudbury Basin based on the inhomogeneity of glass fragments in the Onaping Formation. Also, the massive amount of fall-back breccia that would be formed by an impact is missing. If formation and rotation of the basement and sandstone slabs required that they were at or near surface, then subsequent erosion could not have amounted to much. Even several hundred metres of erosion since the formation of the structure should not be enough to erase all evidence of a fall-back breccia for a crater almost 40 km in diameter. It is suggested, therefore, that most of the breccias were derived from below and did not form as a thick deposit overlying the area.

Split pebbles, which have been ascribed to meteorite impacts, have also been described from several other localities in the Athabasca Basin. Godfrey (1980) has noted "the presence of fractured, offset, and recemented quartzite pebbles in the Athabasca basal conglomerate just west of Fidler Point" (p. 17) and suggests they have been formed by post-Athabasca faulting. They have also been noted in the Maurice Bay area and the Rabbit Lake pit (P. Ramaekers, 1983, pers. commun.).

Shatter cones, which are widely accepted as forming only at meteorite impact sites, have been noted on Fir Island in proximity to the Black Bay Fault. These structures have been confirmed to be shatter cones by geologists from the Sudbury area (P. Ramaekers, 1983, pers. commun.).

The lack of shock features in the overlying dolomites, a good rock type for preserving such features, and the presence of shock features in the underlying rock types suggest that the explosive event originated from below, not above.

A sequence of events is proposed for this model. The clastic sediments of the Athabasca Group were deposited with the Carswell area located on the basin's E-W synclinal axis. Sedimentation took place from about 1600 Ma to not later than 1260 Ma, the age of isotope equilibrium of clay minerals in the sandstones (Clauer et al., 1985). Next, the Douglas and Carswell Formations were deposited as a platform with indications of N-S palaeocurrents. Continuous, but minor, tectonic activity was present in the Carswell area during sedimentation. This activity consisted of NE-SW and NW-SE faults, hydrothermal activity, formation of sedimentary breccias in the dolomites, and the intrusion of a local diabase dyke. Cryptovolcanic activity or a diapiric mechanism during Cambro-Ordovician times caused a quick vertical emplacement of the basement complex into the overlying sediments. Friction formed the fragmentation-derived breccias, cut through by the melt-derived breccias. The dolomites were downfaulted, rotated, and arcuately faulted. Ring faults developed at the boundary of the structure. Shock features were formed by rapid emplacement of the basement complex, and split pebbles developed by shearing along the contact. Continued tectonic activity is evidenced by repeated intrusions of the breccias. All the remaining breccias in the structure indicate an origin from below with an upward flow movement. There is no evidence of any fall-back breccias or similar deposits. Subsequent erosion removed the strongly brecciated sediments which once overlay the basement complex. The original NE and NW faults were reactivated and transected the structure, locally offsetting its circular pattern.

DISCUSSION

Although the Carswell area exhibits a continuous history of tectonic activity, it is possible that such activity has taken place throughout the Athabasca Basin. One of the major problems related with the endogenic model is that it relies on terrestrial processes which are not completely understood. However, Brock (1972) describes similar processes which led to the formation of the Vredefort circular structure in South Africa. Also, Snyder and Gerdemann (1965) have concluded that the Decaturville structure in Missouri was formed by a cryptoexplosion, and Bucher (1963) ascribes the Ries and Vredefort structures and associated "shock" features to cryptoexplosion events in structurally unique areas.

Age dating of the Cluff breccias suggests a sequence of events from 365 to 515 Ma (Bell, 1985), not just one instantaneous tectonic event. If the structure is due to a meteorite impact, the oldest date may represent the age of the event, and the younger dates may be due to subsequent tectonic activity triggered by the impact.

Implications For Uranium Mineralization

According to the impact theory, the presence of uranium deposits in the Carswell structure and impact at Cambro-Ordovician time is a coincidence. The only effect the structures' formation had was to have moved the uranium deposits from a depth of 2.1 km to the surface. During the impact event, mechanical dislocations of the deposits occurred and coffinite was locally precipitated along with the older mineralization, in the Cluff breccias (Ruhlmann, 1985).

In the endogenic model, it has been suggested (e.g., Wheatley, 1982) that the Cluff Lake uranium deposits were the result of concentration along structural and chemical traps during basin-wide mineralization events. These traps were present well before the mineralization events and indicate that the Carswell area may have been structurally unique before the Carswell event. The traps were later reactivated and were overprinted by the deformation resulting from the diapiric-cryptoexplosion Carswell event, which aided in the further remobilization of uranium. The Cambro-Ordovician impact of a meteorite on such structurally controlled Helikian uranium deposits, as in the Carswell area, seems most fortuitious in light of the present data.

ACKNOWLEDGEMENTS

Amok Ltd. provided the financial support for this paper. The authors appreciate the helpful comments and criticism of K.L. Currie, R. Lainé, and D.H. Roussel on an earlier version of this manuscript. We would also like to thank all geologists, past and present, involved in obtaining the data we compiled. Lastly, we would like to thank meteorites and cryptovolcanic explosions everywhere for forcing us to admit that we don't know everything.

REFERENCES

Alonso, D. and Rouve, B., 1973, Rapport de Fin de Mission, La Bordure Orientale du Dome de Carswell: Amok Internal Report.

Bassaget, J.P. and Camps, P., 1973, Rapport de Fin de Mission, Quadrand Nord-Ouest du Noyau de Socle, Géologie et Prospection: Amok Internal Report.

Bell, K., 1985, The Geochronology of the Carswell Area, Northern Saskatchewan: in Lainé, R., Alonso, D., and Svab, M., eds., The Carswell Structure Uranium Deposits, Saskatchewan: Geological Association of Canada Special Paper 29.

Blaise, J.R., and Koning, E., 1985, Mineralogical and Structural Aspects of the Dominique-Peter Uranium Deposits: in Lainé, R., Alonso, D., and Svab, M., eds., The Carswell Structure Uranium Deposits, Saskatchewan: Geological Association of Canada Special Paper 29.

Brock, B.B., 1972, A Global Approach to Geology: Cape Town, A.A. Balkema, p. 212-222.

Bucher, W.H., 1963, Cryptoexplosion Structures, Cause from Without or From Within The Earth? ("Astroblemes" or "Geoblemes"): American Journal of Science, v. 261, p. 597-649.

Clauer, N., Ey, F., and Gauthier-Lafaye, F., 1985, K-Ar Dating of Different Rock Types from the Cluff Lake Uranium Ore Deposits (Saskatchewan-Canada): in Lainé, R., Alonso, D., and Svab, M., eds., The Carswell Structure Uranium Deposits, Saskatchewan: Geological Association of Canada Special Paper 29.

Currie, K.L., 1969, Geological Notes on the Carswell Circular Structure, Saskatchewan: Geological Survey of Canada Paper 67-32, 60 p.

Dence, M.R., 1971, Impact Melts: Journal of Geophysical Research, v. 76, p. 5552-5565.

von Einsiedel, C.A., 1981, Petrography and Geochemistry of the Cluff Lake Breccias, Carswell Structure, Northern Saskatchewan: Unpublished B.Sc. Thesis, Carleton University, Ottawa, 44 p.

Ey, F., 1984, Un Example de Gisement d'Uranium Sous Discordance: Les Minéralisations Protérozoïques de Cluff Lake, Saskatchewan, Canada, Thèse d'Etat, l'Université Louis Pasteur, 171 p.

Gibb, R.A., Thomas, M.D., Lapointe, P.L., and Mukhopadhyay, M., 1983, Geophysics of Proposed Proterozoic Sutures in Canada: Precambrian Research, v. 19, p. 349-384.

Godfrey, J.D., 1980, Geology of the Alexander-Wylie Lakes District, Alberta: Earth Sciences Report 78-1, Alberta Research Council, 26 p.

Harper, C.T., 1983, The Geology and Uranium Deposits of the Central Part of the Carswell Structure, Northern Saskatchewan, Canada: Unpublished Ph.D. Thesis, Colorado School of Mines, Golden, Colorado, 337 p.

Hendry, H.E., and Wheatley, K.L., 1985, The Carswell Formation, Northern Saskatchewan: Stratigraphy, Sedimentology, and Structure: in Lainé, R., Alonso, D., and Svab, M., eds., The Carswell Structure Uranium Deposits, Saskatchewan: Geological Association of Canada Special Paper 29.

Hobson, G.D., and MacAulay, H.A., 1969, A Seismic Reconnaissance Survey of the Athabasca Formation, Alberta and Saskatchewan: Geological Survey of Canada Paper 69-18, 23 p.

Innes, J.J.S., 1964, Recent Advances in Meteorite Crater Research at the Dominion Observatory, Ottawa, Canada: Meteoritics 2, p. 219-249.

Lainé, R., 1983, Some Reflections on the Carswell Structure: Amok Internal Report.

Lambert, P., 1977, Les Effets des Ondes de Choc Naturelles et Artificielles, et le Cratère d'Impact de Rochechouart (Limousin, France), Thèse d'Etat, Université de Paris Sud, 313 p.

Lambert, P., and Pagel, M., 1977, Sur les Elements Planaires des Quartz Provenant des Structures de Carswell et Charlevoix (Canada) et Rochechouart (France): Comptes Rendus Académie des Sciences, t, 284, Série D, p. 1623-1626.

Lillié, F., 1982, Analyse Tectonique de Gisement Claude (Cluff Lake, Saskatchewan): Amok Internal Report.

Pacquet, A. and McNamara, S., 1985, The Study of the Basal Athabasca Succession in the D, E, L, F, and S Areas of the Carswell Structure: in Lainé, R., Alonso, D., and Svab, M., eds., The Carswell Structure Uranium Deposits, Saskatchewan: Geological Association of Canada Special Paper 29.

Pagel, M., 1975, Cadre Géologique des Gisements d'Uranium dans la Structure de Carswell (Saskatchewan, Canada): Etude des Phases Fluides: Thèse de 3é cycle, Universite de Nancy I, 157 p.

Pagel, M. and Poty, B., 1975, Fluid Inclusion Studies in Rocks of the Charlevoix Structure (Quebec, Canada): Fortschrift Mineralogie, v. 52, p. 479-489.

Pagel, M., and Svab, M., 1985, Petrographic and Geochemical Variations within the Carswell Structure Metamorphic Core and their Implications with Respect to Uranium Mineralization: in Lainé, R., Alonso, D., and Svab, M., eds., The Carswell Structure Uranium Deposits, Saskatchewan: Geological Association of Canada Special Paper 29.

Peredery, W.V., 1972, Chemistry of Fluidal Glasses and Melt Bodies in the Onaping Formation: in Guy-Bray, J.V., ed., New Developments in Sudbury Geology: Geological Association of Canada Special Paper 10.

Ramaekers, P., 1981, Hudsonian and Helikian Basins of the Athabasca Region, Northern Saskatchewan: in Campbell, F.H.A., ed., Proterozoic Basins of Canada: Geological Survey of Canada Paper 81-10, p. 219-233.

Ruhlmann, F., 1985, Mineralogy and Metallogeny of the Uraniferous Occurrences in the Carswell Structure: in Lainé, R., Alonso, D., and Svab, M., eds., The Carswell Structure Uranium Deposits, Saskatchewan: Geological Association of Canada Special Paper 29.

Short, N.M., 1968, Petrographic Study of Shocked Rocks from the Steen River Structure, Alberta: in French, B.M., and Short, N.M., eds., Shock Metamorphism of Natural Materials, p. 374-378.

Snyder, F.G., and Gerdemann, P.E., 1965, Explosive Igneous Activity Along an Illinois-Missouri-Kansas Axis: American Journal of Science, v. 263, p. 465-493.

Stoffler, D., 1971, Coesite and Stishovite in Shocked Crystalline Rocks: Journal of Geophysical Research, v. 76, No. 23, p. 5474-5488.

Tona, F., Alonso, D., and Svab, M., 1985, Geology and Mineralization in the Carswell Structure — A General Approach: in Lainé, R., Alonso, D., and Svab, M., eds., The Carswell Structure Uranium Deposits, Saskatchewan: Geological Association of Canada Special Paper 29.

Wanless, R.K., Stevens, R.D., Lachance, G.R., and Edmonds, C.M., 1968, Age Determinations and Geological Studies, K-Ar Isotopic Ages, Report 8: Geological Survey of Canada Paper 67-2A, 141 p.

Wheatley, K., 1982, Mineral Lease 5270-5271, 1982 Final Report: Amok Internal Report.

The Carswell Structure Uranium Deposits, Saskatchewan,
edited by R. Lainé, D. Alonso and M. Svab,
Geological Association of Canada Special Paper 29, 1985

CONCLUSION:
THE CARSWELL URANIUM DEPOSITS — AN EXAMPLE OF NOT SO UNIQUE UNCONFORMITY-RELATED URANIUM MINERALIZATION

R. Lainé
Amok Ltd., 817-825 45th Street W., Box 9204, Saskatoon, Saskatchewan S7K 3X5

INTRODUCTION

This paper summarizes some of the material presented in this GAC Special Paper and compares the Carswell structure uranium deposits with the other Athabasca Basin uranium deposits. The reader is encouraged to read the other papers because of the brevity of this summary.

Much of the Special Paper is centred around the previously unpublished descriptions of the geology of the deposits and the geology of the Carswell structure. This volume is a landmark in our progress toward gaining a fuller understanding of uranium mineralization in this part of Saskatchewan.

THE MAIN CHARACTERISTICS OF THE CARSWELL STRUCTURE

Major characteristics of the Carswell structure may be summarized as follows:

1) The oldest known rocks which comprise the basement gneisses in the Carswell structure have been separated lithostratigraphically, chemically, and petrographically into an upper unit called the Peter River gneisses, derived from shales, and a lower unit called the Earl River complex, derived from felsic and mafic volcano-sedimentary rocks (Pagel and Svab, 1985). Metamorphism reached the granulite facies, possibly during late Archean, however, no age older than 2320 Ma (Rb-Sr and U-Pb) has yet been recorded (Bell, 1985). During the Hudsonian orogeny, between 1900 and 1760 Ma (Rb-Sr), rocks were subjected to amphibolite facies metamorphism. The Earl River complex rocks underwent some anatexis, locally intruded the Peter River gneisses, and formed domal structures in the southern part of the structure.

In the Carswell area, granitoid rocks from the Earl River complex are slightly enriched in uranium, which may have served as a possible source for uranium in the basement.

2) In the main area of mineralization, along the southern contact with the basement, post-Hudsonian sedimentation was different from elsewhere around the structure and shows evidence of pre- or early Athabasca Group sedimentation in small, local depressions (Pacquet and McNamara, 1985). The Athabasca Group sediments covered the basement gneisses and the localized sediments during Helikian time.

3) Toward the end of the Helikian, the Carswell and Douglas Formations were deposited over the Carswell structure (Hendry and Wheatley, 1985).

4) Clay mineralogy has been utilized to evaluate the stratigraphy of the Athabasca Group both inside and around the Carswell structure. On this basis, most of the sandstones within the Carswell structure have been assigned to the upper part of the Athabasca Group (Otherside and Locker Lake Formations). Using lithologic and petrographic studies combined with the clay mineralogy, it appears that at the overturned sandstone-basement contact, the sandstones belong to the lower part of the Athabasca Group: the Fair Point and/or Lazenby Lake Formations (Hoeve et al., 1985).

5) The main alteration stage clay minerals from the D deposit give a K-Ar date of around 1220 Ma (Clauer et al., 1985). A similar age has been obtained from the Douglas Formation pelites exposed along the outer sedimentary rim of the Carswell structure, thus establishing a temporal link between diagenesis and mineralization.

6) Uranium-lead ages indicate a late Hudsonian magmatic emplacement of uraninite and monazite at about 1800 Ma, and the main uranium mineralization between 1150 Ma and 1050 Ma. Remobilized uranium yields ages of 900 Ma, 380 Ma, and 200 Ma (Bell, 1985).

7) Uranium mineralization in the basement is controlled by a flat-lying thrust fault (mylonite zone) and late, steeply dipping faults. The presence of regolith and Athabasca Group sandstone in normal contact at the Dominique-Peter and OP orebodies, and regolith alone at such deposits as the N orebody, suggests that the basement deposits are linked to the unconformity (Blaise and Koning, 1985).

8) Structural studies suggest a link between pre-Athabasca Group (probably Hudsonian) structures and uranium deposits in the sandstone at the unconformity as, for example, at the D deposit (Ey et al., 1985).

9) Ore assemblages are very simple: pitchblende and uraninite constitute most of the ore in all deposits (Ruhlmann, 1985). Te, Se, Au, and Bi accompany the main ore stage, forming an assemblage which closely resembles that of some ore deposits from the Beaverlodge area (e.g., Nicholson Mine, Dahlkamp and Adams, 1981). There are no significant amounts of Ni, As, Cr, or Co as are found in most deposits at the east side of the basin.

10) Organic matter has been observed in association with basement- and sandstone-hosted mineralization. Studies using the NMR ^{13}C (Nuclear Magnetic Resonance) method indicate that the carbonaceous material is the same in all samples and is best defined as a bitumen (Landais and Dereppe, 1985).

11) The Carswell event was the physical intrusion of the basement plug through at least 1200m of Athabasca Group sandstones which created radial and concentric faults (not entirely overprinting Hudsonian fractures), slabbing of the present sandstone-basement contact, and overturning of those slabs along the edge of the basement core (Pagel *et al.*, 1985). Isostatic adjustments created further movement, possibly along radial and circular faults. Late tectonic movements reactivated pre-Athabasca faults and displaced the outermost circular faults by as much as 3 km.

12) Features related to the formation of the Carswell structure have been discussed by Pagel *et al.* (1985). Associated volcanic-like breccias yield dates from 515 to 365 Ma (Bell, 1985) and are possibly contemporaneous with uranium remobilization at 380 Ma.

COMPARISON OF THE DEPOSITS IN THE CARSWELL STRUCTURE WITH THOSE ELSEWHERE IN NORTHERN SASKATCHEWAN

A comparison with some Saskatchewan uranium deposits in the Beaverlodge area and east Athabasca Basin confirms that the Cluff Lake deposits in the Carswell structure are geologically very similar and probably have the same metallogenic history.

Table I summarizes features of the Carswell structure deposits, some Beaverlodge vein-type and east Athabasca Basin unconformity-type deposits. The following characteristic features of uranium in the Carswell area have also been noted in other northern Saskatchewan deposits (e.g., Tremblay, 1982; Dahlkamp and Adams, 1981; and Hoeve *et al.*, 1980.)

1) Uranium deposits are found either in the basement or in the Athabasca Group sandstones.

2) Uranium mineralization in the basement and/or the Athabasca Group sandstones is associated with the unconformity, as illustrated by the D deposit and by sandstone and regolith remnants in proximity to the basement-hosted deposits.

3) Uranium mineralization in the Athabasca Group sandstones is linked to reactivation and extension into the sandstone cover of Hudsonian basement tectonic structures. Uranium in the basement is linked to Hudsonian metamorphic structures and reactivation of Hudsonian tectonic structures.

4) The main ore minerals are pitchblende and uraninite in both types of deposits hosted by the basement and Athabasca Group sandstones.

5) It is suggested in the literature that uranium deposits occur close to Archean granitic domes and possibly Aphebian granitic intrusions (Taylor and Rowntree, 1980; Clark and Burrill, 1981). In the Carswell area, the Dominique-Peter deposit is a good illustration of this spatial association (Powell *et al.*, 1985; Blaise and Koning, 1985).

6) In the Carswell area, as well as in the Beaverlodge area, there is evidence of a late Hudsonian syngenetic magmatic mineralization event characterized by a uraninite-monazite assemblage.

7) Extensive zones of chloritic alteration encompass basement- and sandstone-hosted mineralization. The chlorite within and close to mineralization is a Mg-rich trioctahedral variety, suggesting Mg metasomatism. In the basement gneisses, less than 20 m from the mineralization, that chlorite is replaced by an Al-rich, di-trioctahedral variety and a highly aluminous illite. Away from the mineralization, only the 1M illite polytype occurs in sandstone as well as in bleached basement (Ey *et al.*, 1985).

8) The temperature of the ore forming fluids was less than 200° C. Fluid inclusions suggest that they were very saline but not always saturated (30% NaCl equivalent), and the presence of hematite platelets indicates that the fluids were also oxidizing (Pagel *et al.*, 1980). Pagel (pers. commun., 1983) has observed that fluid inclusions linked to diagenesis and those linked to mineralization have the same characteristics.

SUMMARY

A number of questions remain unanswered at this point and are fruitful areas for further research.

What is the source of the organic matter found in the mineralization? What role, if any, did the organic matter play in the deposition of uranium?

What is the source of uranium? There are indications of enrichments in granitoids and in the Dominique-Peter mylonite which could have acted as a protore.

Are the mafic rocks from the Earl River complex in fact komatiites? Evidence comes only from geochemistry; there are no indications of lava flows or relict minerals that could confirm the origin of these high Mg rocks.

Why are U-Pb dates for mineralization younger than those for clay minerals in the alteration haloes? Is the clay alteration related to ore emplacement or only to diagenesis, or are the two indistinguishable? Is there an evolution from diagenetic clays into mineralization clays?

Why are no true Archean dates obtained in rocks that show relic mineral assemblages of granulite facies metamorphism?

The geochronologic history of the Carswell structure fits very well with the Canadian Shield geological history (Douglas, 1980), particularly for Saskatchewan (Beck, 1969). Outside of a few old dates (2300 Ma to 2000 Ma), rocks have recorded the main Hudsonian phase of around 1870 Ma. The economic mineralization is dated at 1150 Ma to 1050 Ma with

TABLE I
CHARACTERISTICS OF SOME URANIUM DEPOSITS IN NORTHERN SASKATCHEWAN

DEPOSIT	UNCONFORMITY - WELL DEFINED	UNCONFORMITY - OBSCURED BY ALTERATION	REGOLITHIC ALTERATION - STRONG	REGOLITHIC ALTERATION - WEAK	BASEMENT ROCK - TYPE	BASEMENT ROCK - AGE	HOST ROCKS - IN BASEMENT	HOST ROCKS - IN ATHABASCA GROUP	ORGANIC MATTER	ALTERATION - CHLORITE	ALTERATION - ILLITE	LOCALIZATION OF ORE - AT U/C	LOCALIZATION OF ORE - BELOW U/C	LOCALIZATION OF ORE - ABOVE U/C	BASEMENT FAULT CONTROL	ASSOCIATED ELEMENTS - MAJOR	ASSOCIATED ELEMENTS - MINOR	REFERENCES
RABBIT LAKE	Eroded		x		M	Ap	Q,C,G,F		x	xMg			x		x		V,Mo,Cu,Ni	1
COLLINS BAY A		x	x		M	Ap,	F,G		x		x	x	x	x	x	Ni,As	Au,Ag,Pb	1 & 6
COLLINS BAY B	x			x	M,Gr	Ap,Ar	G,C	Sm	x		x	x	x	x	x	Ni,As		1 & 6
EAGLE POINT	Eroded			?	M,F	Ap,Ar?	F,C?,Q		x	xMg	x		x		x			3
MIDWEST LAKE		x	x		Gr	Ap,Ar	G,C,Q	Sm,Sb,P	x	xMg	x	x	x	x	x	Ni,As,Co	Ag,Pb	1 & 3
DAWN LAKE	x			x	M	Ap	G,C	Sb	x	x	x	x	?	x	x	?	?	1
McLEAN LAKE	x			x	M,Gr	Ap,Ar	Q,C	Sm	x	x	x	x	x	x		Ni,Co,As	V,Mo,Cu	4
KEY LAKE		x	x		Gr	Ap,Ar	G	Sb	x	x	x	x	x	x	x	Ni,Co,As		1 & 5
CLUFF D	x		x		M	Ap	Q	P,Sm,Sb	x	xMg	x	x	x	x	x	Au,Te,Se	Co,Ni,Bi	2
CLUFF CLAUDE	Eroded			x	M	Ap	G,F		x	x	x		x	x	x		Mo,Pb	2
CLUFF DOMINIQUE - PETER	Eroded in part		x		M	Ap	G,F		x	xMg	x		x	x	x	Te,Se	Au,Mo	2
CLUFF N	Eroded		x		Gr	Ap	Q,F		x	x	x		x	x	x		Cu,Pb,Au	2
CLUFF OP	Eroded in part		x		M	Ap	G,F		x		x		x	x	x		B,Mo,Cu	2
MAURICE BAY		x	x		Gr	Ar,Ap	F	Sm	x			x	x	x	x		Au,Cu	1 & 7
STEWART ISLAND	x			x	T	Ar,Ap		Sb			?	?	x	x	?		Ni,Co,Cr	1 & 7
FOND du LAC	x		x		Gr	Ar,Ap		Sm	x					x	x			1 & 7
BEAVERLODGE POLYMETALLIC	Eroded				Q,F	Ar	T		x	x			x		x	Au,Se	Ni,Co,As	5

LEGEND

Q — QUARTZITE, FELDSPATHIC QUARTZITE
C — CALC-SILICATE ROCKS
G — GNEISS, GRAPHITIC
F — QUARTZOFELDSPAR GRANITOID
M — METASEDIMENTARY
Gr — GRANITOID & METASEDIMENTARY

T — TAZIN ROCKS
A — APHEBIAN
Ar — ARCHEAN
P — MUDSTONE
Sm — SANDSTONE (MULTICOLOURED)
Sb — SANDSTONE (BLACK)

REFERENCES

1 TREMBLAY (1982)
2 TONA et al (1985)
3 SOPUCK et al (1983)
4 WALLIS et al (1983)
5 DAHLKAMP & ADAMS (1981)
6 JONES (1980)
7 HOMENIUK & CLARK (1983)

228 LAINÉ

TABLE II
GEOLOGIC HISTORY OF THE CARSWELL AREA AND AGE OF MINERALIZATION IN THE ATHABASCA BASIN AND ADJACENT AREA (TIME DIVISIONS OF OROGENIC EVENTS AFTER DOUGLAS, 1980)

Ma	AGE	ORGENY	DEFORMATION	SEDIMENTATION	METAMORPHIC & IGNEOUS ACTIVITY, & ALTERATION	CARSWELL STRUCTURE MINERALIZATION	GEOCHRONOLOGY IN THE CARSWELL AREA	ATHABASCA BASIN & ADJACENT MINERALIZATION
-100	PHANEROZOIC		D_6 POST EVENT FAULTING			COFFINITE ? (D,N) REMOBILIZATION		200 Ma RABBIT LAKE (KNIPPING, 1974)
-200							-250±Ma D MIN'N, U-Pb (BELLON et al., 1976)	215 Ma FOND DU LAC (HOEMENIUK & CLARK, 1983)
-300						PITCHBLENDE & CARBONATE		270±20 Ma KEY LAKE (WENDT et al., 1978)
	HADRYNIAN		D_5 MAIN CARSWELL EVENT		CLUFF BRECCIAS	MINOR REMOBILIZATION HEMATITE & PITCHBLENDE (DONNA AREA) MECHANICAL REMOBILIZATION	365 to 515 Ma CLUFF BRECCIAS Ar-Ar, (BELL, 1985) ca 380 Ma DONNA BOULDERS MIN'N, U-Pb,(BELL, 1985)	290 Ma MAURICE BAY (HOHNDORF et al.,1981) 318 Ma STEWART ISLAND (HOEMENIUK & CLARK, 1983)
-400							467±28 Ma] CLUFF BRECCIAS K-Ar, (WANLESS et al.,1968) 486±55 Ma	338 Ma MIDWEST LAKE (WORDEN et al., 1981) 418 Ma STEWART ISLAND (FAHRIG, 1961)
-500							500±50 Ma ALTERED BASEMENT GNEISS, Ar-Ar, (BELL, 1985)	-450 Ma COLLINS BAY - B (JONES, 1980) -494±19 Ma RABBIT LAKE (CUMMING & RIMSAITE, 1979)
-600								
-700								
-800							820 to 890 Ma OP MIN'N, U-Pb, (BELL, 1985)	-860 Ma RABBIT LAKE (KNIPPING, 1974)
-900	HELIKIAN		D_4 NW-SE TANGENTIAL FAULTING			2nd STAGE MINERALIZATION DOMINIQUE - PETER, OP	945±33 Ma DOMINIQUE - PETER MIN'N, Pb-Pb, (BELL, 1985) 949±33 Ma DIABASE DYKE K-Ar, (GANCARZ, 1979)	918 Ma KEY LAKE (WENDT et al., 1978)
-1000		GRENVILLE					1050±30 Ma NUMAC AREA MIN'N, Pb-Pb, (BELL, 1985)	1075 Ma RABBIT LAKE (KNIPPING, 1974)
-1100						1st STAGE MINERALIZATION (U, B, Mo, Ni, Co, As, Se, Te, Au) CLAUDE, N, D	1095±95 Ma CHLORITE SCHISTS (FROM D), K-Ar, (WANLESS et al., 1979)	1110±28 Ma MIDWEST LAKE (BAADSGAARD et al.,1984) 1110±50 Ma URANIUM CITY AREA (KOEPPEL, 1968)
-1200				END OF SEDIMENTATION CARSWELL FM (celites dated 1222±52 Ma, K-Ar, CLAUER et al., 1985)	(K-Ar, BELL, 1981, WANLESS et al., 1979, WORDEN et al., 1981) Mg CHLORITIC HALOS HYDROTHERMAL ALTERATION & DIAGENESIS AFFECTING BOTH BASEMENT & SANDSTONE (1293±36 Ma, K-Ar, CLAUER et al., 1985)	DOMINIQUE - PETER PROTORE II (NUMAC AREA)	1150±25 Ma D MIN'N, U-Pb,(BELLON et al.,1976)	1150±50 Ma FOND DU LAC (HOEMENIUK & CLARK, 1983) 1194±34 Ma KEY LAKE (WENDT et al., 1978)
-1300			D_3 FAULTING OF THE APHEBIAN BASEMENT, AFFECTING THE ATHABASCA COVER				-1293±36 Ma HYDROTHERMAL CLAYS K-Ar, (CLAUER et al., 1985)	1281±11 Ma RABBIT LAKE (CUMMING & RIMSAITE, 1979)
-1400				WOLVERINE PT & FM (volcanic material dated 1513±24 Ma, Rb-Sr, BELL & BLENKINSOP, 1980)			1330±30 Ma NUMAC AREA MIN'N, U-Pb, (GANCARZ, 1979)	1320 Ma COLLINS BAY - B (JONES, 1980) 1326±17 Ma MIDWEST LAKE (BAADSGAARD et al., 1984)
-1500					ACID VOLCANISM			
-1600			D_2 N140° THRUST FAULTS (MYLONITES) TANGENTIAL NE-SW FAULTING SCHISTOSITY & FOLDING				ca 1000 to 1600 Ma ALTERATION PRODUCTS OF BASEMENT GNEISSES, Ar/Ar, (BELL, 1985)	
-1700	APHEBIAN	PENOKEAN			RETROGRADE METAMORPHISM AMPHIBOLITE/GREENSCHIST	U, Ti LIBERATION		-1780±20 Ma PITCHBLENDE VEINS URANIUM CITY AREA (KOEPPEL, 1968)
-1800		HUDSONIAN	D_1 EARL RIVER COMPLEX DOMING BY DENSITY DIFFERENCE		AMPHIBOLITE GRADE	PROTORE I (URANINITE, MONAZITE IN SOPHIE & LAURE(?) AREAS)	ca 1760 Ma METAPELITIC PETER RIVER GNEISSES, Rb-Sr, (BELL, 1985)	-1930±40 Ma URANINITE IN URANIUM CITY AREA (KOEPPEL, 1968)
-1900					MIGMATIZATION (GRANITOIDS)		ca 1870 Ma FELDSPATHIC GNEISSES, Rb-Sr, (BELL, 1985)	
-2000				SHALES	GRANULITE GRADE			
-2100		BLEZARDIAN	?	ARKOSES & GREYWACKES	MAFIC GNEISS/AMPHIBOLITES (KOMATIITES??)		ca 2130 Ma GRANITOIDS, U-Pb, (BELL, 1985) (Discordant zircons)	
-2200				?	CALC - ALKALINE MAGMATISM			
-2300					GRANITOIDS NE PART OF DOME?		ca 2320 Ma GRANITOIDS, U-Pb, (BELL, 1985) (Discordant zircons)	

remobilization events at 900 Ma, 380 Ma, and 200 Ma, comparable in age to events recorded in uranium deposits in other parts of the basin (Table II). At this time, no dates older than 1150 Ma have been determined for the economic mineralization as noted in the Beaverlodge district by Tremblay (1982).

The Carswell event, responsible for the intrusion of the basement plug through the sandstones, is unique to the Athabasca Basin and had no direct influence in the emplacement of uranium mineralization. Presently, available data favours either an exogenic or an endogenic mechanism for emplacement.

Many models have been proposed for the genesis of the Saskatchewan uranium deposits, a surficial per descensum origin, (Knipping, 1974; Langford, 1977), a hypogene origin (Little, 1974), and a diagenetic hydrothermal origin (Hoeve and Sibbald, 1978). The data presented in this volume best fit the diagenetic hydrothermal model. Thus, as has been proposed for the Saskatchewan uranium deposits, the genetic history of the Carswell deposits "commences very early in the geological development of the Precambrian rocks of Northern Saskatchewan, and several stages of uranium concentration were required to produce significant deposits" (Tremblay, 1982, p. 50).

REFERENCES

Baadsgaard, H., Cumming, G.L., and Worden, J.H., 1984, U-Pb Geochronology of Minerals from the Midwest Uranium Deposit, Northern Saskatchewan: Canadian Journal of Earth Sciences, v. 21, p. 642-648.

Beck, L.S., 1969, Uranium Deposits of the Athabasca Region, Saskatchewan: Saskatchewan Department of Mineral Resources Report NB 126, 139 p.

Bell, K., 1981, A Review of the Geochronology of the Precambrian of Saskatchewan — Some Clues to Uranium Mineralization: Mineralogical Magazine, v. 44, p.371-378.

_____, 1985, Geochronology of the Carswell Area, Northern Saskatchewan: in Lainé, R., Alonso, D., and Svab, M., eds., The Carswell Structure Uranium Deposits, Saskatchewan: Geological Association of Canada Special Paper 29.

Bell, K., and Blenkinsop, J., 1980, Saskatchewan Geochronology Project: in Summary of Investigations 1980, Saskatchewan Geological Survey: Saskatchewan Mineral Resources Miscellaneous Report 80-4, p. 18.

Bellon, H., Devillers, C., Hagemann, R., and Touray, J.C., 1976, Dater les Minéralisations: Mémoires Hors Séries de la Société Géologique de France, v. 7, p. 265-268.

Blaise, J.R., and Koning, E., 1985, Mineralogical and Structural Aspects of the Dominique-Peter Uranium Deposit: in Lainé, R., Alonso, D., and Svab, M., eds., The Carswell Structure Uranium Deposits, Saskatchewan: Geological Association of Canada Special Paper 29.

Clark, L.A., and Burrill, G.H.R., 1981, Unconformity-Related Uranium Deposits, Athabasca Area, Saskatchewan and East Alligator Rivers Area, Northern Territory, Australia: Canadian Institute of Mining Bulletin, v. 74, no. 831, p. 63-72.

Clauer, N., Ey, F., and Gauthier-Lafaye, F., 1985, K-Ar Dating of Different Rock Types from the Cluff Lake Uranium Ore Deposits (Saskatchewan-Canada): in Lainé, R., Alonso, D., and Svab, M., eds., The Carswell Structure Uranium Deposits, Saskatchewan: Geological Association of Canada Special Paper 29.

Cumming, G.L., and Rimsaite, J., 1979, Isotopic Studies of Lead Depleted Pitchblende, Secondary Radioactive Minerals and Sulfides from the Rabbit Lake Uranium Deposit, Saskatchewan: Canadian Journal of Earth Sciences, v. 16, p. 1702-1715.

Dahlkamp, F.J., and Adams, S., 1981, Geology and Recognition Criteria for Vein-Like Uranium Deposits of the Lower to Middle Proterozoic Unconformity and Strata-Related Types: United States of America, Department of Energy Report GJBX-5 (81), 254 p.

Douglas, R.J.W., 1980, Proposal for Time Classification and Correlation of Precambrian Rocks and Events in Canada and Adjacent Areas of the Canadian Shield Part 2: A Provisional Standard For Correlation Precambrian Rocks: Geological Survey of Canada Paper 80-24, 19 p.

Ey, F., Gauthier-Lafaye, F., Lillié, F., and Weber, F., 1985, A Uranium Unconformity Deposit: The Geological Setting of the D Orebody (Saskatchewan-Canada): in Lainé, R., Alonso, D., and Svab, M., eds., The Carswell Structure Uranium Deposits, Saskatchewan: Geological Association of Canada Special Paper 29.

Fahrig, W.F., 1961, The Geology of the Athabasca Formation: Geological Survey of Canada Bulletin 68, 41 p.

Gancarz, A.J., 1979, Chronology of the Cluff Lake Uranium Deposit, Saskatchewan, Canada: International Uranium Symposium on the Pine Creek Geosyncline, Extended Abstract, p. 91-94.

Hendry, H.E., and Wheatley, K.L., 1985, The Carswell Formation, Northern Saskatchewan: Stratigraphy, Sedimentology, and Structure: in Lainé, R., Alonso, D., and Svab M., eds., The Carswell Structure Uranium Deposits, Saskatchewan: Geological Association of Canada Special Paper 29.

Hoeve, J., and Sibbald, T.I.I., 1978, On the Genesis of the Rabbit Lake and other Unconformity-Type Uranium Deposits in Northern Saskatchewan, Canada: Economic Geology, v. 73, p. 1450-1473.

Hoeve, J., Sibbald, T.I.I., Ramaekers, P., and Lewry, J.F., 1980, Athabasca Basin Unconformity-Type Uranium Deposits: A Special Case of Sandstone-Type Deposits: in Ferguson, John and Coleby, A.B., eds., Uranium in the Pine Creek Geosyncline, International Atomic Energy Association, Vienna, p. 575-594.

Hoeve, J., Quirt, D., and Alonso, D., 1985, Clay Mineral Stratigraphy of the Athabasca Group: Correlation Inside and Outside the Carswell Structure: in Lainé, R., Alonso, D., and Svab, M., eds., The Carswell Structure Uranium Deposits, Saskatchewan: Geological Association of Canada Special Paper 29.

Hohndorf, F.A., Voultsidis, V., and von Pechmann, E., 1981, U/Pb Isotopic Investigations of the Maurice Bay Uranium Deposits, Lake Athabasca (Preliminary Results): Canadian Institute of Mining Uranium Symposium, September 8-13, Saskatoon, Technical Program, p. 23-24.

Homeniuk, L.A., and Clark, R.J., 1985, North Rim Deposits Athabasca Basin: Canadian Institute of Mining Bulletin, Special Volume on Uranium Deposits of Canada (unpublished).

Jones, B.E., 1980 The Geology of the Collins Bay Uranium Deposits, Saskatchewan: Canadian Institute of Mining Bulletin, v. 73, no. 818, p. 84-90.

Knipping, H., 1974, The Concepts of Supergene Versus Hypogene Emplacement of Uranium at Rabbit Lake, Saskatchewan, Canada: in Formation of Uranium Ore Deposits, Proceedings of a Symposium, Athens, International Atomic Energy Agency, Vienna, p. 531-549.

Koeppel, V., 1968, Age and History of the Uranium Mineralization of the Beaverlodge Area, Saskatchewan: Geological Survey of Canada Paper 67-31, 111 p.

Landais, P., and Dereppe, J.M., 1985, A Chemical Study of the Carbonaceous Material from the Carswell Structure: *in* Lainé, R., Alonso, D., and Svab, M., eds., The Carswell Structure Uranium Deposits, Saskatchewan: Geological Association of Canada Special Paper 29.

Langford, F.F., 1977, Surficial Origin of North American Pitchblende and Related Uranium Deposits: The American Association of Petroleum Geologists Bulletin, v. 61, p. 28-42.

Little, H.W., 1974, Uranium in Canada: Report of Activities, Part A, Geological Survey of Canada Paper 74-1, p. 137-139.

Pacquet, A., and McNamara, S., 1985, The Study of the Basal Athabasca Succession in the D, E, L, F, and S Areas of the Carswell Structure: *in* Lainé, R., Alonso, D., and Svab, M., eds., The Carswell Structure Uranium Deposits, Saskatchewan: Geological Association of Canada Special Paper 29.

Pagel, M., and Svab, M., 1985, Petrographic and Geochemical Variations within the Carswell Structure Metamorphic Core and their Implications with Respect to Uranium Mineralization: *in* Lainé, R., Alonso, D., and Svab, M., eds., The Carswell Structure Uranium Deposits, Saskatchewan: Geological Association of Canada Special Paper 29.

Pagel, M., Poty, B., and Sheppard, S.M.F., 1981, Contribution to some Saskatchewan Uranium Deposits Mainly from Fluid Inclusion and Isotopic Data: *in* Ferguson, John and Coleby, Ann, B., eds., Uranium in the Pine Creek Geosyncline, International Atomic Energy Agency, Vienna, p. 639-654.

Pagel, M., Wheatley, K.L., and Ey, F., 1985, The Origin of the Carswell Circular Structure: *in* Lainé, R., Alonso, D., and Svab, M., eds., The Carswell Structure Uranium Deposits, Saskatchewan: Geological Association of Canada Special Paper 29.

Powell, B., Koning, E., and Lainé, R., 1985, Geophysical Mapping of Gneiss Domes in the Carswell Structure and their Relationship to Uranium Mineralization: *in* Lainé, R., Alonso, D., and Svab, M., eds., The Carswell Structure Uranium Deposits, Saskatchewan: Geological Association of Canada Special Paper 29.

Ruhlmann, F., 1985, Mineralogy and Metallogeny of Uraniferous Occurrences in the Carswell Structure: *in* Lainé, R., Alonso, D., and Svab, M., eds., The Carswell Structure Uranium Deposits, Saskatchewan: Geological Association of Canada Special Paper 29.

Sopuck, V.J., de Carle, A., Wray, E.M., and Cooper, B., 1983, Application of Lithogeochemistry to the Search for Unconformity-Type Uranium Deposits in the Athabasca Basin, Saskatchewan, Canada: Geological Survey of Canada Paper 82-11, p. 191-206.

Taylor, G.H., and Rowntree, J.C., 1981, The Symposium – Ferguson, J., and Coleby, A.B., eds., Uranium in the Pine Creek Geosyncline: International Atomic Energy Agency, Vienna, p. 751-758.

Tona, F., Alonso, D., and Svab, M., 1985, Geology and Mineralization in the Carswell Structure - A General Approach: *in* Lainé, R., Alonso, D., and Svab, M., eds., The Carswell Structure Uranium Deposits, Saskatchewan: Geological Association of Canada Special Paper 29.

Tremblay, L.P., 1982, Geology of the Uranium Deposits Related to the Sub-Athabasca Unconformity, Saskatchewan: Geological Survey of Canada Paper 81-20, 56 p.

Wallis, R.H., Saracoglu, N., Brummer, J.J., and Golightly, J.P., 1983, Geology of the McLean Lake Uranium Deposits: *in* Cameron, E., ed., Uranium Exploration in the Athabasca Basin, Saskatchewan, Canada: Geological Survey of Canada Paper 82-11, p. 71-110.

Wanless, R.K., Stevens, R.D., Lachance, G.R., and Edmonds, C.M., 1968, Age Determinations and Geological Studies - K/Ar Isotopic Ages, Report 8: Geological Survey of Canada Paper 67-2, 141 p.

Wanless, R.K., Stevens, R.D., Lachance, G.R., and Delabio, R.N., 1979, Age Determinations and Geological Studies - K/Ar Isotopic Ages: Geological Survey of Canada Paper 79-2, 67 p.

Wendt, L., Hohndorf, A., Lenz, H., and Voultsidis, V., 1978, Radiometric Age Determination on Samples of Key Lake Uranium Deposits: *in* Short Paper of the Fourth International Conference on Geochronology, Cosmochronology, and Isotope Geology: United States Geological Survey Open File Report 78-701, p. 448-449.

Worden, J.W., Cumming, G.L. and Baadsgaard, M., 1981, Geochronology, Setting, and Mineralization Ages of the Midwest Uranium Deposit, Northern Saskatchewan: Canadian Institute of Mining Uranium Symposium, Saskatchewan (unpublished).

GEOLOGICAL ASSOCIATION OF CANADA
ASSOCIATION GEOLOGIQUE DU CANADA

SPECIAL PAPERS**

20 The Continental Crust and Its Mineral Deposits
Edited by D.W. Strangway, 1980, 804 p., $24/$30*

21 Cretaceous Rocks and Their Foraminifera in the Manitoba Escarpment
Edited by D.H. McNeil and W.G.E. Caldwell, 1981, 439 p., $24/$30*

22 The Buchans Orebodies: Fifty Years of Geology and Mining
Edited by E.A. Swanson, D.F. Strong and J.G. Thurlow, 1981, 350 p. (2 multi-coloured geological maps in separate binder), $29/$36*

23 Sedimentation and Tectonics in Alluvial Basins
Edited by A.D. Miall, 1981, 272 p., $22/$26*

24 Major Structural Zones and Faults of the Northern Appalachians
Edited by P. St. Julien and J. Beland, 1982, 250 p., $24/$29*

25 Precambrian Sulphide Deposits
(H.S. Robinson Memorial Volume), Edited by R.W. Hutchinson, C.D. Spence and J.M. Franklin, 1982, 791 p., $47/$57*

26 Glacial Lake Agassiz
Edited by J.T. Teller and L. Clayton, 1983, 451 p. (+ 2 maps in pocket), $28/$34*

27 Jurassic-Cretaceous Biochronology and Paleogeography of North America
Edited by G.E.G. Westermann, 1984, 315 p., $30/$36*

28 Evolution of Archean Supracrustal Sequences
Edited by L.D. Ayres, P.C. Thurston, K.D. Card, and W. Weber, 1985, 380 p., $35/$42*

29 The Carswell Structure Uranium Deposits, Saskatchewan
Edited by R. Lainé, D. Alonso, and M. Svab, 1985, 230 p., $35/$42*

*Members/Non-members
**For handling and postage to any address add $3.00 per publication.

Payments must accompany orders. **Make cheques payable to Geological Association of Canada.** Payments may also be made in U.S. funds. No refunds or exchanges.

Mail orders to: GAC Publications
Business and Economic Service Ltd.
111 Peter Street, Suite 509
Toronto, Ontario M5V 2H1
CANADA